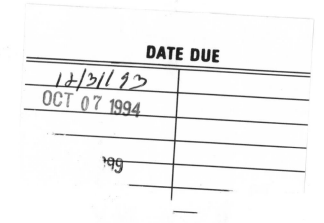

Bacterial Growth
and Division

To Sandi, who has shared over half my life, and who is now a part of me; to Sandi, who has kept my life balanced; to Sandi, who has taught me how to live life; to Sandi, whose presence makes each day a joy.

Bacterial Growth and Division

Biochemistry and Regulation of Prokaryotic and Eukaryotic Division Cycles

Stephen Cooper
Department of Microbiology and Immunology
University of Michigan Medical School
Ann Arbor, Michigan

ACADEMIC PRESS, INC.
Harcourt Brace Jovanovich, Publishers
San Diego   New York   Boston   London   Sydney   Tokyo   Toronto

This book is printed on acid-free paper. ∞

Academic Press, Inc.
San Diego, California 92101

United Kingdom Edition published by
Academic Press Limited
24–28 Oval Road, London NW1 7DX

Library of Congress Cataloging-in-Publication Data

Cooper, Stephen, date
    Bacterial growth and division : biochemistry and regulation of prokaryotic and eukaryotic division cycles / Stephen Cooper.
        p.   cm.
    Includes bibliographical references.
    Includes indexes.
    ISBN 0-12-187905-4
    1. Microbial differentiation.   2. Bacterial growth.   I. Title.
    [DNLM: 1. Baterial--growth & development.   2. Cell Cycle.   3.

Cell Division.   4. Cells.   5. Gene Expression Regulation, Bacterial.   QW
51 C778b]
QR73.5.C66   1991
589.9'08761--dc20
DNLM/DLC
for Library of Congress   90-14426
                                                              CIP

PRINTED IN THE UNITED STATES OF AMERICA
91 92 93 94   9 8 7 6 5 4 3 2 1

Contents

Acknowledgments   xix
Illustrations   xxi
Tables   xxv

Prologue   1

    I. The Field of Division-Cycle Studies   1
    II. The Unity of Biology   3
    III. The History of Cell-Cycle Studies   4
    IV. Theory and Experiment in Biology   4
        Notes   5

1. Bacterial Growth   7

    I. The Study of Bacterial Growth   7
        A. Exponential Growth   7
        B. Balanced Growth   8
        C. The Age Distribution   9
        D. The Classic Life Cycle of Bacteria   12
    II. The Fundamental Experiment of Bacterial Physiology   13
        A. Steady-State Growth   13
        B. The Shift-Up   15
        C. The Copenhagen School   16
        Notes   17

2. A General Model of the Bacterial Division Cycle   18

    I. Theory, Experiment, and Biological Understanding   18
    II. The Aggregation Problem   19
        A. Cytoplasm Synthesis   20
        B. DNA Synthesis   21
        C. Surface Synthesis   22
    III. Interrelationships between Cytoplasm, DNA, and Surface
        Synthesis   22
    IV. Passive and Independent Regulation   23
    V. The Logic of the Division Cycle   24
        Notes   24

3. Experimental Analysis of the Bacterial Division Cycle

	3. Experimental Analysis of the Bacterial Division Cycle	25
I.	The Experimental Analysis of the Division Cycle	25
II.	Synchronization versus Nonsynchronization Methods	26
	A. Criteria for Synchronization Methods	26
	B. Indexing of Synchrony Experiments	29
	C. Differential versus Integral Methods of Cell-Cycle Analysis	30
III.	Synchronization Methods	32
	A. Comparison of Selective and Nonselective Methods	32
	B. Nonselective Methods for Synchronization	33
	1. Heat Treatments and Heat Shock	33
	2. Starvation or Stationary-Phase Methods	34
	3. Starvation for Specific Compounds	35
	4. Entrainment Methods	36
	5. Specific Inhibitors	37
	6. Spore Germination	37
	7. Synchronization by Phage Infection	37
	C. Selective Methods for Synchronization	38
	1. Centrifugation	38
	2. Elutriation	40
	3. Filtration	40
	4. The Membrane-Elution Technique	41
IV.	Nonsynchrony Methods	44
	A. Total Population Analysis	44
	1. Direct Size Measurements	44
	2. Differential Methods for Total Population Analysis	46
	3. Methods Specific to the Analysis of DNA Replication during the Division Cycle	48
	4. Execution Points	48
	B. Backwards Methods	49
	1. Membrane-Elution Technique for Cell-Cycle Analysis	50
	2. Analysis of a Membrane-Elution Experiment	53
	3. Criteria for a Membrane-Elution Experiment	55
	4. Comments on the Membrane-Elution Method	55
V.	A Comment on Methods for Analyzing the Division Cycle	58
	Notes	59
	4. Cytoplasm Synthesis during the Division Cycle	63
I.	A Priori Considerations of Cytoplasm Synthesis	64
II.	Experimental Analysis of Cytoplasm Synthesis during the Division Cycle	69
	A. Total Protein Synthesis during the Division Cycle Analyzed by the Membrane-Elution Method	69

B. Autoradiographic Analysis of Protein Synthesis during the
        Division Cycle   71
C. RNA Synthesis during the Division Cycle   72
D. Size Distribution of Exponential Populations   72
E. Analysis of Specific Protein Synthesis during the Division
        Cycle   74
F. Direct Evidence for the Absence of Cell-Cycle–Specific
        Enzyme Synthesis   76
III. Alternative Proposals for Cytoplasm Synthesis during the
        Division Cycle   76
A. Linear Synthesis during the Division Cycle   77
B. Variation in Specific Protein Synthesis during the Division
        Cycle   80
C. Flagella Synthesis during the Division Cycle   80
D. Variation in the Capacity to Synthesize Specific Proteins   81
E. Specific RNA Synthesis during the Division Cycle   82
F. Soluble Components during the Division Cycle   82
        1. Cation Concentration   82
        2. Nucleotides   82
        3. Calcium as a Trigger of the Division Cycle   83
G. Comment on Specific Synthesis during the Division Cycle   84
IV. A Functional Model for Cell-Cycle Regulation of Specific
        Protein Synthesis   84
V. Determination of the Growth Rate by the Medium   85
A. Physiological States and Growth Rates Produced by
        Different Media   85
B. The Theoretic Maximal Growth Rate   88
VI. Icon for Mass Synthesis during the Division Cycle   89
VII. Events during the Division Cycle   89
        Notes   90

5. DNA Replication during the Bacterial Division Cycle   94

I. The Problem of Schaechter, Maaløe, and Kjeldgaard   94
II. DNA Synthesis during the Division Cycle at Moderate and
        Fast Growth Rates   94
III. Early Studies on the Pattern of DNA Replication   103
A. DNA Synthesis in Synchronized Cells   104
B. Autoradiography   104
C. Number of Growing Points in a Chromosome   105
D. The Meselson–Stahl Experiment   106
E. The Model of Lark and Lark   106
IV. Analysis of DNA Replication Using the Membrane-Elution
        Technique   107

    A. Rate of DNA Synthesis during the Division Cycle   107
    B. DNA Replication at Different Temperatures   109
    C. DNA Content at Different Rates of Growth   110
    D. DNA Content of Individual Cells Using Flow Cytometry   111
    E. Autoradiographic Analysis of the Number of Growing
       Points in a Cell   113
  V. Icon for DNA Replication   114
 VI. A Priori Considerations on the Regulation of Chromosome
     Replication   114
    A. Cell Mass and the Initiation of Replication   114
    B. Implications of the Schaechter–Maaløe–Kjeldgaard
       Experiment   117
    C. The I + C + D Concept   119
 VII. Experimental Analysis of Regulation of DNA Synthesis   119
    A. Cell Mass at Different Growth Rates   119
    B. Inhibition of Mass Synthesis   121
    C. Inhibition of DNA Synthesis   123
    D. The Maaløe–Hanawalt Experiment   125
    E. Thymine Starvation and Premature Initiation   126
    F. Cell Division after Inhibition of DNA Synthesis   127
    G. The Shift-Up   127
    H. Deviations from Rate Maintenance   130
    I. The Shift-Down   132
VIII. Icon for Regulation of DNA Replication   133
 IX. Further Analysis of DNA Replication during the Division
     Cycle   133
    A. Other Methods for Measuring the C Period   133
      1. Increment of DNA Synthesis   134
      2. Increment of DNA Synthesis Using the Total Curve   134
      3. Rate Stimulation   134
      4. Mass-to-DNA Ratios   135
      5. Autoradiographic Analysis of DNA Chain Extension   136
      6. Analysis of Gene Frequency   137
      7. Synchronized Cultures   138
      8. Synchronization of DNA Replication   139
      9. Measuring the C Period by Transductional Analysis   139
     10. DNA Step Times as a Measure of the C Period   139
    B. Modifications of the Rate of DNA Replication   140
      1. Effect of Thymine Concentration   140
      2. The rep Mutation   140
      3. Effect of Inhibitors on Rate of DNA Replication   140
    C. Nucleoid Division in Relation to DNA Synthesis   140
      1. Nucleoid Content during Steady-State Growth and
        Transitions   141
      2. Electron Microscopy of Nucleoids   142

        D. *Determination of the D Period*   142
            1. Residual Cell Division after Inhibition of DNA Synthesis   142
            2. Residual Cell Division after Inhibition of Protein Synthesis 143
    X. DNA Replication during the Division Cycle of
        Slow-Growing Cells   144
        A. *A Priori Considerations on Slow-Growing Cells*   144
        B. *Membrane-Elution Analysis of DNA Synthesis during
            the Division Cycle of Slow-Growing Cells*   144
        C. *Alternative Results and Models*   146
            1. Constant C and D at Slow Growth Rates   146
            2. Marker Ratios during Slow Growth   147
            3. Is C Constant in Slow-Growing Cells?   148
            4. Synchrony Analysis   148
            5. Autoradiography of Slow-Growing Cells   149
            6. Flow Cytometric Analysis of Slow-Growing Cells   149
            7. A Conjecture Regarding DNA Replication in
                Slow-Growing Cells   149
   XI. A Comment on the Chemostat   150
  XII. B, T, and U Periods during the Division Cycle   151
 XIII. DNA Concentration as a Function of Growth Rate   152
  XIV. The Order of Replication of the Genome   152
        A. *Early Work on the Order of Genome Replication*   152
        B. *Bidirectional Replication*   153
        C. *On the Order of the Genes in the Genome*   155
        D. *Is There a Unique Order of Replication?*   156
   XV. The Initiation Problem   157
        A. *Does Cell Division or Termination Regulate
            Initiation of DNA Synthesis?*   157
        B. *Is Initiation Regulated by Positive or Negative Controls?* 158
        C. *Is Initiator Consumed at Initiation?*   160
  XVI. Events Leading to Initiation   160
        A. *The Proposal of Bacterial Restriction Points*   160
        B. *Criterion for Demonstrating Restriction Points*   161
        C. *Accumulation of Initiator in the Absence of DNA Synthesis* 162
 XVII. Plasmid Replication during the Division Cycle   162
        A. *A Priori Ideas on Plasmid Replication during the Division
            Cycle*   162
        B. *Plasmid Replication during the Division Cycle*   163
            1. Flac   163
            2. Resistance Plasmids   167
            3. Minichromosomes   167
            4. Colicinogenic Plasmids   167
            5. P1 Prophage   167
            6. Summary of Plasmid Replication during the Division Cycle 168
XVIII. The Initiation and Termination of DNA Synthesis   169

A. Initiation of DNA Replication   169
B. Replication of DNA   169
C. Termination of DNA Replication   169
XIX. DNA Replication during the Division Cycle of Bacteria   170
Notes   170

6. Synthesis of the Cell Surface during the Division Cycle   177

I. The Structure of the Cell Surface of Gram-Negative
Bacteria   177
A. Peptidoglycan Structure   178
1. Composition and Cross-Linking of the Peptidoglycan
Subunits   178
2. Amount of Peptidoglycan on a Cell   179
3. Heterogeneity of Peptidoglycan Structure   181
B. Arrangement of Peptidoglycan Strands on the Cell Surface   183
II. Biochemistry of Peptidoglycan Synthesis   183
A. Synthesis of the Pentapeptide Precursor   183
B. Transfer of the Pentapeptide to the Growing Chain   184
C. Cross-Linking of Peptidoglycan Chains   184
III. The Relationship of Mass Synthesis to Peptidoglycan
Synthesis   184
IV. Early Studies on the Pattern of Cell-Surface Synthesis
during the Division Cycle   187
A. Early Studies on the Location of Cell Surface Synthesis   187
1. Experimental Support for Zones of Wall Synthesis   188
2. The Unit-Cell Model   189
3. Evidence against Zonal Growth   190
B. Early Studies on Rate of Peptidoglycan Synthesis during
the Division Cycle   190
C. Summary of Early Work on the Rate of Cell-Surface
Synthesis   191
V. Rate and Topography of Peptidoglycan Synthesis during
the Division Cycle   191
A. Quantitative Analysis of Wall Growth during the Division
Cycle   194
B. Experimental Support for the Pressure Model of
Cell-Surface Synthesis   197
1. Ratio of Peptidoglycan Synthesis to Protein Synthesis
during the Division Cycle Using Membrane Elution   197
2. Cell Density   198
3. Segregation of Peptidoglycan   200
4. Rate of Cell-Length Growth   200
5. Location of Newly Synthesized Peptidoglycan   200
6. Leading-Edge Model of Pole Growth   201

7. Variation or Constancy of Diameter during the Division
Cycle 202
8. Stability and Turnover of Peptidoglycan 202
VI. Control Mechanisms for Wall Synthesis 203
A. Stringent Regulation of Wall Synthesis 203
B. Regulation of Surface Synthesis by Cytoplasmic Signals 204
C. Regulation of Wall Synthesis by Peptidoglycan Structure 204
D. Heat Shock and Cell Division 205
E. Cell Division Related to Pattern of Cytoplasmic Protein
Synthesis 205
F. Control of the Cell Cycle by Phospholipid Flip-Out 205
VII. Maturing of Peptidoglylcan 206
A. Change in Acceptors and Donors 206
B. Increase in Cross-Linking 206
VIII. Regulation of Constriction in Rod-Shaped Cells 209
A. Nucleoid-Occlusion Model 209
B. The Periseptal Annulus 209
C. Variable-T Model for Pole Synthesis 211
D. When Does Invagination Occur? 211
IX. Membrane Synthesis during the Division Cycle 212
A. Lipid Synthesis during the Division Cycle 212
B. Membrane Protein Synthesis during the Division Cycle 213
X. The Shape of Rod-Shaped, Gram-Negative Bacteria 213
A. Constant Shape of Cells at Different Growth Rates 215
B. Energetic Argument for Constant Shape 218
C. Exceptions to Constant Shape 219
D. Cell-Width Variability within a Culture 219
E. Cell Length and Width in a Culture 220
F. Apparent Variation of Cell Diameter during the Division
Cycle 222
G. Width Variability and Peptidoglycan Synthesis during
the Division Cycle 224
H. Width Changes after EDTA Treatment or Amino Acid
Starvation 225
I. Central Zone Surface Synthesis during the Division Cycle 225
J. Length Distribution of Newborn Cells 227
K. Molecular Mechanisms of Shape Determination 228
XI. The Volume, Surface Area, and Dimensions of Cells at
Different Growth Rates 230
XII. The Constrained-Hoop Model 230
XIII. Is There a Minimal or Critical Cell Length? 232
XIV. Genetic Analysis of the Division Cycle 234
A. A Priori Considerations of Genetic Analysis 235
B. Mutations in Simple and Complex Processes 235
C. Classification of Mutants Affecting the Cell Surface and
Division 238

    XV. On Laws, Critical Tests, Predictions, and Exceptions   239
   XVI. Icon for Cell-Wall Synthesis during the Division Cycle   240
       Notes   241

7. Density and Turgor during the Division Cycle   247

     I. Density during the Division Cycle   247
        A. A Priori *Considerations of the Meaning of Cell Density*   247
        B. *Experimental Determination of Cell Density*   249
        C. *Cell Density during the Division Cycle*   249
        D. *The Meaning of Isodensity*   250
    II. Turgor during the Division Cycle   250
   III. Are Density and Turgor Regulated during the Division
        Cycle?   252
       Notes   252

8. Variability of the Division Cycle   253

     I. Observed Variation during Cell Growth and Division   253
        A. *Comparing Variations*   253
        B. *Size Variations*   254
        C. *Temporal Variations*   254
    II. Elements of Variation during the Division Cycle   255
        A. *Model for Variability during the Division Cycle*   255
        B. A Priori *Qualitative Analysis of Variability*   257
          1. No Variation   257
          2. Variation in Replication–Segregation Sequence, the
             C and D Periods   257
          3. Variation in the Rate of Mass Synthesis   258
          4. Variation in Initiation Mass   258
          5. Variation in Symmetry at Division   259
          6. Variation in Total Interdivision Time   259
        C. *Size Homeostasis*   259
        D. *Cumulative and Noncumulative Variation*   259
   III. Variation in Equality of Division   262
   IV. Correlations among Different Variables   263
        A. *Correlations between Different Interdivision Times*   263
        B. *Relationship between Size at Initiation and Size at Division*   264
        C. *Relationship of DNA Synthesis and Cell Division*   264
     V. The Inverse Age Distribution   265
    VI. Theory of Backwards Analysis of the Division Cycle   270
   VII. Age–Size Structure of a Bacterial Culture   273
  VIII. Probability and Determinism   276
       Notes   276

9. The Segregation of DNA and the Cell Surface 279

    I. Early Studies on Macromolecule Segregation in
       Gram-Negative Bacteria 279
       A. The Results of Van Tubergen and Setlow 279
       B. Segregation in Microcolonies 280
    II. A Priori Considerations of Chromosome Segregation 281
    III. Methods of Analyzing Segregation of DNA 286
       A. The Methocel Method 286
       B. The Membrane-Elution Method 288
    IV. The Observation and Explanation of Nonrandom
       Segregation Patterns 288
       A. Random Segregation of DNA 288
       B. Nonrandom Segregation of DNA 288
          1. Model of Pierucci and Zuchowski 289
          2. Strand-Inertia Model 289
    V. The Segregation Model of Helmstetter and Leonard 292
    VI. The Alternate-Segregation Model 301
    VII. Alternation of Generations Forbidden 302
    VIII. Segregation of Plasmids 302
       A. Minichromosomes—Origin, Replication, and Significance 302
       B. The Paradox of High Minichromosome Copy Number 303
       C. The Passive-Segregation, Nonequipartition Model of
         Plasmid Maintenance 303
       D. Copy Number and Stability of Minichromosomes 306
       E. Directed Nonrandom Segregation of Minichromosomes 307
    IX. Mechanical Segregation Models 307
    X. Segregation of Cytoplasm 308
    XI. Segregation of the Bacterial Surface 308
       A. Unit-Cell Models of Cell Growth and Division 309
       B. Segregation of Cell Membranes 309
       Notes 310

10. Transitions and the Bacterial Life Cycle 313

    I. A Short History of the Bacterial Life Cycle 313
    II. The Bacterial Life Cycle as Shift-Ups and Shift-Downs 315
       Notes 317

11. The Division Cycle of Caulobacter crescentus 318

    I. The Growth and Division Pattern of Caulobacter
       crescentus 318

A. The Division Cycle of Caulobacter   319
B. DNA Replication during the Division Cycle of Caulobacter   319
   1. DNA Synthesis during the Division Cycle of Stalked and
      Swarmer Cells   319
   2. Is DNA Regulation in Caulobacter Circular?   323
C. Specific Protein Synthesis during the Division Cycle of
   Caulobacter   325
   1. Synthesis of Flagella-Related Proteins during the
      Division Cycle   325
   2. Synthesis of Other Proteins during the Division Cycle   325
D. The Pattern of Pole Development—The Predestined Path
   to the Stalk   326
   1. Linear Model of the Caulobacter Division Cycle   326
   2. Arguments against the Linear Model of Regulation of
      Periodic Protein Synthesis   331
E. Icon for the Caulobacter Division Cycle   333
II. Applications of the Caulobacter Division Cycle   334
   A. Methodological Implications of Caulobacter Growth   334
   B. The Age and Size Distributions of Caulobacter crescentus   334
   C. DNA Segregation in Caulobacter crescentus   325
III. Caulobacter as a Gram-Negative Rod   336
   Notes   337

12. Growth and Division of *Streptococcus*   340

I. The Division Pattern of Streptococcus   340
   A. Surface Growth in Streptococcus   340
   B. An Icon for Cell-Wall Growth of Streptococcus faecium   343
   C. DNA Synthesis during the Division Cycle of Streptococcus
      faecium   343
   D. The Synthesis of Mass   346
   E. Analysis of a Shift-Up Experiment   347
II. The Proposal of the Fundamental Cell   347
III. Events in Surface Growth   350
IV. The Surface-Stress Model and Streptococcal Growth   352
V. Segregation of DNA in Streptococcus   353
VI. Density during the Division Cycle of Streptococcus   353
VII. Comments on the Growth Pattern of Streptococcus   353
   Notes   355

13. Growth and Division of *Bacillus*   358

I. Surface Growth during the Division Cycle of Bacillus
   subtilis   358

A. Structure of the Bacillus subtilis *Cell Wall*   358
B. Turnover of Peptidoglycan   358
C. Inside-to-Outside Growth of Peptidoglycan   359
D. Zonal or Diffuse Surface Growth   360
E. Topography of Surface Growth   362
F. Variability of Growth and Filamentation in Bacillus subtilis   362
G. Multiseptate Growth of Bacillus subtilis   362
II. Cytoplasm Synthesis during the Division Cycle of *Bacillus subtilis*   363
   A. The Synthesis of Enzymes during the Division Cycle of
      Bacilli   363
   B. Macromolecular Synthesis and Cell Growth during the
      Division Cycle of Bacillus subtilis   365

III. DNA Synthesis during the Division Cycle of *Bacillus subtilis*   365
   A. Dichotomous Replication in Bacillus subtilis   365
   B. Bidirectionality of Bacillus subtilis *DNA Replication*   366
   C. Rate of DNA Synthesis during the Division Cycle of Bacilli
      in Balanced Growth   366
      1. Determinations of the C Period   366
      2. The D Period in *Bacillus*   368
      3. Determination of Cell Mass at Different Growth Rates   369
IV. The Surface-Stress Model and Growth of *Bacillus*   371
V. The Segregation of Cell Wall and DNA in *Bacillus*   371
VI. Growth and Regulation during the Division Cycle of
   Bacillus   371
VII. A Unified View of Bacterial Growth during the Division
    Cycle   372
    Notes   372

14. The Growth Law and Other Topics   375

  I. The Cellular Growth Law   375
 II. The Length Growth Law   380
III. Regulation of Synthesis at Initiation   381
IV. The Fundamental Experiment of Bacterial Physiology
    Reanalyzed   383
    Notes   387

15. The Eukaryotic Division Cycle   389

  I. The Eukaryotic Division Cycle   389
 II. An Alternate Analysis of the Eukaryotic Division Cycle   391

    A. The Definition of a G1 Cell   391
    B. The Continuum Model   392
    C. Formal Statement of the Continuum Model   395
    D. The Continuum of Division-Cycle Patterns   395
    E. Progression through the Eukaryotic Cell Cycle   396
    F. Analysis of the Variation in G1-Phase   397
    G. Analysis of G1 Arrest   399
    H. Release from G1 Arrest   400
    I. Analysis of G(0) Cells   402
    J. The G(0) Model of Zetterberg and Larsson   403
    K. Analysis of c-myc Synthesis during the Division Cycle   406
    L. Criteria to Demonstrate Cell-Cycle Regulation   408
    M. The Cloning and Identification of G1 Genes   410
    N. The Restriction Point   411
    O. The Transition-Probability Model   414
    P. Mitogenesis or Cytogenesis   417
    Q. Competence and Progression   417
    R. The Future of the Animal Cell-Division Cycle   419
 III. The Division Cycle of *Schizosaccharomyces pombe*   420
    A. Periodicities in the Growth of Schizosaccharomyces pombe   420
    B. A Critique of the Proposal of Periodicities in
       Schizosaccharomyces pombe   421
      1. Synchronization and the Experimental Evidence   421
      2. The Proposal of Quasi-Linear Synthesis   421
      3. Protein Synthesis during the Division Cycle of
        Schizosaccharomyces pombe   422
      4. Evolution and Periodicities   422
      5. Problems of Mechanism   422
      6. Historical Considerations   423
      7. The Division Cycle of *Saccharomyces cerevisiae*   423
 IV. Backwards Analysis of the Eukaryotic Division Cycle   424
  V. Terminology of the Division Cycle in Prokaryotes and
    Eukaryotes   425
 VI. The Eukaryotic Division-Cycle Icon   426
    Notes   426

16. Conservation Laws of the Division Cycle   429

   I. Conservation of Cell Age Order   429
  II. Conservation of Size Distribution   431
    Notes   431

Epilogue 432

   I. The Unity of Cell Biology 432
  II. Biosynthesis during the Division Cycle 433
 III. The Logic of the Division Cycle 434
 IV. The Remaining Problems 434
     A. *The Initiation Problem* 434
     B. *The Termination Problem* 434
     C. *The Invagination Problem* 435
     D. *The Division Problem* 435
  V. Cartesian and Deductive Science 435
 VI. The Baby Machine 435
     Notes 436

Bibliography 437
Author Index 473
Subject Index 483

Acknowledgments

A book does not come into being in the time it takes to write it. For me, this book is the product of a scientific lifetime. During that lifetime I have discussed these ideas with many people, and I have learned science from many others.

One of my earliest teachers was Norton Zinder. I thank him for introducing me to the excitement of science, and for setting high standards that have guided my work.

Ole Maaløe is another teacher to whom I am indebted, but it took me many years to understand the depth at which Ole thought about biological problems. His work is ever present in this volume, and his suggestions to look, and not touch, or to touch very gently, are some of the lessons I wish to propagate.

Many people have discussed these ideas with me, and among them I would note and thank Arthur Koch, Michael Savageau, Conrad Woldringh, Nanne Nanninga, Olga Pierucci, Frederick Neidhardt, Arieh Zaritzky, Robert Bender, Alan Leonard, Moselio Schaechter, Uli Schwarz, Jochen Höltje, and Jay Keasling. In my laboratory over the last 25 years, a number of associates have worked beyond the call of duty on various experiments, and I want to thank Therese Ruettinger, Sara Scanlon, and Ming-Lin Hsieh for their efforts.

Some have been kind enough to read and comment on various chapters. Thanks for this go to Austin Newton, Bert Ely, Michael Higgins, and David Dicker. Edward Birge read the entire manuscript, and I thank him for his care and insight which have greatly improved this book. Chuck Arthur, my editor at Academic Press, has been supportive and helpful throughout the book's gestation.

I save a special paragraph for Chick (Charles E.) Helmstetter. We met in Copenhagen in 1963, and through some divine intervention wound up with jobs in Buffalo some years later. The ensuing collaboration was one of the most exciting periods of my scientific career. For over a quarter of a century Chick and I have talked, written, and discussed the ideas in this book. I thank him for all of those phone calls and letters that he has received, his comments on this book and on other ideas, and for his friendship.

I thank Harold Winer for his help with the cover design.

But above all, I would note the special help of my wife, Sandi, who has discussed words and syntax and ideas and communication and writing for more hours than she would like to believe. Her help has been invaluable, and I want her to know that she is the driving force behind this creation.

Stephen Cooper

Illustrations

Figure

1-1.	Balanced Growth of a Bacterial Culture	8
1-2.	Age Distribution during Balanced Growth	10
1-3.	Graphic Proof of the Age Distribution	11
1-4.	Life Cycle of a Bacterial Culture	12
1-5.	The Schaechter–Maaløe–Kjeldgaard Experiment: The Fundamental Experiment of Bacterial Physiology	14
1-6.	The Shift-Up	15
2-1.	Icon for Regulation of Growth and Division	23
3-1.	Comparison of Differential and Integral Methods of Cell-Cycle Analysis	31
3-2.	Selective and Nonselective (Batch) Methods for Synchronization	32
3-3.	Size Distribution of a Bacterial Culture	39
3-4.	The Membrane-Elution Apparatus	42
3-5.	Synchronized Culture Produced by Membrane-Elution	43
3-6.	The Technique of Flow Cytometry	47
3-7.	Theory of Backwards Methods of Cell-Cycle Analysis	50
3-8.	The Membrane-Elution Method for Backwards Analysis of the Division Cycle	51
3-9.	Cell-Elution Pattern of a Membrane-Elution Experiment	53
3-10.	Analysis of Biosynthetic Patterns from a Membrane-Elution Experiment	54
4-1.	Protein Synthesis during the Division Cycle Using the Membrane-Elution Method	70
4-2.	Growth during the Division Cycle Determined from the Size Distribution	73
4-3.	Linear or Exponential Accumulation of Mass during the Division Cycle	78
4-4.	Proposed Patterns of Enzyme Synthesis during the Division Cycle	81
4-5.	Semilogarithmic Plots and Cell-Cycle–Specific Protein Synthesis	85
4-6.	Variation in Rate of Mass Synthesis by Repression of Auxiliary Proteins	87
4-7.	Icon for Mass Synthesis during the Division Cycle	90
5-1.	DNA Replication during the Division Cycle	96
5-2.	Continuous Variation of Cell Age at Initiation and Termination	102
5-3.	Continuous Variation in Times of Initiation and Termination Related to Cell Division	103
5-4.	Rate of DNA Synthesis during the Division Cycle Determined with the Membrane-Elution Apparatus	108
5-5.	Values of C and D at Different Growth Rates	110
5-6.	DNA Contents of Bacteria by Flow Cytometry	112
5-7.	Icon for DNA Replication during the Division Cycle	114
5-8.	Derivation of Cell Size at Different Growth Rates	116
5-9.	I + C + D Model of Chromosome Replication	120
5-10.	DNA Synthesis and Cell Division during Inhibition of Mass Synthesis	122
5-11.	Growth and Cell Division during Inhibition of DNA Synthesis	123
5-12.	The Maaløe–Hanawalt Experiment	126

5-13.   Chromosome Patterns following a Shift-Up   128
5-14.   Deviations from Rate Maintenance   131
5-15.   Icon for Regulation of DNA Replication   133
5-16.   Gene-Frequency Analysis of DNA Replication   138
5-17.   Nucleoid Separation and Production   141
5-18.   DNA Replication in Slow-Growing Cells   145
5-19.   Minichromosome and Plasmid Replication during the Division Cycle Using
        the Membrane-Elution Method   168
6-1.   Cellular and Molecular Structure of the Gram-Negative Bacterial Cell Wall   178
6-2.   Three-Dimensional Representation of Peptidoglycan Structure   181
6-3.   Growth of Peptidoglycan Area by Cutting of Stretched Bonds   182
6-4.   Zonal and Diffuse Cell-Wall Synthesis   188
6-5.   Rate and Topography of Peptidoglycan Synthesis during the Division Cycle   193
6-6.   Membrane-Elution Analysis of Surface and Mass Synthesis during the
        Division Cycle   198
6-7.   Why Peaks in the Ratio of Peptidoglycan-to-Cytoplasm Synthesis Occur
        Only Twice   199
6-8.   Increase in Cross-Linking by Natural Selection   208
6-9.   Definition of Cell Shape   214
6-10.   Constant Shape of Rod-Shaped Cells of Varying Size   216
6-11.   Development of Variable Widths during Bacterial Growth   221
6-12.   Length and Width Distribution in a Cell Population   223
6-13.   Peptidoglycan Synthesis in Septate and Nonseptate Cells of the Same Length   226
6-14.   Membrane-Peptidoglycan Interaction for Shape Maintenance   229
6-15.   Volume, Surface Area, and Length of Cells at Different Growth Rates   231
6-16.   The Constrained-Hoop Model   233
6-17.   Surface Synthesis during a Shift-Up   234
6-18.   Icon for Surface Synthesis   240
8-1.   Elements of Variation during the Division Cycle   256
8-2.   Size Homeostasis   260
8-3.   Cumulative and Noncumulative Variation   261
8-4.   DNA Synthesis as a Function of Cell Length   267
8-5.   Comparison of Classical and Inverse Age Distributions   268
8-6.   Comparison of Division-Cycle Analysis by Forward and Backwards Methods   272
8-7.   Age–Size Structure of a Bacterial Culture   274
9-1.   Formal Analysis of DNA Segregation   282
9-2.   The Methocel Method   287
9-3.   Advantages of Presegregation with the Methocel Method   291
9-4.   Comparison of Segregation Determined by the Membrane-Elution Method
        and the Methocel Method   293
9-5.   The Helmstetter–Leonard Surface-Area Model for Nonrandom DNA
        Segregation   294
9-6.   Why Segregation Randomness Varies with Growth Rate Using Methocel and
        Does Not Vary Using the Membrane-Elution Technique   295
9-7.   Analysis of Segregation by the Membrane-Elution Method   296
9-8.   Analysis of Segregation by the Methocel Method   298
9-9.   Nonequipartition Model of Minichromosome Segregation   304
10-1.   The Life Cycle of a Culture as Shift-Ups and Shift-Downs   316
11-1.   Classical Division Cycle of *Caulobacter crescentus*   320
11-2.   Chromosome Pattern during the Division Cycle of *Caulobacter crescentus*   321

11-3.	Alternative View of the Division Cycle of *Caulobacter crescentus*	328
11-4.	Schematic Analysis of *Caulobacter* Growth and Division	331
11-5.	Icon of the *Caulobacter* Division Cycle	333
11-6.	Age Distribution for Cells with Unequal Interdivision Times in Unequal Progeny	335
12-1.	Growth of *Streptococcus* Cell Surface with Overlapping Rounds of Wall Growth	341
12-2.	Icon for Streptococcal Cell-Wall Growth	343
12-3.	Chromosome Replication during the Division Cycle of *Streptococcus*	344
12-4.	Concept of the Fundamental Cell	349
13-1.	Inside-to-Outside Growth of *Bacillus subtilis* Cell Wall	359
13-2.	Side Wall and Pole Growth of *Bacillus subtilis*	361
13-3.	Presumed Chromosome Configuration in Growing *Bacillus subtilis*	367
13-4.	Residual Division in a Culture with a Variable D Period	369
13-5.	Cell Mass as a Function of Growth Rate in *Bacillus subtilis*	370
14-1.	Synthesis of Cell Components during the Division Cycle	379
14-2.	Calculation of Cell-Length Distribution with Variable Cell Widths	382
14-3.	Microscale Pattern of Protein Synthesis during the Division Cycle	384
14-4.	Idealization of the Schaechter, Maaløe, and Kjeldgaard Experiment	386
15-1.	G1 Arrest	393
15-2.	The Continuum Model	394
15-3.	Icon for the Classical Eukaryotic Division Cycle	396
15-4.	Comparison of the Eukaryotic and Prokaryotic Views of G1-Phase Variability	398
15-5.	Explanation of Complementation of G1 Mutants	399
15-6.	G1 Arrest according to the Continuum Model	400
15-7.	Schematic Description of the G(0) Model of Zetterberg and Larsson	403
15-8.	Reanalysis of a G(0) Model	405
15-9.	The Transition-Probability Model	415
15-10.	External Conditions Induce DNA Synthesis by First Activating Mass Synthesis	418
15-11.	The Frequency-of-Labeled-Mitoses Method: A Backwards Method for Eukaryotic Cell-Cycle Analysis	425
15-12.	Continuum-Model Icon for the Eukaryotic Division Cycle	426

Tables

Table

5-1.	Replication Point Determination by Autoradiography (after Bird, Louarn, Martuscelli, and Curo)	113
5-2.	Mass and DNA Content of Cells Growing at Different Rates	117
6-1.	Dimensions of Cells Growing at Different Rates	217
6-2.	Classification of Morphogenes (according to Donachie, Begg, and Sullivan)	236
8-1.	Cell Sizes and Their Variation at Particular Events during the Division Cycle	254
9-1.	The Segregation Results of Pierucci–Zuchowski and Its Comparison with Different Models	290
9-2.	Selection for High Plasmid Copy Number	306
12-1.	Cell Composition Expected for *Streptococcus faecium* at Different Growth Rates	354
14-1.	Rates of Chain Extension for Macromolecules	384
15-1.	Terminology of Cell-Cycle Phases in Prokaryotes and Eukaryotes	425

Prologue

And since then, I have tended to class certain research procedures with military operations: the will to conquer; applying a strategy and a tactic; the necessity of choosing a terrain; developing a plan of attack; concentrating one's forces on a particular sector, focusing and modifying the initial plan according to the reactions obtained. In short, going on the offensive on all fronts.

F. Jacob, *The Statue Within*, 1988.

I. THE FIELD OF DIVISION-CYCLE STUDIES

Growth of cells can be considered in two different but complementary ways. We can consider the growth of any property of a population such as the change in the total number or total weight of a group of individuals or cells. Alternatively, we can study the growth of individual members of the population. In the human population, an individual begins life weighing about seven pounds and grows rapidly over the next decade. The increase in weight then slows. There is no necessary relationship between the growth pattern of a population and the growth pattern of its individual members, as can be seen by our current population explosion. The growth pattern of an individual could be different. We could remain the same weight for 10 years, grow to 300 pounds by the age of 20, and then drop to 150 pounds; this would have no effect on the growth of the population.

This book deals primarily with bacterial growth at the individual cell level, and only slightly at the population level. It is the study of the bacterium during its cell cycle or division cycle. Though there have been a number of excellent treatises on the *growth* of bacteria, such as the text by Ingraham, Maaløe and Neidhardt,[1] its successor by Ingraham, Neidhardt and Schaechter,[2] or the encyclopedic treatment of *Escherichia*

1

coli and the related *Salmonella typhimurium*,[3] the division cycle of bacteria has rarely been given the central role. The seminal volume by Mitchison[4] is an outstanding exception.[5]

This volume describes and attempts to unify this important area of research. I feel that the field of the bacterial division cycle is not well understood by a large number of researchers outside of the field. When I have turned to the section on the division cycle in many microbiology texts, the ideas on the bacterial division cycle are usually misrepresented. It is *de rigueur*, for example, to present one method for synchronizing cells (usually size separation by filtration or centrifugation). The description of the synchrony method is usually followed by a general statement that the method allows the determination of the sequence of events that compose the division cycle. Even the simple notions that cell synchronization is the best approach to investigate the division cycle, or that there *are* events to be understood are, at best, not to be taken as gospel and are, at worst, wrong.

The study of the division cycle may serve as a model of progress in science. There have been many false leads, many incorrect results, and many difficulties. This field has had more than its share of scientific argument. But if the past is buried by the knowledge of the present we are doomed to repeat past errors. Only by seeing the past and interpreting it in the light of our current knowledge can we understand how to avoid the same problems today. This book deals with past arguments, controversies, and ideas that have dominated the field of division-cycle studies at various times in its development. There is a tendency to present science as a collection of known facts, in which progress follows a smooth and untroubled road. This perception not only takes away the human aspect of science, but it also deprives the student of the historical material that should be studied to see how the current ideas arose. The lessons of the study of the division cycle should find application in the study of other scientific areas.

The boundaries of this book are determined by the bacterial division cycle. Growth and biosynthesis of the cell is not treated except as it impinges on the division cycle. Many of the phenomena of bacterial growth are presented as black boxes, to be accepted and used and not further analyzed. For example, we accept that DNA replicates in a certain manner, using certain enzymes and precursors. Precisely how DNA replicates is not necessary for our understanding of the bacterial division cycle. How DNA replication is regulated and initiated, however, is of concern. When we limit our focus by this definition, we still come upon an enormous body of experimental and theoretical work. The approach has therefore been to search out and describe the general principles that govern the division cycle, and thus make the presentation of this body of work more manageable.

Most of the book will be devoted to a discussion of the rod-shaped, gram-negative *Enterobacteriaceae*, *Escherichia coli*, and *Salmonella typhimurium*. For historical reasons, these organisms have the best-understood division cycles. When other bacteria or cells are discussed, it will be within the framework of the *Escherichia coli* model of the division cycle. Although we will deal with other bacteria, such as the bacilli, streptococci, and stalked bacteria, these organisms will be used to study particular points or to emphasize differences or similarities of interest. I hope that future work will concentrate on these other bacterial systems, bringing our understanding of their division cycles to the same level as those of the *Enterobacteriaceae*. These other organisms will be better served in the future by experimentalists applying the knowledge of the gram-negative cells, rather than looking at each system as a new division cycle with its own particular principles and logic.

II. THE UNITY OF BIOLOGY

Bacterial models have been useful paradigms for understanding more complex systems. The application of bacterial virus studies to animal viruses is a well-known example. Similar observations could be made for studies of the genetic code, mutagenesis, cell genetics, protein synthesis, and DNA synthesis. During the last decade, the use of bacteria as models for higher systems has diminished. Because a number of phenomena present in animal cells—exons, introns, splicing, development, and immune responses—are not found in bacterial cells, there has been a tendency to proceed in new directions without looking back to work in bacterial cells. As part of this trend, in cell-cycle studies there has been a strong separation between ideas about bacterial cells and eukaryotic cells. There is a direct applicability (Chapter 15) of bacterial work to the analysis of the animal cell cycle. It is important that investigators of both types of cells appreciate this. If bacteriologists see this connection, their work will have an extra measure of relevance. If cell biologists see the connection, they will be able to take a fresh look at their past work, their models, and the future direction of experimental analysis.

The unity of biochemistry has been a keystone in its development. I suggest that a unity of cell biology, with regard to the division cycle, should also be recognized as a valuable principle.

III. THE HISTORY OF CELL-CYCLE STUDIES

Although the genesis of cell-cycle studies can be traced back almost 100 years,[6] the practical beginning is more recent. The observation of Hotchkiss,[7] that cells synchronized by heat shock were maximally transformed by DNA during a specific portion of the division cycle may be considered the start of the field. Other work during the 1950s dealt with attempts to produce synchronized cultures and to measure biochemical processes during the division cycle. This was merely a prelude to the "Fundamental Experiment of Bacterial Physiology."[8] This experiment of Schaechter, Maaløe and Kjeldgaard is taken as the end of the beginning and perhaps the beginning of the end. We shall have many opportunities to refer to this experiment throughout the book; in many ways, it is the foundation upon which the study of the bacterial division cycle rests.

In the decade following the Fundamental Experiment,[9] the division cycle problem was further refined by the work of Maaløe and his colleagues. They studied the biosynthesis of various molecules during steady-state growth and following simple perturbations of growth. One experiment in particular, the Maaløe-Hanawalt experiment,[10] introduced the concept of a round of DNA replication and thus defined the problem of the regulation and initiation of DNA synthesis. By the end of the sixties, the basic outline of the bacterial division cycle—at least with regard to DNA synthesis—was understood. In the next two decades, there was much exciting work on the bacterial division cycle. The lesson of this work is that the division cycle is a *simple* process. Many of the detailed molecular processes are complex and not yet understood, but we do understand the division cycle's broad framework.

IV. THEORY AND EXPERIMENT IN BIOLOGY

Biology, and microbiology in particular, are primarily experimental sciences. In other sciences this is not so. Theoretical approaches have grown up within which experiments can be discussed and critisized. Soon after Einstein published his Special Theory of Relativity, the eminent physicist Walter Kaufmann described an experiment that contradicted its predictions. Einstein blithely (if one can imagine a blithe Einstein) ignored this result and went confidently on with his work. Ten years later it was demonstrated that there were leaks in Kaufmann's vacuum systems; this defect invalidated his results. The Law of Conservation of Energy provides an even more common example. There is a long history of perpetual motion machines. If such machines existed, they would invalidate the law of conservation of energy. The latest was the patent application for a machine that had an input of milliwatts and an output of watts (Martin Gardner in *Science Digest*, October, 1985). When such reports are published, we do not revise the law of conservation of energy; rather, we criticize the experiment. This relationship of experiment and theory has been good for physics, but it does not exist, except in rare instances, in biology. There are times, however, when the theory is better than the experiment. To report conflicting results, with a simple "On the one hand . . . and on the other hand . . ." style prevents us from getting beyond the experimental results. Theory can help us make judgements about experimental work. One can, and should, make judgements about competing experiments.

There is an even more immediate and important reason to take a theoretical or conceptual view of the field; it aids pedagogy and memory. Consider the universally taught operon model of Jacob and Monod. One of the experimental supports of this model is a table of enzyme activities in cells containing one or more copies of the DNA coding for the structural and regulatory genes in their original or mutant form, and growing in the absence or presence of an inducer of the enzyme. The original table may be a dim memory to those who have read the original proposal, but the ideas and model remain clear. The various circuits of induction, by inactivation of a negative repressor, are found in all contemporary textbooks of microbiology and biochemistry. With this textbook description of the model, we can now re-create the original table. A model allows us to remember and understand an array of facts in what would otherwise be an unintelligible morass of meaningless numbers. So it is with the division cycle of bacteria. There are enormous numbers of facts, many contradictory and irreconcilable. How should we treat the facts? Are the facts all equal and presentable? Only by having a concept can we arrange the results in an understandable manner and make judgements as to which experiments are more likely to be correct.

NOTES

1. Ingraham, Maaløe, and Neidhardt, 1983.
2. Neidhardt, Ingraham, and Schaechter, 1990.
3. Neidhardt, Ingraham, Low, Magasanik, Schaechter, and Umbarger, 1987.
4. Mitchison, 1971.
5. The books by Lloyd, Poole, and Edwards (1982) and Edwards (1981) examine the division cycle specifically, but in my opinion do not do justice to the subject. Many of the ideas in those books are the opposite of the ideas presented here.

6. Ward, 1895; see Chapter 14 for an extensive history.

7. Hotchkiss, 1954.

8. Schaechter, Maaløe, and Kjeldgaard, 1958.

9. The two papers that belong to this fundamental experiment are Schaechter, Maaløe, and Kjeldgaard, 1958, and Kjeldgaard, Maaløe, and Schaechter, 1958.

10. Maaløe and Hanawalt, 1961.

1

Bacterial Growth

I. THE STUDY OF BACTERIAL GROWTH

Bacteria grow in different ecological niches and have varied patterns of growth. This book proposes that a single, archetypal description of the pattern and regulation of cell growth and division can accommodate these myriad patterns. In order to present and apply such a model, we must first have a system that can be fully described. The defined system we shall adopt is the growth of bacteria in a laboratory culture. Many call growth in the laboratory artificial and unrepresentative; they argue that most bacteria growing in nature are starving, and usually adhering to surfaces. Exponential growth in the laboratory with unlimited medium is therefore unrepresentative. The answer to this criticism is that cell growth under laboratory conditions is analyzable and reproducible. Further, the natural situation can be explained by ideas generated by laboratory growth, but the reverse is more difficult and less common.

A. Exponential Growth

When a bacterial culture growing in unlimited medium is kept below a given concentration by dilution at suitable intervals, the culture grows continuously and exponentially. If we plot the number of cells per volume (or any other property per volume) against time, a straight line is produced on semilogarithmic paper. The growth curve is specified by:

$$N_t = N_0 \cdot 2^{t/\tau}$$

where N_0 is the cell number at zero time, N_t the cell number at any time t, and τ is the doubling time of the culture. The cell number doubles every τ minutes. The line for any other cell property increases with the same doubling time as cell number; therefore, the plotted lines are parallel (Fig. 1–1). The average properties of the culture are constant with time. The *extensive* properties of the culture—the amount and the number—increase. The *intensive* properties—the average cell size, the RNA per cell, the DNA per cell, and so forth—remain constant and invariant with time. In practice, a culture can be kept growing exponentially for many hours to ensure that it is in a constant state of exponential growth. It is

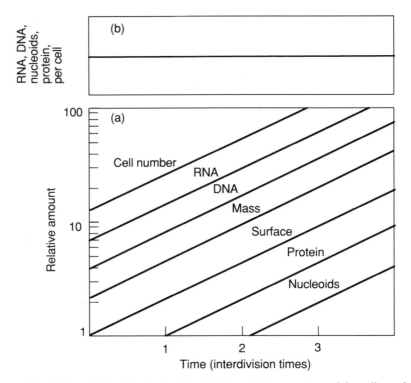

Figure 1.1 Balanced Growth of a Bacterial Culture. Determinations of the cell number, DNA, RNA, protein, or any property of cells in balanced growth give straight, parallel lines when plotted on semilogarithmic graph paper [panel (a)]. As the rates of increase of all cell properties are the same at all times, the average cell composition is constant. This is plotted in the upper graph where the cell composition is constant [panel (b)].

not unusual to have cells growing exponentially for up to 15 hours before performing an experiment.

B. Balanced Growth

Such an exponentially growing culture can be said to be in balanced growth.[1] In most textbooks on bacteriology, the growth of bacteria is usually presented as a curve in which an overgrown culture is inoculated into fresh medium. There is lag phase before cell number begins to increase, and then a period of time when the number increases, referred to as early-, middle-, and late-log[2] phases. The culture stops growing as it enters stationary phase, and a final death phase may follow. This growth pattern is not an example of balanced growth. Balanced growth occurs only in that middle phase of the classic growth cycle in which cells are growing exponentially with constant properties. One of the benefits of considering balanced growth is that cultures in balanced growth have constant properties. One need not consider the problem of obtaining a reproducible physiological state for each experiment. In balanced exponential growth, the properties of the cells are *a*historical. The cell properties are independent of the age of the culture. Any results obtained are independent of the precise time when samples were removed from a culture.

The lesson of this discussion should not be missed. The terms early-log, mid-log, and late-log phase are phrases that should be eliminated from the scientific literature. Such terms may be satisfactory merely to describe harvested cells irrespective of their physiological state. In experiments where the physiological state of the bacteria is important, however, such terms are ill-defined and irreproducible. Physiological experiments should use cells in balanced growth. Experiments in which the precise physiology of the cells is unimportant could describe the precise optical density or cell density when cells were harvested or analyzed; their position in the life cycle would be unimportant; growing cells for enzyme production would be one example. Only by being careful with the growth of bacteria will the concept of log, or balanced, growth be made a rigorous and useful concept.

C. The Age Distribution

Consider a bacterial culture in which all cells are growing with precisely the same interdivision time; there is no variability in the interdivision times. The doubling time, τ, of a culture is obtained by measuring the time required for any of the properties in Fig. 1–1 to double. The time for this culture to double is the same as the time between cell divisions. A newborn cell, usually referred to as a daughter or baby cell, originates by division of a mother cell. A baby cell has an age of 0.0 and a mother cell, an age of 1.0. Cells of intermediate age are referred to by their fractional age with a cell halfway between birth and division having an age of 0.5.

What is the age distribution of an exponentially growing culture? How many cells of each age are found in an exponentially growing culture? The age distribution is a plot of the cell frequency as a function of age. We might initially think that the age distribution of a growing culture is random or uniform. There are, however, twice as many newborn cells (age 0.0) as dividing cells (age 1.0), and there is a smooth distribution of cells between these ages. The age distribution is described by:[3]

$$F_\alpha = 2^{(1-\alpha)}$$

where F_α is the fraction of cells at age α during the division cycle.

The ideal age distribution is plotted in Fig. 1–2. At age zero the relative frequency is 2.0, at age 1.0 the relative frequency is 1.0, and at age 0.5 the relative frequency is 1.41. A simple proof of this distribution is illustrated in Fig. 1–3 where growth is plotted on a semilogarithmic graph. At time zero, there is a population of cells that will divide within the next doubling time. Consider the cell number increase during a short interval δt_1 at the start of the plot. This time interval is associated with a cell number increase δn_1. In the last time interval, δt_2, there is a cell increase δn_2. Inspection of semilogarithmic graph paper indicates that the increase in cell number at the end of the doubling time is approximately twice that at the start of growth. At the limit ($\delta t \rightarrow 0$), $\delta n_2 = 2 \cdot \delta n_1$. The cell increase at the beginning of the doubling time is due to the existing cells that are just about to divide in the original culture at time zero, and the cell increase at the end of the doubling time is due to the existing cells that were just born in the culture, since these newborn cells must wait for one doubling time before they again divide. It follows that there must have been twice as many newborn cells as mother cells in the original culture at time zero. As the properties of a culture in balanced growth do not change with time, during balanced exponential growth the number of newborn cells is twice that of the mother cells.

The age distribution is time-invariant, and therefore the age of the *average* cell is constant through time. As a simple demonstration, assume

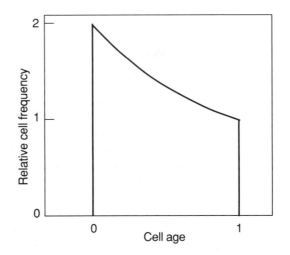

Figure 1.2 Age Distribution during Balanced Growth. The relative frequency of cells of different ages is plotted against cell age. In an ideal culture, where all interdivision times are equal, there are exactly twice as many young cells as old cells.

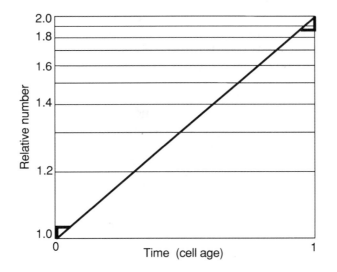

Figure 1.3 Graphic Proof of the Age Distribution. The growth of a cell culture is plotted over one interdivision time. At the left and right of the graph the rise in cell number for a short time is shown. The absolute vertical increase for the two triangles is different, with the increase at the end of the growth period twice as large as that at the start of the growth period. This is owing to the logarithmic scale. The increase at the end of the growth period is due to the existing younger or newborn cells in the culture at time zero, and the increase at the start of the growth period is due to the oldest cells that were just about to divide, so it can be seen that there must have been twice as many young cells as old cells at time zero.

that the age distribution was chosen to be artificially uniform, with all cell ages represented equally. The average age of the cells would be 0.5. An instant later each of the cells would move up from one age interval to the next. For all cells except the youngest cells, the number of cells in the interval would remain the same. The oldest cells divide to yield two young cells, so the number of cells in the youngest interval would be twice the number of cells in any other interval. The average age of the cells in the culture would be slightly younger than 0.5. This is an example of a time-variant age distribution. If such time variation existed, we would not have a culture in balanced growth. Each time cells were taken, the average cell age would be different, and experiments would not be reproducible. This is not the case for the exponential age distribution.

The age distribution means that no matter what the pattern of synthesis of a particular cell component is during the division cycle, the increase of that component in a culture growing in balanced growth is exponential. This is because the proportion of cells of each age does not change during exponential growth. The exponential age distribution described here is the only distribution that gives a population whose age distribution and characteristics do not vary with time (but see Chapter 8, where statistical variation is analyzed).

D. The Classic Life Cycle of Bacteria

Since the beginning of this century the bacterial life cycle has been a central part of bacteriology textbooks (Fig. 1–4). When an overgrown culture is diluted into fresh medium, an initial lag phase is observed, giving way to the log phase, and ending with a stationary phase. This pattern defined bacterial growth.[4] Experiments were described in terms of this life cycle. Stationary phase cells, early-log, mid-log, or even late-log phase cells were taken and studied. Henrici[5] measured cell sizes during the life cycle of a culture and showed that during these phases, there was a change in the size and shape of bacteria. The initial cells were small; they grew larger during lag phase, were largest during the log phase, and then became smaller during the late-log and stationary

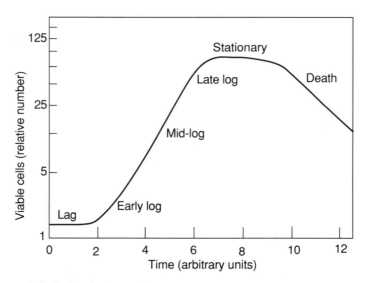

Figure 1.4 Life Cycle of a Bacterial Culture. The bacterial growth rate as measured by viable cells exhibits a precise sequence of changes when an overgrown culture is diluted into fresh medium. There is a lag before any cell-number increase is observed. Then there is an increase in cell number that gives a straight line when plotted on semilogarithmic paper (log growth). The log phase has been subdivided into early-, mid-, and late-log phases. The culture stops growing as the cells enter stationary phase. At extended periods, there is a decrease in cell number as cells die and lose colony-forming ability.

phases. The interpretation of these observations was that bacteria passed through a series of stages in a life history. The life cycle of bacteria could be studied from birth to death, in the same way that we might study the life cycle of a vertebrate organism. The cells described by Henrici were not in balanced growth. The properties of the cells were not constant, with the size and composition changing with time. When a cell is in balanced growth, the cell properties are constant and time-invariant. As we shall see, the life cycle of a bacterial cell (Fig. 1–4) is not a necessary process, and can be dispensed with when we study cells in continuous, steady-state, balanced growth.

II. THE FUNDAMENTAL EXPERIMENT
OF BACTERIAL PHYSIOLOGY

A. Steady-State Growth

A new interpretation of the classic growth curve was introduced by Schaechter, Maaløe, and Kjeldgaard,[6] who studied bacteria in steady-state growth. As a culture grew, the bacteria were diluted back at intervals, so they never achieved a cell concentration greater than some value. Unlike the cells in the classic life-cycle culture, such cells had a constant size. Cells sampled at any time in a steady-state culture were identical to cells sampled at any other time.

When cells were grown in media that allowed widely divergent growth rates, from cells with a 25-minute doubling time to cells with a 2-hour doubling time, Schaechter, Maaløe, and Kjeldgaard found that each doubling time defined a particular physiological state of the cell. As can be seen in Fig. 1–5, the composition of a cell, in terms of RNA, DNA, cell mass, or cell nucleoids,[7] was determined by its particular rate of growth, and was independent of the means used to obtain that growth rate. If a particular bacterial growth rate was obtained by adding 10 amino acids to minimal medium, and the same growth rate was achieved by adding six nucleosides to minimal medium, the size and composition of the bacteria were the same in the two different media. The physiological state was determined by the growth rate and not by the composition of the medium. Slow-growing cells were smaller, had less DNA and RNA, and had fewer nucleoids per cell than fast-growing cells. This observation now explained, in part, the classic life cycle (see Chapter 10 for a detailed analysis). Stationary-phase cells are slow-growing cells with an infinite interdivision time. These are the smallest cells; rapidly growing cells are the largest. Cells diluted into fresh medium grow larger and

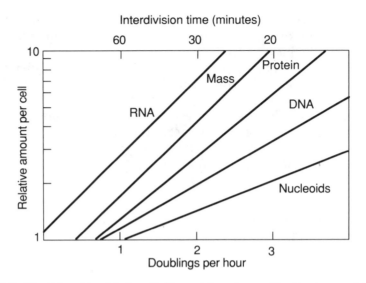

Figure 1.5 The Schaechter–Maaløe–Kjeldgaard Experiment: The Fundamental Experiment of Bacterial Physiology. When bacteria are grown at a number of different growth rates by adding different carbon sources or different supplements of amino acids, vitamins, or nucleosides, the macromolecular composition of the cells changes. The abscissa gives the doublings per hour, with a 3 indicating rapidly growing cells, with a 20-minute doubling time (three doublings per hour) and a 1 indicating slower cells with a 60-minute doubling time. The experimental lines are straight lines on semilogarithmic paper, but the slopes of the lines are different, indicating that the ratio of the different components changes at each growth rate. The number of nucleoids per cell is an average number, as cells have either one or two nucleoids.

achieve their largest size during the period of exponential or log-phase growth. As the culture enters the stationary phase, the cells slow their growth and become smaller.

Schaechter, Maaløe, and Kjeldgaard were primarily concerned with the numerical relationships of RNA composition and content. At the time of their work, the relationship of the ribosome to protein synthesis was just being explored. For our purposes, the DNA content of bacteria growing at different rates will be examined more closely. The amount of DNA per cell varies over a fourfold range, and there appears to be a continuous variation in DNA content with growth rate. How can this be? Does a bacterial cell growing at rapid rates have 100 chromosomes or genomes? As growth rate decreases, do the cells have successively 99, or 98, or fewer and fewer chromosomes per cell? How is the RNA content, or protein content, or cell size, regulated at different growth rates?[8]

B. The Shift-Up

These initial and elegant studies of balanced growth were followed by an analysis of the dynamic changes in cell composition as one growth rate was changed to another.[9] Because the experiment involved an increase in growth rate (by adding additional nutrients to bacteria growing in a minimal medium), this experiment is generally referred to as a nutritional shift-up. Typical results of a shift-up are shown in Fig. 1–6. There is an essentially immediate change in the rate of increase of cell mass and a slower change in the rate of increase of DNA. Most curiously, the rate of cell division continues at the pre-shift rate for approximately 60 minutes; then the rate of cell division shifts abruptly to the new growth rate. The phenomenon of continued cell division at the pre-shift rate for approximately 60 minutes is called *rate maintenance*.

The shift-up results fit, and in a sense explain, the steady-state composition results. Because the rate of mass, RNA, and protein synthesis increases first, followed by DNA synthesis, and only then by cell division, the resulting cells are larger and have more RNA, protein, and DNA after a shift-up. Although the phenomenon of rate maintenance could not have been predicted from the steady-state results, it could have been

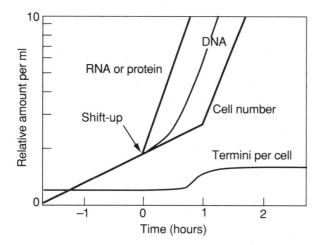

Figure 1.6 The Shift-Up. Cells growing in a poor or minimal medium are shifted to a richer medium (by the addition of more nutrients), and the cell number increase and macromolecular syntheses followed. For approximately 60 minutes, the rate of cell increase remains unchanged (rate maintenance) before it abruptly changes to the new growth rate. The change in the rate of mass increase is essentially instantaneous, while there is a delay in the rate of increase of DNA synthesis.

predicted that the cell number would be the last to achieve a new steady-state rate of increase following a shift-up.

We can now understand the dominance of the classical life cycle before the fundamental experiment. In the early days of bacteriology, spectrophotometers were not available.[10] The main method of measuring bacterial growth was counting colonies. There was rate maintenance of the nongrowing cells when they were diluted into fresh medium, so there was no observed growth for at least 60 minutes. This is the classic lag phase. When mass measurements were made on cells placed in a fresh medium, growth was observed during the lag phase.[11] The lag phase thus corresponded to the immediate increase in the rate of mass synthesis during the shift-up, with a period of *rate maintenance* for cell number.

The history of bacterial growth, as typified by the life-cycle concept, is neatly explained as a succession of shift-ups and shift-downs, with the corresponding changes in cell size. In Chapter 5, the biochemical basis of the phenomenon of rate maintenance will be explained. In Chapter 10, the life cycle of a culture will be analyzed in more detail.

C. The Copenhagen School

Because the original work on steady-state growth and shift-ups was performed in Copenhagen, and many of the early workers in the field studied in the Maaløe laboratory, the ideas generated by these experiments have been collectively referred to as the Copenhagen School.[12] By *School*, I do not mean a particular institution that teaches a body of knowledge, but rather a mode of thinking, a way of doing experiments, an approach to bacterial growth exemplified by the work initiated in Copenhagen. In addition to a particular mode of thought, there is a particular set of ideas and paradigmatic experiments that are central to the discussions of the Copenhagen School. Many who have never been to Copenhagen feel they are a part of this tradition.

One of the central tenets of the Copenhagen School is that the best experimental approach to a biological system is one that perturbs the system least or not at all. Experiments in Copenhagen treated the systems gently, and any approach that might perturb the cells was either used cautiously, or discarded in favor of less disturbing methods. "Look, don't touch," may be one way of summarizing the Copenhagen School. Make measurements on cells in undisturbed states the approach suggests, and by making these measurements, try to understand the natural state of affairs. This is seen in its clearest form in the measurements on cells grown at different rates. As we shall see in Chapter 3, the perturbation of cell growth is a real and ever-present problem.

In addition, the Copenhagen School formulated a quantitative theoretical perspective. It allowed us to think clearly and accurately about the growth of the bacterial cell, and to draw subtle conclusions from experiments. Calculations based on quantitative measurements gave insight into the mechanisms of bacterial cell growth. The same approach can be applied to the analysis of the bacterial division cycle.

NOTES

1. Campbell, 1957.
2. The term log arose because the data for bacterial growth give a straight line on semilogarithmic graph paper. The equation of growth is more properly described as exponential. For historical reasons, the term log phase has become synonymous with exponential growth.
3. The equation should be $F_\alpha = \ln2 \cdot 2^{(1-\alpha)}$. The integral of this equation gives a unit amount of cells in the culture. This does not change the idea implicit in the equation, that there are twice as many young cells as old cells.
4. Buchanan, 1918.
5. Henrici, 1928.
6. Schaechter, Maaløe, and Kjeldgaard, 1958.
7. The concentrations of DNA in the bacterial cell are referred to as nucleoids to distinguish them from the membrane enclosed genomes found in eukaryotic cells.
8. The original paper by Schaechter, Maaløe, and Kjeldgaard is notable for a number of other ideas. It may be the first explicit statement regarding the difficulty of deciding between a linear function and a logarithmic function when the data vary over a factor of two. They also used the chemostat to obtain very slow growth rates, and defined the difference between restricted and unrestricted growth.
9. Kjeldgaard, Maaløe, and Schaechter, 1958.
10. Longsworth, 1936.
11. Hershey, 1938, 1939, 1940; Hershey and Bronfenbrenner, 1937, 1938. This is the same Alfred Hershey who won a Nobel Prize for the analysis of T-even phage growth. It is forgotten that he also had an important place in defining the nature of bacterial growth.
12. The book by Maaløe and Kjeldgaard (1966) is a superb summary of the original ideas that form the core of the Copenhagen School.

2

A General Model of the
Bacterial Division Cycle

I. THEORY, EXPERIMENT, AND BIOLOGICAL UNDERSTANDING

The history of biology is in the Baconian tradition; it is an inductive science. Researchers look at biological systems, amass experimental results and observations, and from this array of data, they bring forth general theorems and models. An opposite tradition is the deductive or Cartesian approach. In this tradition, models are conceived and predictions derived; experiments are then used to test the proposed theories or models. This approach has been more successful in physics than in biology.

Even though biology is a firmly rooted experimental science, and generally proceeds through inductive processes, our best understanding of biological phenomena comes when we can interpret a wide array of experimental results in terms of an idealized model. When facts are so interpreted, the facts are seen as derived from the model itself. One well-known example is the operon model of Jacob and Monod. Measurements of enzyme production in different conditions, with combinations of different structural and regulatory gene mutations placed on the same or different strands of DNA, form the experimental base of the Jacob–Monod model. We may not remember the original published results, but they may be reconstructed by working out the predictions of the model portrayed in textbooks. Our understanding of the results comes from having the model before us. Without that model the results are confusing, unintelligible, and forgettable.

Confusion is possible in the study of the bacterial division cycle because the body of experimental results is large, and largely contradictory. Some order may be brought to this field by considering a general model of the division cycle and looking at experiments in terms of this model. The model presented in this chapter will be supported by experiments described in later chapters.

II. THE AGGREGATION PROBLEM

In economics, the aggregation problem concerns how to combine various sectors of an economy in order to understand and predict the overall behavior of an economy. Should the figures for the production of capital machinery be combined with those for the production of consumer goods? Is paper produced for boxes in the same economic category as stationery? It is difficult to treat each item in an economy individually; some aggregation is necessary in order to understand the whole system. For example, the economy of an individual country is often aggregated into a single number, the gross national product. We must now consider the aggregation problem for the analysis of the bacterial division cycle.[1]

In the analysis of the bacterial division cycle, the question arises as to how we aggregate the different parts of the cell in order to achieve an understanding of the biochemistry of growth and division. Is there a unique pattern of synthesis during the division cycle for each enzyme? Or are there a limited number of patterns with different enzymes or molecules synthesized according to any one of these patterns? Are there ways of grouping proteins or RNA molecules so that we can consider classes of molecules rather than individual molecular species? Should we consider the cell membrane a different category from that of peptidoglycan? Are the enzymes involved in macromolecule metabolism synthesized differently during the division cycle from those involved in small-molecule metabolism? There are approximately a thousand proteins in the growing cell, and if there were a unique cell-cycle synthetic pattern for each protein, or if there were only a few enzymes exhibiting any particular pattern, we would have an insuperable task describing the biosynthesis of the cell during the division cycle.

With regard to synthesis during the bacterial division cycle, there are only three categories of molecules, each of which is synthesized with a unique pattern. The growth pattern of the cell is the sum of these three biosynthetic patterns. The first category is the cytoplasm, the entire accumulation of enzymes, proteins, RNA molecules, ribosomes, small molecules, water, and ions that makes up the bulk of the bacterial cell. It is the material enclosed within the cell surface. The second category is the genome, the one-dimensional linear DNA structure. For our understanding of biosynthesis, the linear aspect of DNA is important, although the folded genome is a three-dimensional object. The third category is the cell surface, which encloses the cytoplasm and the genome. The surface is composed of peptidoglycan, membranes, and membrane-associated proteins and polysaccharides. Everything in the cell fits into one of the three categories, and each category has a different pattern of synthesis during the division cycle. These three patterns are simple to understand as they can be derived from our current knowledge of the principles involved in the biosynthesis of cytoplasm, genome, and cell surface.

A. Cytoplasm Synthesis

The cytoplasm of the cell is composed of all macromolecular and low-molecular-weight material that is not part of the genome or cell surface. To a large extent, it is composed of the elements of the protein-synthesizing system. Consider a unit amount of bacterial cytoplasm. The cytoplasm assimilates nutrients from its environment, metabolizes them, and prepares low-molecular-weight precursors. The cytoplasm then synthesizes all of the enzymes, ribosomes, and macromolecules that make up the cytoplasm of the cell. How does this unit of cytoplasm increase? It increases exponentially. As an example, consider a unit amount of cytoplasm. Assume there are 1000 ribosomes in this unit of cytoplasm, and the rate of protein synthesis, including ribosomal protein synthesis, is proportional to the number of ribosomes. Soon there will be 1001 ribosomes. The rate of protein synthesis is now increased by 0.1%. The next ribosome is made in a period that is slightly shorter than that for the first ribosome. As more and more ribosomes are made, the rate of protein synthesis increases; eventually the number of ribosomes is doubled. The doubling time of the cytoplasm is the time taken to have 1000 ribosomes make 1000 new ribosomes. This time is less than 1000 times the time it took for the first 1000 ribosomes to make one ribosome, because the newly made ribosomes participate in the biosynthetic reactions immediately after being made, and the rate of ribosome synthesis continuously increases.

In the same manner as ribosomes, enzymes also increase exponentially during the division cycle. The enzymes present at the start of the measurements make the precursors for more enzymes, and the newly synthesized enzymes participate immediately in the biosynthetic processes. In this way, the entire cytoplasm of the cell grows exponentially. Every part of the cytoplasm increases exponentially, so the composition of the cytoplasm is invariant during the division cycle. As shown in Chapter 1, for exponential growth of a culture of cells, the composition of the culture does not change with time. In the same way, the composition of the cytoplasm of each individual cell does not change with time during the division cycle.

The time for the synthesis of cytoplasmic molecules is short compared to the bacterial interdivision time. For this reason, the synthesis of the cytoplasm is continuous. There are no events that mark cytoplasm synthesis during the division cycle, as the relative rate of cytoplasm synthesis (i.e., the rate of accumulation *per* existing amount of cytoplasm) is constant and invariant. We could argue that there are events occurring at a microscopic level. Each synthesis of a ribosome could be considered a unique event (see Chapter 14 for a detailed analysis). We could measure these events by observing when each new ribosome appears. The time scale of these events is insignificant when compared with the time scale of the division cycle. In a bacterial cell with 20,000 ribosomes, 20,000 new ribosomes will be made within 3600 seconds; this is approximately 6 ribosomes per second. It is not possible to resolve these individual synthetic events, so we do not consider these molecular events relevant to the division cycle.

Linear synthesis of cytoplasm has also been proposed. In this case, new cell cytoplasm is not produced at constant efficiency; the new components act as though they are unable to join immediately in new biosynthetic reactions. In Chapter 4, we will argue that the exponential mode of synthesis is not only the simplest, but is also the actual mode of cytoplasm synthesis.

B. DNA Synthesis

The synthesis of DNA during the division cycle is different from cytoplasm synthesis. DNA synthesis during the division cycle is not exponential. Let us assume that the unit cytoplasm of the previous section is associated with a single piece of DNA or bacterial genome. At some time during cytoplasm synthesis, the DNA, in some unknown way, senses that there is a particular amount of cytoplasm, let us say that associated with 20,000 ribosomes. At that time the DNA begins to replicate. Replication occurs at a constant rate, from one end of the DNA to the other. Assume that the time for DNA to replicate is less than the doubling time of the cytoplasm. During that synthetic period, the DNA will double in amount, and the rate of DNA synthesis will be constant. A constant synthetic rate means that the rate of DNA chain elongation is invariant and does not change as the DNA replicates. After the completion of DNA replication, there are two DNA molecules. The DNA does not replicate again because the two DNA molecules share the extant cytoplasm. The amount of cytoplasm per DNA molecule is less than the amount of cytoplasm necessary for initiation, so the two DNA molecules are not initiated again until the cytoplasm doubles, that is, when there are 40,000 ribosomes present, 20,000 for each DNA molecule.[2]

If DNA synthesis ends before another initiation event, there may be periods of the division cycle devoid of DNA synthesis. If DNA synthesis is initiated before prior replication points are terminated, there may be complex, multiforked replicating structures. At all growth rates, however, the pattern of DNA synthesis during the bacterial division cycle is composed of constant rates of DNA synthesis. These constant rates are explained by the presence of an integral number of replication points at which DNA is synthesized at any time during the division cycle. Thus, as with cytoplasm, the pattern of DNA biosynthesis during the division cycle can be explained in terms of the molecular events involved in the DNA biosynthesis.

C. Surface Synthesis

The final component of the bacterial cell, the cell surface, has an altogether different mode of synthesis. The cell surface grows to just enclose the cytoplasm of the cell. The pattern of growth is not simple to describe, but for rod-shaped cells we can give a general description of the process. Consider that the amorphous cytoplasm described above is contained in an open cylinder. Assume that the cylindrical cell surface increases to just enclose the newly synthesized cytoplasm. The length of the cylinder doubles in exactly the doubling time of the cytoplasm. In this case, the rate of surface area synthesis is strictly proportional to the rate of cell cytoplasm synthesis, and is exponential. This picture of a bacterial cell surface does not exist, as in reality there are ends to the cylindrical cell wall. Because new ends of the cylinder are not made throughout the division cycle, this exponential pattern of wall synthesis does not exist. The volume of the cell increases exponentially, and the cell has a tubelike shape with hemispherical ends, thus cell-surface synthesis cannot be exponential. As we shall see, the cell surface grows so that the density of the cell is constant, and there is just enough surface to cover the existing cytoplasm. This means that the cell volume increases exponentially. Although the rate of surface synthesis is close to exponential, it is not precisely exponential; therefore, the cell surface is in a different category from that of the cytoplasm.

III. INTERRELATIONSHIPS BETWEEN CYTOPLASM, DNA, AND SURFACE SYNTHESIS

There is a simple relationship between the three major components of the cell. The cytoplasm accumulates as fast as it can in a given environment. It does not outpace the rate of DNA synthesis because DNA initiation, the regulatory point of DNA synthesis, is regulated by cytoplasm synthesis and accumulation. If cytoplasm is synthesized more rapidly, then the

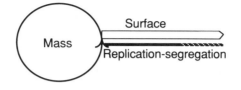

Figure 2.1 Icon for Regulation of Growth and Division. An icon is a simplified representation of a rule in pictorial form. The circle represents the continuous increase in cell mass—without a beginning and without an end—and the two lines indicate that the mass is driving both the replication–segregation sequence (DNA synthesis) and cell-surface synthesis. As we shall see, the requirement for mass synthesis is continuous for surface synthesis, while it is required only for the initiation of DNA replication, not for continued DNA replication.

cells initiate DNA synthesis more frequently; thus cytoplasm synthesis never outpaces the synthesis of DNA. Similarly, surface is synthesized to just enclose the cytoplasm. Cytoplasm synthesis is the driving force regulating the synthesis of the other cell components (Fig. 2–1).

IV. PASSIVE AND INDEPENDENT REGULATION

In the model presented here, one idea is implied by what the model does not propose. That idea is that the relationships among the different components of the cell are regulated by passive, local mechanisms. The different components act as independent entities that are synthesized in response to local signals. There is no superregulatory mechanism coordinating the biosyntheses. A timer that runs throughout the cell cycle might be envisioned in this model; the synthesis of various cell components would respond to, and follow, the dictates of the central timing mechanism. The master timer tells each cell component when to start synthesis or change the rate of synthesis.

In contrast, it is proposed that there is no single regulating or organizing mechanism. The cytoplasm increases in response to the interaction of the nutrients, precursors, energy sources, and environmental factors. DNA synthesis responds to cytoplasm synthesis; it is synthesized at a particular rate during the division cycle without regard to the continued synthesis of cell cytoplasm, or when DNA synthesis occurs during the division cycle. The cell surface grows or stretches to accommodate the increased cytoplasm or cell mass.

V. THE LOGIC OF THE DIVISION CYCLE

The first rule of cell division is that a cell shall not divide unless two genomes are present. This requires some means or mechanism to ensure that cell division does not occur, or start to occur, unless the genome has replicated once. If cell division occurred with fewer than two genomes present, at least one of the daughter cells would not survive. The second rule of the division cycle is that a cell shall not start DNA synthesis unless the cell has enough cytoplasm. As we shall see, once DNA synthesis starts, cell division shall usually follow. This second rule ensures that if division followed the initiation of DNA synthesis, and even if there were no more cytoplasm synthesis following the initiation of DNA synthesis, each of the daughter cells would have the minimal amount of cytoplasm to enable it to survive. For example, it would be unproductive for a cell to divide with only one RNA polymerase molecule; one of the daughter cells would be dead, as it would not be able to make any RNA in order to make even the first RNA polymerase molecule. By ensuring that DNA synthesis is not started unless some minimal amount of cytoplasm is present, then even if cytoplasm synthesis ceased for the duration of the division cycle, the cytoplasm in the two daughter cells would be sufficient for survival.

NOTES

1. Cooper, 1990b.
2. This nonexponential synthesis of DNA in a single cell should not be confused with the exponential synthesis of DNA during the balanced growth of a bacterial population consisting of many cells. The exponential synthesis is a property of the population and is independent of the synthetic pattern during the division cycle.

3

Experimental Analysis of the Bacterial Division Cycle

There is always a question as to whether experimental methods should be discussed before or after the experiments. If methods were precise techniques allowing us to understand natural events, as they are generally portrayed, then it would make sense to present the methods first. We would describe how methods are used to understand reality. But the situation is not that simple; we must be able to judge methods and choose between competing approaches. We need a historical view of methods to see which have worked in practice and not merely in theory, and to see which have led to greater understanding. This suggests that we see how the experimental results were obtained, and then note which experimental approaches work. For practical reasons, it is important to have an introduction to the methods used to study the division cycle. Without such an introduction, much of the discussion in the following chapters would be confusing.

I. THE EXPERIMENTAL ANALYSIS OF THE DIVISION CYCLE

Experimental approaches to the study of the division cycle may be separated into two distinct categories, synchrony and nonsynchrony methods. Because cells are so small, a large number of them are required to make a chemical measurement. In order to have a large number of cells at a particular age or time during the division cycle, many investigators have developed methods to produce synchronized cultures; such cultures are believed to contain cells that are all the same age. With time, the cells progress as a cohort through the different ages of the division cycle. By sampling the culture at different times we can measure the composition of the cells or the activity of a particular function, and determine the rate of change of cell properties as a function of cell age.

Because there are problems with synchrony methods, other complementary, nonsynchrony approaches have been developed. The main benefit of these nonsynchrony methods is that the investigated cells may

be unperturbed. Because there are no naturally occurring cell populations that are synchronized in a manner suitable for cell-cycle studies, we
always have to do something to produce a synchronized culture. Although the possibility is usually noted that such synchronization procedures may produce alterations to the normal cell cycle, this caveat has
been more honored in the breach than in practice.

II. SYNCHRONIZATION VERSUS
NONSYNCHRONIZATION METHODS

If a survey asked biologists, "What is the most appropriate approach to
studying the division cycle?" synchrony methods would be, without
doubt, the most popular answer. Because chemical measurements on
single cells are difficult, and because many measurements have to be
made as cells pass through the division cycle, it is obvious that a synchronized culture could produce enough material for analysis.

Nonsynchrony methods are also available, and they are superior to
synchrony methods. It is ironic, considering the effort expended on
synchronization methods and synchronized cultures, that synchrony
studies have added little to our understanding of the bacterial division
cycle. It may be that synchronized cultures will be used to obtain some
new result in the future, but the most that can be said now is that
synchronization can only confirm known results. This critique may come
as a surprise, but there are historical as well as theoretical reasons for
eliminating synchrony as a method for the study of the division cycle.

A. Criteria for Synchronization Methods

There are many methods proposed for synchronizing cultures. In order
to compare different methods we need a set of criteria for judging the
different methods. The following criteria are proposed:

1. *If the selected cells are newborn cells, there should be no cell division for a
   significant fraction of the generation time.* Assume that a method was
   developed that produced synchronized newborn cells. At the start of
   the experiment the slope of the plot of cell number against time
   should be as close to zero as possible. If the slope were significantly
   above zero, that would mean that some cells were dividing. These
   dividing cells would contaminate the younger cells. It is assumed that
   there is some minimal time between birth and the next division, and
   that the minimal time is some significant fraction of the interdivision
   time. It is difficult to propose quantitative limits, as all experiments

have their own set of limitations. A particular degree of contamination or division might be acceptable in one experiment but not in another.

2. *The time required for the increase in cell number at the time of synchronous division should be a small fraction of the interdivision time.* Division should occupy a relatively small time span. This is the most important criterion for judging synchrony methods; it determines the resolution with which different cell cycle phenomena can be distinguished during the division cycle. That the time for the increase in cell number should be as short as possible is obvious, but this criterion is usually not discussed in synchrony experiments. In a perfectly synchronized culture, this time would be negligible. The longer the increase time, the more the culture is not synchronized, and the more the culture is like an exponentially growing culture. If the starting culture had a doubling time of 60 minutes, a well-synchronized culture would have a very short increase time—i.e., the period during which the majority, if not all, of the the cells divide once—possibly as short as 5 to 10 minutes. This is generally not found; in most cases the increase time is as broad as 20 minutes. This means that the resolution of the synchrony method is only 10 to 20 minutes. It is difficult to make firm rules and set quantitative limits for the allowable initial increase and the slope during synchronous division, but investigators are encouraged to at least characterize their experiments by noting any initial increase and the increase time.

3. *After one doubling time (measured in the exponentially growing culture), the cell number should double; successive generations should be of equal length.* No specific limits are proposed that eliminate a synchrony experiment. At a minimum, the cell increase factor should be noted in any synchrony experiment (i.e., 95% increase, or a 1.95-fold increase in cell number). A variable generation time indicates some perturbation of the cells. In an ideal synchrony experiment, the time for half of the cells to divide equals the time for the cells to double in exponential or log-phase growth.

4. *There should be at least two independent cycles of synchronized growth; any experimental results should occur similarly in the two successive cycles of synchrony.* Any results obtained in the first cycle could be due to a perturbation of the cells, the most common and most often overlooked problem with synchronized cells. Many reported results demonstrating particular patterns of growth or biosynthesis during the division cycle are due to perturbations caused by synchronization; the first generation results are not reproducible in the second and succeeding generations.

5. *When plotted on a graph to demonstrate synchrony, the cell numbers, as well*

as any measurements of biosynthesis, should show plateaus and rises without the need to add lines. Any enhancement of the results by added lines should be unnecessary. Often it is possible to take a published synchrony curve or synchrony result, trace the points leaving out the interpretive line, show it to independent observers, and discover that they do not see the synchrony. Enough points should be taken to allow experimental values to stand alone.

6. *The cell number should be obtained by a method that eliminates investigator bias.* The use of a Coulter Counter to determine cell number is superior to the use of a haemocytometer. The results are more accurate, and there can be no unconscious investigator bias in the determination of the cell number.

   The Coulter Counter is an electronic apparatus that measures the sizes and numbers of particles in solution. A glass tube with a small orifice is placed in a solution, and a fixed volume of fluid is drawn through the orifice. Electrodes placed astride the orifice allow the resistance to be measured across the orifice. Any particles entering the orifice cause an increase in resistance in proportion to their size. Measuring the distribution of resistance, and hence the sizes, reveals the size distribution of a population of particles. The number indicating the concentration of particles is objective, and not subject to operator error. The numbers are accurate, reproducible within 1%, and simple to obtain.

7. *Any results obtained by a synchrony method should be consistent with available observations on cells growing in unperturbed exponential growth.* For example, if some process can be shown to occur in cells of all ages during unperturbed exponential growth, and after synchronization the process occurs only during a narrow age interval, then the synchrony method is not good for studying the normal (i.e., unperturbed) division cycle.[1]

8. *If some aspect of the cell cycle is known, then a synchrony method should confirm the result.* If a biosynthetic pattern in the division cycle is known from other nonsynchrony methods, then the synchronized culture should yield a confirmation of that result. Performing this control ensures that the synchronization procedure does not perturb the cells and obscure a known result.

9. *The synchrony should decay.* There is a variation in the interdivision times of cells in a culture, and this variation has a cumulative component (see Chapter 8). This means that variation is additive, and synchronization decays. If a synchronized culture maintains synchrony for an extended number of generations without decay (i.e., the synchronous divisions do not become less and less distinct), this is

evidence for growth different from that in the original exponentially growing culture. That a particular cell culture had a smaller amount of variation than other cultures would have to be demonstrated independently.

It is also important to have at least a theoretical understanding of the mechanism by which a synchronized culture is produced. For most methods, this is a trivial exercise; smaller cells are assumed to be younger cells. For other methods, particularly those relying on batch treatments, such an understanding is elusive. As we shall see in Chapter 16, there are theoretical reasons to expect that batch treatment methods will not produce a synchronized culture.

The criteria presented here are based on the assumption that we wish to study the normal, unperturbed cell cycle of growing cells. To study synchronized cultures irrespective of whether they are normal, or if perturbations from normality have no bearing on the work (i.e., to get a significant fraction of bacteria with septa, or animal cells in mitosis), then these criteria might not be important. To understand the normal cell cycle, discard these criteria only at the risk of obtaining artifactual results.

B. Indexing of Synchrony Experiments

Various proposed formulae give a numerical measure of the extent to which a culture is synchronized. One example is[2]

$$F = (N/N_0) - 2^{t/\tau}$$

N/N_0 is the extent that a particular synchronized culture doubles in cell number, t the time for the increase in cell number, and τ is the doubling time. For a perfectly synchronized culture, the time for doubling the cell number is zero (t = 0). When the culture doubles, F = 1. For an exponential, perfectly nonsynchronous culture, F = 0. This occurs when t = τ. A culture may have a value between zero and one as an index of the synchronization. The higher the index, the better the synchronization.

Another proposed index, with the same ideas but a slightly different mathematical form is

$$F = (N/N_0 - 1) \cdot (1 - t/\tau)$$

In this case, at the extremes (when t = 0 or t = τ), the values of the indices are the same. Different values are obtained for intermediate levels of synchronization.

There are two major problems with these indices, or in fact any single value proposed as a measure of synchronization. First, these two formulae mix two incommensurate aspects of cell growth, the cell number increase and the duration of the increase. A culture with a sharp increase and a poor fraction of cells dividing may have the same value as a culture with a broad increase with a perfect doubling in cell number. It is not clear that these two results are indicative of the same degree of synchrony. The second criticism is that there are many other aspects of the synchrony problem not included in a single value. Many of the properties of a synchronized culture are nonscalar, and are difficult to reduce to a numerical value. For example, should the reproducibility of the interdivision times between two different increases in some variable be considered? Are aspects like scatter, sharpness of the start of the increase, reproducibility of the methods and the results, and the other properties of a synchrony experiment important?

Another approach to indexing synchronization experiments is to describe a population as though it were composed of a perfectly synchronized population contaminated with an exponentially growing, unsynchronized population.[3] For this index, the F value is the fraction of the population that acts perfectly synchronized. A value of 0.9 means that a synchronized culture appears to contain 90% perfectly synchronized cells and 10% exponentially growing cells. (This indexing method will be analyzed in more detail at the end of this chapter.)

These numerical measures of synchronization are flawed, and may produce a high value even if the synchrony is poor. A high value for the first term (more than a cell doubling) can compensate for a large t value. As suggested in the criteria above, the culture must only double, no more, no less. The value of the synchrony is then given by the t value.

Each method of synchronization must be judged on its own. The final judgement must be the utility of the method and its ability to produce a result that advances understanding of the division cycle.

C. Differential versus Integral Methods of Cell-Cycle Analysis

In addition to the dichotomy between synchrony and nonsynchrony methods, there is another important pair of approaches, those using differential methods of analysis and those using integral methods of analysis. The patterns of synthesis during the division cycle are best considered and understood as *rates*, rather than varying *amounts* during the division cycle, and the differential methods provide the rates directly.

Consider an experiment to decide whether a particular molecule is synthesized linearly or exponentially during the division cycle. Linear means that the rate of synthesis is constant during the division cycle; the amount of the molecule increases linearly. A straight line is obtained when the amount per unit time is plotted on a rectangular graph. Exponential synthesis means that the amount of material per cell gives a straight line when plotted on semilogarithmic coordinates. Exponential synthesis is generally due to an autocatalytic increase in the rate of synthesis. Exponential synthesis does not have a constant rate of increase during the division cycle, as the rate of synthesis also increases exponentially.

As can be seen in Fig. 3–1a, the amount of material in the middle of the division cycle differs at most by 6% between linear and exponential synthesis. Extremely accurate measurements of the amount of material in the newborn, midcycle, and dividing cells would have to be made to distinguish these two patterns. In contrast, as can be seen in Fig. 3–1b, there is a marked difference in the predicted rates of synthesis during the division cycle for these two patterns. In the linear case, the rate of synthesis is constant, and in the case of exponential synthesis, the rate of synthesis is exponential.

The use of rate measurements will be termed the differential approach to the division cycle, and the use of total amounts, the integral approach or method. As a general rule, differential methods are preferred to integral methods because they are more sensitive to slight differences in the rates of synthesis during the division cycle. There are many integral methods that can, in theory, give important information about the division cycle. In practice, however, such methods have not been successful. Perhaps all integral methods are inherently less sensitive than differential methods. An analogy might be to consider comparing two large numbers directly or measuring the difference between two large numbers. The difference method is more accurate. Given the alternative of measuring the amount of DNA per cell in a synchronized culture, or measuring the rate of DNA synthesis by pulse-labeling at different times,

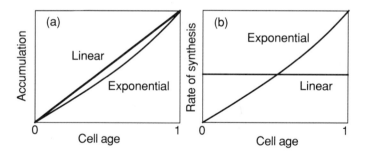

Figure 3.1 Comparison of Differential and Integral Methods of Cell-Cycle Analysis. The cumulative values for exponential and linear synthesis are plotted in panel (a). The rates of synthesis are plotted in panel (b). An exponential increase gives an increasing rate, while a linear increase gives a constant rate.

the more accurate, reproducible, and interpretable result would come from the pulse-labeling measurement.

III. SYNCHRONIZATION METHODS

A. Comparison of Selective and Nonselective Methods

Synchrony methods can be separated into two distinct categories, selective and nonselective methods (Fig. 3–2).[4] Selective methods are those that take all of the cells of a particular age from a population and produce a synchronized population. Only a fraction of the cells are removed from

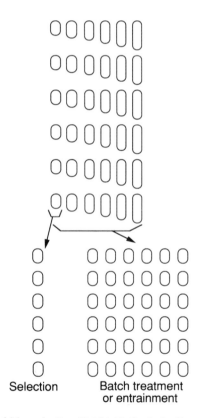

Selection   Batch treatment
                                  or entrainment

Figure 3.2 Selective and Nonselective (Batch) Methods for Synchronization. With selective methods, only a small fraction of the original cell population is retained for analysis. The defining characteristic of batch methods of synchronization is that all of the cells in the population are aligned to produce a synchronized culture.

a culture; the remainder are discarded. Separating all of the newborn cells, and placing them in a new culture, would result in a synchronized culture. The cells in such a culture would proceed through the division cycle and divide as a cohort after one interdivision time. The cells would then proceed through another division cycle. Nonselective methods produce a synchronized culture by some manipulation of all cells. No selection of part of the population occurs. The theoretical bases for selective techniques are clear, whereas the theoretical bases for nonselective techniques are not clear. Nonselective methods assume that there is some master reaction, or some cell-cycle event, at which the cells are arrested by a treatment. When the cells are released from this treatment, a synchronized culture is produced.[5]

It is theoretically unexpected for any batch treatment of cells to yield a synchronized culture. The law of conservation of cell age order (presented in detail in Chapter 16) specifically states that it is impossible to synchronize all cells in a culture by batch treatment. It is possible, however, to produce a pseudosynchronous culture by batch treatment; a pseudosynchronous culture is synchronized for only a single aspect of the cell cycle (see Chapter 5). The cells in this type of culture, while apparently synchronized, are not representative of cells in a normal, unperturbed culture. The cells still remember their age order, and after a single pseudosynchronous cell division or event, the cells express their differences and resume growing in a nonsynchronous manner.

B. Nonselective Methods for Synchronization

1. Heat Treatments and Heat Shock

One of the first methods producing bacterial synchrony used temperature shifts to synchronize or entrain bacteria. This approach follows Hotchkiss's success using temperature entrainment to "synchronize" *Pneumococcus*. Hotchkiss produced oscillations in the transformability of this organism.[6] It was the oscillations in transformability, rather than any particular cell-cycle parameter, that implied synchronization. Soon Lark and Maaløe[7] described a series of heat treatments that appeared to synchronize *Salmonella typhimurium*.[8] As they were working shortly after the finding of midcycle DNA synthesis in eukaryotic cells[9] by Howard and Pelc,[10] it was of interest to see whether DNA synthesis in bacteria was also confined to a particular phase of the division cycle. Lark and Maaløe observed a doubling in the amount of DNA during a short time in their cultures; this implied that DNA was synthesized only during a portion of the division cycle. This method contains the central cautionary tale that applies to all work in synchrony, that synchrony may produce

artifacts. Schaechter, Maaløe, and Bentzon[11] studied unperturbed, exponentially growing *Salmonella typhimurium*, and found that almost all of
the cells had incorporated thymidine during a pulse label.[12] In unperturbed cells, DNA synthesis was continuous and not confined to any
portion of the division cycle. This led Maaløe to his insight on *normality*,
and the notion that perturbation of cell growth and division was something to be avoided. The heat-shock method produced perturbations and
artifacts, and was therefore not useful in the study of the division cycle.
The important insight was that rather than accept the original result at
face value, the implications of the initial result were considered and
tested in an unperturbed culture. This is incorporated into Criterion 7 for
judging synchrony techniques. These results exemplify the problem with
synchronization: synchronization can lead to perturbations and artifacts.

There have been many attempts to use temperature shifts, both single
and multiple, to produce synchronized cells.[13] No idea or conclusion
from temperature-shift synchronization has become an established part
of the literature. Until there is a full understanding of how temperature
shifts affect cells, such methods for synchronization should not be used.
In addition to heat shock, some have proposed that cold shocks also
produce synchronized cultures.[14]

The use of temperature shifts is based on the assumption that we can
block cells at some point in the division cycle, at some *master reaction*.
After the cells have accumulated at this point, they are allowed to resume
growth. When growth is restarted, the cells proceed synchronously
through the division cycle. If the cell were a simple clock[15] with only one
value determining the place of the cell in the division cycle, the method
could work. But the cell is not a simple clock. The reason that cells cannot
be truly synchronized using temperature shocks is that only one aspect of
the cell cycle is affected by temperature. Consider a culture in which
chromosomes of the cells are all different ages and configurations. Treat
the cells so that only cell separation or division is inhibited. DNA synthesis and the initiation of DNA synthesis, as well as mass synthesis, are not
affected by the temperature treatment. When division is allowed to
resume, we might expect to see a short period of rapid division because
the cells accumulated at the point just before cell division. But the chromosomes in the cells of the culture still remain in different stages of
replication. The pattern of DNA synthesis in such a culture could not be
studied because DNA synthesis would not be synchronized.

2. Starvation or Stationary-Phase Methods

Starvation methods assume that cells proceed to a particular point at
which they are arrested when the cells are starved. Starvation methods

are analogous to the heat-shock methods for synchronization. When the cells are all at this point, the missing nutrient is added back to the culture, and the culture is synchronized. While this method has a great appeal, and there have been many reports of success with starvation, it is not clear that starvation can produce a synchronized culture.

Cutler and Evans[16] published curves demonstrating that an overgrown culture diluted into fresh medium produced a synchronized culture. The method was used to demonstrate that RNA synthesis during the division cycle was periodic.[17] The results of Dennis[18] demonstrating exponential, continuous, aperiodic RNA synthesis during the division cycle suggest that the starvation method caused perturbations. All of the criticisms of temperature shifts can be applied to starvation techniques. There are now theoretical and experimental insights that allow us to state that starvation affects various parts of the cell differently. It is now known that when a culture is starved so that cytoplasm synthesis is inhibited, rounds of DNA replication in progress are not inhibited (Chapter 5). Thus the mass or protein of the cell is fixed in the prestarvation state while the DNA is allowed to achieve a state never observed in normal, unperturbed, unsynchronized cells. The overgrowth of cultures has even been proposed to produce a synchronized culture before the completion of starvation.[19] When a culture is allowed to overgrow, it is reported that there is a step in the growth curve that is a synchronized division.

3. Starvation for Specific Compounds

In addition to starvation of cultures by overgrowth, starvation for specific compounds has been used. For example, thymine starvation[20] has been used to produce a culture for the study of the division cycle. Others have used amino acid starvation[21] to produce a synchronized culture. The criticisms described for general starvation apply also to these methods. Despite the continuous failure of starvation methods to produce any palpable result, there is something so seductive about the idea that the method will probably go on forever. In a recent review,[22] it was suggested that in " . . . stationary cultures . . . the (newborn) cells . . ." would be useful for the study of the division cycle.

In Chapter 5 the phenomenon of pseudosynchrony will be described, in which a series of starvation regimens produces a synchronous burst of cell division. This presents the appearance of synchrony, but the cells are not actually synchronized, because the DNA replication patterns are different in cells of different ages. When synchrony means that all cells are of the same age, then starvation methods do not appear to be able to synchronize cells.

4. Entrainment Methods

Entrainment methods are an extension of single treatment methods. They are a subvariant of starvation or heat-shock methods; rather than one shock or treatment, there is a sequence of shocks or treatments. The culture is slowly brought into a synchronous mode of growth and division by these treatments. The idea is that if one treatment does not work, then applying the same treatment many times may work. A simple analogy of entrainment is a child on a swing. Whether the child pumps to increase the swing of the arc, or the parent pushes the swing at the appropriate time, the addition of a little extra momentum adds to each swing. Rather than one big push, we get the swing going by little pushes at just the right time. If there is a slight synchrony effect, then repeated treatments add incrementally to the synchronization of the culture, until the culture is synchronized.

Goodwin and Kepes used a protocol in which a series of repetitive starvations for phosphate were used to produce a synchronized culture.[23] Kepes and Kepes[24] set up a mechanical starvation apparatus to produce a sequence of up to 16 starvations and refeedings of phosphate, and a similar approach has used glucose starvation for synchronization.[25] Koch[26] has analyzed this type of procedure and has come up with a rationale to explain the synchrony. Koch proposed that there is a particular relationship between the uptake of a solute, in this case phosphate, and the stage of the division cycle. If there were such a relationship, we could get a synchronized culture. First note that there is no independent demonstration that the particular assumptions made to justify this method are correct. Second, and more important, Koch's theory requires an exact dilution by a factor of two at each doubling. If there is any deviation from such a protocol, there is a decided falling off in the degree of synchronization. Rather than support the phosphate entrainment method, this analysis says that the method should not work. Koch's theory actually states that it is almost impossible to synchronize cultures using entrainment methods.

Another model explaining entrainment is that of Pritchard, analyzed in detail by Grover.[27] Pritchard proposed that larger cells would take up the limited phosphate in the medium faster than the smaller cells because of their larger surface area. They would grow and divide, and accumulate at the smaller size. After repeated cycles of starvation and refeeding, the cells would become synchronized. Grover pointed out that this cannot work because of the relationship of DNA synthesis to cell division (see Chapter 5). Also, there is no evidence that cells take up material in a cell-cycle–specific manner (see Chapter 4).

The automatic phosphate starvation method of Kepes and Kepes[28] requires a special comment. The data supporting synchronization by repeated phosphate starvations are fabulous. There are precise steps, and the increases and steps repeat for a number of generations, up to 15 in one published experiment. This result has been used to support the idea that division cycles are very regular, that decay of synchronization is not a necessary process, and that it is possible to do batch synchronization. It is hard to explain the published data,[29] and unfortunately the method has never been reproduced in other laboratories.[30] Automatic phosphate starvation has never yielded a result that has been confirmed by other, alternative methods. The continuous phosphate starvation method does not appear to be a valid method for studying the bacterial division cycle. The very fact that the synchronization is so perfect and long lasting is less a support for the value of the method than a criticism of the method (see Criterion 9 above). (See Chapter 8 for a discussion of the decay of synchronization, and Chapter 16 for a more general critique of such entrainment methods.)

5. Specific Inhibitors

Treatment of cells with inhibitors such as p-fluorophenylalanine[31] or phenylethanol[32] have been proposed to synchronize various organisms. These inhibitors may be equivalent to methods that starve cells for an amino acid or a nucleotide. It might be assumed that there are compounds that stop cells at particular points in the division cycle. As explained for temperature manipulations and starvation, and in more detail in Chapter 4, such an assumption is not valid.

6. Spore Germination

For those organisms that form spores or microcysts, it may be possible to get a synchronous culture by the simultaneous germination of a spore population. While this has a firm logical base, and there have been many reports of synchronized events following germination,[33] the main question to be asked is whether these events are related to the normal cell cycle. There have been no extensions of the analysis to a second synchronous cycle in most of the germination experiments. Therefore it is difficult to distinguish between the results due to the germination process and those specific to the processes occurring between cell divisions.

7. Synchronization by Phage Infection

It has been reported that infecting Escherichia coli with a certain mutant phage induces synchrony in cell division.[34] This synchrony is proposed to last for several cycles, and is not due to selection because over 80% of the bacteria survive the infection. The mechanism of synchronization is not known. This method does not fit many of the criteria presented above; synchrony does not decay, interdivision times vary, the mechanism is mysterious, and it is not clear that the data are incompatible with exponential growth. In my opinion this is not a method for synchronizing bacteria.

C. Selective Methods for Synchronization

Selective methods produce a synchronized culture by removing or selecting a portion of cells all of the same age. This contrasts with nonselective methods, which attempt to synchronize an entire culture by a particular treatment or series of treatments. Although it is theoretically impossible to synchronize a culture by batch techniques (see Chapter 16), selective methods can, in theory, produce a synchronized culture.

1. Centrifugation

One of the first methods for producing a synchronous culture by selection was the use of centrifugation to separate cells by size. Cells of a narrow size class were selected to produce the synchronized culture (the Mitchison–Vincent technique).[35] This technique has proved successful in providing synchronized cultures of both yeast and bacteria. The method consists of layering a concentrated cell sample from an exponentially growing culture on top of a linear sucrose gradient, centrifuging the gradient, and removing the upper layer of cells. This layer contains the smallest cells and produces a synchronized culture.

Unfortunately there are two problems with the centrifugation method. One is the generic problem of whether the manipulations of the cells, no matter how gentle and benign, perturb the normal functioning of the cell. Many who have followed the Mitchison–Vincent approach have worked hard on this point. They carefully made sure that there were no temperature fluctuations and that there were no harsh osmotic shocks. The other problem, one that is inherent to the method, can be seen from inspection of the cell size distribution of cells from an exponentially growing culture and newborn cells (Fig. 3–3). Whenever newborn cells are analyzed for their size distribution, and the size distribution is compared with the size distribution of the exponential culture, there is a large overlap of cell sizes. This means that newborn cells have sizes that cover a large portion of the range of the total size distribution. The smallest cells in the culture are not representative of the newborn cells, but rather of cells smaller than the average newborn cell. These cells may give results because they are very small; cells may be perturbed, injured,

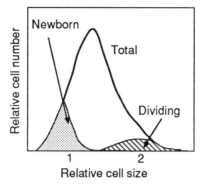

Figure 3.3 Size Distribution of a Bacterial Culture. A comparison of the size distribution of cells from an exponentially growing population with the size distribution of newborn and dividing cells. If the cells divided exactly in half at each division, the newborn distribution would be the same as the dividing distribution, with all values halved.

dead, or have other problems that contribute to their being exceptionally small. On the other hand, selection of a larger size distribution would give newborn cells contaminated with cells of other ages. This contamination would vitiate any results obtained.

In the midst of all of this negative commentary about synchronization, there is something to be said in favor of the method. It is possible to use the imperfections of the gradient-centrifugation technique to obtain useful results. This is illustrated by experiments that used differential centrifugation in a gradient to select the smallest cells in a culture.[36] DNA synthesis during the division cycle was analyzed, and it was found (anticipating the analysis of DNA replication in Chapter 5) that there was a delay in the youngest cells until DNA synthesis started. This result can be understood (assuming that in general such a delay does not occur) by considering that in the normal population there is a distribution in the sizes of the newborn cells. In this newborn population there is a spread among the population such that half of the cells have started DNA synthesis and half have not (i.e., the cells start DNA synthesis, on the average, at the time of cell division). If the population selected by centrifugation were smaller than average, the selected population would be enriched with cells that are not big enough to start DNA synthesis. We will return to this in Chapter 5, but suffice it to say that such a method could be utilized to study particular aspects of the division cycle, even though the result is not representative of the normal, unperturbed culture.

2. Elutriation

Elutriation is a special type of centrifugation method that allows the production of a synchronized culture with fewer perturbations than other methods. In this method, the sedimentation of bacteria is counteracted by a continuous flow of medium upward through the centrifuge cell. At a particular flow rate, the smallest cells are washed out of the centrifuge chamber and are collected as a synchronized culture.[37] The bacteria are introduced into the rotor in growth medium, and the isolation of a fraction of small cells can be achieved in about 10 minutes. When this method was applied to *Escherichia coli*, a synchronized culture was produced[38] that had a reported purity factor (i.e., the fraction of the cells that were newborn) of 90%.

3. Filtration

Filtration to select cells of a particular size[39] is an alternate approach to producing a synchronized culture by selection. This method is so intuitively simple, and so simple to explain, that it has become the dominant textbook explanation for synchronizing cultures. The basic idea is to pass a culture through a filter or a series of filters, and to take the few cells that initially come through as the start of a synchronized culture. The first successful method used a stack of filter papers as the selection apparatus; the cells were concentrated before filtration. Abbo and Pardee[40] eliminated the concentration steps in order to minimize perturbations.

There are some problems with this technique. First, it is not a reproducible method. The conditions have to be worked out anew each time it is used on a different organism. Second, although much effort, similar to the effort with centrifugation methods, has gone into making sure that the methods do not perturb normal cell metabolism, in practice significant distortions can arise. Conflicting results have been obtained by different workers using the same methodology. Maruyama[41] found a step-wise increase in DNA content during the division cycle of *Escherichia coli*, while Abbo and Pardee,[42] Cummings,[43] and Nagata,[44] found a continuous increase in the DNA content. The third problem is that there has yet to be an independent confirmation that this method can work.

It is important to place the method of Abbo and Pardee in its proper historical context; their filtration method is one of the archetypal methods for producing synchronized cells. The rate of DNA synthesis during the division cycle, measured using radioactive thymidine and a differential method, indicated that DNA synthesis was exponential during the division cycle. But the rate of incorporation of thymidine is not expected to be exponential (Chapter 5). This prediction, however, was not known in 1960 when the experiments were performed. Also, their cells were analyzed a long time after filtration (50-120 minutes) and this makes it difficult to judge the state of the cells immediately after synchronization. It is now known that the cumulative variability of cell interdivision times would lead to the decay of synchrony by the time the cells were studied (see Chapter 8). Reanalysis of the published synchrony curves (e.g., plotting the points without lines, analyzing differences in successive division cycles) suggests that the cells may not have been synchronized. But past experiments should not be judged based on the standards of today. As time passes, the state of the art changes. Abbo and Pardee's experiments and ideas were excellent ones, particularly as they were the first to note the problems with *forced synchrony*. These investigators were aware of the possibility that the cells might be disturbed by the treatments used to produce the *synchronized* culture. It was for this reason that they tried to use an *unforced synchrony* method to produce a *synchronous* culture.

4. The Membrane-Elution Technique

A different approach to obtaining synchronized cultures stems from the ability of various strains of bacteria (*Escherichia coli* B/r, *Salmonella typhimurium, Bacillus subtilis*) to bind to nitrocellulose membranes, to grow on these membranes, and to release newborn cells to the medium flowing over the membranes. This technique, first described by Helmstetter and Cummings,[45] forms such an integral part of this book that the origin of the method and its subsequent development will be described in detail.

The membrane-elution method involves taking exponentially growing cells and filtering them onto a membrane, inverting the membrane, and pumping medium through the filter. Cells, presumably attached by one end,[46] grow on the filter and newborn cells are released from the membrane (Fig. 3–4). By looking at synchrony curves and size distributions with a Coulter Counter, it was shown that the released cells are newborn. The size distributions indicate that cells released from the membrane are small and that they grow in size. After approximately one doubling time, a bimodal size distribution is observed as small cells arise by the division of the larger cells (Fig. 3–5).

This method followed from the filtration synchronization method of Yanagita and Abbo and Pardee. In their method, cells were filtered through a mat of filter paper (see above). After continued filtration, newborn cells were found in the eluate. With further elution with warm growth medium, newborn cells continued to appear in the eluate. It was concluded that the cells were growing on the filter. Reducing the number of filter papers improved the method, and the best results involved only a single piece of filter paper. It was thus realized that the synchronized

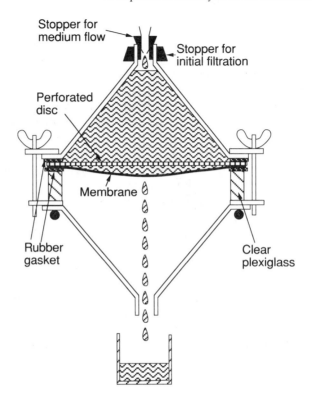

Figure 3.4 The Membrane-Elution Apparatus. The membrane-elution apparatus is composed of those pieces held together by the wing-nut bracket. The apparatus is initially inverted and placed on a side-arm suction flask. The nitrocellulose membrane is supported by a perforated disc. Suction is applied, and the cells are poured onto the membrane. The cells are washed with additional volumes of prewarmed medium, and the entire apparatus is inverted and placed on a cone-shaped support resting on a ring stand. The entire apparatus is in a full-view incubator permitting work at room temperature while keeping the experimental apparatus at 37°C. Medium is poured into the apparatus, and a tube from a metering pump keeps a steady flow of medium through the membrane. The drops fall directly through the supporting cone into a beaker, where they are collected at intervals.

Figure 3.5 Synchronized Culture Produced by Membrane-Elution. Cell attachment is visualized as though cells were bound to the nitrocellulose membrane at their poles. When the membrane is inverted, and the cells grow, the first newborn cells eluted come from the oldest cells (age 1.0) at the start of the experiment. Whenever cells are collected, the newborn cells represent a synchronized culture. Hence, they grow as illustrated at the bottom of the figure. The synchronized cultures in any fraction come from cells of different initial ages at the start of the experiment.

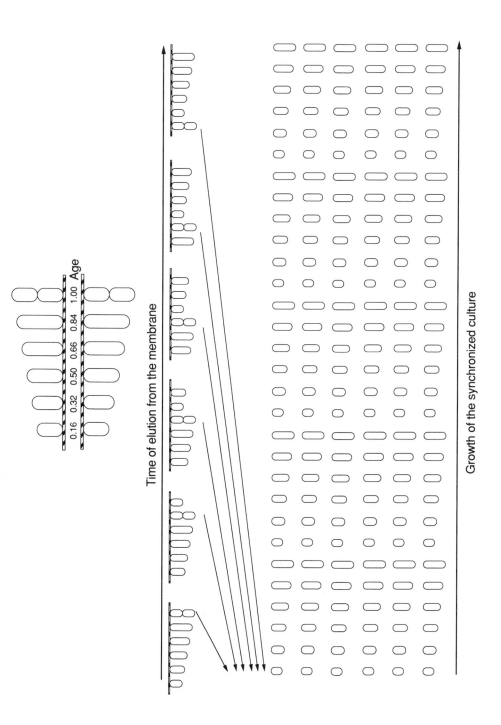

Age

0.16 0.32 0.50 0.66 0.84 1.00

Time of elution from the membrane

Growth of the synchronized culture

cultures obtained were not due to the differential filtration properties of the paper, but rather to the growth of cells on the filter paper. An improved method was developed by substituting a nitrocellulose filter for the paper.[47] Since that time this has been the basis for the membrane-elution method. While this method has been used to produce synchronized cells,[48] the most important use of this method is to analyze cells by a completely different, *backwards* method.

The membrane-elution method is simple to reproduce, and it is described in detail by Helmstetter.[49] The method is not generally applicable to all bacteria or all strains of *Escherichia coli*. Very few bacterial species have been tested on the membrane system; therefore the method is still open to expanded use. Although the membrane-elution method yields reproducible cells that fit essentially all of the criteria for synchrony described above, the method is open to the criticism that manipulation of the cells on the membrane—even with extreme precautions against perturbations such as constant temperature environment, washing with conditioned, prewarmed medium, and so forth—may alter the normal cell cycle.

IV. NONSYNCHRONY METHODS

It is interesting to consider that the division cycle may be analyzed without ever preparing a synchronized culture, simply by studying exponentially growing cells and manipulating them in various ways. The time during the division cycle that an antibiotic acted in preventing cell division can be determined by synchronizing a culture and adding the antibiotic at different times during the division cycle.[50] If cells receiving the antibiotic within 15 minutes of division are still able to divide, while cells receiving the antibiotic earlier than 15 minutes before division are inhibited and do not divide, then we could conclude that 15 minutes before division, the cells pass the point at which they are sensitive to the antibiotic. This same experiment is more easily accomplished by adding the antibiotic to an exponentially growing culture and seeing that cell division continued for 15 minutes. This experiment gives exactly the same result as the synchrony experiment.

A. Total Population Analysis

1. Direct Size Measurements

Examination of a total population of cells can give information about the division cycle; synchronization is not necessary to analyze the division cycle.

a. The Collins–Richmond Principle   Collins and Richmond[51] derived a relationship between the growth rate of cells and cell size using three distributions: the distribution of the cell sizes in the growing population, the distribution of the cell sizes of dividing cells, and the distribution of the cell sizes of newborn cells. If the variation in the sizes of dividing cells were exactly twice the variation of the sizes in the newborn distribution (this means that cells divided exactly in half at division), then only two experimentally derived distributions are needed, the total distribution and either the newborn or the dividing distribution. To understand this method, consider a growing and dividing cell population with no size variation at birth and division. Assume that all the cells are between sizes 1.0 and 2.0. If cells grow linearly during the division cycle, the size of the cells at each cell age will be proportional to the cell age. The size distribution will therefore be exactly the same as the age distribution (Chapter 1). If the measured size distribution were plotted and a pattern obtained that was the same as the age distribution (i.e., twice as many small cells as large cells with a smooth distribution between those two frequencies), then the pattern of growth during the division cycle would be linear. As we shall see in Chapter 4, cell growth is actually exponential, and the size distribution is not the same as the age distribution. What should be seen here is that we can obtain the growth pattern of cells during the division cycle directly from the size distribution of cells in a growing culture.

b. Microscopic Examination   Because bacterial cells, particularly those of the rod-shaped, gram-negative class, have a rigid shape, it is possible to measure the size of such cells accurately.[52] Size measurements of three different cultures demonstrated that whatever the pattern of size increase during the division cycle might be, it is the same in three different gram-negative, rod-shaped cells.[53] (A detailed analysis of this method is presented in Chapter 14.)

c. Electronic Particle Sizing   The invention of the electronic particle-sizing instrument has added a new approach to the study of bacterial sizes. Cells in solution are placed around a glass tube with a small orifice (30μ or less). These cells are drawn through the orifice by a slight vacuum. The electrical resistance across the orifice is measured. As each cell passes through the orifice, an increase in resistance occurs, and this is converted into an electrical pulse proportional to the size of the cell in the orifice. The data can be accumulated in a computer. This distribution could be used to determine the growth pattern of cells using the Collins–Richmond approach.

While the use of such an instrument[54] can accurately determine cell number, the measurement of the size distribution is more problematic.[55] The size distribution is dependent on the rate of liquid flow through the orifice, cell concentration, current across the aperture, and the electronic components in the instrument.[56] Cells resident along the orifice wall have a slower passage and a slightly higher apparent size. This problem can be accommodated by slowing down the flow rate so that there is less shear. When this is done, all of the cells pass through the orifice at the same rate, no matter where in the orifice they may be located.

d. Time-Lapse Analysis　　The temporal pattern of cell division can be studied by observing the pattern of division of cells either in a microscope, as in the classic studies of Powell,[57] or by the use of film-recording techniques and time-lapse photography.[58] Although it is extremely difficult to determine the pattern of size increase in living cells during the division cycle, we can see when division occurs and determine the interdivision time.

2. Differential Methods for Total Population Analysis

a. Autoradiography　　The light-microscopic and electron-microscopic methods for whole-population analysis can be improved by adding autoradiography. After a culture is pulse-labeled and fixed, autoradiography involves placing the cells against a photosensitive emulsion and allowing the β-particles to produce grains in the emulsion. The autoradiographic grain count in cells as a function of cell size gives a measure of the pattern of biosynthesis during the division cycle. By classifying the cells according to size, which in practice means according to length, and making some assumptions about the relationship of cell size to cell age, we can determine the rate of synthesis of a particular compound or the rate of incorporation of a particular precursor as a function of cell age. If the grain counts were proportional to the size of the cell—the larger cells had more grains than the smaller cells—then an exponential mode of growth could be presumed. If the grain counts were constant and independent of cell size, this would support a linear mode of synthesis. A fundamental problem with this approach is that cell size, or length in the case of rod-shaped bacteria, is not a good measure of cell age. This is because the length of the cell is dependent upon the width of the cell. Because, as we shall see in detail in Chapter 6, the volume or mass of the cell most likely determines the regulation of cell-cycle events, cell length is a very poor measure of cell age.

b. Flow Cytometry　　Flow cytometry is the group of methods that allows the measurement of the chemical, immunological, and other properties of single cells using combinations of laser, computer, and elegant liquid-handling techniques. Flow cytometry has revolutionized the study of the cell cycle of eukaryotes. It is also possible to apply the flow cytometry principles to the study of bacteria.[59] Steen[60] has described a modified microscope that allows the measurement of the DNA content of individual bacteria (Fig. 3–6). Cells in liquid are spread out on a slide after being expelled through an orifice under high pressure. The laminar flow of the liquid across the slide allows individual cells to enter the microscope field of view where the fluorescence activation of the DNA is measured.[61] In addition to fluorescence at two different wavelengths, light-scattering measurements are also possible. Flow cytometry has been particularly important in the study of the bacterial division cycle because it has provided an independent confirmation of the pattern of DNA synthesis during the division cycle that was discovered using the membrane-elution method. Thus, flow cytometry provides the link that allows us to choose between different methods, and enables us to select the method that can give the correct patterns of biosynthesis during the division cycle.

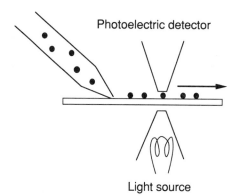

Figure 3.6 The Technique of Flow Cytometry. The content or composition of single cells can be measured by applying a specific fluorescent dye and allowing the cells to pass over an optical system that measures cell fluorescence. The fluorescence is proportional to the content of DNA or any other cell component. In contrast to the standard flow-cytometric method, where cells are held in individual drops, the bacterial cells are passed over the optical system in a continuous stream.

3. Methods Specific to the Analysis of DNA Replication during the Division Cycle

a. Gene-Frequency Analysis   It is possible to measure the concentration of genes in a preparation by various genetic and biochemical methods. For example, we can use transformation, the introduction of new genetic material into a bacterial cell, to measure the concentration of a particular gene.[62] Similarly, we can measure the relative transducing activity of a phage preparation for different genetic markers. These differences in gene concentration are indicators of the pattern of DNA replication. Similar studies can now be undertaken with cloned fragments.[63]

b. Density-Shift Analysis   It is possible to incorporate nucleotides with a different density from that of normal nucleotides into DNA. Using isopycnic centrifugation techniques, dense DNA can be separated from light DNA. By combining this approach with radioactive labeling, some understanding of the replication of DNA may be gained.[64]

c. Radioactive Decay   Labeling DNA with radioactive precursors that kill the cell as the label decays shows what fraction of the population is resistant to such killing.[65] The resistant fraction indicates cells that were not involved in DNA synthesis during the labeling period (see Chapter 5).

4. Execution Points

Assume that there is a particular point during the division cycle at which cells can be inhibited by a particular environmental treatment. Cells that are beyond that point when the treatment is applied will be able to divide, while cells that are before that point will be prevented from dividing. It is thus possible to operationally determine such a point where the particular event occurs. This point has been called the *execution point*. Taking into account the age distribution of an exponentially growing population, it is possible to calculate the time at which an event is executed during the division cycle. By determining the amount of residual division after treating the cells with an inhibitor of cell division, we can calculate the execution point, EP, in minutes before cell division:[66]

$$EP = \tau \cdot \frac{\ln(N/N_0)}{\ln 2}$$

In this equation N_0 is the initial cell number, N is the cell number after cell division has ceased, and τ is the doubling time of the culture.

The determination of execution points is attractive because it is simple to use, requires measurements of cell number that are easy to obtain using an electronic cell counter, and has a straightforward theory. It is not clear, however, that the method can lead to the discovery of any cell-cycle event. It has usually worked only when the quantitative aspects of a particular cell-cycle event that were already known were confirmed. The identification of unknown aspects of the division cycle has not yet been achieved using execution points (see Chapters 5, 6, 12, and 15 for more examples). For these reasons, this method should be used sparingly, and when used to identify new properties of the division cycle, the results obtained should be accepted with caution. The main problem with the execution-point method is that leakage of cell division in the presence of the inhibitor allows some cell number increase; this leakage operationally determines a time at which some event appears to have happened.

B. Backwards Methods

Consider a population of cells growing in a culture with all cell ages represented. The cells are in balanced growth and have a normal age distribution. Some molecule is synthesized during the division cycle with a constant rate during the first half of the division cycle (age 0.0-0.5), and this rate doubles to a new constant rate during the last half of the division cycle (ages 0.5-1.0). This is illustrated in Fig. 3–7. A radioactive label that is specifically incorporated into the molecule of interest is added. After a short labeling period, the cells below middle age have 100 counts per minute (cpm) per cell and cells above middle age have 200 cpm per cell. Take a "magic forceps," reach into the culture and remove all cells of ages 0.0 to 0.1 and place them in one tube. Place the cells of ages 0.1 to 0.2 in a second tube, and continue placing all the cells into ten tubes according to age. Now measure the radioactivity per cell in the different tubes. The radioactivity per cell in the cells from ages 0.0 to 0.5 will be half that of the cells of ages 0.5 to 1.0. We conclude that the rate of synthesis was constant during the first half of the division cycle and doubled in the middle of the division cycle. The cells are labeled in an exponentially growing and unperturbed state; the only perturbation, if any, is adding the label. While it may be argued that even this manipulation may perturb the cells under study, this is less a perturbation than any synchronization procedure; it is therefore more likely to yield results of value. Furthermore, the experimental results obtained with this approach suggest that it gives results unaffected by perturbation artifacts. The important question left unanswered in this discussion is what are the "magic forceps" that allow us to sort cells by age.

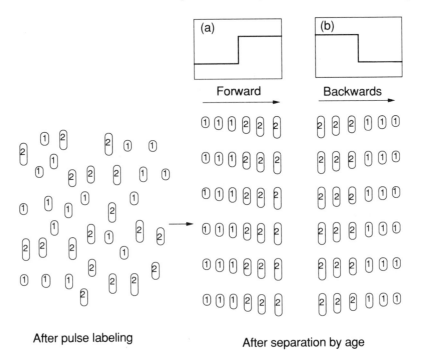

Figure 3.7   Theory of the Backwards Methods of Cell-Cycle Analysis. A forward synchronization analysis is presented in (a) and a backwards analysis is presented in (b). The forward method is a standard synchronization approach, which may be the selection of newborn cells by either membrane-elution or centrifugation. In backwards methods (b) cells are separated after a nonperturbing treatment (enzyme induction, labeling for a short time) and arranged according to their ages at the time of treatment. Cells can be arranged in either ascending or descending order of cell age. The membrane-elution method releases cells in reverse order to our usual way of considering cells, i.e., the cells released first come from the oldest cells and then younger and younger cells are represented, so the method is called a backwards method. The size of the cells is representative of their ages at the time of labeling, and the numbers (indicating, for example, the amount of radioactivity per cell) indicate that there is a doubling in the rate of synthesis of some component in the middle of the division cycle.

1. Membrane-Elution Technique for Cell-Cycle Analysis

A schematic diagram of the membrane-elution procedure for determining the rate of synthesis of any particular molecule during the division cycle is presented in Fig. 3–8. Six representative cells from an exponential-phase culture are followed through the steps of the procedure. The ages of the cells are shown as fractions of an interdivision period. The cells are first pulse-labeled with a radioactive precursor of the particular molecule of interest, for example, thymidine for DNA or leucine for

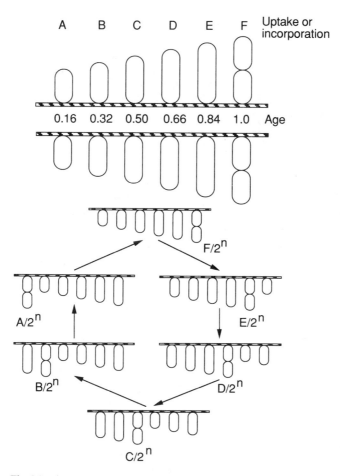

Figure 3.8 The Membrane-Elution Method for Backwards Analysis of the Division Cycle. Cells of different ages are labeled in an exponentially growing culture. The label is incorporated into the different cells as indicated by the letters A to F. The cells are filtered onto a membrane which is then inverted, and the elution of newborn cells begins. The first cells eluted contain half the label of the oldest cell placed on the membrane. After one-sixth of a generation, the next oldest cells grow up and produce newborn cells with half the label of the original cells. This continues for one complete interdivision time, at which point the original oldest cells, with label in the amount F, produce cells with one-quarter of the original label. The amount of label in the initial cells can be determined from the label in the progeny cells eluted from the membrane.

protein. The amount of label incorporated into the cells of different ages is indicated by a letter above each cell. At the time of labeling, the cells are growing in balanced exponential growth, and the only perturbation is the addition of label. The cells are then filtered onto a nitrocellulose membrane, and the radioactivity is removed by washing with fresh, prewarmed medium. Assume that whatever label has been incorporated into the cells remains in the cell from this point forward, i.e., there is no turnover or release of label into the medium. The membrane is inverted, and fresh medium is pumped through the membrane so that bound cells can continue to grow and divide. The division stages of the cells at six distinct times during each generation of continuous elution are shown progressing clockwise (Fig. 3–8).

The cells bind to the membrane as though they are attached at one pole. There is no evidence as to how attachment occurs. All that is required is that when cell division occurs, at least some of the daughter cells are released from the membrane. The release of cells does not have to be complete, as the radioactivity *per cell* in the eluate will be determined. If there are fewer cells released, there will be less radioactivity, but the radioactivity per cell will remain the same.

The letter n (Fig. 3–8) indicates the generational relationship between the eluted cells and the original bound cells; during the first generation of elution, n equals one; during the second, it equals two; and so forth. At the beginning of elution, the newborn cells are daughters of cells that were attached when they were age 1.0; these cells contain F/2 units of label. (That this is so can be seen by asking the following question: Which cells in the original culture give the first newborn cells into the eluate? The answer is, those cells closest to cell division, the almost-mother cells in the original culture.) The bound cell divides in half, so only half of the original radioactivity is present per cell in the eluate. After 0.16 generations of elution each cell has progressed through the cycle by one-sixth of an interdivision time. The newborn cells eluted from the membrane at this time are daughters of the cells that were labeled and attached at age 0.84. These cells contain E/2 units of label. This sequence of events continues until all of the cells attached to the membrane have divided once. This repeats during the second and succeeding generations of elution.

For clarity, it has been assumed that when a cell divides on the membrane, one daughter is eluted and the other remains attached. This has been indicated by showing the cells attached to the membrane by one end. If cells were attached longitudinally, and both poles were attached to the membrane, no daughter cell would be released into the medium. If upon further growth the cells released newborn cells into the medium, then the number of cells coming off the membrane in later generations would be greater than in the first generation. Occasionally this is observed in practice; therefore, it is possible that some cells do not release progeny during the first generation. (See Chapter 9 for an analysis of this point.)

It is important to note that the only requirement of the membrane-elution technique is that cells are eluted in chronological order for one generation. The progeny cells could be perturbed for subsequent study and analysis, but this would not hinder the determination of the original incorporation pattern during the division cycle. The only situation precluding analysis is when the molecule of interest is unstable or turns over. A considerable time must elapse before obtaining the newborn cells from a particular age cell, so such turnover may make the results uninterpretable. As we shall see in Chapters 4, 5, and 6, such turnover is negligible, and thus the membrane-elution method is broadly applicable.

2. Analysis of a Membrane-Elution Experiment

Two points must be considered when analyzing and interpreting a membrane-elution experiment. One is how the time for elution of one generation of cells is determined; the second is how the patterns of radioactivity per cell are interpreted. There is an age distribution for cells growing in unperturbed culture (Chapter 1), with twice as many newborn cells present as dividing cells. If the cells of different ages were bound with equal efficiency to the membrane, we would find a sawtooth pattern in the cell elution pattern as shown in Fig. 3–9. The number of cells in the eluate is the mirror image of the age distribution. The first cells are proportional to the number of dividing cells and the last cells in a generation of elution are proportional to the number of newborn cells. The end of the first generation of elution is indicated by a sudden decrease in cell numbers; subsequent generations are indicated by de-

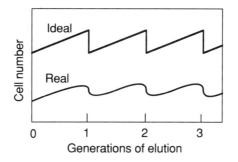

Figure 3.9   Cell-Elution Pattern of a Membrane-Elution Experiment. A sawtooth pattern is observed in the cell-elution curve because the first cells eluted from the membrane are the progeny of the oldest cells in the original culture. After one generation, the eluted cells are the progeny of the youngest cells in the original culture. This elution pattern is attributable to the fact that in a culture there are twice as many newborn cells as cells ready to divide (Chapter 1). Because of cell-cycle variability, this sawtooth pattern is smoothed in an actual experiment. The midpoint of the decrease is taken at the time of one generation of elution.

creases in cell number. The cells eluted within each generation proceed from the oldest to the youngest cells bound to the membrane. The elution curve during each generation of elution is the reverse of the age distribution.[67]

What radioactivity pattern in the eluate would be expected from different incorporation rates? Two different incorporation patterns are presented and analyzed in Fig. 3–10. An exponential incorporation pattern and a step-wise pattern with the rate doubling at midcycle are presented. The expected elution patterns, in radioactivity per cell, are shown in Fig. 3–10a. For the step-wise incorporation pattern, the initial measure of radioactivity per cell is constant. After one-half generation of elution there is a decrease in the radioactivity per cell as the cells from the younger half of the division cycle now contribute their progeny to the eluate. The same pattern is repeated in the second generation of elution, with the radioactivity per effluent cell half that of the first generation. An exponentially increasing pattern of incorporation into the original cells gives an exponentially decreasing pattern of radioactivity per cell. The patterns from the elution experiment are *backwards*, as we normally think about division cycles from young to old. In order to see the actual pattern, we could plot the elution curve in reverse, and present the data from youngest to oldest in the usual fashion (Fig. 3–10b).

There is a degree of variability in cell interdivision times, and the sawtooth patterns, while clear and evident, are not so sharp as in the

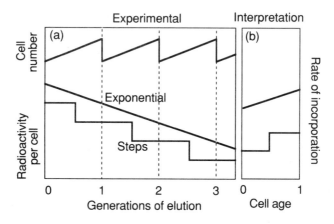

Figure 3.10   Analysis of Biosynthetic Patterns from a Membrane-Elution Experiment. An exponential-labeling pattern and a step-labeling pattern are illustrated in panel (a). These are redrawn in panel (b) in terms of the rate of incorporation during the division cycle. Reading from left to right in panel (b), cell age increases from age 0.0 to age 1.0.

idealized patterns shown here. For this reason, the midpoint of the decrease in cell number is used to determine when one generation of elution has occurred. Because the generation time on the membrane may not be the same as that in the unperturbed growing culture, it is important to have an independent method for determining the growth rate on the membrane. Assuming cellular material is partitioned equally to daughter cells, the time it takes for the radioactivity per cell to decrease by half is an independent measure of the generation time of a culture.

3. Criteria for a Membrane-Elution Experiment

This chapter began with a set of criteria to be used in judging whether a synchronization method is a good one for division-cycle analysis. We can now list criteria for a membrane-elution experiment.

1. *The label used for measuring the rate of synthesis of a molecule should be stably incorporated into the cell. No turnover leading to release of the molecule should take place.* If there were any turnover and release from the bound cells, then cells from other age cohorts could incorporate label after being bound on the membrane. In addition, the release would modify the amount of label in the cells of a particular age, changing the pattern of synthesis.
2. *The cells that attach to the membrane should divide once and in order.* As will be seen in the analysis in Chapter 5, this requirement is satisfied by the analysis of DNA synthesis during the division cycle. The success of the analysis of DNA synthesis, and its confirmation by a number of other methods, means that cells do divide at least once and in order. As the method is reproducible (and if necessary we could include a DNA synthesis control in each experiment), the membrane-elution method satisfies this requirement.

No requirement is made for the cells growing on the membrane to be normal and to divide normally after release from the membrane. If the cells are perturbed after release from the membrane, this does not reduce the utility of the membrane-elution method.[68] It will also be demonstrated in Chapters 6 and 9 that there is no flipping of the cells on the membrane; the cells act as though they are firmly bound by one pole throughout many generations of elution.

4. Comments on the Membrane-Elution Method

It is sometimes suggested, when the purpose of the experiment is to measure synthesis, that the membrane-elution method measures uptake, not synthesis. Similarly, the method has been described as measuring synthesis, not uptake. What is usually sought is a measure of macromolecular synthesis during the division cycle. When radioactive labeling is used, the membrane-elution method actually measures only uptake. When the label is added for 1 or 2 minutes before placing the cells on the membrane, the label that is taken up into the cells is what is measured in the eluted cells. In subsequent chapters, we shall see that this uptake, in all cases studied, reflects the rate of macromolecular synthesis. In particular, as we shall see in the Chapter 5 on DNA synthesis, a large number of experiments have supported the observations obtained by the backwards membrane-elution method. The pool sizes are so small relative to the amount of uptake and synthesis taking place at any time, that uptake is equivalent to synthesis.

A misconception regarding the membrane-elution method is that the cells may be perturbed after membrane-elution. There are reports that cells are altered by the filtration onto the membrane.[69] Even if the cells are altered, the radioactivity the cells have taken up before filtration is still reflected in the eluting cells. It does not matter how altered the cells are as long as the cells divide at least once and maintain their age order. The successful results with DNA synthesis support the idea that cells do divide in order on the membrane and continue to do so for a number of generations (Chapter 5). It has been suggested that the doubling times of the cells eluted from a membrane are not the same as the doubling times before placement on the membrane. In some cases, the generation time of the cells bound to the membrane-elution apparatus, or the cells eluted from the apparatus, may be different from the growth rate prior to filtration.[70] The *relative* position of an event within the division cycle is measured, so this problem is not critical, though we should be aware of it. For example, consider that cells with a 60-minute interdivision time are placed on the membrane, the cells grow on the membrane with a 30-minute interdivision time, and the eluted cells have a 40-minute interdivision time. If we find that an event occurs in the middle of the division cycle, i.e., 15 minutes after elution starts, we would still conclude that the event occurred in the middle of the 60-minute division cycle. In all cases, the results are referred to the original, unperturbed cells that were labeled before being placed on the membrane.

Pools must be considered in the analysis of the membrane-elution method. In the early days of this method, cells were precipitated and washed. In recent years, however, this step has been eliminated. The introduction of water-miscible scintillation fluids allows the counting of radioactivity in cells without precipitation, filtration, or washing. The change in methodology does not alter the method or its success. Once the cells have been placed on the membrane, all of the radioactive medium is washed away. Radioactivity within the cell is then incorporated into macromolecules. The pools of the precursors that have been used are relatively small, generally less than 2 minutes' worth of soluble material in the cell at the time of filtration, so within a few minutes all of the soluble radioactivity in the cell is rendered insoluble.

One limitation of the membrane-elution method is that it only works with one or two strains. The method has been shown to work with *Escherichia coli* B/r strains, *Salmonella*, *Bacillus*, and even K12 strains of *Escherichia coli* (although not always successfully with the latter); additional organisms should be tried on the membrane-elution apparatus.

A very important point regards whether we can study synchronized cells eluted from the membrane-elution apparatus as well as using the prelabeling method. Even if we had a perfectly unperturbed collection of synchronized cells coming from the membrane, the membrane-elution method and synchronization methods are not equivalent. In Chapter 8 it will be shown that the backwards membrane-elution method (or any backwards method) may be inherently better than any synchronization method.[71]

As very few cells are bound to the membrane, it has been suggested that the results obtained may be due to a selected group of cells. The cell elution curves from the membrane-elution apparatus indicate that a constant fraction of cells of all ages are bound to the membrane; there is no indication that any particular age of cells is preferentially bound. Even if there were such preferential binding, the results would still be satisfactory, as all calculations are made on a per cell basis; if there were a variation in the cell-elution curve, the ratio would not be affected. More conclusive is the observation that the correct division-cycle DNA pattern was obtained by the backwards membrane-elution method (Chapter 5). This indicates that the bound cells, whatever fraction of the total is represented, are able to show what is happening in an entire culture.

It has been proposed that the eluted cells are contaminated with a large number of randomly eluted cells, thus making interpretation of a membrane-elution experiment difficult. In one set of reported experiments, where the cells eluting from a membrane were observed in the electron microscope, a significant number of cells had constrictions. Assuming that newborn cells are not constricted, and that the constricted cells came from the release of cells of all ages, the randomly released fraction of the cells from the membrane-elution apparatus could be calculated. A synchronization index, F, the fraction of eluted cells that arise by division, was proposed.[72] F values of 0.77, 0.68, and even values as low as 0.43 were reported. This last value means that only 43% of the cells coming from the membrane were newborn cells, with 57% of the cells coming from the bound population at random. If this were the case, then the number of cells on the membrane should be continuously decreasing; the number of cells eluting in subsequent generations should be lower than in the first generation. It is generally found that the number of cells eluted in later generations is higher than that in the initial generation. Further, studies in which the division of cells on the membrane was inhibited give an extremely large decrease in the number of eluted cells when cell division ceased.[73] This means that the cells eluted from the membrane are essentially all newborn cells. An alternative possibility to explain the results is that newborn cells can invaginate, and that invagination may be widely spread over the division cycle.

Finally, two successive generations of elution are not equivalent to two successive synchronous generations in a synchrony experiment. In a membrane-elution experiment each generation of elution is actually an analysis of the same division cycle. In general, any elution pattern must be repeated in successive generations. (For exceptions, see Chapters 6 and 9.)

Finally, one very important advantage of the membrane-elution method should be emphasized. In a synchrony experiment, with cells being labeled at different times during a period of growth, there is the possibility that each of the labeling periods has some variation. In contrast, with the membrane-elution method, all cells are labeled for precisely the same time. It may be recalled that the label is added to all of the cells, and then the label is removed from the cells by filtration and washing of all of the cells at the same time. This difference in labeling protocols may improve the determinations of incorporation during the division cycle.

V. A COMMENT ON METHODS FOR ANALYZING THE DIVISION CYCLE

This chapter has reviewed a number of methods used to obtain information about the bacterial division cycle. In the remainder of this book, it will be seen that the membrane-elution method has given the most permanent understanding of the bacterial division cycle. When we analyze DNA synthesis, cytoplasm synthesis, cell-wall synthesis, or the segregation of DNA, we shall see this method playing a central role. My experience suggests that the membrane-elution method is not only the best method in practice, but also is the best method in theory as well (see Chapter 8). In contrast, synchrony has not fared so well as a method for the analysis of the division cycle. No doubt this method will continue to be developed, studied, used, and analyzed. Anyone using synchronization to study the division cycle should take note of the history of synchrony, the criteria for judging synchronization, and the results obtained by other methods.

NOTES

1. An example of this criterion is seen in the analysis of the heat-shock synchronization procedure of Lark and Maaløe (1954) in which subsequent analysis demonstrated that the periodic synthesis of DNA was not found in unperturbed cells.

2. Blumenthal and Zahler, 1962.

3. Figdor, Olijhoek, Klencke, Nanninga, and Bont, 1981.

4. Maaløe (1962) has reviewed the theories of both selective and nonselective synchronization. For an exhaustive summary of synchronization methods, see Evans (1974).

5. For an analysis of synchronization by manipulation of a population of cells, see Campbell (1964).

6. Hotchkiss, 1954.

7. Lark and Maaløe, 1954, 1956.

8. Another influence on the development of this methodology was the apparent success in the use of temperature changes in synchronizing a protozoan, *Tetrahymena* (Zeuthen, 1964).

9. DNA synthesis, in animal cells, is confined to a central portion of the division cycle. There is no DNA synthesis in a newly divided cell, and after a period of time (the G1 phase) DNA synthesis (the S phase) starts. The S phase is finished some time before mitosis or division, and thus another phase of the division cycle, the G2 phase, is defined.

10. Howard and Pelc, 1951a,b.

11. Schaechter, Bentzon, and Maaløe, 1959.

12. See also Pachler, Koch, and Schaechter, 1965.

13. Single shifts have been used by a number of workers on different microorganisms (Hegarty and Weeks, 1940; Falcone and Szybalski, 1956; Hunter-Szybalska, Szybalski, and DeLamater, 1956; Scott and Chu, 1958; Starka and Koza, 1959; Doudney, 1960; Zeuthen, 1964), but the apparently most successful methods were those using a number of temperature shifts in succession (Lark and Maaløe, 1954; Zeuthen and Scherbaum, 1954; Zeuthen, 1958; Padilla and Cameron, 1964).

14. Starka and Koza, 1959; Kunicki-Goldfinger and Mycielski, 1966; Zaitseva, 1963.

15. Campbell, 1957. A simple clock is one in which there is a single value that determines the age of the cell, and all parameters of the cell are determined by this single value.

16. Cutler and Evans, 1966.

17. Cutler and Evans, 1966.

18. Dennis, 1971, 1972.

19. Ricciuti, 1972.

20. Barner and Cohen, 1956; Scott and Chu, 1958; Burns, 1959.

21. Matney and Suit, 1966; Ron, Rozenhak, and Grossman, 1975; Ron, Grossman, and Helmstetter, 1977.

22. Holland, 1987.

23. For this approach see Goodwin, 1969; Kepes and Kepes, 1980, 1981; Joseleau-Petit, Kepes, Peutat, D'Ari, and Kepes, 1987. The specific use of these methods to study either enzyme synthesis or lipid synthesis during the division cycle will be considered in Chapters 4 and 6.

24. Kepes and Kepes, 1980, 1981.

25. Anagnostopoulos, 1971.

26. Koch, 1986.

27. Grover, 1988.

28. Kepes and Kepes, 1980, 1981.

29. In a recent paper, where the synthesis of DNA during the division cycle was published, the data for only the third division cycle of synchrony were published (Robin, Joseleau-Petit, and D'Ari, 1990). We can only speculate that there were problems with the synthesis of DNA during the first two cycles. The DNA synthesis curves do not appear to support the conclusions of a step in DNA synthesis.

30. De Jonge, 1989.

31. Brostrom and Binkley, 1969.

32. Altenbern, 1968.

33. See, for example, Sussman and Halvorson, 1966; Young and Fitz-James, 1959.

34. Paolozzi, Nicosia, Liebart, and Ghelardini, 1989.

35. Centrifugation methods were first used with yeast cells and therefore the method is sometimes referred to as the Mitchison–Vincent Technique (Mitchison and Vincent, 1965). It has been used many times to produce synchronized bacterial cultures (Maruyama and Yanagita, 1956; Manor and Haselkorn, 1967; Kubitschek, Bendigkeit, and Loken, 1967; Shapiro and Agabian-Keshishian, 1970; Baldwin and Wegener, 1976).

36. Gudas and Pardee, 1974.

37. Poole, 1977a,b.

38. Figdor, Olijhoek, Klencke, Nanninga, and Bont, 1981. See also Lloyd, John, Edwards, and Chagla (1975) and Lloyd, John, Hamill, Phillips, Kader, and Edwards (1975).

39. Other workers have reported success with the filtration method (Anderson and Pettijohn, 1960; Lark, 1958; Lark and Lark, 1960; Nagata, 1963; Sargent, 1973). This method has been used to study nucleic acid synthesis (Maruyama, 1956; Maruyama and Lark, 1959, 1962; Abbo and Pardee, 1960; Nagata, 1962, 1963; Masters, Kuempel, and Pardee, 1964; Rudner, Rejman, and Chargaff, 1965), protein synthesis (Maruyama, 1956; Abbo and Pardee, 1960; Masters, Kuempel, and Pardee, 1964; Kuempel, Masters, and Pardee, 1965; Nishi and

Kogoma, 1965), the sequence of replication of the bacterial genome (Nagata, 1962), and studies with radiation and other chemical treatments (Yanagita, Maruyama, and Takebe, 1958; Helmstetter and Uretz, 1963; Ryan and Cetrulo, 1963).

40. Abbo and Pardee, 1960.

41. Maruyama, 1956.

42. Abbo and Pardee, 1960.

43. Cummings, 1965.

44. Nagata, 1962, 1963.

45. Helmstetter and Cummings, 1963, 1964.

46. The proof of this assumption, which is not crucial to the method, is presented in Chapter 6.

47. Helmstetter and Cummings, 1963.

48. Helmstetter and Cummings, 1963, 1964; Cummings, 1965, 1970; Clark and Maaløe, 1967.

49. Helmstetter, 1969.

50. Gmeiner, Sarnow, and Milde, 1985.

51. Collins and Richmond, 1962.

52. Most of the elegant work in this area has been done by Nanninga and Woldringh and their associates; Nanninga and Woldringh, 1985; Nanninga, Woldringh, and Koppes, 1982; Woldringh, 1976; Woldringh, Binnerts, and Mans, 1981; Woldringh, De Jong, van den Berg, and Koppes, 1977; Woldringh, Huls, Pas, Brakenhoff, and Nanninga, 1987; Olijhoek, Klencke, Pas, Nanninga, and Schwarz, 1982).

53. Trueba, Neijssel, and Woldringh, 1982.

54. The instrument is generically called a Coulter Counter, although other companies manufacture similar instruments.

55. Grover, Naaman, Ben-Sasson, and Doljanski, 1969; Kubitschek, 1971b. For further analysis of the use of a Coulter Counter, see Kubitschek, 1958, 1960, 1962a; Kubitschek and Friske, 1986.

56. Harvey and Marr, 1966.

57. Powell, 1956.

58. Hoffman and Frank, 1965; Adler and Hardigree, 1964; Adolph and Bayne-Jones, 1932; Bayne-Jones and Adolf, 1933.

59. Boye, Steen, and Skarstad, 1988; Steen, 1980, 1983; Steen and Boye, 1980; Steen and Lindmo, 1979.

60. Steen, 1980.

61. Lindmo and Steen, 1979.

62. Chandler, Bird, and Caro, 1975.

63. Lane and Denhardt, 1974, 1975.

64. Lark, Repko, and Hoffman, 1963; Bird and Lark, 1970.

65. Pachler, Koch, and Schaechter, 1965; Kubitschek and Newman, 1978.

66. Canepari, Del Mar Lléo, Satta, Fontana, Shockman, and Daneo-Moore, 1983.

67. Bremer and Churchward, 1978.

68. Although suggestions have been made that the cells eluted from a mem-

brane are perturbed, our cells divide normally after elution, and there is no evidence of any major perturbation. The major conclusion is that the cells are normal, but even if they were not normal, the membrane-elution method is satisfactory.

69. One example is given by Boyd and Holland, 1977.
70. Nanninga and Woldringh, 1985.
71. Cooper, 1990a.
72. Koppes, Meyer, Oonk, De Jong, and Nanninga, 1980.
73. Helmstetter and Pierucci, 1968.

4

Cytoplasm Synthesis during the Division Cycle

The cytoplasm of the bacterial cell is composed of all the enzymes, RNA molecules, ribosomes, solutes, and other low-molecular-weight molecules responsible for the biosynthesis of new cell material. An additional definition of cytoplasm is the material not associated with the cell surface or the genome. In bacteria, a number of proteins are found as part of the cytoplasmic and outer membranes; these proteins will be regarded as surface material, not part of the cytoplasm.

In this chapter we will consider how the cytoplasm is made during the division cycle. The conclusion of this chapter, that the synthesis of cytoplasm is a simple process devoid of sequential events during the progress of the cell from birth to division, is contrary to much of the experimental evidence. If this were a standard review article, the experimental evidence on cytoplasm synthesis would be presented, and some problems would be noted; the final conclusion would be left to the reader. The appropriate analysis of cytoplasm synthesis is contained in fewer than ten articles, not in the over one hundred or more proposing a complex sequence or pattern of cytoplasm synthesis. I shall first present what I consider to be the better description of cytoplasm synthesis during the division cycle. This will be followed by a description of the alternative proposals.

The discussion of the growth of the bacterial cell during the division cycle begins with an analysis of cytoplasm synthesis. Cytoplasm synthesis is the driving force behind the synthesis of DNA and the cell surface. It stands temporally before these syntheses. It is a simple and easily understood pattern, and the analysis of cytoplasm synthesis introduces us to the membrane-elution method. Most important, it introduces us to some simple *a priori* thoughts about biosynthesis. The idea of an uneventful process during the division cycle is repeated in the following analyses of DNA and cell-surface synthesis.

I. *A PRIORI* CONSIDERATIONS OF CYTOPLASM SYNTHESIS

There are two different views of cytoplasm synthesis. At one end there is a cascade of events during cytoplasm synthesis. This cascade is like a series of falling dominoes. Enzyme synthesis, or an enzyme activity, varies during the division cycle. One molecule accumulates, and at some concentration, triggers the synthesis of another molecule. The progress of the cell through the division cycle is a succession of these syntheses. This view proposes that there is a precise and reproducible historical sequence to the events occurring during the division cycle.[1] At the other end there is the proposal that cytoplasm synthesis is invariant during the division cycle; i.e., there are no sequential variations in protein synthesis or cytoplasm concentration during the division cycle. Cytoplasm synthesis is *a*historical. There is no marker during cytoplasm synthesis that can inform an observer when a particular synthesis is taking place during the division cycle. All cytoplasm syntheses are taking place at all times.

The function of cytoplasm—with regard to cell growth—is to make more cytoplasm. The cytoplasm does this by synthesizing all of the enzymes and molecules that metabolize external nutrients, converting these nutrients into precursors of macromolecules, and then synthesizing the macromolecules of the cell. If we assume that a newly made portion of cytoplasm is able to immediately function at the same efficiency as the cytoplasm that was involved in its production, then the expected mode of cytoplasm synthesis is exponential. Mathematically, this is expressed by saying that the rate of mass synthesis is proportional to the amount of mass, X, present at any time, t; therefore:

$$dX/dt = k \cdot X$$

which after integration gives us

$$X = X_o e^{k \cdot t}$$

where X_o is the initial mass and X is the mass at any time, t.

This pattern of synthesis is referred to as exponential because of the form of the rate equation. It is analogous to the exponential pattern of cell increase during balanced growth. An exponential rate of mass synthesis during the division cycle means that the absolute rate of synthesis is continuously increasing; the amount of mass synthesized in a unit time is not constant throughout the division cycle. Mathematically, this is due to the consideration that the differential (i.e., the rate of change) of an exponential function (mass increase) is an exponential function.

Let us consider cytoplasm synthesis at a more intuitive level. Take one gram of cytoplasm. If in 1 minute this initial gram made 1 milligram of new cytoplasm (i.e., there was a 0.1% increase in the extant cytoplasm), we would have 1.001 grams of cytoplasm. In the next minute, there would be somewhat more cytoplasm made than in the first minute because there is an extra milligram of cytoplasm making new cytoplasm. Therefore it takes approximately 0.1% less time to make the next milligram of cytoplasm. If the initial gram of cytoplasm made 1 milligram of cytoplasm every minute, it would take 1000 minutes to make the second gram of cytoplasm. Exponential synthesis means that rather than taking the 1000 minutes to produce the next gram of cytoplasm, it takes less time. The cytoplasm would double in 693 minutes rather than the 1000 minutes that would be required if there were a constant amount of cytoplasm added each minute.

Other modes of increase are conceivable, and have been proposed. For example, if cytoplasm increased linearly during the division cycle, there would be a constant rate of cytoplasm synthesis. Mathematically, the rate of mass synthesis would be constant:

$$dX/dt = k$$

and, after integration,

$$X = kt$$

What mechanisms could be imagined to produce a linear rate of cytoplasm synthesis during the division cycle? A linear rate of mass synthesis could arise if the newly made cytoplasm was not immediately able to synthesize more cytoplasm. If cytoplasm synthesis were linear between birth and division, the new cytoplasm made during the division cycle would be inactive until cell division, and the act of cell division would presumably activate the cytoplasm made during the preceding division cycle. An alternative mechanism for linear synthesis would be that there was a steady decrease in the efficiency with which a given amount of cytoplasm made new cytoplasm during the division cycle. For example, as more and more cytoplasm was made, each part of the cytoplasm would work at a continuously lower efficiency. A third possible mechanism is to envision a fixed number of functioning molecules involved in the uptake of nutrients. If new molecules involved in the incorporation of nutrients were not activated until a cell division, then there would be a constant amount of uptake during the division cycle. This would give a linear pattern of synthesis during the division cycle.

It is also possible that an exponential pattern of total cytoplasm synthesis could be accompanied by variations in the synthesis of individual molecules. We could imagine successive waves of the synthesis of individual proteins taking place at particular times during the division cycle. These waves would combine to give exponential synthesis of the cytoplasm. If a single gene gave a cell-cycle–specific period of synthesis of an enzyme, then when that enzyme was synthesized, there would be an infinitesimal reduction in the synthesis of all of the other proteins in the cell. These reductions in synthesis would not be measurable, but the total of all syntheses in the cell would continue to be exponential. Thus, exponential cytoplasm synthesis is not necessarily incompatible with cell-cycle–specific biosynthetic events. Another way of phrasing this idea is to say that cytoplasm is limiting, with regard to cytoplasm synthesis, and the variations of other cell components do not alter the exponential accumulation of total cytoplasm.

A large number of enzymes have been reported to be synthesized in a cell-cycle–specific manner.[2] There have also been reports of cell-cycle–specific synthesis of various RNA molecules.[3] The patterns of synthesis are varied, and have been found for different enzymes, different organisms, and different situations. Despite voluminous evidence for cell-cycle–specific biosynthesis, I suggest that there are no cell-cycle–specific syntheses of cytoplasm components. Let us look at some simple *a priori* arguments explaining why we would not expect these synthetic patterns.

From a logical or evolutionary point of view, cell-cycle–specific protein synthesis during the division cycle would not be expected. One particular pattern often found is the synthesis of an enzyme or protein at a particular time during the division cycle. Consider a certain protein that is made over a short period at a particular time in the division cycle. During that short window of synthesis, the protein is made in sufficient quantity for the needs of the cell during the next interdivision time; enough protein or enzyme is made until the next window of synthesis appears. A moment before the short period of biosynthesis, the cell had enough of that protein; immediately after synthesis, it had an excess. The protein is in excess because it must now last, without additional synthesis, until the next division cycle and the next window of synthesis. This is an inefficient use of resources. The resources of the cell would be better used synthesizing the protein continuously during the cycle and devoting its saved energy for additional ribosome and protein synthesis. This would give that cell an evolutionary advantage over a cell that retained a cycle-specific synthetic pattern.[4]

Although the argument above was couched in terms of the synthesis of a particular protein at a particular time during the division cycle, any pattern that is not exponential would lead to inefficiency. If total cytoplasm synthesis were exponential, and a particular protein had a bilinear pattern of synthesis (i.e., its rate of synthesis doubled at some point in the division cycle), there would necessarily be periods of enzyme excess during the division cycle. Only the exponential synthesis of each component of the cell cytoplasm leads to the most efficient use of cell resources.

There have been numerous proposals of cycle-specific variations in enzyme synthesis. For example, one review suggested ". . . that based on our present knowledge, the pattern of enzyme regulation in synchronous cultures of both prokaryotic and eukaryotic cells is complex."[5] My proposal is that synthesis of enzymes during the division cycle is simple as well as efficient; cytoplasm synthesis is unrelated to the division cycle.

The idea of a sequence of events during the division cycle, leading from cell birth to cell division, may come from the description of sequential metabolic pathways in biochemistry. These are presented as events that follow one after another, as in:

$$A \to B \to C \to D \to E$$

Although we draw the metabolic sequence as proceeding from left to right, with B made from A, and so on, in a metabolic system all of the reactions are proceeding at all times. There is no differentiation in time such that all A is converted to B and then all B is converted to C. The overwhelming success of biochemistry and this description of sequential pathways led to the search for similar sequential patterns during the division cycle. While there may be sequential synthetic patterns of cytoplasm synthesis, there is no temporally sequential pattern of synthesis.

Other systems have served as models for sequential pathways during the division cycle. In bacteriophage λ there are a large number of positive and negative controls that govern transcription. During phage development, these controls work in a sequential manner to regulate the synthesis of enzymes and proteins that lead to an efficient production of new λ bacteriophage. It seems logical that in a more complex system, like the cell, there would be controls at least as complex. These complex levels of transcription control in bacteriophage λ are not synchronized and occur in a continuous and overlapping manner.

There is another source of the idea of sequential events during the division cycle; this is the metaphor of life. The life cycle of a cell, or more specifically the division cycle of a cell, is commonly envisioned as a microscopic version of the life pattern of an individual. Just as individuals have their lives marked by events—birth, school, graduation, marriage, children, grandchildren, retirement, and death—so the cell's life history is marked by events. These events flow from birth to division. A cell is born, then performs a sequence of events culminating in the final event, cell division. In contrast to this sequential view of the cell, the model proposed here is that of a division cycle with no events in cytoplasm synthesis. It may be difficult at first to imagine an uneventful life history of a cell, marked only by simple events such as the start and finish of DNA replication and the start and finish of cell division. We must reject the idea of discrete, temporally sequential events occurring during the division cycle and replace it with the view of cytoplasm as an organic, growing, homogeneous, and time-invariant entity.

Having proposed that there are no cycle-specific variations in enzyme synthesis, we may note one type of exception. Consider a cell where DNA is synthesized during a relatively short period in the middle of the division cycle (see Chapter 5 for further details of DNA synthesis). If, at the instant DNA synthesis was initiated, there were a sudden reduction in the pool of deoxythymidine triphosphate (TTP, a precursor of DNA), we might expect a relaxation of repression of the enzymes producing this molecule. A surge in the synthesis of these enzymes would follow. As the concentration of TTP returns to normal, repression reduces the rate of enzyme synthesis. Even if such a synthetic pattern were too small to be experimentally measured, it is conceivable that such a variation in cytoplasm synthesis does exist. Does the proposal of an uneventful cytoplasm synthesis mean we should not consider this wave of enzyme synthesis a cell-cycle event? This change in synthetic rate is not significant for the regulation of the division cycle. When such a change in the synthesis of an enzyme occurs, there is an infinitesimal decrease in the synthesis of all the other proteins in the cell. Some ribosomes now switch over to the translation of RNA messages determining the synthesis of enzymes related to TTP synthesis; but there is no change in the total rate of cytoplasm synthesis. The ribosomes that make the enzymes related to TTP synthesis stop synthesizing the other cell proteins. Should we consider the decreases in these syntheses to be cell-cycle–specific events? It is proposed here that any of these minor fluctuations in biosynthesis are not related to the regulation of the cell cycle. They are symptomatic responses to the other changes that are occurring during the division cycle.

It has been proposed that another source of variation in individual enzyme synthesis during the division cycle is gene dosage. If this were the case, then at the instant a gene doubled by replication there would be a sudden change in the rate of synthesis of a particular protein.[6] Because this change could happen for any enzyme, we could argue that there are myriad separate events during the division cycle. This conclusion should be reversed. The very multiplicity of these changes renders each one so minor as to be unimportant with regard to the regulation of the division cycle.

II. EXPERIMENTAL ANALYSIS OF CYTOPLASM SYNTHESIS DURING THE DIVISION CYCLE

A growing cell is composed primarily of RNA and protein, and the synthesis of the cell cytoplasm or mass may be loosely associated with the synthesis of the major components (see Chapter 14 for a detailed analysis). The total mass of the cell should include other cell components such as cell surface and DNA, but because these components are such a small fraction of the total mass, the RNA and protein are essentially equivalent to the cell mass.

A. Total Protein Synthesis during the Division Cycle Analyzed by the Membrane-Elution Method

Membrane-elution is the best method for measuring the rate of synthesis of protein or mass during the division cycle.[7] Let us first look at this approach, and then see how other methods have contributed to the analysis of protein and mass synthesis during the division cycle.

The results of a membrane-elution experiment supporting exponential protein synthesis are schematically illustrated in Fig. 4–1. A radioactive amino acid is added to an exponentially growing culture, the culture is filtered onto a membrane, and the newborn cells are eluted as described in Chapter 3. A plot of the radioactivity per cell in the eluate is a straight decreasing line on semilogarithmic paper.[8] The analysis of this curve entails looking at it in reverse, from right to left, over one division cycle, so the *rate* of incorporation of the radioactive amino acid is exponential.

In this experiment, the pool size of the amino acid used, leucine, was so small that the uptake of leucine into the cell was an accurate measure of the synthesis of protein. The same exponential result was obtained with different amino acids. The conclusion is that protein synthesis is exponential during the division cycle, with no indication of any change in the rate of leucine incorporation at the instant of cell division. In Fig. 4–1, the expected results for alternative patterns of biosynthesis, such as linear and bilinear synthesis, are plotted. (Bilinear synthesis is a pattern in which two linear synthetic rates combine to produce the total synthesis during the division cycle; the break in the rates may occur at any time. Linear synthesis is a special case of bilinear synthesis with the change in rate occurring at division. This leads to a single linear rate during the division cycle.) The results clearly eliminate linear and bilinear patterns of amino acid uptake and protein synthesis during the division cycle. This experiment measures the rate of synthesis of protein during the division cycle, and therefore it is a differential method as defined in Chapter 3. Below we shall discuss an alternative approach that used an

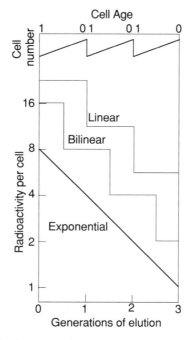

Figure 4.1 Protein Synthesis during the Division Cycle Using the Membrane-Elution Method. This is a schematic illustration of the results obtained on the biosynthesis of protein during the division cycle. A radioactive amino acid is added to an exponentially growing culture for a short time, the cells are filtered onto a membrane, the membrane is inverted, and the newborn cells produced by division are collected. The upper line is the concentration of cells in the eluate, and the lowest line is the observed result for radioactivity per cell. The two thinner lines are the expected results for either linear or bilinear protein synthesis. Linear synthesis is a special case of bilinear synthesis, with the break in the rate of synthesis occurring precisely at cell division. The observed exponential result is very different from that expected from either linear or bilinear synthesis. There are no flat regions in the elution curve that would indicate any constant rate of protein synthesis.

integral method for the analysis of mass synthesis during the division cycle.

This discussion was started by proposing that protein synthesis is exponential and finished by concluding that the incorporation of an amino acid is exponential. What is the difference between these two ways of looking at the division cycle? This difference is important and has played a crucial role in either aiding, or hindering, our understanding. Suppose that there were some other pattern of protein synthesis during the division cycle, for example, linear. There would be a constant rate of protein synthesis during the division cycle. But in the membrane-elution method, all that is measured is the uptake of the amino acid during the division cycle. It can be imagined that the uptake of radioactive amino acid could be exponential—that is, the rate of incorporation of the amino acid into the cell increases during the division cycle—yet the uptake may not reflect the actual rate of protein synthesis. If this were the case, then we would have an accumulation of amino acid in the cell during those portions of the cycle when the synthesis was less than the uptake, and a decrease when synthesis was greater than the uptake. Once they have taken in the labeled precursor, the cells on the membrane will be able to incorporate it at a later time because the cells grow on the membrane. Because the newborn cells are not released from the membrane for some time after the cessation of the pulse-labeling period, the cells will be able to incorporate all of the pool material into insoluble protein by the time the cells are available for analysis.

How can we analyze this membrane-elution experiment demonstrating exponential uptake of an amino acid in view of the pool problem? One answer is that the soluble pool is negligible. The time for residual incorporation of amino acid into the cell cytoplasm is less than one minute following filtration onto the membrane. As the pool is less than 2% of the incorporated material after a very short pulse, the uptake of the radioactive amino acid is a good measure of protein synthesis. As we shall see in the next chapter, the pattern of DNA replication derived from experiments with the membrane-elution method is completely supported by a completely independent method.[9] This result indicates that the membrane-elution method can be used to determine the pattern of biosynthesis during the division cycle. Although in one case, such as DNA synthesis, the pool problem is negligible, in other cases the pool may be important. It is proposed here (and in more detail in Chapter 5) that the membrane-elution method is the only method that has had any division-cycle result confirmed by an independent method.

B. Autoradiographic Analysis of Protein Synthesis during the Division Cycle

Exponential synthesis has been supported by the autoradiographic analysis of amino acid incorporation as a function of cell size.[10] An exponentially growing culture was pulse-labeled with radioactive leucine, and the cells were fixed onto slides. After covering the cells with photographic emulsion and storing for autoradiography, the number of grains per cell was determined as a function of cell size. When the cells were grouped according to length, there was a linear relationship between the length of the cells and the number of grains per cell. The length of the cell is a good first approximation of the cell size (but see Chapter 6 for another view).

As the incorporation of radioactivity was a proportional to cell size, this indicates that synthesis is exponential.

In this experiment, the cells were washed before autoradiography, so the pool was removed. What was therefore determined was the incorporation into cell protein. At the extreme sizes there were deviations from exponentiality. Very large and very small cells may be abnormal and not representative of the rules by which cells grow during the division cycle.[11] It should also be realized that very large cells may be an extremely small fraction of the total population; searching through enough cells may produce a statistically valid determination of the grain count for these cells. Yet as a fraction of the total population, this incorporation would be negligible, and would not change the major result, that protein synthesis is exponential during the division cycle.

C. RNA Synthesis during the Division Cycle

The membrane-elution method has also been used to determine the pattern of RNA synthesis during the division cycle. RNA synthesis during the division cycle is exponential.[12] It is interesting that the initial measurements of RNA synthesis using the membrane-elution method indicated that there were perturbations, and the result was not as clear as that for protein synthesis. It was then shown that these perturbations were due to the incorporation of the precursors of RNA into DNA, and that there were variations in the rate of DNA synthesis during the division cycle. When these DNA-related variations were considered (see Chapter 5), it became clear that RNA synthesis during the division cycle was exponential.

D. Size Distribution of Exponential Populations

If the pattern of mass synthesis during the division cycle is exponential, we can derive the expected size distribution of an exponentially growing population. While the age distribution is such that there are twice as many newborn cells as dividing cells (Chapter 1), when we consider the size distribution, the predicted result depends upon the pattern of cell growth. Assume that a cell grows linearly (i.e., with constant rate) throughout the division cycle. This means that the cell size is directly proportional to cell age, and the cell-size distribution should be the same as the cell-age distribution. At each age increment between ages 0.0 and 1.0, cell size increases linearly between sizes 1.0 and 2.0. The predicted result for linear synthesis is illustrated by the dashed line in Fig. 4–2. It is the same as the age distribution, because there is a one-to-one correspondence between cell size and cell age. In contrast, the size distribution is different if mass synthesis is exponential during the division cycle. If the cells grow exponentially during the division cycle, there will be *four*

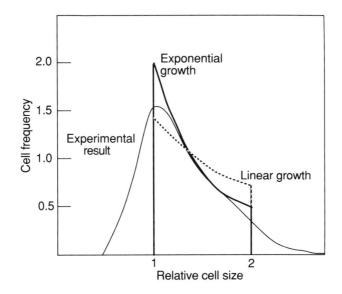

Figure 4.2 Growth during the Division Cycle Determined from the Size Distribution. The expected size distribution patterns for either a linear or exponential increase in cell size during the division cycle are illustrated. For the linear case, a factor of two is expected between the frequency of the largest and smallest cells, while for exponential synthesis a factor of four is expected; this means that the exponential pattern yields a much steeper graph of the size distribution. The thin line is a generalized illustration of the typical result obtained for the size distribution of cells during exponential growth. The middle of the experimental distribution fits the exponential pattern rather than the linear pattern.

times as many small cells as large cells. How can this be, if there are only twice as many newborn cells as dividing cells? Consider that the newborn cells have a size of 1.0 and the dividing cells a size of 2.0. Consider that the newborn cells go from size 1.0 to size 1.0 + δs in an instant; δs is a small increment in size. The number of cells within this interval is a measure of the number of cells of size 1.0. Now consider the cells of size 2.0 − δs. These are the cells that are within δs of dividing. The largest cells will pass through this size interval at twice the rate that the newborn cells will pass through δs because the rate of mass synthesis is nearly twice as great in the large cells as in the small cells. At the limit (δs → 0) cells move through the large sizes at twice the rate of cells moving through the small sizes. Exponential growth means that the older cells are growing at twice the rate of newborn cells, because these oldest cells are twice the size of the youngest cells. If we now consider that the newborn cells are twice as numerous as the dividing cells (by the age distribution), and pass through the various size intervals at half the rate of older cells, we find that there are four times more of the smallest cells than the largest cells. (A graphic proof is given in Fig. 8–7.) This argument can be reapplied to cells growing linearly during the division cycle. In this case the cells in the smallest and oldest size classes are resident for the same amount of time in each of their size classes. Therefore the size distribution is the same as the age distribution. The size distribution for exponential mass increase during the division cycle is given by the bold line in Fig. 4–2.

A typical experimental result for the size distribution is given by the thin line in Fig. 4–2. In the middle range of cell sizes the observations fit the exponential pattern better than the linear pattern. At the extremes of cell size, there is less a fit because of variability in the sizes of cells at division and birth (see Chapter 8). This type of experiment will be analyzed in more detail in Chapter 14 where the general growth law of bacteria is presented. As we shall see, cell size measurements fail to unequivocally support exponential growth and eliminate other models that are approximately similar to exponential growth. What *can* be concluded is that the size distribution of an exponentially growing culture does fit the pattern expected by exponential growth of the cell during the division cycle, and does not fit the pattern expected for linear growth during the division cycle.[13]

As described in Chapter 3, Collins and Richmond[14] have derived an equation that allows determination of the growth law of a bacterial culture from the existing size distributions, if the distributions at birth and division are known. When *Bacillus cereus* was studied, Collins and Richmond found that the cells in the middle ranges were increasing their length (size) exponentially. At the ends of the size distribution, there were complications owing to the difficulty of getting the size distribution of cells at the instant of division and at the instant of birth. The method is satisfactory for the middle range of cell sizes. It would be impossible to have linear mass synthesis give the midrange steepness of the experimental line (Fig. 4–2). Such analyses have supported the exponential growth law more than the linear growth law, although other growth laws are consistent with the observed size distributions.[15] This means that measurements of the size distributions of exponentially growing cells are consistent with exponential mass synthesis, and inconsistent with linear synthesis. Other models of synthesis that may be close to exponential synthesis (such as various forms of bilinear or even trilinear synthesis) cannot be eliminated by total-population methods.

E. Analysis of Specific Protein Synthesis during the Division Cycle

It is possible to measure specific protein synthesis during the division cycle using the membrane-elution method. The experiment would be similar to the analysis of total protein; however, the analysis of each of the fractions eluted from the membrane would be different. An exponentially growing culture would be pulse-labeled with a precursor of protein. The cells would be placed on the membrane and newborn cells eluted from the membrane. The protein in each of these fractions would be subjected to two-dimensional electrophoretic separation, autoradiography, and finally the radioactivity in each of the spots would be quantitatively analyzed. If there were, for example, a period of time when a particular protein was synthesized, then only a portion of the eluted fractions would reveal this labeled protein. At this time, I know of no published experiment using this approach,[16] though an analogous approach has been used.

Lutkenhaus, Moore, Masters, and Donachie[17] have looked at the proteins of *Escherichia coli* by two-dimensional gel electrophoresis as a function of cell-cycle age using both synchronized cells and cells separated by size after labeling. The second approach is analogous to the membrane-elution method in that the cells are labeled in an unperturbed state, and there is no further incorporation during the period of size separation of cells. In their study of over 750 different proteins, none of the proteins appeared to be synthesized in a cycle-specific manner. Some minor protein may be synthesized in a cycle-specific manner, but for various reasons—it was covered over by another major protein, or it was too minor to be detected—it was not measured or discovered. It is impossible to eliminate, by this experimental approach, each and every possibility of a cycle-specific synthetic process. This type of criticism is made from the assumption that there is some cell-cycle–specific protein synthesis.

A slightly more interesting experiment, because there was a likelihood that there were cycle-specific syntheses and the number of proteins studied was finite and clearly defined, is that of Wientjes, Olijhoek, Schwarz, and Nanninga,[18] who studied the penicillin-binding proteins. They showed that the penicillin-binding proteins were synthesized at constant rates throughout the cell cycle. There is circumstantial evidence suggesting[19] that there are different peptidoglycan-synthesizing systems active at different times during the cell cycle (one for cylinder extension and another for pole growth; see Chapter 6 for details). Therefore we might expect a pattern of synthesis of these penicillin-binding proteins reflecting the changes in activity during the cycle. As it turned out, in this well-designed and -executed experiment, there was no measurable change in the quantity of any penicillin-binding protein as a function of the division cycle.

The analysis of the penicillin-binding proteins used synchronized

cultures produced by a carefully controlled elutriation technique. Because the assay of the penicillin-binding proteins affects cell growth and
division, this type of experiment cannot be performed using the
membrane-elution method. Adding the assay molecules to label the
binding proteins would affect cell division and thus alter the subsequent
elution of the cells from the membrane.

F. Direct Evidence for the Absence of Cell-Cycle–Specific Enzyme Synthesis

In a series of model experiments on the synthesis of enzymes during the
division cycle, Bellino[20] demonstrated that aspartate transcarbamylase
was made continuously and exponentially during the division cycle. No
evidence of cycle-specific synthesis was observed, even though previous
work had suggested that this enzyme was made in a cell-cycle–specific
manner. Any experiments that purport to demonstrate cell-cycle–
specific synthesis of any enzyme or protein should use Bellino's work
(which used the membrane-elution method) as a model. Of particular
note is Bellino's analysis of the possible disturbances of cellular metabolism caused by synchronization.

III. ALTERNATIVE PROPOSALS FOR CYTOPLASM SYNTHESIS DURING THE DIVISION CYCLE

In the previous sections, it was argued that the synthesis of cytoplasm is
continuous, ahistorical, invariant, and exponential; the synthesis of cytoplasm is independent of the division cycle. Nevertheless, an enormous
number of papers propose just the opposite. There are proposals for
different modes of mass synthesis, specific synthetic patterns during the division cycle for various enzymes and proteins, and there are
even cell-cycle–specific relationships proposed for the accumulation of
cations.

Science is not democratic. Science does not progress by majority vote.
No matter how many articles oppose the postulate of invariant, exponential synthesis during the division cycle, the idea of a simple, organic
increase of all cytoplasmic components is still correct. For every experiment involving other models of cytoplasm synthesis, the results are
either not convincing, or subject to criticism based upon the criteria for
cell synchrony analysis as presented in Chapter 3.

Is it possible to eliminate *all* proposed nonexponential patterns of
synthesis during the division cycle? Even one experiment that proved the
existence of a periodic synthetic pattern would invalidate the entire

proposal. In order to prove exponential synthesis of cytoplasm we would have to show what was wrong with each experiment in the enormous literature on cell-cycle–specific variations in the rate and pattern of cytoplasm synthesis. Instead, we will analyze some representative examples of experiments that have been used to support the alternative position, and the general rules discovered by this analysis can then be applied to any other results.

One other point should be noted. No example of cell-cycle–specific cytoplasmic synthesis has served as a starting point for a detailed analysis of the mechanism of the phenomenon. For example, once the phenomenon of induction of β-galactosidase was described, a large number of workers could confirm the phenomenon and work on the problem of the mechanism of induction. This work eventually led to the isolation of molecules that were involved in the regulation of this enzyme. No cytoplasmic cell-cycle–specific event approaches this level of analysis. Cell-cycle–specific syntheses and events may be reported, but further work usually demonstrates that they do not exist.

A. Linear Synthesis during the Division Cycle

An alternative model of cell-mass synthesis during the division cycle has been proposed by Kubitschek. His main thesis is that mass synthesis during the division cycle is linear. This means that the rate of mass synthesis is constant during the division cycle and does not increase as proposed by the exponential model of cytoplasm synthesis. The initial proposal for linear synthesis stems from measurements of the uptake (or incorporation) of a large number of compounds (glycine, leucine, glucose, acetate, sulfate, phosphate, and thymidine) as a function of the cell cycle. Kubitschek observed that the uptake of all of these compounds was constant during the division cycle.[21] This result was interpreted as indicating that the total uptake of nutrients from the medium is constant during the division cycle. The net accumulation of mass, which is dependent on this uptake, is therefore proposed to be linear. Direct measurements of cell size have been put forward to support linear synthesis.[22] These results are part of a general model proposing that in all cells, growth during the division cycle is linear.[23] The latest support for linear growth is an analysis of cell sizes following a shift-up.[24]

There have been other reports of linear synthesis during the division cycle.[25] Although a doubling in the rate of mass synthesis was reported to coincide with the initiation of DNA replication, the same doublings in the rate of mass synthesis were observed when DNA synthesis was inhibited. It was concluded that the observed doublings in the rate of increase in cell size were not correlated with, or related to, the initiation of DNA replication.

The linear-synthesis model of Kubitschek implies that there is a global control system that regulates mass synthesis. This control system makes mass increase linearly during the division cycle. If macromolecule synthesis is essentially exponential,[26] this global control system adjusts the uptake of molecules during the division cycle so that the total mass increase is linear.

In contrast, the passive, exponential model proposed at the beginning of this chapter (and in Chapter 2) does not propose any global control system. Each of the components of the cell—protein, DNA, peptidoglycan, lipid, and RNA—is synthesized according to signals given to it by the state of the cell; there is no mechanism that adjusts mass synthesis to give a linear mass increase during the division cycle. The mass increase of the cell is the sum of the masses of its individual components, and although it is essentially exponential, in detail it is a complex function of the sum of the rates of synthesis of the different macromolecular and micromolecular components of the cell (see Chapter 14).

The main problem in determining whether cell growth is linear or exponential during the division cycle is that the two models differ by at most 6% when looked at in a direct manner (e.g., by looking at the size of synchronized cells). The problem is illustrated in Fig. 4–3. Cells are synchronized using a centrifugation technique. The smallest cells are collected from the gradient. The size of cells as they grow during the

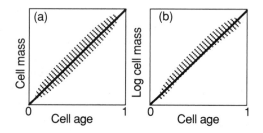

Figure 4.3 Linear or Exponential Accumulation of Mass during the Division Cycle. Panel (a) is a schematic illustration of the increase in cell size during the division cycle (shaded area indicates the experimental points). Small cells selected from a sucrose gradient are allowed to grow, and cell size is determined electronically using a Coulter Counter. In panel (b), the same experimental points have been plotted on a semilogarithmic graph. The straight lines are the expected results for either linear mass synthesis (a) or exponential mass synthesis (b). Although there is some curving of the data when plotted on semilogarithmic paper (b), statistical analysis reveals that the different models cannot be distinguished on the basis of this cumulative experiment.

division cycle is determined using a Coulter Counter. As cell density is constant during the division cycle (see Chapter 7), the size of the cell is proportional to mass. A plot of the results, schematically illustrated by the shaded area in Fig. 4–3a, is consistent with linear synthesis. But when the experimental points are replotted (Fig. 4–3b), the points fit the exponential pattern equally well.[27] Statistical analysis cannot distinguish between the two models. In contrast, as noted above and in Chapter 3, the predictions of the two models when studied in a differential manner—e.g., pulse-labeling cells at different ages during the division cycle so that the *rate* of synthesis is measured—are quite different. The linear model predicts a constant rate of mass synthesis; the exponential model predicts an exponential increase in the rate of mass synthesis. The more conclusive experiment—that is, the differential analysis—supports exponential synthesis of macromolecules (RNA as well as protein), while the less conclusive experiment—direct measurement of cell sizes—cannot distinguish between the different models.

One major difference between linear mass synthesis and exponential mass synthesis lies in what the two models predict for the variation in uptake mechanisms during the cycle and at division. The exponential model proposes that as the cell grows, there is a continuous increase in the ability of the cell to take up molecules from the medium, and there is no sudden change in, or activation of, any uptake mechanism at any time during the division cycle. The linear model predicts that there is a sudden activation of the uptake mechanism.[28] Kubitschek has proposed such a constancy of active uptake sites during the division cycle and their duplication before or during division.

The major difference between the experiment of Kubitschek (Fig. 4–3) and the membrane-elution experiment (Fig. 4–1) is that with sucrose gradient selection the cells are handled to produce the smaller cells. In contrast, the actual labeling of cells in the membrane-elution method is the only perturbation of the cells. All of the other manipulations of the cells occur after the labeling period, and thus are not causes of perturbation error (see Chapter 3).

As the smallest cells in the gradient were selected for the synchronized culture, it is also possible that there was not a random selection of young cells. That is, the cells selected were possibly smaller than an average newborn cell. Without an independent knowledge of the size distribution of newborn cells (which would entail independent knowledge of the rate of mass synthesis during the division cycle), we cannot know that the smallest cells selected are an accurate reflection of the youngest cells in the exponential culture.

The method of measuring the size of cells, the Coulter Counter size distribution, is also a problem. As the cells grow and divide, the baseline provided by the newly forming and growing baby cells makes it very difficult to extract the rate of size increase for cells near the end of the division cycle. If there were a large separation between the newborn cells and the largest dividing cells,[29] this might not be a problem. But the published graphs indicate that the newborn cell distribution does carry over to the dividing cell distribution. It should also be noted that the difference between the smallest cell-size peak and the largest cell-size peak was only 16 channels on the pulse height analyzer. A change in peak location of only one channel could account for the difference between linear and exponential growth.

One explanation for linear mass synthesis and exponential macromolecule synthesis has been offered.[30] It has been proposed that the difference between the two patterns is due to the accumulation of soluble precursors in the middle of the division cycle. The requirements for such a pool are that it is at least 6% of the total cell mass, and thus approximately threefold larger than the pool available. Variation in pool size cannot allow a linear increase in mass synthesis during the division cycle.

The method of Kubitschek could be tested by demonstrating, with the synchronized cells obtained by sucrose gradient selection, that the uptake of thymidine varies in the manner predicted by the current model of DNA synthesis during the division cycle (Chapter 5). If this were found, it would support the method of Kubitschek for the analysis of the division cycle. Until this control is done, there is always the possibility that the method does perturb the cells.

B. Variation in Specific Protein Synthesis during the Division Cycle

There is a considerable amount of evidence supporting the variation of specific enzyme synthesis during the division cycle. This type of variation has been reported not only for *Escherichia coli*,[31] but also for many other organisms.[32] A summary of the types of syntheses that have been proposed to exist is given in Fig. 4–4.

C. Flagella Synthesis during the Division Cycle

The synthesis of flagella during the division cycle is a special case. If, for example, flagella were made at a pole after a division process, then one would expect to find periodic synthesis of flagella protein. In *Bacillus subtilis* such a periodic synthesis has been reported.[33] In one bacterium, *Caulobacter crescentus*, there is clear and definitive evidence that there is a cell-cycle–specific synthesis of a number of proteins (flagellin is one) that are precursors to the visible flagellum. In Chapter 11 it will be proposed that such cell-cycle–specific syntheses are not really related to the cell

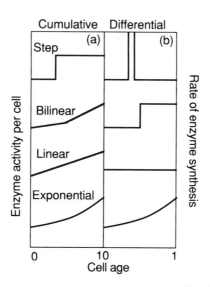

Figure 4.4 Proposed Patterns of Enzyme Synthesis during the Division Cycle. Four different types of enzyme accumulation during the division cycle are illustrated in (a). The
exponential synthesis gives a straight line when plotted on semilogarithmic paper, although the line is curved on rectangular paper. In (b), the differential rates of enzyme
synthesis are plotted. The differential plots, indicating the rates of synthesis, are more
distinguishing than the total amount present during the division cycle.

cycle or to cytoplasm synthesis during the division cycle. This cell-cycle–
specific synthesis has a different interpretation and does not support the
idea of cell-cycle–specific syntheses in the cytoplasm.

D. Variation in the Capacity to Synthesize Specific Proteins

A distinction must be made between the variation in protein synthesis
during the division cycle during steady-state growth, and the variation
during the division cycle in the capacity to synthesize a particular
protein. If a specific inducer of an enzyme is added to a growing culture,
there is a cell-cycle–specific variation in the ability of the cell to synthesize
that enzyme. For example, it has been shown that addition of an inducer
of β-galactosidase results in a doubling in the amount of enzyme produced in cells over a certain cell-cycle age.[34] This age is related to the time
when the gene for the enzyme has doubled (Chapter 5). When there is a
sudden change in the synthesis of an enzyme, from negligible synthesis
to maximal synthesis by the addition of a specific inducer, it appears as if
the genetic material, the number of genes, is the limiting factor. A cell
with twice as many genes will make twice as much enzyme. This result,

however, should not be compared to a situation in which the cells are growing continuously and with no change in the external conditions. Such sudden change in the capacity to synthesize an enzyme does not mean that the rate of enzyme synthesis changes during steady-state growth conditions.

E. Specific RNA Synthesis during the Division Cycle

Cell-cycle–specific synthesis of RNA molecules has been reported.[35] Experiments using the membrane-elution method have demonstrated that RNA synthesis during the division cycle is not periodic but is exponential.[36]

F. Soluble Components during the Division Cycle

1. Cation Concentration

There has been one report that the zinc content of synchronized cultures varies during the division cycle,[37] whereas other cationic components (potassium, magnesium, and calcium) do not vary. A reanalysis of the data reveals that the results for zinc do not support a cycle-specific variation. The data are quite consistent with a continuous and exponential increase in these ionic components during the division cycle. Others, rather than looking at the contents of the cells, have looked at the rate of uptake of various ions, both cations and anions,[38] and have found cell-cycle–specific variations.

2. Nucleotides

There is a long history of the variation of nucleotide synthesis during the division cycle and the possibility that such variation may be a clue to the regulation of the cell cycle. For example, stepwise changes in the synthesis of various phosphorus-containing molecules were reported during the division cycle of *Alcaligenes fecalis*.[39] It was also reported that there were variations in the nucleoside triphosphate pools during the division cycle of *Escherichia coli*.[40] The latest chapter in this history is the proposal that the tetraphosphorylated nucleotide (Ap$_4$A; diadenosine $5',5'''$-P^1,P^4-tetraphosphate) is somehow involved in the initiation of DNA synthesis.[41] Reports of its binding to the DNA replication system and its use as a primer for DNA synthesis[42] supported the idea, as well as measurements in synchronized animal cells showing that DNA synthesis was accompanied by cell-cycle–specific increases in the concentration of Ap$_4$A.[43] In order to see whether such variations in nucleotide concentration occurred during the bacterial division cycle, measurements of its concentration were made in synchronized cultures.[44] It was found that no variation in any tetranucleotide occurred (there are other analogues as well: Ap_4U, Ap_4G, Ap_4C), indicating that at least for these nucleotides, there was no cell-cycle relationship for their appearance or synthesis.[45]

It has been proposed that cyclic guanosine-3'5'-monophosphate is a regulator of cell division.[46] The evidence for cell-cycle–specific synthesis is extremely striking, with a sudden pulse of appearance of this nucleotide at a particular time during the division cycle. In this particular synchrony experiment, the increase in cell division is at least an order of magnitude longer than the time for the specific appearance of the nucleotide. This result is difficult to explain, for we would expect the synthetic period to be at least as long as the increase time of a synchronized culture. This suggests that the variation in nucleotide content has more to do with the perturbation of the cells than the timing of cell division.

It has been proposed that cyclic AMP can affect the morphology of bacteria[47] and thus may be a trigger of events during the division cycle. Direct measurements of the concentration of cAMP during the division cycle indicates that the amount of cAMP increases exponentially during the division cycle.[48] There does not appear to be any cell-cycle–specific synthesis of cAMP.

3. Calcium as a Trigger of the Division Cycle

A proposal was recently made that variations in the calcium content of a cell can be a general trigger of cell-cycle events.[49] This proposal is based on the conclusions of a series of experiments demonstrating there are cytoskeletal-like proteins in bacteria, and that these proteins are involved in regulating the division cycle.[50] The reasoning continues with the observation that calcium ions play an important role in regulating certain cell-cycle events in higher organisms, where these cytoskeletal proteins have a definite function, and that during the bacterial division cycle, a major change in the distribution of calcium from the cell envelope to the cytoplasm occurs.[51]

One problem with this line of reasoning is that it rests heavily on unproven experimental results. The measurements of the variation in cytoplasmic calcium during the division cycle were based on a static analysis of cells classified according to length. The larger invaginating cells could have been an abnormal fraction of the population. No evidence was presented of the variation in calcium as cells moved through the division cycle. Further, the identification of proteins in bacteria that crossreact with myosin does not mean that the protein so identified has any cytoskeletal function.

G. Comment on Specific Synthesis during the Division Cycle

My conclusion from the literature proposing variations in concentration or synthetic rates of various components of the cytoplasm is that there is yet to be proof of such variation. The criteria for a successful analysis of the cell cycle using synchronization (Chapter 3) have not yet been satisfied by these experiments. When the actual data are studied, there is just as good a fit to a random or exponential pattern of synthesis as to some regular cell cycle variation.

The history of the analysis of cytoplasm synthesis during the division cycle may be a result of the psychological aspects of science. Negative results are not valued as much as positive results. It is always better and more rewarding to find something than to find nothing. In the examples cited in this chapter, two conditions are uniformly found. First, if an objective choice were to be made between an event model and a nonevent model at the time of the original experiment, the original result could not distinguish between the two alternatives. And second, the main reason many of the secondary, refuting experiments were done was specifically to refute the original experiments. Alone, they were negative results. But in combination with the previous work, they could be published because they changed a prior idea. It is interesting to speculate how many experiments may lie buried in notebooks because the investigators were not able to show a cell-cycle–specific event. The cumulative effect of such evidence would be to support the nonevent model of the division cycle.

IV. A FUNCTIONAL MODEL FOR CELL-CYCLE REGULATION OF SPECIFIC PROTEIN SYNTHESIS

It is illuminating to look at one particular model of specific protein synthesis during the division cycle because of the instructive nature of the experiment. When spores of *Bacillus subtilis* germinate, DNA replication proceeds from one end of the genome to the other (see Chapter 13). It has been proposed that when various genes are replicated, there is a specific start of the synthesis of gene products from those genes. That is, when the first gene is replicated, there is a specific stimulation of that gene product, and as other genes are replicated, the other products are synthesized. This model[52] suggests that there is a sequential program of ordered syntheses of gene products that is advantageous for the germination process. Such a model could be adapted for the synthesis of gene products during the normal division cycle by proposing that when genes replicate there is a specific synthesis of gene products at particular times during the division cycle.

The evidence in support of this model is based on an error in the plotting of the data. The enzyme activities during germination were plotted on a semilogarithmic plot. As can be seen in Fig. 4–5, when straight-line data, with no lag, are plotted on a semilogarithmic plot, there is an apparent time sequence of enzyme synthesis. The synthetic pattern is caused more by the units used, than by the values obtained. If the values were changed by a factor of 10 relative to the other enzymes, the order of appearance of the enzymes would be changed. There is no evidence that there is a sequential appearance of enzymes during spore germination.

V. DETERMINATION OF THE GROWTH RATE BY THE MEDIUM

A. Physiological States and Growth Rates Produced by Different Media

Up to this point, we have been concerned with the pattern of mass synthesis during the division cycle. It is also of interest to consider how the absolute rate of mass synthesis is determined. This question can be considered completely independent of the particular pattern of synthesis during the division cycle.

It is not clear that bacteria growing in different media should have different rates of growth or different rates of mass synthesis. In the sad personal experience of many, eating much less does not make one thinner. Human beings have a homeostatic mechanism that moderates the effect of large variations in ingestion. Bacteria, in contrast, have a large

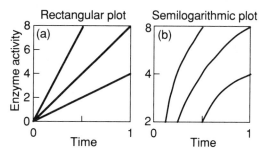

Figure 4.5 Semilogarithmic Plots and Cell-Cycle–Specific Protein Synthesis. Consider three rates of enzyme synthesis with slopes of 0.5, 1.0, and 2.0, shown in panel (a). When these same results are plotted on a semilogarithmic plot as in panel (b), it appears as if there is a sequential synthesis of enzyme activity.

number of growth rates and sizes precisely determined by the growth medium (see Chapter 1). How are these growth rates determined? (Chapter 5 will analyze how different cell sizes are determined by different growth rates.)

This problem is an area of active research, and the final answer is not in. Presented here is a way of thinking about the problem, a minimal answer upon which more complex aspects may be embroidered as they are discovered.[53]

Consider a cell growing in minimal medium with glucose as the sole source of carbon. For simplicity, consider only the involvement of the ribosomes in cell growth and cytoplasm synthesis. The ribosomes are involved in making the ribosomal proteins as well as all of the other proteins of the cell. The other proteins means all of those proteins involved in the metabolism of glucose or a carbon source, and the synthesis of the amino acids, nucleotides, and sugars responsible for cell growth. Let us say that 80% of the protein synthesized by the ribosomes is other protein and 20% is ribosomal protein. This means that for each doubling of cell mass, the other 80% of the ribosomes make the other 80% of the proteins, and 20% of the ribosomes, considered as a class, must work until they have made enough ribosomal protein equal to the total amount of ribosomal protein at the beginning of the division cycle. On average, it takes 8 minutes for a ribosome to make all of the protein required for a ribosome (see below), so we can consider that in 40 minutes the ribosome content will be doubled.[54] In these 40 minutes, the other four-fifths of the ribosomes will make an amount of protein that is four times as great as the ribosomal protein. At this growth rate, the amount of non-ribosomal protein should be approximately 80% of the total protein.

Now imagine that we add to the medium an amino acid that represses or inhibits the synthesis of a particular set of amino acid-synthesizing proteins. Within a few minutes ribosomes will exist that would, in the absence of the added amino acid, have been synthesizing those proteins; these ribosomes are now able to synthesize ribosomal protein. The time required for the ribosomes as a group to synthesize a cell's worth of ribosomes is slightly shortened. As we add more and more amino acids, the amount of material devoted to nonribosomal protein synthesis is decreased, and that devoted to ribosomal protein synthesis is increased.

As in the cell with a 40-minute doubling time, if two-fifths of the ribosomes are involved in making ribosomes and only three-fifths of the ribosomes are involved in making all of the nonribosomal proteins, we would expect the ribosomes to double in number in approximately 24 minutes. This is the time for two-fifths of the ribosomes to double the number of ribosomes.

The situation is not really as simple as portrayed, because as we reduce the amino acid requirements, and the cell grows faster, the rate of synthesis of the remaining proteins will increase slightly. This would lead to a slight lowering of the growth rate based only on the prediction of how much protein would be spared by adding a particular amino acid. The time required for the ribosomes as a group, now enlarged by the extra ribosomes, to double the number of ribosomes will not be predicted simply by the partition model, but will be a complex function of the entire regulatory system of the cell. A schematic illustration of the regulation of growth rate is presented in Fig. 4–6.

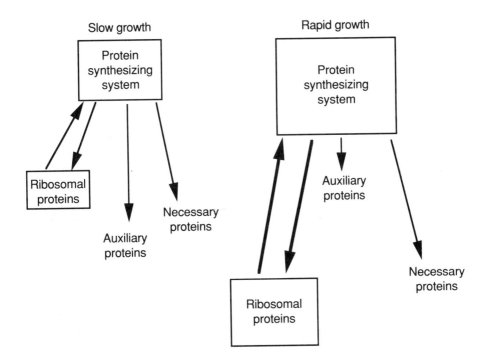

Figure 4.6 Variation in Rate of Mass Synthesis by Repression of Auxiliary Proteins. During slow growth in minimal medium, a number of auxiliary or dispensible proteins are made. These are proteins that will cease being made when they are not necessary for growth. When cells are placed in a richer medium, reducing the synthesis of auxiliary proteins, there is an increase in the proportion of the protein-synthesizing system devoted to synthesizing the protein-synthesizing system (ribosomes, RNA, polymerase, etc.). The more rapid production of ribosomes, for example, leads to the production of more and more protein-synthesizing system. Thus, a given amount of cell mass can double its mass in a shorter time because more of the mass is devoted to the process of making the machinery that makes more machinery.

Although this description uses ribosomes as an example, it is equally possible to use the partitioning of RNA polymerases among the genes for the protein-synthesizing system and the other genes of the cell. The result of the analysis is the same; the more genes are repressed, the more rapidly will a cell be able to double its cell mass and decrease the doubling time.

Regarding this analysis, there are two questions remaining. First, is the partitioning of the ribosomes among the different protein synthetic requirements of the cell enough to determine the growth rate of the cell? Second, is the partitioning, if it occurs as described, due to the passive responses described here, or is it due to some regulatory mechanism that directs the flow of syntheses such that the proper fraction of ribosomes is utilized to make the right amount of ribosomes in relation to total protein synthesis?

B. The Theoretic Maximal Growth Rate

It is interesting to consider the maximal possible growth rate. It is a truism that *Escherichia coli* has a maximal growth rate of about three doublings per hour, or a 20-minute interdivision time.[55] Consider the possibility that we may be able to add to the medium all of the material that a cell needs to grow, not only the small molecules such as the amino acids, nucleotides, and vitamins, but even the polymerases and synthetases that are used by the cell for macromolecular synthesis. Consider further that when these enzymes are added to the medium, the cell, sensing their presence, represses the synthesis of the respective enzymes. If all of the auxiliary and necessary proteins are added (see Fig. 4–6), then in theory, the cell needs to make only ribosomes.

How long does it take a cell to make a ribosome? It is possible to determine the step times (the time required to add an additional amino acid or a nucleotide to a chain) for RNA and protein synthesis. For amino acids, this is approximately 1/17th of a second,[56] which means that the chain elongation rate is 17 amino acids per second. For RNA the corresponding value is 1/45th of a second or 45 nucleotides per second. As the codon size (nucleotides per amino acid) is 3, this allows for a coordinate synthesis of the protein and the messenger RNA for that protein. The bacterial ribosome has approximately 8000 amino acids.[57] At a translation rate of approximately 17 amino acids per second per ribosome, approximately 8 minutes would be required for a single ribosome to make all of the protein for another ribosome. This is the minimal doubling time of a cell or the maximal growth rate.

This theoretic maximal growth rate of 8-minute doubling time is not actually attainable because the adding of the other enzymes to the cell is not possible; even if it were, there is no reason to expect that the cell would have a mechanism to repress the synthesis of those proteins. It is unavoidable that a portion of the ribosomes must be available for nonribosomal protein synthesis.[58] This portion determines the fraction of ribosomes available for ribosome synthesis. At any particular growth rate a fixed amount of ribosomes is working at ribosome synthesis. For example, if it did take 8 minutes to make a ribosome, then for cells growing with a 24-minute doubling time, approximately one-third of the ribosomes would be available for ribosome synthesis.[59]

This analysis contains an important but hidden message. In some experimental systems, the growth rate of cells is varied by changing the concentration of one particular nutrient. For example, a chemostat can be used to vary the rate of feeding a single limiting nutrient, or in batch growth, a particular amino acid can be added at different concentrations (see Chapter 12). Different growth rates may be obtained by this experimental manipulation. This variation in growth rate is not necessarily the same as the variation in growth rate obtained by adding different nutrients. If, for example, as a single nutrient were limited, and the rate of chain elongation of every component of the cell changed by the same factor, there would not be a variation of cell sizes as obtained by Schaechter, Maaløe, and Kjeldgaard (Chapter 1). Only those growth-rate changes that give a variation in cell size and composition are valid for studying the different steady-state sizes of cells.

VI. ICON FOR MASS SYNTHESIS DURING THE DIVISION CYCLE

It is possible to pictorially summarize the main idea of this chapter, that mass is synthesized in an undifferentiated manner during the division cycle, in the diagram in Fig. 4–7. This simple circle, with no indication of any sequence of events, indicates the continuous production of cell mass. The next chapters will offer a simple summary of the division cycle whereby all events are integrated.

VII. EVENTS DURING THE DIVISION CYCLE

The main conclusion of the analysis presented here is that the synthesis of the cytoplasm is uneventful. Cytoplasm increases in a simple, continuous, invariant, ahistorical manner between birth and division. No cascade of events or markers changes the rate of cytoplasm synthesis during

Figure 4.7   Icon for Mass Synthesis during the Division Cycle. Mass is synthesized continuously and uneventfully during the division cycle. Mass synthesis is determined by prior mass synthesis. The continuous circle is an icon for mass synthesis.

the division cycle. This proposal will stand as a model for the subsequent chapters, because the argument against events is repeated in various forms for both DNA synthesis and cell-surface growth. Only four events can be distinguished during the division cycle. These are the start and finish of DNA synthesis, and the start and finish of cell pole formation. There are no other events.

NOTES

1. For a summary of this event-signal viewpoint, read the recent review by D'Ari and Bouloc (1990).

2. Halvorson, Carter, and Tauro, 1971; Mitchison, 1969.

3. Rudner, Rejman, and Chargaff, 1965.

4. In the discussion following the article by D'Ari, Maguin, Bouloc, Jaffe, Robin, Liebart, and Joseleau-Petit (1990), the following argument is made: "Cooper's vision of the cell cycle, as described in the present Forum, supposes that the cytoplasm is uniform throughout the cell cycle, thus eliminating cytoplasmic signals such as a "termination protein" (synthesized only after chromosome termination and required for subsequent cell division). Cooper makes the teleological argument that nonuniform synthesis would be wasteful: the product would be either underproduced before the turn-on or overproduced afterwards. This is only true, however, if the product is required constantly throughout the cycle, which of course would not be the case for signal molecules." In the subsequent chapters it will be argued that the division cycle can function without the intervention of signal molecules that are made at any specific time during the division cycle. Signal molecules that regulate the division cycle are accumulated continuously during the division cycle between the events that they initiate.

5. Halvorson, Carter, and Tauro, 1971, p. 98.

6. It is important to distinguish the cell-cycle–specific variation in enzyme synthesis during steady-state, unperturbed growth, and variations in enzyme

synthesis during a sudden pulse of enzyme induction. As will be discussed in detail in Chapter 6, when a gene is duplicated there is a doubling in the capacity of the cell to synthesize the enzyme when the cells are pulse-induced for that enzyme. This nonsteady-state experimental situation should not be used as support for the change in synthetic rate during unperturbed growth conditions.

7. As described in Chapter 3, Abbo and Pardee (1960) used filtration to produce a synchronized or synchronous culture. The pattern of macromolecular synthesis was determined using these synchronous cultures. They observed that protein synthesis was exponential during the division cycle. The conclusion of this experiment is correct, but unfortunately there was no evidence that the method used, filtration synchronization, gave a culture suitable for the analysis of protein synthesis during the division cycle. The internal evidence indicated that these cells were unsuitable for determining the pattern of macromolecular synthesis during the division cycle. Their pattern of thymidine incorporation was also exponential; this pattern would not be expected if the cells were synchronized (See Chapter 5.) In the same historical vein, Perry (1958) was the first to suggest that growth during the division cycle is exponential.

8. Cooper, 1988a.

9. The experimental determination of the pattern of DNA synthesis during the division cycle (Helmstetter and Cooper, 1968; Cooper and Helmstetter, 1968; Cooper, 1969) has been confirmed by a completely independent method, flow cytometry (Skarstad, Steen, and Boye, 1985), that is not affected by changes in the pool sizes of precursors during the division cycle. No synchrony method has received such support.

10. Ecker and Kokaisl, 1969.

11. It should also be noted that there is an inverse correlation between length and width for cells in an exponentially growing population (Chapter 6). The longest cells are thinner than average, and this variation could account for the deviations from exponentiality.

12. Dennis, 1971, 1972.

13. Although it might appear that the experimental line is between the two theoretical lines in Fig. 4–2, and thus the models could not be differentiated, experimental error would lead only to a flattening of the curves. If the cells from adjacent size classes were mixed, the curve would become successively flatter. Thus an experimental result more steep than that expected for linear growth eliminates linear growth, while a line that may be somewhat flatter than expected for exponential growth does not eliminate exponential growth.

14. Collins and Richmond, 1962.

15. Grover, Woldringh, and Koppes, 1987.

16. Preliminary experiments in the author's laboratory did not reveal any cycle-specific protein synthesis when the membrane-eluted fractions were analyzed.

17. Lutkenhaus, Moore, Masters, and Donachie, 1979.

18. Wientjes, Olijhoek, Schwarz, and Nanninga, 1983; Wientjes, Schwarz, Olijhoek, and Nanninga, 1981. This is an example of an experiment that cannot

be done in the backwards methodology, because adding penicillin to the cells affects their cell division.

19. Spratt, 1975.

20. Bellino, 1973.

21. Kubitschek, 1968b.

22. Kubitschek, 1986.

23. Kubitschek, 1970a, 1968a. There have also been proposals that the rate of synthesis is bilinear (Kubitschek, 1981).

24. Kubitschek, 1990a.

25. Ward and Glaser, 1971.

26. Kubitschek, 1981.

27. A complete analysis of the data is given by Cooper (1988a). In Fig. 4–3a, the data of Kubitschek (1986) are plotted on a rectangular graph, with the best straight line (determined by statistical analysis) drawn through the data. A reanalysis of the published data on a semilogarithmic plot is shown in Fig. 4–3b. The relevant statistical parameter, R^2 (R square), which is an indication of how well the data fit the various models, was determined for each regression line and was found to be 0.99251 for the linear model and 0.98559 for the exponential model. The values obtained are so close to 1.0 that it can be concluded there is no statistical difference indicating that the data fit a linear model rather than an exponential model. There is an obvious curvature of the line plotted using the logarithmic transformation. This difference, however, does not allow a statistical elimination of the exponential model because of the insensitivity of the experimental method to distinguish the difference of 6% between the models. Arguments have been put forth in favor of using more sophisticated statistical methods that follow the curvature of the data that are able to distinguish between different models.

28. Kubitschek (1990b) has summarized the evidence for such a cell-cycle–specific activation of various uptake mechanisms. All of these experiments are synchrony experiments, and the original data should be studied before accepting any of the results.

29. As was shown on the data on microspheres used to calibrate the Coulter Counter.

30. Kubitschek and Pai, 1988.

31. Kuempel, Masters, and Pardee, 1965; Goodwin, 1969; Donachie and Masters, 1969; Nishi and Hirose, 1966; Shen and Boos, 1973; Scott, Gibson, and Poole, 1980; Dietzel, Kolb, and Boos, 1978.

32. Some of these cases are discussed in detail in Chapters 11, 12, and 13. See also Wraight, Lueking, and Kaplan, 1978; Wraight, Lueking, Fraley, and Kaplan, 1978.

33. Van Alstyne, Grant, and Simon, 1969. The data are not strong, and such synthesis is not proven.

34. Cooper, 1972.

35. Rudner, Rejman, and Chargaff, 1965; Cutler and Evans, 1967.

36. Dennis, 1971, 1972.

37. Kung, Raymond, and Glaser, 1976.

38. Kubitschek, Freedman, and Silver, 1971; Kubitschek, 1968b. To be precise, the experiments by Kubitschek (1968b) did not find cell-cycle–specific variations, as the finding was that uptake was constant during the division cycle. This means that at division there was a sudden doubling in uptake when cells separated. In the context of the comparison between exponential changes during the division cycle and other proposals, the linear-uptake proposal means that there are cell-cycle–specific variations. In the case of bilinear synthesis, the sudden change in synthesis occurs in the middle of the division cycle rather than at division.

39. Maruyama and Lark, 1962.

40. Huzyk and Clark, 1971; Grummt, 1983; Zamecnik, 1983.

41. Rapaport, Zamecnik, and Baril, 1982.

42. Weinmann-Dorsch, Hedl, Grummt, Albert, Ferdinand, Friss, Pierron, Moll, and Grummt, 1984.

43. Zamecnik, Rapaport, and Baril, 1982.

44. Plateau, Fromant, Kepes, and Blanquet, 1987.

45. Although a synchrony experiment is used here to support the position that there are no cell-cycle events, the author does not feel that even this result indicates that synchrony can work. Even if the synchrony were quite imperfect, and thus close to an exponential growth pattern, an exponential pattern of synthesis would be observed. Thus, imperfectly synchronized cells should give an exponential pattern of synthesis no matter what is studied.

46. Cook, Kalb, Peace, and Bernlohr, 1980.

47. Botsford, 1981.

48. Höltje and Nanninga, 1984.

49. Norris, Seror, Casaregola, and Holland, 1988.

50. Holland, Casaregola, and Norris, 1990.

51. Chang, Shuman, and Somlyo, 1986.

52. Kennett and Seuoka, 1971.

53. Cooper, 1970; Maaløe and Bentzon, 1985.

54. Although throughout this chapter the model of synthesis has been exponential cytoplasm synthesis, this simplified discussion of growth rate considers the ribosome increase without any complications of the calculations due to exponential mass synthesis. In a precise calculation, the time for the synthesis of the ribosomes would be shortened as new ribosomes completed within a short time enter into biosynthesis immediately.

55. Koch (1980) has data to support the growth of *Escherichia coli* at 17.5-minute doubling times.

56. Young and Bremer, 1976; Manor, Goodman, and Stent, 1969; Gausing, 1972.

57. Ingraham, Maaløe, and Neidhardt, 1983.

58. Cooper, 1970.

59. The actual value would be somewhat different because as ribosomes are made, they are able to immediately contribute to the synthesis of ribosomes.

5

DNA Replication during the Bacterial Division Cycle

The pattern of DNA synthesis during the division cycle is relatively complex compared to the simple pattern of mass synthesis. This complexity, however, is only superficial. There is an abundance of experimental and theoretical work on DNA synthesis during the division cycle, and almost all of this work fits into a single, unifying framework. The framework will first be described, and then the supporting experiments will be presented.

I. THE PROBLEM OF SCHAECHTER, MAALØE, AND KJELDGAARD

The problem of DNA replication during the division cycle is highlighted by the observation of Schaechter, Maaløe, and Kjeldgaard that there is a continuous variation in cell DNA content at different growth rates (see Chapter 1). If there were a large number of identical genomes in the cell, we could assume that cells at fast growth rates, with a large DNA content, would have 100 genomes, and as growth slowed, cells would have fewer and fewer genomes. With slight decreases in the growth rate there would be 99 genomes in the cell, then 98, and so on, accounting for the continuous variation in DNA per cell.

II. DNA SYNTHESIS DURING THE DIVISION CYCLE AT MODERATE AND FAST GROWTH RATES

The description of DNA synthesis will first be presented for cells growing at moderate and rapid rates. Slow-growing cells will be discussed separately.

The pattern of DNA synthesis during the division cycle of *Escherichia coli* B/r is illustrated for a number of different growth rates in Fig. 5–1.[1] The length of each of the graphs is drawn (at the left) proportional to the time between divisions. Replication is bidirectional, with DNA replication starting at a particular point on the chromosome and proceeding in different directions around each half of the chromosome. Therefore the chromosome could be considered as two separate chromosomes replicating simultaneously. For simplicity, in Fig. 5–1, replication is drawn as unidirectional. All of the chromosomal figures may be doubled and yield the same results. The passage of a replication point from one end of the chromosome to the other is called a round of replication.

In the cells growing with a 60-minute doubling or interdivision time, the newborn cell contains a chromosome that has just begun to replicate. The DNA content of this newborn cell is defined as one genome equivalent. At the end of the division cycle, the mother cells must contain exactly two genome equivalents of DNA, so that at division the two daughter cells each contain one genome equivalent of DNA. The DNA replicates for 40 minutes at a constant rate with the chromosome replicating one-quarter of its length every 10 minutes. Chromosome replication ends, or terminates, 40 minutes into the division cycle; thus termination occurs 20 minutes before cell division. The pattern of DNA synthesis during the division cycle is shown at the left in Fig. 5–1. In cells with a 60-minute interdivision time, there is a constant rate of DNA synthesis for the first 40 minutes, with no DNA synthesis during the last 20 minutes of the division cycle. If DNA replication began immediately after termination there would be more than two genome equivalents in the mother cell—the two genomes would be half replicated in that final 20 minutes—and at division, each newborn cell would have more than one genome equivalent. This would violate the basic principle of cell-cycle studies, that at the instant of cell division there should be exactly twice as much of each type of material in the mother cell as there is in the daughter cell. If this rule were violated, then the composition of the cells would be continuously varying, and a steady-state growth pattern would not be obtained.

A distinction is made between DNA synthesis and DNA replication. Replication refers to the movement of the growing point along the genome and the extension of the strands of DNA. Synthesis refers to the actual increase in the amount of DNA, irrespective of its arrangement in chromosomes. While the rate of replication may be constant, meaning that the movement of a replication point along the chromosome is constant, the rate of DNA synthesis varies during the division cycle. In the case of cells with a 60-minute interdivision time, there is a 40-minute period with a constant rate of DNA synthesis and then a 20-minute period with a zero rate of DNA synthesis.

Now consider cells growing with a 50-minute interdivision time. The

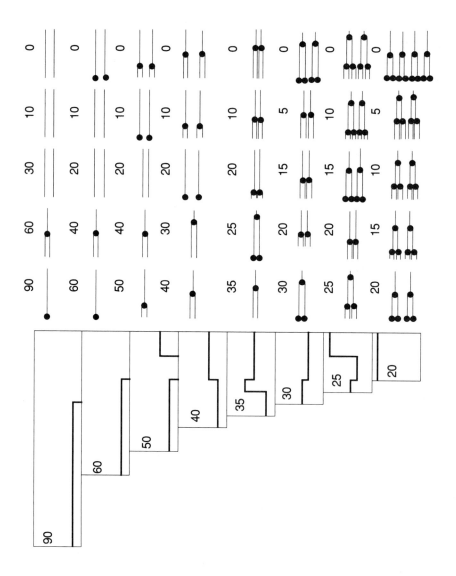

newborn cell contains a chromosome that is one-quarter replicated. At each 10-minute interval, the chromosome replicates another quarter of the chromosome; thus termination of replication occurs at 30 minutes into the division cycle or 20 minutes before cell division. As the newborn cell had a chromosome that had been replicated for 10 minutes, the initiation of replication of the two genomes in the mother cell occurs 10 minutes before cell division, at 40 minutes into the division cycle. At the end of the division cycle, the mother cell contains two genomes, each one-quarter replicated; these genomes are partitioned into the two daughter cells. In cells with a 50-minute interdivision time, two conditions were satisfied: termination of a round of replication occurred at 20

Figure 5.1   DNA Replication during the Division Cycle. DNA replication during the division cycle of *Escherichia coli* is illustrated. At doubling times between 20 minutes and 60 minutes, termination of replication occurs 20 minutes before cell division, and initiation occurs 40 minutes before termination. The lines indicate double-stranded DNA replicating semiconservatively. Although the chromosomes of bacteria are circular, and replication is bidirectional, the lines drawn here are equivalent to the actual biological situation. The pattern of DNA replication during the division cycle is indicated by the stick-figure drawings of replicating or nonreplicating chromosomes. The entire genome is represented by a single straight line. The degree to which a chromosome has replicated is indicated by the length of the single line compared to that of the double line. A chromosome that is half-replicated has the length of the double lines equal to the single, unreplicated line. The replication point, at which DNA replication is occurring, is indicated by a black dot. The numbers associated with the chromosome patterns indicate the minutes before cell division. To the left of the chromosome patterns is a plot of the rate of DNA synthesis during the division cycle. The ordinate indicates the relative rate of DNA synthesis in terms of the number of growing points in a cell. In the cells growing with a 60-minute doubling time, the result is clear, because initiation occurs at cell birth and termination occurs 40 minutes later. There is a period of 20 minutes devoid of DNA synthesis. Another way to describe cells growing with a 60-minute interdivision time is to say that initiation occurs at the end of the division cycle. From this point of view, with faster growth, initiation of replication occurs earlier and earlier in the division cycle. For example, in the cells growing with a 50-minute interdivision time, initiation occurs 10 minutes before cell division and continues for another 30 minutes in the subsequent cycle. Thus, replication takes 40 minutes and terminates 20 minutes before a cell division. In cells growing with 40-minute doubling times, there is no period devoid of DNA synthesis. Cells growing with faster than 40-minute doubling times have multiple forks. Multiple forks occur when initiation occurs on a chromosome that has not yet completed an existing round of replication. At doubling times slower than 60 minutes, the rate of DNA synthesis appears, in some cells, to occupy a constant fraction (two-thirds) of the division cycle. As illustrated by the cells with a 90-minute interdivision, time, the pattern of DNA synthesis in cells growing more slowly than a 60-minute doubling time is a constant pattern, with the first two-thirds of the division cycle occupied by DNA synthesis, and the last one-third devoid of DNA synthesis. Supporting this model are experimental determinations of the rates of DNA synthesis during the division cycle of cells growing at different rates. These determinations are consistent with the idealized representations of DNA synthesis at the left.

minutes before division, and the DNA was replicated at a constant rate of 40 minutes for a complete round of replication. One-quarter of the genome is replicated every 10 minutes. The pattern of DNA synthesis in cells with a 50-minute interdivision time is 30 minutes of constant synthesis at the beginning of the division cycle, a 10-minute "gap" in synthesis, and 10 minutes of DNA synthesis at the end of the cycle at a rate twice the rate of DNA synthesis in the younger cells.

Cells could grow with a 50-minute doubling time where the time for a round of DNA replication is only 30 minutes, or the time between termination of the round of replication and cell division is only 10 minutes. Another alternative would be cells with a period of 33.3 minutes of DNA replication and a period of 16.6 minutes between termination of replication and cell division. In all of these cases, the DNA would initiate at the start of the division cycle in the newborn daughter cell, as in cells with a 60-minute doubling time. There is nothing *a priori* impossible about these proposals. In the last case, the pattern of DNA replication during the division cycle would be exactly the same as in cells growing with a 60-minute interdivision time; there would be DNA replication during the first two-thirds of the division cycle and a gap with no DNA synthesis in the last one-third of the division cycle.

But the experiments support a constant period for a round of DNA replication, now called the C period, and a constant period between termination and cell division, now called the D period. At all growth rates, as seen with the 50- and 60-minute interdivision times, terminations of rounds of replication occur 20 minutes before cell division, and a round of replication—the time from the start of replication at the origin to the end of replication at the terminus—takes 40 minutes. The main difference between the two growth rates is that in one case (60-minute interdivision time), all of the replication of the DNA takes place in one division cycle. In the other (50-minute interdivision time), a round of DNA replication starts in one division cycle and finishes its last 30 minutes of replication in a second division cycle.

In cells growing with a 40-minute interdivision time, the newborn cell contains one half-replicated chromosome that becomes fully replicated 20 minutes later, because the replication point moves half of the length of the chromosome in 20 minutes. Termination occurs 20 minutes before cell division. Because the dividing cell has two half-replicated chromosomes, and it takes 20 minutes of replication to produce a half-replicated chromosome, the replication of the two chromosomes commences immediately upon termination of the previous round of replication. DNA synthesis occurs continuously during the division cycle; and unlike cells growing with 50- and 60-minute interdivision times, in these cells there is no gap in DNA replication. DNA synthesis is continuous during the division cycle. There is no decrease in the rate of DNA synthesis coinciding with termination, as observed in cells with a 50- or 60-minute interdivision time. Rather there is a doubling in the rate of DNA synthesis in the middle of the cycle. This doubling occurs when DNA replication terminates; one replication point disappears and two replication points appear at the origins of the chromosomes. In the cells with a 40-minute interdivision time, as in the cells growing more slowly described above, it takes 40 minutes to replicate the genome, and termination occurs 20 minutes before cell division.

Can the two rules that we have been developing apply to cultures growing faster than 40-minute doubling times? Can we maintain a constant rate of DNA replication, with 40 minutes for DNA to replicate, and still have a culture that divides more often than every 40 minutes? Consider a culture with a doubling time of 35 minutes. The newborn cell has a chromosome that will need 15 minutes of replication before termination. This is because a round of replication terminates 20 minutes before cell division. This chromosome has replicated for 25 minutes before appearing in a newborn cell. In a cell with a 35-minute interdivision time, termination occurs 15 minutes into the division cycle. At 10 minutes into the cycle, 5 minutes before termination, new rounds of replication are started at the origins of the chromosome. This produces chromosomes at division that have been replicating for 25 minutes. There are two chromosomes in the dividing cell with 15 minutes left before termination. In cells with a 35-minute interdivision time, new rounds of replication begin *before* the end of previous rounds of replication; there are replication forks on top of replication forks. These chromosomes are referred to as multifork chromosomes.

What is the rate of DNA synthesis during the division cycle in cells with a 35-minute interdivision time? The rate of DNA synthesis is constant for the first 10 minutes of the division cycle. After 10 minutes, an additional two replication points are inserted into the chromosome, and the cell has a total of three replication points.[2] The rate of DNA synthesis increases threefold. But this rate of synthesis is short lived, because 5 minutes later there is a termination; there are only two replication points in the last 20 minutes of the division cycle. The relative rate of DNA synthesis goes from one, to three, to two during the division cycle.

Similar considerations apply to the division cycle of cells growing with a 30-minute doubling time. Because termination of DNA replication occurs 10 minutes into the division cycle (in order to satisfy the requirement for 20 minutes between termination and the end of the division cycle) the chromosomes in the newborn cells must have a round of replication that will terminate in 10 minutes. The chromosomes will be three-quarters replicated in the newborn cell. How will two three-quarters–replicated chromosomes be produced in the mother cell? There will be an initiation of DNA replication at the start of the division cycle, i.e., 30 minutes before the division cycle that we are analyzing. These two replication points will continue replication, and the replication points will be three-quarters along on two separate chromosomes at cell division. If there is an initiation at the start of one division cycle, then there must be an initiation of division at the start of the next division cycle, because every division cycle is identical. For 10 minutes at the start of the cycle, three replication points will be proceeding down the chromosome. At 10 minutes into the cycle, 20 minutes before cell division, the leading replication point will terminate. Thus the rate of DNA synthesis during this division cycle will go from three to two, a decrease in the rate of DNA synthesis.[3]

In cells with a 25-minute interdivision time, the chromosome in the newborn cell has a replication point 5 minutes from termination. This satisfies the rule that there must be a termination occurring 20 minutes before cell division. In order for this replication point to have only 5 minutes until termination, the replication point must have started 35 minutes earlier. How can this be possible, when the interdivision time is only 25 minutes? Initiation takes place not in the antecedent cycle, but two cycles before the one we are analyzing. In the cycle before the one that the mother cell came from, an initiation took place 10 minutes before cell division. This round of replication was 10 minutes old at division, and this round of replication continued through the next cycle (25 minutes) so that at the next division it is 35 minutes along the chromosome. This gives a replication point that is 5 minutes from termination in the newborn cell. What is the rate of DNA synthesis during the division cycle in such a culture? There is a chromosome in the newborn cell with 5 minutes to replicate at the leading replication point. There is also a 10-minute-old pair of replication points closer to the origin of replication. The rate of replication in this newborn cell is three. This rate continues for the next 5 minutes, at which time there is a termination, and the rate of DNA synthesis goes down to two. There are two chromosomes replicating in the cell. Ten minutes later, at 15 minutes into the division cycle, new replication points are inserted into both chromosomes. Ten minutes before division, four new replication points start at the four origins in the cell. The rate of DNA synthesis triples and goes from two to six. The relative rate of DNA synthesis during the division cycle goes from three, to two, to six.

Two points should be noted. First, compare the rate of DNA synthesis in the two newborn cells to the rate of DNA synthesis in the dividing mother cell. There is no change in the total rate of DNA synthesis at the instant of division. Second, when a cell has a multiply forked chromosome, the leading and the lagging replication points are separated by a time value equal to the interdivision time. This means that successive replication points will reach the end of the chromosome and terminate at intervals equal to the interdivision time. The intertermination time will be equal to the interdivision time, and both of these will be equal to the time between initiations of new rounds of replication.

In cells with a 20-minute interdivision time, there is a termination of DNA replication coincident with an initiation of DNA replication. The newborn cell appears, unlike the previous examples, to have two independent chromosomes. There are two termini throughout the division cycle. There is a termination coincident with cell division so that the newborn cell will have just terminated a round of replication 20 minutes before cell division. The chromosomes in the newborn cell are half-replicated, as there must be a termination at the end of the division cycle. The pattern of DNA replication during the division cycle when all of these requirements are taken into account—a termination during each division cycle at the start of the cycle, and DNA replication proceeding at a constant rate so that a round of replication takes 40 minutes—has the newborn cell with two half-replicated chromosomes, with new rounds of replication starting at the instant of cell division. The rate of DNA synthesis is constant throughout the division cycle. The absolute rate of synthesis in cells with a 20-minute interdivision time is six times the rate of DNA synthesis in cells with a 60-minute interdivision time.

We have derived a set of chromosome patterns during the division cycle for cells growing with wide variations in growth rate using two simple rules: DNA replication proceeds at a constant rate and takes 40 minutes from initiation to termination; and termination of replication occurs 20 minutes before cell division. In each case, the chromosome content and configuration in dividing cells is exactly twice that in each newborn cell.

The patterns of DNA replication can be replotted in terms of cell age at initiation and termination in cells growing at different rates of growth. Four successive division cycles for a range of growth rates are presented in Fig. 5–2. The patterns of DNA synthesis expected for each division cycle are drawn within each division cycle. As growth rate increases, there is a continuous backward movement of the cell age at initiation and the cell age at termination. Termination moves back from near the end of a division cycle to the beginning of a division cycle. The cell age at termination moves from cell age 0.66 to cell age of 0.0 as the doubling rate

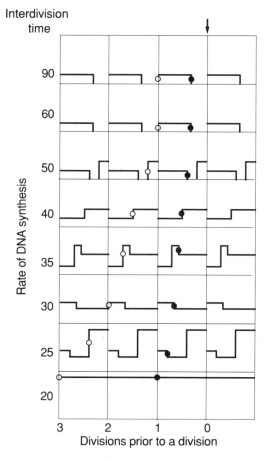

Figure 5.2 Continuous Variation of Cell Age at Initiation and Termination. The patterns of DNA synthesis are aligned to indicate the backward movement, within the division cycle, of the initiation and termination of DNA replication. The closed circles indicate the termination of DNA replication during the division cycle associated with division at the arrow. Termination moves back from the last third of the division cycle to the start of a division cycle. The open circles indicate the initiation of DNA replication needed for the cell division that occurs at the arrow. Initiation occurs at cell division during slow growth and moves back through the division cycle as the growth rate increases. As the growth rate increases above 30-minute doubling times, initiation occurs at the end of a division cycle before continuing to move back within the division cycle.

decreases from 60 to 20 minutes. The cell age at initiation has a more complicated pattern. At doubling times of 30 minutes, there is a discontinuity, with initiation occurring at the end of the division cycle.

The patterns of initiation and termination are clarified when we consider, not the age during the division cycle when an event occurs, but the length of time before a particular cell division that an event occurs. The ages during the division cycle at initiation and termination are plotted as a function of growth rate in Fig. 5–3a. There is a discontinuity at 30-minute interdivision times. When the same data are replotted in terms of division cycle times before a particular division, there is a continuous variation of the division cycle age when initiation occurs (Fig. 5–3b).

The pattern of DNA replication during the division cycle has been presented for cells between 60- and 20-minute interdivision times. What happens outside these extremes? At faster growth rates, with doubling times faster than 20 minutes, the question is whether additional levels of multiple forks exist. Growth rates of 15.1- to 17.5-minute interdivision times have been reported.[4] This growth range has not been explored because extremely rapid growth rates are difficult to achieve and, when achieved, the cells are in such rich medium that labeling experiments are precluded. The pattern of DNA replication in slow-growing cells will be presented in detail below.

III. EARLY STUDIES ON THE PATTERN OF DNA REPLICATION

Many early experiments laid the groundwork for the eventual under-standing of DNA replication, and they may be important in the study of other bacterial systems. The history also illustrates many of the problems that have arisen during the study of the bacterial division cycle.

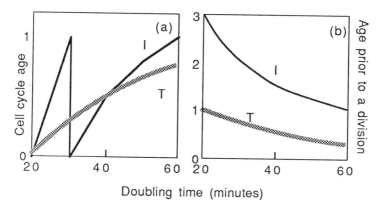

Figure 5.3 Continuous Variation in Times of Initiation and Termination Related to Cell Division. In panel (a) the times for initiation (I, solid line) and termination (T, hatched line) are plotted as times during the division cycle. In panel (b) the same data are plotted as division cycles before a particular cell division.

A. DNA Synthesis in Synchronized Cells

One of the earliest studies on the pattern of DNA synthesis during the division cycle was performed on cells synchronized by heat shocks.[5] At that time, understanding of the eukaryotic division cycle was just beginning to emerge. It was found that DNA replication during the eukaryotic cell cycle occurred in the middle of the division cycle,[6] and the bacterial results were similar to the eukaryotic results. The proposed pattern of DNA replication was then shown not to be correct for unperturbed, exponentially growing cells.[7] Cells were pulse-labeled with tritiated thymidine and then autoradiographed. Almost all cells in the culture were synthesizing DNA. There was a continuous synthetic pattern as would be expected for cells growing with doubling times of 40 minutes or faster. Continuous DNA synthesis was also demonstrated by showing that cells pulse-labeled with tritiated thymidine and stored for many months in a freezer were almost all susceptible to killing by the incorporated isotope.[8] This experiment indicated that over 90% of the cells growing in a glucose-minimal medium (moderate growth rate) are engaged in synthesizing DNA.

This short historical digression is the prototypical finding for much of the work with synchronized cells. First a result is found, and then it is shown that the method was not adequate to the task. A more interesting case is the measurement of the rate of DNA synthesis during the division cycle of cells synchronized by physical selection.[9] As noted in Chapter 3, this method was believed to produce synchronized cells more normal than those produced by drastic treatments such as heat shock. It was found that the rate of DNA synthesis increased exponentially during the division cycle. Aside from the fact that this result is not in accord with later findings, the finding of exponential synthesis is possible if the culture was not synchronized. The exponential increase would be found if the synchrony method were unsatisfactory. This experiment may be turned into a general word of caution—an exponential increase in any cell component using synchronization could indicate a faulty experimental situation; by default, there would be an exponential rate of synthesis.[10]

B. Autoradiography

In a classic paper, Van Tubergen and Setlow[11] analyzed the segregation at division of specifically labeled cell components. When the segregation of DNA was studied using thymidine label, the DNA was conserved in a number of large units that were not halved at each division. This is in accord with the idea that there are a small number of genomes. Even more interesting, they were able to show that their distributions suggested that the DNA was present in both large and small structures. This is consistent with whole chromosomes and partly replicated chromosomes (with less label) in the labeled cells.

Forro[12] advanced this approach by studying the distribution of DNA from labeled cells that were allowed to grow into microcolonies. Microcolonies are groups of cells grown from a single cell, but which retain a monolayer arrangement so that each of the descendents of the original cell may be observed. Two different classes of labeling patterns were found. After pulse-labeling and autoradiography, microcolonies were observed that had either two or four labeled cells. From Fig. 5–1 it can be seen that pulse-labeling cells with a 40-minute interdivision time would produce cells (or microcolonies) with two labeled subunits from the youngest cells, and cells (or microcolonies) with four labeled subunits from the older cells. After long-term labeling, microcolonies would be found with two highly labeled cells and two less-labeled cells. Some microcolonies had four large subunits and four less-labeled subunits. In the cells with a 40-minute interdivision time in Fig. 5–1, we would expect to find, from completely labeled cells, microcolonies with two highly labeled units and two less labeled units (from the younger cells) or four highly labeled units and four less labeled units (from the older cells).[13] The microcolony results fit these predictions.

These studies were able to yield the *pattern* of DNA replication—a general description of the synthesis of DNA during a division cycle; it was difficult, however, to obtain quantitative data from such experiments. The autoradiographic data fit the pattern of DNA replication for one growth rate described in Fig. 5–1. Because the autoradiographic experiments were so time consuming, it was difficult to do the experiments with cells growing at different rates; thus, the pattern of DNA replication was not determined over a wide range of growth rates.

C. Number of Growing Points in a Chromosome

Another early experiment demonstrated that the chromosome replicated at only one point. Bonhoeffer and Gierer[14] labeled cells with 5-bromouracil (a density analogue of thymine) and found that the lengths of densely labeled DNA produced after particular labeling times were similar to those expected if the DNA of the cell replicated at only one point. If the cell had more than one replication point per chromosome, a bromouracil-labeled strand of DNA would be shorter than was found. It is likely that the cells were not in an enriched medium and were growing with configurations similar to those of cells growing more slowly than doubling times of 40 minutes.

D. The Meselson–Stahl Experiment

Although the well-known Meselson–Stahl[15] experiment proved DNA replication is semiconservative, this result is independent of the replication of DNA during the division cycle. Even cells growing with a 20-minute interdivision time and multiple forks will give the expected Meselson–Stahl result. With regard to this chapter, this experiment demonstrated that each segment of DNA replicated only once during the division cycle. This result put constraints on models of DNA replication. All models with a random selection for replication from a large number of DNA molecules could be eliminated. This experiment demonstrated that DNA replication was an orderly process, with each portion of the genome replicated once during the division cycle.

E. The Model of Lark and Lark

Lark and Lark[16] studied cells growing on glucose and on succinate. Lark and Lark found, in accordance with Schaechter, Maaløe, and Kjeldgaard, that there was less DNA per cell in the slower-growing succinate cells compared to the faster glucose-grown cells. Autoradiographic experiments on the replication of DNA indicated that there were two chromosomes in newborn cells at each growth rate. They proposed that both chromosomes replicated simultaneously during the division cycle in the glucose-grown cells, but sequentially in the succinate-grown cells. This model was consistent with their finding that there were more replication points in the faster growing cells. They proposed that glucose-grown cells have two replication points throughout the cycle, and succinate-grown cells have one replication point during the cycle. Unfortunately, the expected amount of DNA per cell for these chromosome patterns is the same for both growth rates, at all points during the division cycle. The newborn cells would have two genome equivalents, middle-aged cells would have three genome equivalents, and dividing cells would have four. The model of Lark and Lark cannot, therefore, account for the different DNA contents of cells at slow and moderate rates of growth.

Koch and Pachler[17] tested the alternation model directly. They labeled cells growing in succinate for either one-half generation or two generations with high-specific-activity tritiated thymine. When the cells were stored, the kinetics of survival was the same for both cultures. A first-order inactivation was observed. This result eliminated the possibility that the cells possessed two identical chromosomes that replicated alternately.

IV. ANALYSIS OF DNA REPLICATION USING THE MEMBRANE-ELUTION TECHNIQUE

A breakthrough in the determination of the pattern of DNA replication came with the introduction of the membrane-elution method with pre-labeling: the backwards, membrane-elution method (see Chapter 3). For a number of years the membrane-elution method (the baby machine) was used to produce synchronized cells. These cells were analyzed extensively, yet no conclusive results were obtained. Although the method clearly produced synchronized cultures, the experiments were unable to give a clear pattern of DNA synthesis during the division cycle.[18] In contrast, the backwards approach successfully gave the pattern of DNA synthesis during the division cycle. This pattern fully supported the description of chromosome replication presented in Fig. 5–1.

A. Rate of DNA Synthesis during the Division Cycle

Our understanding of the pattern of DNA replication during the division cycle came from a series of experiments, beginning in 1967, in which the membrane-elution technique was used to measure the rates of DNA synthesis during the division cycle of cells growing at different rates. The first results obtained were those on the pattern of DNA synthesis during the division cycle of slow-growing cells.[19] One of the cultures studied was a glucose-grown culture with an interdivision time of approximately 40 minutes. It was observed that in the middle of the division cycle there was a doubling in the rate of DNA synthesis. Inspection of Fig. 5–1 indicates a doubling in the rate of DNA synthesis is expected for this growth rate. Cells growing slower than the glucose culture had gaps in synthesis as expected for cells growing slower than 40-minute interdivision times. The gap in synthesis was found only in the older cells, as would be expected for cells growing with 50- and 60-minute interdivision times (Fig. 5–1).

Following the experiments with slow-growing cells, the pattern in more rapidly growing cells was determined.[20] Figure 5–4 illustrates the rate of thymidine incorporation during the division cycle of cells of *Escherichia coli* B/r growing at different rates as determined using the membrane-elution technique. Because of the degree of variability of interdivision times of cells, there is some dispersion in the results (Fig. 5–4). Nevertheless, where such dispersion does not eliminate the increases and decreases in the rate of DNA synthesis, the patterns of DNA replication as illustrated in Fig. 5–4 do fit the model in Fig. 5–1. In slow-growing cells (between 40- and 60-minute interdivision times), it is possible to see a gap

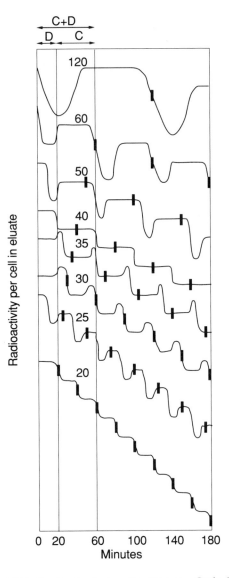

Figure 5.4 Rate of DNA Synthesis during the Division Cycle Determined with the Membrane-Elution Apparatus. The elution patterns for a number of membrane-elution experiments with cells growing over a wide range of growth rates is illustrated. The short vertical bar delimits individual division cycles eluted from the membrane. Reading from right to left (i.e., from a bar to the left axis) gives the rate of DNA replication during the division cycle. The cells with a 40-minute interdivision time have a constant rate of incorporation that doubles in the middle of the division cycle. There is a gap in synthesis in the 50- and 60-minute interdivision cells. The peak values in the cells growing faster than 40-

in DNA synthesis in the older cells. At faster growth rates, a peak is seen where there is an expected peak in incorporation (see the 35-minute cells in Fig. 5–1). A simpler approach to the experimental analysis is to note that at all growth rates there is a drop in the incorporation pattern at approximately 60 minutes of elution (Fig. 5–4). This drop is interpreted as the time of initiation of new rounds of replication. It is expected that these initiations will always occur C+D minutes (i.e., 60 minutes) before a division.

Based on this interpretation of the data, the time for a round of chromosome replication (C), the time between the end of a round of replication and the following cell division (D), and the sum (C+D) over a wide range of growth rates can be determined. These parameters are plotted versus the growth rate in Fig. 5–5. Between growth rates of 1 and 3 doublings per hour, the values for C, D, and C+D were fairly constant and equal to about 41, 22, and 63 minutes, respectively. In some strains of *Escherichia coli* B/r, at growth rates below one doubling per hour, the experimental values cluster along values such that (C+D) = τ [i.e., rounds start at division, C = $\frac{2}{3}\tau$, and D = $\frac{1}{3}\tau$]. At slow growth rates, the chromosome is synthesized during the first two-thirds of the division cycle; at doubling times slower than 60 minutes, DNA is synthesized with the same pattern as in cells with a 60-minute doubling time. This type of pattern is shown for the cells with a 90 minute interdivision time in Fig. 5–1.

B. DNA Replication at Different Temperatures

The pattern of DNA replication in a given medium is the same when cells are grown at different absolute rates of growth by varying the temperature.[21] It is not the absolute growth rate that determines the pattern of DNA replication. Rather, for a given medium, if the growth rate is changed by varying the temperature, the pattern of DNA synthesis does not change. This is consistent with the original observation of Schaechter, Maaløe, and Kjeldgaard[22] that variations in temperature do

---

minute interdivision times are due to the short periods of multiple forks. These patterns are the same as those illustrated in Fig. 5.1. These results are plotted in real time (not in terms of generations of elution), and at C+D minutes of elution, there is a drop in the radioactivity per cell (at least for cells where C+D is constant), for cultures growing between 20- and 60-minute doubling times. This is because 60 minutes earlier, when the cells were labeled, there was an initiation of DNA synthesis related to the dividing of the cells 60 minutes later. In practice we can see the labeling patterns for approximately three to four generations of elution.

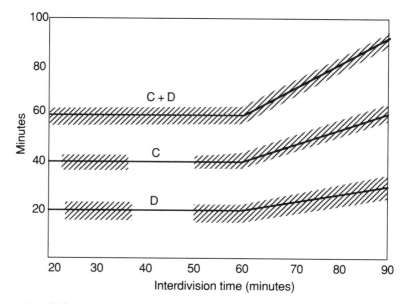

Figure 5.5 Values of C and D at Different Growth Rates. The values of C and D for a number of growth rates are summarized. The shaded area indicates that there is some variability in the determinations. When the membrane-elution technique is used, it is possible to measure the C+D values at all growth rates, because the drop in radioactivity at C + D minutes is always observed. At intermediate and rapid growth rates, when initiation and termination coincide, it is more difficult to determine C and D independently. Other methods have shown that the interpolation of the results across these gaps is justified. C and D are constant over a wide range of growth rates and increase at longer interdivision times.

not change the cell composition (Chapter 1). If every aspect of the cell increases in the same proportion as the temperature changes, the pattern of DNA replication would be expected, as a function of the division cycle, to remain the same.

C. DNA Content at Different Rates of Growth

Inspection of Fig. 5–1 reveals that the average chromosome content increases with increasing growth rate. If an unreplicated genome is one genome equivalent, the newborn cells have one genome equivalent and the dividing cells have two genome equivalents in cells growing with a 60-minute interdivision time. The average number of genome equivalents is between one and two per cell. In the cells with a 20-minute interdivision time, the newborn cell has three genome equivalents per cell, and a dividing cell has six genome equivalents. The average genome content is between these two values. It is intuitively obvious that the

DNA content per cell increases with increasing growth rate. The genome content in terms of C, D, and the doubling time[23] is given by the formula

$$G_{avg} = \frac{\tau}{C \ln 2}(2^{(C+D)/\tau} - 2^{D/\tau})$$

G_{avg} is the average number of chromosome equivalents per cell. A chromosome equivalent is that mass of DNA corresponding to a single non-replicating chromosome. This equation takes into consideration the age distribution of the culture (Chapter 1). The DNA contents of cells at different growth rates are consistent with the patterns of DNA replication determined by membrane-elution (Fig. 5–1).[24]

Not only did the cell DNA contents confirm and support the proposed replication patterns during the division cycle, but the DNA measurements also allowed a calculation of the molecular weight of the chromosome. With a knowledge of the number of genome equivalents in a cell derived from the membrane-elution method, a chemical determination of the cellular DNA content can yield the DNA per genome equivalent. The value determined was 2.5×10^9 daltons per genome equivalent.[25] This value is consistent with other determinations of the molecular weight of the bacterial genome.[26]

Measurements of the absolute amount of DNA per cell[27] also indicate that the chromosome pattern drawn in Fig. 5–1 is correct. An alternative could be to have a multiple of such chromosome configurations present. Any multiple of chromosomes that replicated in synchrony would give the same DNA synthesis pattern, but the absolute amount of DNA expected from independent measurements of genome size would be different. The DNA contents predicted for cells growing in different media was the same as that determined by Schaechter, Maaløe, and Kjeldgaard. This means that in *Salmonella typhimurium*, the organism they studied, the same model with constant C and D values could account for the variation in DNA content with growth rate. The pattern of DNA replication measured by the membrane-elution method is similar to that of *Escherichia coli*.[28] Thus we come to the explanation of the problem posed by the results of Schaechter, Maaløe, and Kjeldgaard: there are a small number of genomes or genome equivalents per cell; the continuous variation in DNA content is owing to the continuous variation in cell age at initiation and termination as described in Figs. 5–1 and 5–2.

D. DNA Content of Individual Cells Using Flow Cytometry

A completely independent support of the the constant C and D model (Fig. 5–1) has been provided by measurements of the DNA contents of individual cells growing at different growth rates.[29] These measurements were made by adapting the flow cytometry method, which has been used so successfully for measuring DNA contents of animal cells, to the analysis of bacteria (see Chapter 3). In Fig. 5–6 the predicted and observed results are compared for a few selected cases. The experiment is extremely simple. Cells are moved past a source of light, and the fluorescence of the cells is measured. The fluorescence is proportional to the amount of DNA present in a cell. The cells are unperturbed and unsynchronized. The results, plotted as histograms, are then compared to the idealized results expected on the basis of a constant C and D. The predicted results are also modified to account for variations in the actual measurement of the DNA in a particular cell. The results agree with the predictions.

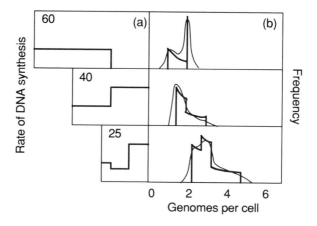

Figure 5.6 DNA Contents of Bacteria by Flow Cytometry. At the left [panel (a)] are patterns of DNA synthesis at various growth rates. At the right [panel (b)] are the expected flow-cytometric results for the particular synthetic pattern. When there is a gap in synthesis, a spike appears; because of instrumental variation, the actual result is spread out. When the DNA content of exponentially growing bacteria is analyzed by flow cytometry, a frequency distribution of DNA concentrations is obtained. Comparing such a result with the expected theoretical result gives independent support for the patterns of DNA replication during the division cycle. The correlation is stronger when we assume that there is a broadening of the measurements due to variability in the method. For the cells with a period devoid of DNA synthesis, all of the older cells in the 60-minute culture have a genome content of two. This leads to the spike in the frequency distribution at values of two. The other part of the ideal curve is the result of the age distribution and the linear rate of synthesis of DNA during the synthetic period. The thin lines superimposed on the thick-lined, ideal pattern, are the observed results. The observed results are due to a broadening of the ideal results caused by experimental instrument variations.

The importance of the flow cytometry results should not be underestimated. They provide a crucial link in the analysis of the division cycle. The flow cytometry results demonstrate that the membrane-elution method can be successfully used to measure the rates of biosynthesis of different cell components during the division cycle. While other experiments have substantially supported the initial membrane-elution results, the flow cytometry results determine the pattern of DNA replication without any perturbations of the cell. With the flow cytometry results, we know that the growth of the cells on the membrane is as described in Chapter 3, and that the backwards membrane-elution method is a good method for cell-cycle analysis.

E. Autoradiographic Analysis of the Number of Growing Points in a Cell

A simple approach to the confirmation of the model for DNA synthesis presented here—one that is applicable to a large number of different organisms, whether synchronizable or not—is that used by Caro and his colleagues[30] to analyze the number of growing points in a cell. A cell is pulse-labeled for a short time and then allowed to grow in unlabeled medium until a number of unlabeled cells appear. One can get the average number of grains per cell, correct for the unlabeled cells, and obtain the fraction of cells labeled (Table 5–1). From the increase in cell number, it is possible to calculate how many labeled units were in an average cell at the time of labeling. The number of labeled units is then interpreted in terms of the number of growing points per cell. The results agree with the predictions of the model presented in Fig. 5–1.

Table 5.1 Replication Point Determination by Autoradiography (after Bird, Louarn, Martuscelli, and Caro)

	Medium and Generation Time			
	L-broth 25 minutes		M9 with Casamino Acids 34 minutes	
Growth in cold medium after a tritiated pulse (increase in cell number)	8.0×	16.1×	7.3×	12.66×
Average grain count per cell	2.29	1.21	2.20	1.31
Average grain count per labeled cell	3.19	2.93	3.64	3.20
Fraction of cells labeled	0.678	0.337	0.550	0.349
Number of strands labeled at t=0 (average)	5.42	5.42	4.02	4.42
Number of forks or pairs of forks per cell	2.71	2.71	2.01	2.2

V. ICON FOR DNA REPLICATION

At some time during the division cycle there is an initiation of DNA synthesis. DNA replication then proceeds in a linear manner to the end of a round of replication, and cell division follows. The act of cell division has no regulatory relationship to the initiation of DNA synthesis. This idea is summarized in an icon for DNA synthesis presented in Fig. 5–7. In the following sections, we shall combine this icon with the icon for mass synthesis (Chapter 4) and arrive at an icon for the regulation of DNA synthesis during the division cycle.

VI. *A PRIORI* CONSIDERATIONS ON THE REGULATION OF CHROMOSOME REPLICATION

A. Cell Mass and the Initiation of Replication

Given the pattern of DNA synthesis during the division cycle at different rates of growth, the question arises as to what regulates the pattern so that the cell initiates DNA replication at a particular time during the division cycle. We could frame the question in a very simple way, by asking why the cell, when growing with an interdivision time of 60 minutes, does not immediately initiate new rounds of replication at the instant of termination, as is seen in cells growing with a 40-minute interdivision time. How does a cell or a genome *know* to wait in cells doubling with a 60-minute interdivision time, while not waiting in cells growing with a 40-minute interdivision time?

Assume that the initiation of DNA replication occurs when the amount of a particular material is present in a unit amount per origin of DNA. For simplicity, assume that this is the mass of the cell. Initiation occurs when the mass of the cell is present at some unit amount per origin. Inspection of the patterns of cells growing at different rates between 60- and 20-minute doubling times (Fig. 5–1) indicates that the number of origins in each cell at initiation in the cells with 60-, 50-, 40-, 35-, 30-, 25-, and 20-minute doubling times is 1, 2, 2, 2, 2, 4, and 4,

Figure 5.7 Icon for DNA Replication during the Division Cycle. DNA replication, the C period, is indicated by a straight line followed by a D period, indicating the time until cell division. The linear aspect of DNA replication indicates that the end of DNA replication does not feed back or regulate the synthesis of DNA.

respectively. The assumption of a constant mass per origin at the time of initiation means that the relative cell mass for a newborn cell with a 60-minute doubling time must be 1 and for the newborn cell growing with a 20-minute doubling time, must be 4. Because there must be exactly twice the amount of material in the mother cell compared to the newborn cell, the cell doubling in 60 minutes must increase in relative size from 1 to 2 during the division cycle, and the cell doubling in 20 minutes must increase in relative size from 4 to 8 during the division cycle. Similarly, the newborn cells from a culture growing with a 30-minute interdivision time must be size 2 and must grow to size 4 during the division cycle.

How can we determine the size of cells at other growth rates? For example, cells with a 40-minute doubling time initiate DNA replication in the middle of the division cycle. How can we make comparable estimates of the size of the newborn and dividing cells? Because the mass of the cell increases exponentially during the division cycle (Chapter 4), these values can be determined graphically. In Fig. 5–8b, a series of parallel lines is drawn, increasing over a vertical distance of 2 on a semilogarithmic scale. A single point in each graph indicates the mass that exists at that particular time during the division cycle. Because the cultures growing with 50-, 40-, 35-, and 25-minute doubling times initiate at cell ages of 0.80, 0.50, 0.29, and 0.60 during the division cycle, these points are placed at the time of initiation of DNA replication. Although the cells with 25- and 40-minute interdivision times initiate at similar times during the division cycle (ages 0.50 and 0.60), the cells with a 25-minute interdivision time have a mass of 4 at initiation, and the cells with a 40-minute interdivision time have a mass of 2 at the midpoint of the division cycle. These points determine, for cells growing exponentially in mass during the division cycle, the mass at all other times during the division cycle,[31] including the mass at the time of cell birth and cell division. The size of the newborn cells at any interdivision time can be determined from the graph.

As the cell age distribution is the same for cells growing with any doubling time, the mass in the newborn cells is proportional to the average cell mass. When the cell sizes determined in Fig. 5–8b are replotted against the reciprocal of the interdivision time, a straight line is obtained on a semilogarithmic plot. Cells with doubling times of 60, 30, and 20 minutes have reciprocal doubling times of 1, 2, and 3 doublings per hour. For each of these cultures, the relative cell sizes are 1, 2, and 4. The complete set of calculations is presented in Table 5–2.

Consider cells with a 120-minute doubling time (not included in Fig. 5–8). Assume that in this cell, DNA replication follows the constant C and D rules, and replication initiates 60 minutes before cell division. Such

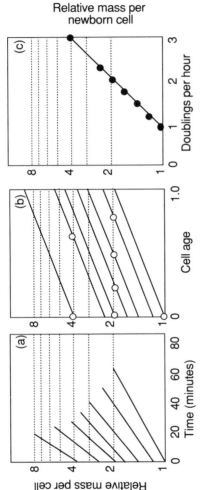

Figure 5.8 Derivation of Cell Size at Different Growth Rates. At the left [panel (a)] is a graph of the growth of cells from a newborn size to a dividing size at different doubling times. The smaller cells with a 60-minute doubling time grow from size 1 to size 2, while the cells with a 20-minute doubling time grow from size 4 to size 8. When these size values are plotted over a normalized cell-cycle time, as shown in panel (b), we can use the sizes at the time of initiation of DNA replication (open circles) to place the lines. For example, the cells with a 60-minute interdivision time initiate DNA replication at age 1.0 (i.e., at division), and the cells with a 40-minute interdivision time initiate DNA replication in the middle of the division cycle. For cells between 30- and 60-minute interdivision times, two replication points are initiated. Thus initiation takes place at size 2 in these cases. The relative sizes of cells in the culture are indicated by the size of each of the cells at the start of the division cycle. When these sizes are replotted in panel (c) against the reciprocal of the doubling time, there is a straight line on a semilogarithmic graph. (The reciprocal of the doubling time is the value of 60 over the doubling time, giving the number of doublings per hour.) The average size per cell increases the growth rate if the pattern of mass synthesis during the division cycle is the same at all growth rates and the mass per origin at the time of initiation is constant.

Table 5.2 Mass and DNA Content of Cells Growing at Different Rates

Doubling Time (Minutes)	Doublings Per Hour	Origins at Initiation	Mass at Initiation	Cell age at Initiation	Newborn Mass
20	3.0	4(8)	4(8)	0.0	4.0
25	2.4	4	4	0.60	2.64
30	2.0	2(4)	2(4)	0.0	2.0
35	1.7	2	2	0.29	1.64
40	1.5	2	2	0.5	1.42
50	1.2	2	2	0.8	1.15
60	1.0	1(2)	1(2)	0.0	1.0
120	0.5	1	1	0.5	0.76
∞	0.0	1	1	1.0	0.5

a cell will have a period of 60 minutes with no DNA synthesis at the start of the division cycle. DNA synthesis will take place between 60 and 100 minutes, and there will be a period devoid of DNA synthesis at the end of the cycle, corresponding to the 20 minutes between termination and cell division. The reciprocal doubling time is 0.5 doubling per hour. This cell will have a unit initiation mass in the middle of the division cycle and the newborn cell will have a mass of 0.76. These cells also fall on the straight line determined by the growth rates between 20- and 60-minute doubling times. This result can be extrapolated to extremely large doubling times. These doubling times are cells with approximately zero doublings per hour. If we assume that C+D remains 60 minutes even at extremely long interdivision times, cells near the end of the division cycle will have a size of 1.0. This is because initiation of one nonreplicating chromosome will occur at the end of the division cycle. Thus, newborn cells will have a size of 0.5. The smallest set of cell sizes will be cells born with a size of 0.5 and increasing until they divide at size 1.0. This is the minimal cell size, obtained for cells with a zero growth rate. If we consider the set of growth rates with 0, 1, 2, and 3 doublings per hour, the relative sizes of the newborn cells are 0.5, 1.0, 2.0, and 4.0 as illustrated in Fig. 5–8. The cells with a growth rate of 0.0 doublings per hour are equivalent to stationary phase, nongrowing cells, and will be discussed in detail in Chapter 10.

B. Implications of the Schaechter–Maaløe–Kjeldgaard Experiment

The mathematical analysis of DNA regulation to be presented now is more rigorous and interesting. Assume that there is a constant size or mass per origin at initiation.[32] Mathematically this is represented as:

$$M_i/N_i = K$$

The cell mass at the time of division, M_d, is given by:[33]

$$M_d = K \cdot e^{(C+D)/\tau}$$

In the original derivation, it was assumed that mass increased exponentially during the division cycle (see Chapter 4), but this requirement may be relaxed; we need assume only that the pattern of mass synthesis, whatever it is, is the same at all growth rates. If this be the case, it is easy to show that the average cell size at a particular growth rate is proportional to the size at any particular age—in this case at birth or at division—and thus we arrive at the result that

$$M_{avg} = K' \cdot e^{(C+D)/\tau}$$

The average mass of a cell is an exponential function of the inverse of the doubling time. The abscissa of the original Schaechter, Maaløe, and Kjeldgaard experiment (Chapter 1, Fig. 1–5) is the inverse of the interdivision time. If the determination of the mass per cell at different growth rates gives a straight line against the reciprocal of the doubling time, and if the slope is equal to $C+D$, then this is evidence that the cell mass can be the substance the cell measures to determine when to initiate new rounds of replication. Conversely, if it were known that mass is the initiating substance, and initiation takes place when there is a constant amount of mass per origin, then the slope of the mass per cell in a Schaechter, Maaløe, and Kjeldgaard experiment (Chapter 1) may be used to calculate the value of $C+D$ for a particular cell. If $C+D$ is not constant, then one would expect the values for the initiator per cell, whatever that value was, not to give a straight line in the Schaechter, Maaløe, and Kjeldgaard experiment. The slope for the mass per cell is indicative of a value of $C+D$ of 60 minutes for *Salmonella typhimurium* as well as *Escherichia coli*.

Let us consider this result in a more intuitive way. If $C+D$ were extremely small, then cells at all growth rates would divide shortly after initiation of a round of replication. The sizes at division would be approximately the same for all cells over a wide range of growth rates. Therefore the slope, in a Schaechter, Maaløe, and Kjeldgaard experiment, would be very close to zero. If $C+D$ were large, then we would get a large variation in the size per cell with growth rate. We can intuitively see that the slope of the mass per cell at different growth rates is an indication of the value of $C+D$. In the three decades since the measurements of Schaechter, Maaløe, and Kjeldgaard, there have been essentially no repetitions of these measurements with other cells. Given the simplicity of this experimental approach to the measurement of $C+D$ values, further studies of a wide range of organisms by this very simple approach will yield an understanding of the possible ranges of $C+D$ values in diverse bacteria.

C. The I + C + D Concept

An alternative approach for deriving the various DNA synthesis patterns for cells growing at different rates can now be presented. This approach is based on a three-step process: I+C+D, where I is the time for accumulation of an initiator complement of constant size per origin (a unit of initiator), C is the time for a round of replication, and D is the time between termination and cell division. The process indicates that cell division follows I+C+D minutes after the start of initiator synthesis. By means of this linkage, the rate of initiator synthesis governs both the frequency of initiation and the rate of cell division. The important point is that complex division cycles are due to the overlapping of the simpler I+C+D sequences.[34]

The pattern of chromosome replication during the division cycle can be determined by applying the I+C+D formulation to a hypothetical cell containing a single chromosome. Constructions for five values of I (that is, for cells with different growth rates, as I=τ, where τ is the interdivision or the doubling time) are shown in Fig. 5–9. The C and D periods are shown at the top of Fig. 5–9; D is half of C, as has been found experimentally. In each case, replication begins at the origin of a single chromosome, and new replication points (filled circles) are introduced at the origins of the chromosomes every I minutes. In Fig. 5–9 the rounds of replication are initiated at zero minutes in all cases, so a cell division takes place C+D minutes later as indicated by the rectangles enclosing the chromosome configurations at the right. A division cycle is composed of the cells and their DNA configurations starting I minutes earlier and continuing until the next division at the right in Fig. 5–9. One point is worth noting: the diagrams in Fig. 5–9 are not dependent upon the constancy of the C and D periods. C and D could vary in different media, and this formulation is still valid. For any steady-state growth condition, the I+C+D rule would be followed.

VII. EXPERIMENTAL ANALYSIS OF REGULATION OF DNA SYNTHESIS

There have been a number of different approaches to the analysis of the mechanism of regulation of DNA synthesis. The general concept that the initiation of DNA synthesis occurs in response to the accumulation of mass, or some molecule that acts or is synthesized in the same manner as mass, has been supported by many different experiments.

A. Cell Mass at Different Growth Rates

The best support for the identity of mass with the initiator is the basic result of Schaechter, Maaløe, and Kjeldgaard (Chapter 1). The straight

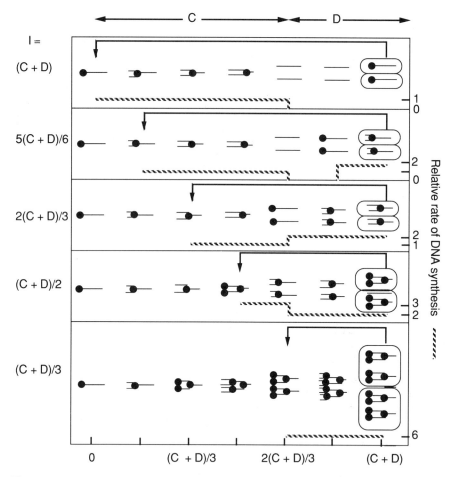

Figure 5.9 I+C+D Model of Chromosome Replication. A cell with a single chromosome is allowed to grow at different doubling times represented as fractions of C+D. It is assumed that C is always twice D, but this restriction is not required. The chromosome grows according to the C period, and initiation of new rounds of replication is allowed to occur at the expected times according to the specified interdivision or I period. When a division occurs, we look back one interdivision time and specify the pattern of DNA replication during the division cycle. The hatched lines indicate the predicted relative rate of DNA synthesis during the division cycle in units of replication points per cell.

line, when the log of the cell mass is plotted against the inverse of the doubling time, indicates that mass per cell varies in the proper manner for cells with constant C and D periods. Calculations of the C+D time using the original data of Schaechter, Maaløe, and Kjeldgaard gives values similar to those determined directly by the membrane-elution method.

It is important to distinguish two different ideas about initiation of DNA replication. Up to this point, we have used the idea of initiation mass very loosely. Let us consider two extreme possibilities within this framework. First, if the cell did measure just the global amount of mass, even down to the point of total dry or wet weight of the cytoplasm, then a measurement of the mass per cell should give a result consistent with a slope equal to C+D. Now let us consider that the actual initiator is a specific molecule that is a fixed fraction of cell mass. The cell measures, in some way, the amount of initiator present in a cell, and when there is a unit amount of this particular initiator per origin, initiation occurs. If, as assumed, the initiator were present in a fixed amount per total mass at all growth rates, size determinations at different growth rates could not distinguish between mass in general or some specific part of the cell mass as being the initiating factor. On a more practical level, if the specific initiator were not present as a perfectly constant fraction of total mass at different growth rates, the experimental measurements could not distinguish the different models. If the initiator, at a particular growth rate, were present at a 10% lower concentration or fraction of total mass (from some trivial metabolic cause), we would expect a 10% increase in the mass per cell at that growth rate. The experimental determinations have at least that much variability. Thus, when the term mass or size is used, a specific metabolic interaction's being the proximal cause of initiation of DNA synthesis is not excluded.

B. Inhibition of Mass Synthesis

Consider an exponentially growing culture to which an inhibitor of protein or RNA synthesis is added. Because there is no further accumulation of mass, those chromosomes in the process of replication complete the round of replication, but no new rounds of replication start. Assume that cells with two genomes are able to divide even in the absence of mass synthesis. During the period of mass-synthesis inhibition, and in the period following resumption of mass synthesis, what is expected for DNA synthesis and cell division? The pattern of mass synthesis, DNA synthesis, and cell division during this period of inhibition and regrowth is shown in Fig. 5–10. When mass synthesis stops, DNA replication continues, with a steady slowing in the rate of DNA synthesis as the replication points in progress terminate replication. If the rate of replication point movement were unaffected by the inhibition of mass, then after exactly C minutes, DNA synthesis ceases. Cell division continues for C+D minutes at which time an abrupt cessation of cell division is observed. Cell division continues for C+D minutes because those initiation points less than C+D minutes from the cell division will continue to

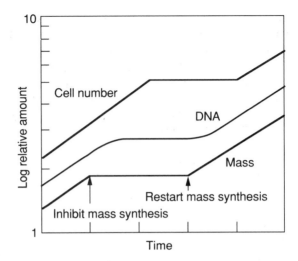

Figure 5.10 DNA Synthesis and Cell Division during Inhibition of Mass Synthesis. Mass synthesis is inhibited for approximately one generation time. When inhibition ceases, it is assumed that mass synthesis starts again at the same rate as before inhibition. DNA synthesis and cell division continue for 40 and 60 minutes respectively, although the rate of DNA synthesis slows, while cell number increase stops abruptly. When mass synthesis resumes, there is a delay period before any cell division is observed. DNA synthesis resumes slowly after mass synthesis resumes.

determine the cell divisions at the appropriate times, even during an inhibition of mass synthesis.

We shall assume that when the inhibitor of mass synthesis is removed, there is an immediate resumption of mass synthesis at the rate before the start of inhibition. In contrast to mass synthesis, there is a slow resumption of DNA replication as the mass in different cells achieves the amount required for initiation of DNA replication. To understand this asynchronous initiation of DNA synthesis,[35] consider a cell just about to initiate DNA replication when inhibition began, and another cell that had just initiated DNA replication before inhibition. The cell that was just about to initiate DNA replication will fulfill the mass requirements for initiation in a short time, while the cell that had already initiated will require one interdivision time before an initiation will occur. There was a doubling in the number of origins in the latter cell when initiation occurred. All of the other genomes will be initiated between these two extremes, and after one interdivision time the rate of DNA synthesis will be back to the original rate. Cell division follows a slightly different course. During inhibition of mass synthesis, cell division continues for C+D minutes at the original rate and then stops. When initiation of DNA synthesis resumes, there will be a period of C+D minutes before cell division resumes at the initial rate. Any initiations of new rounds of replication occurring after the resumption of mass synthesis will not lead to cell divisions until C+D minutes later. At this time the cell will have a composition equal to those cells before the period of inhibition, and growth will resume at the original steady-state composition.

C. Inhibition of DNA Synthesis

If DNA synthesis is inhibited, the expected result is shown Fig. 5–11. Assuming that mass continues at the original rate, after a period of time (if DNA synthesis was inhibited for exactly one interdivision time) all of the chromosomes would have enough mass associated with them to have initiated a new round of replication at all of the origins in the cell.

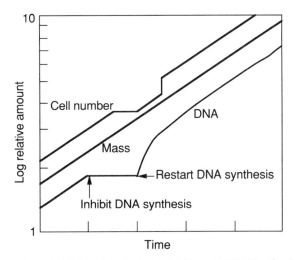

Figure 5.11 Growth and Cell Division during Inhibition of DNA Synthesis. When DNA synthesis is inhibited for one interdivision time, then allowed to resume, the rate of DNA synthesis at the resumption of DNA replication is higher than the rate of DNA synthesis when inhibition started (rate stimulation). It is assumed that mass synthesis continues exponentially during this period of DNA inhibition. Cell number continues to increase for D minutes after DNA replication is inhibited. When DNA synthesis resumes, there is a delay of D minutes before cell division resumes. Cell division resumes at the initial rate as growing points caught in the middle of a round of replication reach the terminus. Cell division follows D minutes later. At 60 minutes following resumption of DNA synthesis, there is a sudden doubling in cell number. This is a *pseudosynchronous* division, as the division is synchronized, but the underlying DNA synthesis pattern is not synchronized. The chromosomes in the synchronously dividing population are in all stages of replication; there is no second round of synchronous cell division.

When DNA synthesis is allowed to resume, the initial rate of DNA synthesis will be greater than its rate at the time it was inhibited. DNA synthesis increases owing to the insertion of new replication points. To see this, consider a culture that has one replication point per chromosome, as is found in our ideal cells growing with a 40-minute interdivision time (Fig. 5–1). If we insert new replication points into each of the chromosomes, there will be a tripling in the number of replication points. When DNA synthesis is allowed to resume, the rate of DNA replication shows a threefold increase. When the synthesis of DNA is plotted on a semilogarithmic graph, the DNA synthesis curve has a continuously decreasing slope that eventually, after C minutes, is the same as the original rate of DNA synthesis. After C minutes, the group of synchronously starting replication points reaches the end of the genome, and the only replication points in place are those resulting from the normal response to continued mass synthesis.

It is interesting to consider cell division during and following inhibition of DNA synthesis. Cell division will continue for D minutes after inhibition of DNA synthesis, as cells within the D period will have two complete genomes and will continue to divide. Cell division then abruptly ceases. When DNA synthesis resumes, there will be a delay of D minutes before the resumption of cell division. Any replication points frozen in place during inhibition come to the end of the genome, and after a D-minute delay, produce a cell division.[36] Cell division now occurs at the normal, preinhibition rate for C minutes. This is related to the termination of those replication points frozen in place during the inhibition of DNA synthesis. There is an abrupt increase in cell number C+D minutes after resumption of DNA synthesis. This *synchronous division* is determined by the initiation points inserted in all chromosomes when DNA synthesis was inhibited. These replication points are presumed to have started replication synchronously, and after passing through a C and a D period lead to a synchronous cell division. This division, while occurring in all cells of the population simultaneously, is not a true synchronous division, and is not an indication that the culture is synchronized. This division in all cells of the population is more properly called quasi-synchronous or pseudo-synchronous, because it occurs only once. The culture is not synchronized, but cell division is synchronized. At the instant of this increase in cell number there is a recovery of the conditions of balanced growth that prevailed before the inhibition. When DNA synthesis is inhibited, and mass continues at the normal rate, there is an increase in the mass per DNA and the mass per cell. After resumption of DNA synthesis, the pattern of DNA accumulation and cell division is such that they catch up to the mass, either by the immediate increase in the rate of DNA synthesis or by the sudden doubling in cell number C+D minutes after release of inhibition. Cell growth then resumes with its normal composition, size, and steady-state growth rate.

D. The Maaløe–Hanawalt Experiment

One of the fundamental experiments illuminating the mechanism of regulation of chromosome replication was performed by Maaløe and Hanawalt.[37] They inhibited protein synthesis by various starvation regimens in auxotrophic bacteria and measured the amount of residual DNA synthesis. During starvation the amount of DNA increased approximately 36% over that initially present. This is precisely the amount of DNA synthesis expected if the time for a round of replication is equal to the doubling time of the culture. This is demonstrated by the cells with a 40-minute interdivision time in Fig. 5–1. There are between 1.5 and 3.0 genome equivalents per cell; the precise value is 2.04 if the age distribution is taken into account. After completion of all rounds of replication in progress (with no new rounds of replication starting), there will be between 2.0 and 4.0 genome equivalents per cell. If the age distribution is taken into account, the value is 2.79 genome equivalents. This is a 36.8% increase in the amount of DNA.

The residual increase in the amount of DNA during inhibition of mass synthesis depends on the growth rate. The faster the growth rate, the greater the residual synthesis (Fig. 5–12). It was fortuitous that the cultures studied by Maaløe and Hanawalt gave the expectation for a single growing point on all chromosomes. If they had cells growing faster or slower, the values would have been higher or lower. The percent increase in DNA per cell (%ΔG) is given by

$$\%\Delta G = \left(\frac{C(\ln 2)(2^{C/\tau})}{\tau(2^{C/\tau} - 1)} - 1 \right) \cdot 100$$

For the case where $C = \tau$, the result reduces to $(2\ln 2 - 1) \cdot 100$, or 36.8%. The higher the growth rate, the greater the percentage increase in the amount of DNA synthesis during inhibition of mass synthesis (Fig. 5–12). The Maaløe–Hanawalt approach, though not utilized very often, has been used to confirm the basic idea that at faster growth rates, there is a larger increase in the runout of DNA replication.[38] Others have used this method to advantage.[39]

The most important result of the Maaløe–Hanawalt experiment is that it distinguished the process of initiation of DNA replication from the process of chain elongation. The Maaløe–Hanawalt experiment demonstrated that once started, DNA replication proceeds to the end of the

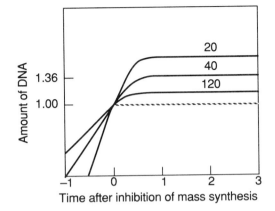

Figure 5.12 The Maaløe–Hanawalt Experiment. When mass synthesis is inhibited at time zero, there is a continuation of DNA replication until all extant rounds of replication are completed. The amount of residual incorporation is dependent upon the ratio of the C value to the interdivision time. The interdivision times are indicated on top of each line. As cells grow faster, the amount of residual synthesis increases. The dashed line is the expected amount of DNA synthesis if there is no residual increase in DNA.

genome. Only the initiation process is inhibited when mass synthesis is inhibited.

E. Thymine Starvation and Premature Initiation

When DNA synthesis was inhibited by removal of thymine from a thymine-requiring cell, there were premature initiations of new rounds of replication.[40] The insertion of new growing points was observed at one of the two arms of the replicating chromosome. It was suggested that removing thymine had, in some way, deregulated initiation so that replication points were inserted before previous rounds of replication terminated; the new replication points were termed premature. We can now look at this result in a different way. The initiation of new rounds of replication before the old ones had finished was not premature, but was a normal response to the increase in mass during the period of thymine starvation. The initiations occurred at the correct time and were not premature. The increase in the rate of DNA synthesis was due to the increase in the number of replication points. Further studies on the rate of DNA synthesis following a period of inhibition of DNA synthesis quantitatively fit the prediction that mass accumulation regulated the initiation of DNA replication (see section on Rate Stimulation, below).

F. Cell Division after Inhibition of DNA Synthesis

The rate of cell division following thymine starvation has also been studied, and the results are consistent with the analysis in Fig. 5–11. When cells resume growth following a period of inhibition of DNA synthesis, there is a sudden synchronous division C+D minutes after the resumption of DNA synthesis.[41]

G. The Shift-Up

In a companion paper to their analysis of the steady-state composition of bacteria at different growth rates, Kjeldgaard, Maaløe, and Schaechter[42] analyzed the physiology of the transition between steady states. As they knew that cells growing at faster rates were larger than the genetically identical cells growing at slower rates, they investigated how cells changed their size when they were transferred from a poor to a rich medium. This experiment has been called a shift-up. An idealized summary of what Kjeldgaard, Maaløe, and Schaechter observed is shown in Fig. 1–6. The rate of mass synthesis increased to the new, postshift rate almost immediately. DNA synthesis followed at a continuously increasing rate. Cell division continued at the initial, preshift rate for 60 minutes, at which time there was a sudden and sharp change to the new growth rate. This observation of rate maintenance is found for a number of shift-ups.[43]

The model of DNA replication and its regulation described here can explain these observations. The maintenance of the rate of cell division at the preshift rate for 60 minutes can be simply explained by the constancy of the C and D periods. Because new replication points inserted into the chromosomes do not express themselves in new cell divisions until C+D minutes later, only those replication points within 60 minutes of division will determine the first 60 minutes of cell division. The distribution of replication points gives the continued rate of cell division as though the cells were not shifted to a richer medium.[44] After 60 minutes, the abrupt change to a new rate of cell division is due to a change in the pattern of insertion of new growing points into the chromosome at the time of the shift-up.

Let us consider a shift from a medium supporting a doubling time of 40 minutes to a medium supporting a doubling time of 20 minutes. The expected patterns of chromosome replication are illustrated in Fig. 5–13. Two representative cells from a 40-minutes culture are shown at the left. At 0 minutes the cells of ages 0.0 and 0.5 contain 0.5 and 0.0 units of initiator per chromosome origin. The unit of initiator is that quantity required to begin the replication of a chromosome. In this simple case,

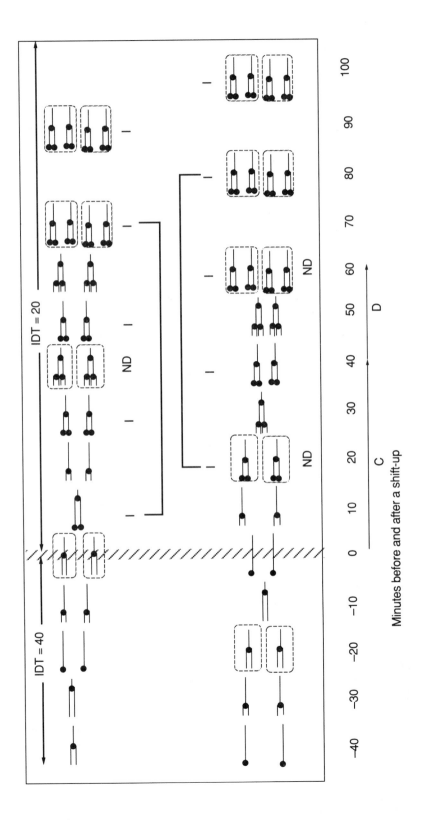

Minutes before and after a shift-up

where we assume that in one-quarter of an interdivision time the cell makes one-quarter of the required initiation material, the amount of initiator is (for 40-minute cells) equivalent to the degree of replication of the chromosome. If the shift had not occurred, initiation of DNA synthesis would occur at 20 and 40 minutes. These are the times required to synthesize 0.5 and 1.0 units of initiator in cells growing with a doubling time of 40 minutes. Assume that there is an immediate change in the rate of initiator synthesis after a shift-up. In the case of cells with a 40-minute interdivision time shifted to a 20-minute interdivision time there would be a doubling in the rate of synthesis. Only 10 and 20 minutes will be required until the next initiations. These times are the fractions of the 20-minute doubling time required for the synthesis of the remaining 0.5 and 1.0 units of initiator.

A shift-up can be characterized by the definitive pre- and postshift doubling times, t_1 and t_2. If it is assumed that the definitive postshift rate of synthesis of initiator is established immediately, the time (T_2) of the first act of initiation after the shift is calculated as follows:

$$T_2 = T_1 \cdot (t_2/t_1)$$

where T_1 is the interval between the time of the shift and the time at which the new initiation would have occurred in the cell according to the preshift rhythm. Whatever time remains until initiation at the slow growth rate is shortened in proportion to the ratio of the growth rates before and after the shift. Subsequently, new acts of initiation occur every t_2 minutes.

In this analysis, the synthesis and amount of initiator is presented with the idea that initiator increases from zero to the threshold amount, or that it is consumed at the time of initiation. This consumption of initiator is a formal concept and should not imply that previously synthesized initiator is destroyed at initiation, but that a full complement of

---

Figure 5.13 Chromosome Patterns following a Shift-up. The division cycles of two cells growing with a 40-minute interdivision time are illustrated. The dotted lines around the chromosomes indicate cells about to divide. At zero time, when the cells are shifted to a richer medium, one cell has just divided and one cell is in the middle of the division cycle. The newborn cells at zero time would normally initiate (NI) at 20 minutes after the shift-up, but because the rate of initiator synthesis is doubled, this initiation now takes place at 10 minutes. Sixty minutes later (indicated by the bracket), there is a cell division at 70 minutes. This cell division would not have occurred if there had not been a shift-up, and division would normally (ND) be at 80 minutes. Initiation takes place 20 minutes after the shift-up in the other cell; this initiation would normally occur at 40 minutes after the shift-up. After the first initiation, initiations occur regularly every 20 minutes. After 60 minutes, cell divisions occur every 20 minutes.

initiator per chromosome origin is synthesized each division cycle.[45] This new complement can be added to the initiator already present in the cell. The origin number has doubled, so it requires one interdivision time to regain the proper amount of initiator per origin. It was also assumed that the initiator increased at a constant rate, i.e. linearly, between initiations. This simplifying assumption is not necessary. Any pattern of initiator synthesis will give the same result, if the pattern of initiator synthesis during the interinitiation interval is the same in the pre- and postshift media. The pattern of initiator synthesis could be linear, bilinear, exponential, or any other pattern. The pattern of mass synthesis during the division cycle must be the same in different media and different growth rates to give the observed shift-up result. As discussed in Chapter 4, it is probable that the initiator is synthesized exponentially during the division cycle at all growth rates.

Whereas there is no change expected in the rate of cell division for at least C+D minutes, there is a slow increase in the rate of DNA synthesis beginning shortly after the shift-up. The cell that is almost at initiation will initiate slightly earlier than if there had not been a shift-up. There will be a continuously increasing rate of DNA synthesis for a time equal to the faster doubling time. In the experiment described in Fig. 5–13, DNA synthesis will achieve its final steady-state rate 40 minutes after a shift-up. The insertion of new replication points earlier than they would ordinarily have occurred is observed as a steadily increasing rate of DNA synthesis following the shift-up. Computer analysis[46] of the shift-up has indicated that the description presented here fits the observations.

H. Deviations from Rate Maintenance

Deviations from rate maintenance can be measured using an electronic cell-counting device to measure the kinetics of cell division following a shift-up.[47] These deviations can also be seen in the original data of Kjeldgaard, Maaløe, and Schaechter.[48] Slight deviations in the rate of cell division following a shift-up can be accurately measured using the membrane-elution apparatus. The cells eluting from the membrane are a measure of the rate of cell division, and this direct rate of cell division is easier to measure than the rate of cell division from the cumulative total number of cells. There are slight deviations from rate maintenance following a shift-up.[49]

If C and D were not constant at different growth rates, or if transient changes in C and D occurred at the time of the shift-up, then deviations from strict rate maintenance would be expected. These deviations could be predicted by appropriate modifications of the model presented above.

Changes in D would alter the rate of cell division during the first D minutes following a shift-up, while changes in C would alter the rate of cell division subsequent to D minutes after a shift-up. A few simple cases are shown diagrammatically in Fig. 5–14. For example, if the D period after the shift-up were shortened, there would be a period of rapid cell division followed by a period of slower cell division as cells that terminated after the shift-up passed through the shortened D period. If the D period lengthened after a shift-up, there would be a corresponding decrease in the rate of cell division. More complex patterns have been analyzed.[50]

There have been a number of reports of shift-ups that give results different from those described here. The differences have been with regard to cell division, with division after the shift-up either being faster or being inhibited.[51] The problem with these results is that there is no analysis of the implications in terms of the basic results of Kjeldgaard, Maaløe, and Schaechter. Whenever some deviation from rate maintenance is reported, the results are not presented with the expected chromosome configurations, the expected amounts of DNA per cell after the shift-up, and other parameters that lead to a complete explanation of the shift-up. The shift-ups are presented as simple results with no attempt to place them within the context of chromosome replication, as has been done with the observation of rate maintenance.[52] Any future proposals of deviations from rate maintenance must be analyzed and placed within a context of steady-state growth before and after the shift-up.

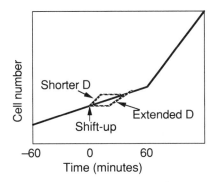

Figure 5.14 Deviations from Rate Maintenance. If the D period changes following a shift-up, deviations from rate maintenance in the rate of cell increase will be observed. If D shortens, then cells in the D period will divide earlier. There will be an increase in the rate of cell division over that expected for rate maintenance. Conversely, an extension of D leads to a slowing of the rate of cell division after a shift-up.

One subtle point about the shift-up has often been missed. During a shift-up there is a recapitulation, at every moment during the transition period, of the continuous series of DNA synthetic patterns found at all growth rates between the two growth rates of the shift-up. Thus, if cells growing with a 60-minute interdivision time were shifted to medium supporting growth at a 20-minute interdivision time, at some time during the 60 minutes it takes for the shift-up, each of the patterns for DNA synthesis at 50-, 40-, 35-, 30-, and 25-minute doubling times would occur.[53] To have, for example, initiation occur at some particular time during the division cycle, the desired pattern can be obtained by choosing a time during the shift-up rather than searching for a particular growth rate.

The strength of the shift-up experiment is that the results distinguish the mass- or initiator-accumulation models from other models. Any models that propose a sudden change in a cell property not associated with an accumulation phenomenon may be hard pressed to account for the shift-up. Any proposals of models for initiation should at a minimum be applied to explaining the shift-up results.

I. The Shift-Down

The shift-down is the converse of the shift-up. Cells growing rapidly are shifted to a medium supporting a slower growth rate. For a shift-down, cells are removed from a medium with a large number of nutrients and placed in a minimal-salts medium. If mass synthesis changed to the new growth rate immediately, there would be a sudden decrease in the rate of mass synthesis. Cell division would continue at the preshift rate. Because the rate of cell division is maintained at the higher, preshift rate (rate maintenance), there would be a decrease in average cell size. The cells would change from one steady-state composition to another.[54]

In practice, the shift-down is not so simple. When cells grow in a rich medium, they cease to synthesize those enzymes responsible for the production of various proteins required for growth. The cells might not have enzymes required for synthesizing amino acids if they were grown in a medium with amino acids. When placed in a medium without these amino acids, the cells cannot make the proteins that are required to synthesize them. There is a bottleneck and a sudden cessation of mass synthesis. Cells are starved until they can synthesize the enzymes required for mass synthesis. As the cells slowly begin to synthesize protein again, mass synthesis begins to increase and return to the faster rate. As cell division continues during the period when mass increase ceases, the size of the cells decreases. Although the shift-down experiment is not the mirror image of the shift-up experiment, normalization of the results for the different kinetics of mass synthesis would show that the shift-down experiment exhibits the inverse pattern of the shift-up experiment. (The importance of this idea is illustrated in the discussion of the bacterial life cycle in Chapter 10.)

VIII. ICON FOR REGULATION OF DNA REPLICATION

In Chapter 4, a simple icon illustrating the synthesis of mass during the division cycle was presented. Mass was represented as a circle, indicating that mass is made continuously without any differentiation in time. An icon for the regulation of DNA synthesis during the division cycle is presented in Fig. 5–15. A straight line indicates the C period when DNA is synthesized, and another straight line, of different texture, indicates the D period. The icon is not circular because the ends of the C or D periods do not regulate or feed back to affect the start of a C or D period. At intervals, a round of DNA replication is initiated; this round of replication leads to cell division. Cell division is the end of a process and the beginning of none.

IX. FURTHER ANALYSIS OF DNA REPLICATION DURING THE DIVISION CYCLE

A. Other Methods for Measuring the C Period

It is important to recognize that there are a number of methods for measuring the C time that do not depend on the ability of a cell to bind to a membrane. The following methods, along with those such as the

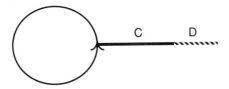

Figure 5.15 Icon for Regulation of DNA Replication. Combining the icons for mass synthesis (Fig. 4.7) with the icon for DNA synthesis (Fig. 5.7) gives an icon for the regulation of DNA synthesis during the division cycle. Initiation of DNA synthesis is regulated by mass synthesis. Neither the end of DNA synthesis nor cell division affects the synthesis of cell mass, which proceeds continuously during the division cycle. The circle represents the continuous synthesis of mass. At regular intervals the synthesis of DNA is initiated, and this leads to cell division. Cell division is the end of a process; no processes begin with cell division.

measurement of residual DNA synthesis after mass inhibition, are applicable to all types of cells and do not require a synchronization procedure.

1. Increment of DNA Synthesis

The C value can be determined by measuring the amount of residual DNA synthesis after inhibiting mass synthesis. This is the Maaløe–Hanawalt experiment described previously. This approach to measuring the value of C is extremely simple, requiring only the ability to measure the amount of DNA in a culture.

2. Increment of DNA Synthesis Using the Total Curve

An improvement on the increment-of-DNA-synthesis method is that of Bremer and Churchward.[55] If the rate of DNA replication changes in the presence of chloramphenicol, an inhibitor of protein synthesis, and it takes more than 40 minutes to complete replication, the formula for the fractional increase is still valid, after all rounds of replication have been completed. In practice, this is difficult because the amount of DNA synthesized under inhibitory conditions asymptotically approaches the final value. A slight continuous increase in the amount of DNA may be missed. Bremer and Churchward avoided this problem by replotting the increase in DNA in the presence of an inhibitor of mass synthesis. A plot of the square root of the amount of synthesis remaining until all rounds of replication have been completed gives a straight line. In this way, all of the values on the DNA curve were used to determine the intercept of DNA synthesis at an infinite time of incubation. This plot also allowed them to determine the precise time that chloramphenicol inhibited the initiation of DNA synthesis; a delay of 6 to 11 minutes was found before chloramphenicol inhibited the initiation of new rounds of DNA replication.[56] When this delay was taken into account, the C values in *Escherichia coli* strain B/r were 39, 41, and 38 minutes in different experiments. Similar results were obtained using rifampicin, an inhibitor of RNA synthesis, to inhibit new rounds of replication. The results also indicate that there is no slowing of replication forks in progress at the time that mass synthesis was inhibited.

3. Rate Stimulation

Zaritsky[57] developed the rate-stimulation method for measuring the value of C in exponentially growing cultures. The rate-stimulation method assumes that as cells grow under conditions in which DNA replication is inhibited, new replication points will be inserted in the extant origins of the cell (Fig. 5–11). Consider the ratio of the number of origins before and after inhibition of DNA synthesis. In Fig. 5–1, we saw that cells growing with a 60-minute interdivision time have approximately 0.6 replication points per cell. If DNA synthesis is inhibited for one doubling time, during which each of the existing origins obtains a replication point, then there would be between two and three replication points per cell. We expect approximately a fourfold increase in the rate of DNA synthesis when DNA replication resumes. This is the ratio of the number of replication points after inhibition to those before inhibition. In a culture with a 20-minute doubling time, there are six replication points in all exponentially growing cells. After one generation of starvation, we would expect the cells to have 14 replication points, because each cell has eight origins that can accept new replication points during the period of starvation. The increase in rate after resumption of DNA synthesis would therefore be slightly more than twofold. For a given value of C, the rate stimulation is greatest when growth is slowest. A formula for the value of stimulation ratio expected for various values of C is given by

$$\%S = \frac{2^{(C/\tau+1)} - 1}{2^{C/\tau} - 1}$$

This method makes a number of assumptions (all origins are available for initiation, the pool of precursors does not change during inhibition of DNA synthesis, and the rate of fork movement is the same before and after starvation); if these assumptions hold, this method can yield values for the C period.

When the rate-stimulation method was applied to strains that required thymine for growth, it was found that the rate of replication—that is, replication fork movement—slowed as the exogenous thymine concentration decreased. For example, as the thymine concentration was lowered from 20 to 1.1μg/ml, the C value increased from 48 minutes to 80 minutes. The increase in C in thymine-requiring strains was first reported by Zaritsky and Pritchard.[58] The rate of fork movement increased as the thymine concentration increased. Measurements of the concentration of thymidine triphosphate, a direct precursor of DNA, in cells growing on limiting concentrations of thymine[59] indicated that the rate of replication was determined largely by the intracellular concentration of TTP. These results point out that while the general picture of DNA replication in thymine nonrequiring strains is as described in Fig. 5–1, there is no requirement that the value of C be constant.

4. Mass-to-DNA Ratios

Inspection of the Schaechter, Maaløe, and Kjeldgaard experiment (Fig. 1–5) shows that there is an exponentially increasing amount of mass of the cell with growth rate, and an exponentially increasing amount of

DNA per cell. The slope of the mass curve is greater than that of the DNA curve. This means that as the growth rate increases, there is a decrease in the amount of DNA per unit mass. The formula for the average DNA per cell as a function of growth rate, divided by the formula for the average mass per cell as a function of growth rate

$$M = k \cdot 2^{(C+D)/\tau}$$

yields a formula for the DNA per mass, sometimes referred to as the concentration of DNA:

$$G/M = \frac{\tau}{k \cdot C \cdot \ln2} (1 - 2^{-C/\tau})$$

If C and k (the mass per origin required for initiation) is constant, and if we know one G/M value, then we can predict the G/M ratio for any other growth rate. If the ratio at different growth rates fits the predictions, this would confirm the constancy of C. Conversely, if we know the G/M ratio and the C value at one growth rate, and the initiation mass is constant, then the value of C at one growth rate can be calculated from the G/M ratio at another growth rate.

5. Autoradiographic Analysis of DNA Chain Extension

Autoradiography was first used to determine the number of DNA subunits in a cell, and the pattern of replication was then determined. With more sophisticated analysis, the number of replication points could be precisely measured. Further work has allowed a direct determination of the rate of replication fork movement along the genome.

The method is conceptually simple. A cell is labeled for a short time with radioactive thymidine. The cell is gently lysed, the DNA is spread out on a slide, and the slide is processed for autoradiography. There are numerous short dashlike arrangements of radioactivity on the slide, due to the presence of short segments of labeled DNA. Even if the chromosome is not spread out in its entirety, if there is enough spreading that the probability is high that the short radioactive segment will be spread out, the length of the radioactive segment, and therefore the rate of fork movement, can be obtained. An additional refinement has been to use two different specific activities of label to determine the directionality of fork movement. Consider that a cell is pulse-labeled for 4 minutes. If the tracks are precisely 100 μm long, and if the length of the genome is 1,000 μm in length, then in 40 minutes the entire genome could be replicated. This experiment would give a C value of 40 minutes. In practice, bidirectional replication would give a length of 50 μm, as there would be twice as many forks responsible for the replication of the DNA.

Even with good statistical data, absolute C-value determinations from this method are still beset by a number of inherent errors. For example, it is difficult to estimate the error due to shrinkage of the DNA during fixation. The first estimates of Cairns[60] gave a length of the genome of 700 to 900 μm, while later measurements[61] extended this to 1,100 to 1,400 μm. Also, for a given pulse, the lengths of the tracks differ by at least a factor of two in any preparation. It is difficult to say whether the short-term replication pattern of DNA can give a measure of the total C value. The most important errors probably stem from changes in the pools of radioactive intermediates that could alter the track length.

6. Analysis of Gene Frequency

Inspection of the pattern of DNA replication at different growth rates (Fig. 5–1) indicates that the ratio of the frequency of different genes in the chromosome varies with growth rate. Consider the cells with a 60-minute doubling time. The unreplicated chromosomes have all genes present in a unit amount. Replicating chromosomes have some genes, those near the origin, present at twice the frequency of those near the terminus. For genes between the origin and terminus, there is a decreasing ratio of gene concentration relative to the genes present at the terminus. Cells growing with a 20-minute doubling time have genes at the origin present four times more frequently than genes at the terminus. As the doubling time relative to the rate of DNA replication increases, there is an increase in the ratio of genes at the origin compared to genes at the terminus (Fig. 5–16).

We can calculate the gene dosage per cell. For a given gene, X, at a position of the chromosome, x, the mean number of copies of a gene in an exponential phase culture is

$$X = 2^{[(1 - x)C + D]/\tau}$$

where C, D, and τ are the chromosome replication time, the division time, and the doubling time of a culture. This equation can be simplified by considering that there is no D period and that we are considering an exponential population of chromosomes replicating in exponential culture. In this case we get

$$X = 2^{[(1 - x)C]/\tau}$$

For a gene at the terminus, position x = 1.0, and the relative value is taken as unity. Consider a gene at the origin of replication, at position x = 0.0. When τ = C (a culture with a 40-minute interdivision time), the relative value is 2.0. When τ = C/2 (a culture with a 20-minute interdivision time), the relative value of the origin to terminus gene content is 4.0.

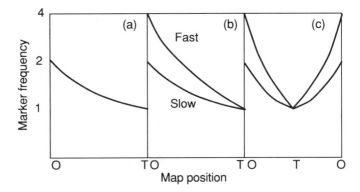

Figure 5.16 Gene-Frequency Analysis of DNA Replication. A comparison of the relative marker or gene frequency from cells with a single chromosome (either stationary-phase cells or spores) and cells that are growing is shown in panel (a). The marker-frequency curve depends on the growth rate. If the time for a round of replication equals the interdivision time, the value changes from 1 to 2, proceeding from the terminus to the origin. These values are only relative, and in practice are normalized to a culture with a defined distribution according to the lower flat line. A comparison of a fast-growing and a slower-growing culture is shown in panel (b). The ratio of the two curves is a function of the growth rate and the C value. Bidirectional replication gives the general result in panel (c).

Measurements of gene ratios have been used to determine the rate of DNA replication during the division cycle.[62] As a bonus, such measurements have also led to the understanding that replication is bidirectional (see below). One of the more common ways to determine gene dosage is to use quantitative nucleic acid hybridization. The use of hybridization has difficulties. For example, nonspecific hybridization may account for 44-48% of the total hybridized radioactivity.[63] Further, Bremer and Churchward noted that the amount of specific hybridization observed can be greater (three- to fourfold) than can be expected from the size of the phage DNA molecules.[64]

7. Synchronized Cultures

Sucrose gradient synchronization (Chapter 3) has been used to study the pattern of DNA replication during the division cycle of strains that did not work with the membrane-elution method.[65] In this way, the pattern of DNA replication of *Escherichia coli* K12 strains could be determined. Although the results were consistent with the general model, one word of caution should be noted. When the cells are taken from the top of a sucrose gradient, and then allowed to grow out, it is possible that the smaller cells are an unrepresentative portion of the newborn cells. In this case, there would be large periods devoid of DNA synthesis in the younger cells. These large gap periods would not necessarily be found if all of the newborn cells of a culture were studied and analyzed. This can be understood by considering that initiation of DNA replication occurs *about* cell division. Some cells initiate before and some cells initiate after division. Smaller cells would be most likely to initiate after division and to thus give a gap in synthesis in the beginning of the division cycle. These smaller cells arise either from a smaller mother cell or by not precisely midcell division. This type of variability is discussed in detail in Chapter 8.

8. Synchronization of DNA Replication

We can use the idea that accumulation of mass causes the initiation of DNA replication to synchronize a round of DNA replication. In one experiment, cells were starved of amino acids to allow all rounds of replication to proceed to the end of the genome; then cells were starved of thymine in the presence of amino acids to induce a new round of replication in all chromosomes. A synchronous round of replication was then allowed to proceed by the addition of thymine.[66] Samples were exposed to radioactive thymidine at intervals, and the time at which various portions of the genome replicated was determined by DNA-DNA hybridization. The C period was 40 minutes in rich medium and slightly longer (52 minutes) in glucose-minimal medium.

It is important not to refer to cultures that have only DNA synthesis synchronized as synchronized cultures. Synchronized cultures should be reserved to describe cells that are proceeding through the division cycle as if they had not been synchronized at all.

9. Measuring the C Period by Transductional Analysis

A unique approach to measuring the C period involves studying the time, after transduction, that cells with recombinant phenotypes begin to increase.[67] Transductants for markers located at different positions on the chromosome begin to increase at different times, in reverse order to that in which they are replicated. The period in which this happens is equal in duration to the time taken to replicate the chromosome. The transductional analysis gave a C period of 39 minutes at 37°C, a value in good agreement with other determinations.

10. DNA Step Times as a Measure of the C Period

Another way to determine the C period is to measure the thymidylate step-time.[68] The steptime is the average time it takes to add one more nucleotide to a growing chain. These are complex measurements to make, requiring extremely short labeling times (6 to 18 seconds) and determinations of the ratio of internal to end label in DNA. The determinations of the chain growth rate were consistent with other measured values of C. The step-time measurements indicated a chain growth rate of 143-250 nucleotides per second at 20°C. When the temperature coefficient for growth is taken into account, a predicted value for 37°C would be 500-700 nucleotides per second, which is in very close agreement with the calculated values of C from the membrane-elution method.

B. Modifications of the Rate of DNA Replication

Although the model presented at the start of this chapter emphasized the constancy of C and D values at intermediate growth rates, there is evidence that C does not have to be constant.

1. Effect of Thymine Concentration

When different thymine-requiring strains were grown with less than optimal thymine concentrations, the rate of DNA synthesis varied with thymine concentration.[69]

2. The *rep* Mutation

A mutant that could not replicate various bacteriophages had altered C values, as shown by comparing the length of the autoradiographic strands after pulse labeling.[70]

3. Effect of Inhibitors on Rate of DNA Replication

DNA fork movement is sensitive to antibiotics that have ribosomes or RNA polymerase as their primary site of action.[71] Chloramphenicol or streptolydigin slow down fork movement. Addition of rifampin, an inhibitor of RNA synthesis, reverses this effect; this suggests that fork movement might be obstructed by transcription complexes temporarily immobilized on the DNA template.

C. Nucleoid Division in Relation to DNA Synthesis

The DNA of the cell is not dispersed randomly throughout the cell volume, but is concentrated in an observable nucleoid. When cells are studied in the light microscope, it is possible to stain the cells and see condensed regions as a result of the DNA. These are called nucleoids, to distinguish them from the membrane-bound organelles that contain the genetic material in higher cells. The question arises, what is the relationship of the cell nucleoids to the initiation, replication, and termination of DNA synthesis during the division cycle?

1. Nucleoid Content during Steady-State Growth and Transitions

Schaechter, Maaløe, and Kjeldgaard (Chapter 1) observed a continuous increase in the number of nucleoids per cell as a function of growth rate. A simple explanation of the nucleoid increase is to assume that each nucleoid is determined by a separate chromosome. This is another way of saying that each nucleoid is determined by an independent terminus. If we look at the patterns of DNA replication during the division cycle (Fig. 5–1), we see that the maximal number of nucleoids per cell should be 2.0 in cells with a 20-minute doubling time and approximately 1.6 in cells with a 60-minute doubling time.

If we assume that termination produces two nucleoids where only one existed before, then we can use the measurements of Schaechter, Maaløe, and Kjeldgaard to measure the D period in *Salmonella typhimurium*. When this is done for the original data, the value for D is too large. By considering, however, that the nucleoids begin to condense into two new nucleoids before actual termination (Fig. 5–17), and by assuming that thin strands connecting these new nucleoids are sometimes not observed, the data can be adjusted to fit the observed D period.[72] When such an adjustment is performed, the proposal fits the data of Schaechter, Maaløe, and Kjeldgaard precisely.

During a shift-up, Kjeldgaard, Maaløe, and Schaechter[73] observed that the number of nucleoids per cell increased approximately 40 minutes

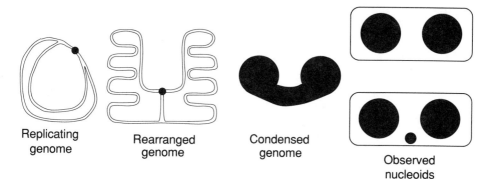

Replicating genome   Rearranged genome   Condensed genome   Observed nucleoids

Figure 5.17 Nucleoid Separation and Production. It is possible to interpret the observed nucleoid content in terms of the separation of replicated DNA. Although toward the end of a round of replication there is still a connection between the two condensed masses of DNA, this strand can sometimes be missed owing to problems with microscopic resolution. Thus it could appear that nucleoid division occurred earlier than a measured time of termination of DNA replication. Upon occasion, small dots of material are seen, which are interpreted as residual connecting material.

after a shift-up. This means that independent of the error in the absolute number of nucleoids per cell, we can associate the nucleoids with termination of DNA replication.

2. Electron Microscopy of Nucleoids

Woldringh[74] analyzed thin sections of bacteria in a growing culture and classified the bacteria according to size. The fraction of bacteria in each size class with one or two separate and condensed regions of DNA was determined. Using this approach, Woldringh was able to show that the mean time of change of one to two nucleoids was equal to the D period. This is a strong support for the idea that when cells terminate DNA replication, they immediately produce two distinct nucleoids. The conclusion of this analysis is that termination is very likely associated with the formation of observable nucleoids in the cell. Brock[75] has presented an insightful historical analysis of our understanding of the bacterial nucleoid.

D. Determination of the D Period

1. Residual Cell Division after Inhibition of DNA Synthesis

If the D period were a period of preparation for cell separation, and if the preparations for cell division were started at the termination of DNA replication, then cells that have terminated DNA replication will be able to divide. Those that have not terminated will be prevented from a subsequent division in the absence of DNA synthesis. If DNA synthesis were inhibited, only cells with two termini, that is, those in the D period, would divide. In this case, the D period could be estimated by measuring the residual cell division after inhibition of DNA replication. An alternative possibility is that there is no necessary relationship between cell division and the termination of DNA replication, but there is a fortuitous coincidence of an independent time of termination and an independent pathway of events leading to cell division, such that the division occurs approximately 20 minutes after termination. While in the first case, termination would be a necessary and sufficient condition for cell division, in the second case, termination would only be a necessary condition for cell division. In either case, if inhibition of DNA replication did not disturb the division process except by lack of termination, then we could estimate the D period by estimating the residual division time.

Helmstetter and Pierucci[76] studied residual division after inhibiting DNA synthesis by three independent methods. Their results suggested that the D period is constant and independent of the doubling time. An interesting experimental point should be noted. Helmstetter and Pierucci did not measure the residual division time by looking at the residual division in a batch culture. Rather, they measured the residual time by determining the differential rate of cell division. This was done by looking at the rate of release of cells from a membrane-bound population. A growing culture was treated either with UV irradiation, mitomycin, or was starved of thymine (all of which inhibit DNA synthesis in ways thought to be specific), and the cells were bound to a membrane. For 20 minutes, there was a normal increase in the eluted cell numbers, after which time there was a dramatic decrease.

This experiment illustrates an important property of the membrane-elution method. The number of cells being eluted from a membrane-elution apparatus at any time is a measure of the *rate* of cell division at any time. For this reason, accurate measurements of the changes in the rate of cell division can be derived following any particular treatment. Even changes so slight that they could not be easily measured using the direct measurement of cell number in a culture can be determined using the membrane-elution method.[77]

2. Residual Cell Division after Inhibition of Protein Synthesis

A slightly different approach was taken using an inhibitor of protein synthesis to determine the D period.[78] If adding an inhibitor to a culture allows only those cells in the D period to divide, then the fractional increase in the number of cells is

$$N/N_0 = 2^{D/\tau}$$

where N and N_0 are the final and the original cell number and D is the period between termination and cell division. When determinations for a number of different growth rates are considered, then

$$\ln(N/N_0) = \ln 2 \cdot (D/\tau)$$

If D is a constant, D/τ approaches zero at slow growth rates.

This method is based on the assumption that protein synthesis is required for the synthesis of the terminal segment of the chromosome in *Escherichia coli*.[79] There are strong theoretical objections to the use of chloramphenicol for such a measurement, as inhibition of protein synthesis may have effects on the division process itself, independent of the initiation of any specific D-period processes (see Chapter 6 for a discussion). When residual division[80] was studied in cells growing at different rates, the D period was constant. At rates slower than 60-minute doubling times, the results were not consistent with a D period's being a constant fraction of the interdivision time.

X. DNA REPLICATION DURING THE DIVISION CYCLE OF SLOW-GROWING CELLS

There is less consensus regarding the pattern of DNA replication during the division cycle of slow-growing cells than of fast-growing cells.

A. *A Priori* Considerations on Slow-Growing Cells

If we consider cells growing at a doubling time greater than 60 minutes, we can visualize four possible patterns of DNA replication (Fig. 5–18). One possibility is that the C and D values remain constant even as the interdivision times approach large values. Another possibility is that the C and D patterns enlarge as the doubling times increase and remain a fixed fraction of the interdivision times. In this case, the C and D periods would be two-thirds and one-third of the interdivision time. In the first case, it might be assumed that the material required for the synthesis of DNA is adequate to support the given C values even though growth becomes extremely slow. The second example assumes that at slow growth rates, the material required for DNA synthesis decreases, thus leading to a slowing of the rate of DNA fork movement. A third possibility is that the C values remain constant and the D values increase. A fourth conceivable example is that D values remain constant and the C values increase.

B. Membrane-Elution Analysis of DNA Synthesis during the Division Cycle of Slow-Growing Cells

A number of reports have appeared showing that there is a period in the division cycle devoid of DNA synthesis; a gap in DNA synthesis exists.[81]

The membrane-elution method was first applied to the study of DNA replication in slow growing cells. Its success there was immediate. The pattern of uptake of radioactive thymidine during the division cycle was consistent with DNA synthesis during the first part of the division cycle and a gap during the last part of the division cycle. There was no gap observed in the beginning of the division cycle even for very slow growing cells. This result fits the idea that at slow growth the sum of C and D is equal to the interdivision time. The pattern of DNA synthesis during the division cycle in slow-growing cells is shown in Fig. 5–2.[82] At the start of elution, there is a sharp dip in the incorporation per cell, then a plateau. The first cells eluted are from the oldest cells in the culture, so this result indicates that there is a gap in DNA synthesis in the oldest cells. In the younger cells—at least those in the first two-thirds of the division cycle— there is a constant rate of DNA replication.

In the examples of slow growth in Fig. 5–1, there is an increase in

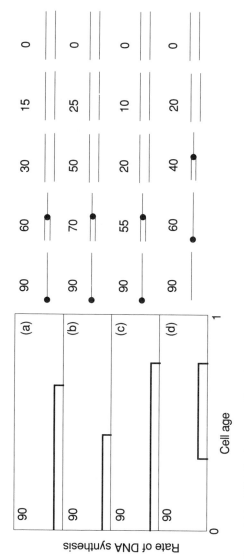

Figure 5.18 DNA Replication in Slow-Growing Cells. Four possible patterns of DNA replication in slow-growing cells are presented. In all cases it is assumed that the cells have a 90-minute interdivision time. In (a) the C and D periods occupy two-thirds and one-third of the division cycle; the C and D periods increase in proportion to the increase in doubling time over 60 minutes; the pattern of DNA replication is the same as in cells with a 60-minute interdivision time. In (b) there is a constant C period (40 minutes) and in (c), a constant D period (20 minutes). In (d) both C and D periods are the same as in growth rates faster than 60-minute interdivision times. The uppermost pattern is the most common one observed, with the C and D periods increasing in proportion to the interdivision time. Pattern (d) presents a gap in the initial part of the division cycle, a B period.

incorporation in the oldest cells.[83] This is indicated by the sharp decrease in radioactivity per cell in the first cells eluted. This reveals a subtle but important idea. As the 60-minute cells were drawn in Fig. 5–1, it was indicated that initiation occurred precisely at the time a cell was born. Consider that there is some variability in the initiation process, and that initiation actually occurs around cell division, either a little before, at, or a little after cell division. This would mean that some cells initiate DNA synthesis in the very oldest cells of the population, and other cells would initiate DNA synthesis after cell division. There would be a slight gap in the synthesis of DNA in some of the youngest cells of the culture. On average, initiation would occur at the time of cell division.

A definite contribution to the slow-growth problem came from the analysis of different strains at slow growth rates. There were strain differences in the pattern of DNA replication during the division cycle. When three different B/r strains (A, F, and K) were compared, it was found that B/rA had no B period at slow growth, while strains B/rF and B/rK did have such a visible gap in the youngest cells.[84] This variation is unexpected, considering that the three strains differ only in their history of storage in various laboratories.

C. Alternative Results and Models

1. Constant C and D at Slow Growth Rates

An alternative to the proposal that C and D increase as the interdivision time increases at slow growth rates was proposed by Kubitschek and Freedman.[85] They measured the DNA contents of cells over a wide range of growth rates and proposed that the constant C and D model is correct even at slow growth rates. At the outset, it should be mentioned that the slow-growing cells in these experiments were all produced by chemostat growth, resulting in a large number of cultures with apparent doubling times of between 1 and 50 hours. Slow-growing cultures were also produced by batch growth, with poor carbon sources giving interdivision times between 1 and 3 hours.

The rate of DNA replication was determined by measuring the amount of DNA per cell at very slow growth rates. This is equivalent to a Schaechter, Maaløe, and Kjeldgaard experiment for slow-growing cells. In this region of growth, the most extreme values in the DNA content expected by different models would not be greater than a factor of two. For example, consider that the DNA all replicates in the beginning of the division cycle or at the end of the division cycle. In the first case, the DNA content per cell would be close to two; in the second case, close to one genome equivalent per cell. For intermediate cases, with DNA replication occurring in the first two-thirds of the division cycle or occurring in the last 60 minutes of the division cycle, the expected differences are even slighter. When such measurements were done,[86] the absolute amount fit with the idea that C and D were constant even at very slow growth rates. This model (the self-consistent model) proposes that even in cells growing with a 50-hour interdivision time, the time for C and D remains 40 and 20 minutes. In these slowest-growing cells, one would expect a period devoid of DNA synthesis for the first 98% of the division cycle.

There are three concerns with this result. First, when the results are replotted on the semilogarithmic coordinates of Schaechter, Maaløe, and Kjeldgaard, the scatter of the points mitigates against any firm conclusion as to the meaning of the DNA contents. Second, the conclusions were based on determinations of the absolute amounts of DNA per cell, and the amount of DNA per genome. Both of these determinations are difficult to obtain. Finally, and more generally, this is an example of an integral type of experiment which, as noted in Chapter 3, is not so sensitive an approach as differential methods (see Chapter 3).

An appreciation of the difficulty in using measurements of the DNA per cell to determine the replication pattern can be gained from the following analysis. The self-consistent (constant C and D at very slow growth rates) model would have the average number of genome equivalents asymptotically approach 1.0. At very slow growth rates, the greater part of the division cycle would be occupied with cells that have one genome equivalent. Only in the last small fraction of the division cycle would there be an increase in the DNA per cell. Averaging over the entire culture would yield a value of 1.0 genome equivalent per cell. With a C and D period occupying two-thirds and one-third of the division cycle respectively, as is proposed by the membrane-elution results, there is a DNA content of approximately 1.63 genome equivalents per cell. This means only absolute measurements of the DNA content per cell can clearly distinguish between these two predictions.

2. Marker Ratios during Slow Growth

A different approach has also suggested that at slow growth rates the values of C are constant. An alternative way of looking at marker frequency as a measure of the C value is to consider the relative frequency of any two markers. The frequency of markers in a culture is given by

$$a/b = 2^{(C \cdot \Delta / \tau)}$$

where a and b are the frequency of two chromosomal markers, Δ is the relative distance between the markers as a fraction of the genome, and C and τ are the replication and doubling times of a culture. If C is constant as the doubling time increases, the ratio between the two markers should decrease. If C is a constant fraction of the doubling time (i.e., it varies with growth rate and is not a constant period of time), then the ratio should be constant.[87] Measurements of the relative numbers of different positions on the chromosome were performed by assaying the relative amounts of different bacteriophages inserted at different points on the genome. When such measurements were performed at slow growth rates, a decreasing ratio was found for growth rates slower than 60 minutes, consistent with a constant C even at very slow growth rates.[88] The differences were over the order of 20–30% at most, and the data, when considered as a part of a larger fraction of interdivision times, are indistinguishable from the model proposed for a variable C at slow growth rates.

These approaches to determining the DNA pattern—DNA contents and marker ratios—are examples of integral methods of measurement. Rather than measuring the rate of DNA replication in the division cycle directly, they measure a result of the rate of change. This measurement is inherently less sensitive than the direct measurement of the rate, done with the membrane-elution method.

3. Is C Constant in Slow-Growing Cells?

The earliest experiments on the membrane elution method[89] indicated that there was a gap in DNA synthesis in the cells at the end of the division cycle. More quantitative work showed that DNA replication in slow-growing cells (slower than 60-minute doubling times) occupied two-thirds of the division cycle, with one-third of the division cycle having a gap. In terms of the general model described above, the C and D periods summed up to the total interdivision time; the C period occupied the first two-thirds and the D period occupied the last one-third of the division cycle. The conclusion of the membrane-elution experiments is that C and D are not constant as the growth rate slows.

4. Synchrony Analysis

The pattern of DNA replication during the division cycle of a number of different bacteria has been studied using sucrose gradients to synchronize cells. The results are all compatible with the general model proposed here, but one aspect of this method should be noted. When the cells were taken[90] from the top of a sucrose gradient and allowed to grow, there was a noticeable gap in the pattern of DNA replication. As mentioned above, this result could be due to the selection of cells smaller than average newborn cells.

5. Autoradiography of Slow-Growing Cells

Another approach was taken by the Dutch workers[91] who pulse-labeled slow-growing cells with tritiated thymidine and subjected the cells to autoradiography. The grains and the size of the cells were measured by electron-microscopy. The fraction of labeled cells for different size classes was then determined. It was found that for strains B/rA and B/rK grown with interdivision times of approximately 100 minutes, there was a dearth of labeled cells among the cells in the smallest size classes. Cells of the largest size classes as well as those in the midsize range were almost completely labeled. This result indicated that the cells had a gap in the synthesis of DNA during the first part of the division cycle.

These experiments relied on classifying cells by length and using length as a measure of cell age. As will be seen in Chapter 6, cell length is a poor measure of cell age, and in the particular case of autoradiography of cells, we can get an exaggerated view of the size of the gap in the shortest cells by using the length measurement. The shortest cells would be only a small fraction of the population. These shortest cells are possibly unrepresentative of the class of newborn cells, so the gap period in the youngest cells may be exaggerated.

6. Flow Cytometric Analysis of Slow-Growing Cells

Flow cytometric measurements of cells growing slowly in chemostat cultures suggested that a B period could be detected. This was clearly indicated by a peak in the frequency of cells with one genome equivalent per cell. In fact, in B/rA (the strain that in the membrane-elution work had no B period) there was a large B period (80% of the division cycle) when the cells had a 17-hour interdivision time. Similar results were found for B/rK.[92]

7. A Conjecture Regarding DNA Replication in Slow-Growing Cells

What are we to make of different results from the same bacteria? Is there some common understanding to be derived from these conflicting results? It is proposed that in cells growing with a doubling time of greater than 60 minutes there is a lengthening of the C period. There are, however, some minor differences in the lengthening of the C period (and the D period) that lead to either the appearance of or absence of a B period; the B period is defined below. There is no theoretical difficulty with this idea because there is no *necessary* relationship or association of the initiation of DNA replication with this or any other cell division. As will be seen (Chapter 6), initiation can be either before or after a particular division.

The extremely large values for D, particularly in the definitive work with flow cytometry, is best understood as resulting from slow growth in a chemostat. When cells are allowed to cease growth—for example, by overgrowth—they continue to divide until the cells have one genome per cell. Cells in the process of replication of DNA, or those in the D period with two genomes, will continue to replicate DNA and divide until all cells have one genome per cell. The same situation applies in a chemostat, with a fraction of the cells acting as though they are starved and not growing, with only a fraction of the cells able to proceed through the initiation process.

XI. A COMMENT ON THE CHEMOSTAT

The chemostat has been widely used by researchers of diverse interests.[93] This machine for growing cells uses the slow introduction of a limited nutrient supply to keep the cells growing continuously at some very slow growth rate. Despite the popularity of the chemostat, and the rapid introduction of this method into any problem where it appears to be applicable, with the analysis of the division cycle the chemostat has been a disappointment. Aside from results on the chemostat itself, no important, interesting, or undeniably correct result has been obtained from the chemostat.

If we want to study bacteria growing over a wide range of growth rates, we can readily obtain rapidly growing cells by adding nutrients to the medium; we obtain more slowly growing cells by adding less to the medium, then adding poorer and poorer carbon sources. When we finally add acetate, we get cells with a two- to three-hour doubling time. We wish to get slower cells but cannot find a poorer carbon source. We turn to the chemostat. We set the dilution rate to change the medium every 25 hours, so that the interdivision time is 25 hours. The question raised is whether this is actually, or physiologically, a very slow-growing cell. Consider that the limiting nutrient is glucose. The cell metabolizes the glucose after each addition. There is a moment of feeding and a moment of starvation; the cell therefore goes through a time of feast and a time of famine. This cell is not the same type of cell as one growing continuously on a carbon source with no period of starvation. Although the mathematics of the chemostat allows us to think of infinitesimally small additions of nutrient, and infinitesimally small intervals between additions, in practice we must give finite drops at finite intervals. This leads to problems. In the chemostat, there is a mixture of cells, those growing relatively fast and those that are starved for relatively long periods.

Because the production of cells and cell mass is linear with time in a chemostat, it is sometimes believed that cells in the chemostat grow linearly. Without going into a complete proof, let it be noted that the cells inside the chemostat grow exponentially, because the age distribution is the same as that in an exponentially growing batch culture. If the age distribution were different, then the culture would not be in steady state, as the age distribution would be continuously changing. If the age distribution were uniform, then an instant later, there would be an excess of young cells as the oldest cohort divide to produce twice as many young cells as any other age group. Just because a culture is diluted at very short intervals, so that the concentration is constant, does not mean that the growth pattern of the cells in the culture is different from that in a culture in which such dilution does not occur.

XII. B, T, AND U PERIODS DURING THE DIVISION CYCLE

In addition to the C and D periods defined above, other periods of the division cycle have been accorded names. The B period is the period between cell division and initiation of DNA synthesis when there is a gap in DNA synthesis at the start of the division cycle. The B period is analogous to the G1 period of animal cells. How does the B period's size vary with growth rate? Does it have any function? What events occur in the B period? The problem with these questions is that the B period has no biological function. The B period can exist, but it has no reality that should be investigated. When some strains grow very slowly, it is possible to see a gap in the start of the division cycle. Call this the B period. Now speed up the growth rate. The B period will get smaller and smaller, and eventually decrease to zero when initiation occurs at the same time as cell division. As the growth rate increases, and initiation occurs before cell division, we get a negative B period. This indicates that the B period has no biological function. The B period, when it exists, is merely part of a larger period, the period between the starts of new rounds of replication. That it is found in some particular cells at some growth rates does not mean that it, as an entity, is worthy of study. The importance of this concept will become apparent in Chapter 15, when the pattern of DNA replication in eukaryotic cells is analyzed.

The T period is the time from the start of invagination until cell division. If there were a relationship between the D period and the T period, it might show that cell division is triggered by termination.

The U period is related to the T period in that it is the time between initiation of DNA synthesis and invagination. The time between initiation of DNA synthesis and cell division is equal to U+T and also C+D. If the T period were equal to the D period, then the C period would be equal to the U period, and this would be evidence that initiation of DNA synthesis is temporally related to the start of invagination as well as termination of DNA synthesis.

XIII. DNA CONCENTRATION AS A FUNCTION OF GROWTH RATE

Some researchers have plotted the classic Schaechter, Maaløe, and Kjeldgaard results using the ratio of DNA to mass as a function of growth rate. The mass increases with a greater slope than DNA (see Chapter 1), so the ratio of DNA to mass must decrease. There is less DNA per mass as the growth rate increases. This way of looking at DNA replication is confusing and should be eliminated. The concentration of DNA per mass is caused by two different effects, cell mass increase and DNA increase due to a relatively constant C and D period. We understand the DNA concentration or amount per cell as stemming from the C and D and initiation rules, so it is not necessary to confuse the issue by considering DNA concentration.

XIV. THE ORDER OF REPLICATION OF THE GENOME

A. Early Work on the Order of Genome Replication

One of the earliest attempts to determine the origin and direction of replication was that of Pato and Glaser,[94] who used the membrane-elution technique both for preparation of synchronized cells and for backwards analysis (see Chapter 3). They assumed that when a particular part of the genome was replicated, there would be a doubling in the inducibility of the genes in that region. It was assumed that the rate of induction of the enzyme was limited by the amount of DNA present, and when the gene replicated, there would be a doubling in the amount of the enzyme synthesized after a short pulse with inducer. When they performed this experiment with three enzymes, they concluded that the origin was between 7 and 8 o'clock on the genetic map with a unidirectional, clockwise direction of replication. Helmstetter[95] reached the same conclusion using the backwards membrane-elution approach with the same three enzymes. He concluded that the origin was at 60 minutes on the standard map, and replication was clockwise.

An analogous approach was taken by Ward and Glaser[96] who used nitrosoguanidine to mutagenize synchronized cultures prepared by membrane-elution. Assuming that nitrosoguanidine acted preferentially at the growing point,[97] they expected an increased reversion frequency in those cells treated when the genome was replicating that particular portion. They found repetitive peaks and concluded that the origin was at 49 minutes with a clockwise direction of replication.

Other attempts were made to determine the origin and direction of replication using elegant approaches combining chromosome alignment, density labeling, and transduction methodology. For example, Wolf, Newman, and Glaser[98] labeled parts of the chromosome with a density label (bromodeoxyuridine) either during completion of chromosome replication in the absence of an amino acid, or after restarting chromosome replication following a procedure of alignment, growth in the absence of replication, and subsequent initiation of DNA replication.[99] Transducing lysates were prepared from these cells, and the fraction of transductants from dense phage or light phage was determined. The results suggested an origin of replication between 54 and 70 minutes on the genetic map, and probably a clockwise direction of replication. Different strains gave different results, and some appeared to have alternative origins with a counterclockwise direction of replication. A more precise location of the origin used cells in which DNA synthesis was synchronized by a sequence of starvations.[100]

These results are interesting in that they are generally compatible with bidirectional replication. In most cases, the number of markers (three enzymes, four reverting loci) were inadequate to make a distinction between unidirectional and bidirectional replication. In any case, the simple demonstration of a doubling in the inducibility of an enzyme at the time of gene replication and the periodic variation in inducibility do survive as correct results.

B. Bidirectional Replication

Shortly after the above work, a number of papers appeared conclusively demonstrating bidirectional replication of the *Escherichia coli* genome. The most elegantly presented set of experiments are those of Bird, Louarn, Martuscelli, and Caro,[101] who used a marker-frequency analysis of exponentially growing cultures to determine the order and mode of replication. They used two phages, one at a fixed location in the genome, and the other inserted at many different locations in the genome. They measured the ratio of the amount of DNA in the two phages in different cultures. The amount of phage at a particular genetic location in all cultures was used to normalize the results. The phage yields were quantitated by determining the ratio of the DNA contents of the two phages rather than the absolute amount of each phage. As illustrated in Fig. 5–16, a gradient of markers revealed the origin and the direction of replication. They found that the origin was at 74 minutes on the map, and proceeded simultaneously in two directions; this is bidirectional replication. Two items contributed to their success. First, they used 14 markers to determine the gradient. Second, they did their measurements at different rates of growth and found, as expected from a constant C period, that the gradient was steeper at faster growth rates (Fig. 5–16). Most interesting, they took the observation that at low thymine concentrations in a thymineless mutant the rate of DNA replication is slowed[102] to produce multiforked cultures without a drastic increase in growth rate. This experiment gave an extremely steep gradient of markers and clearly demonstrated bidirectional replication. This work also confirmed the series of assumptions descending from Maaløe and Hanawalt[103] that when cells were allowed to grow in the absence of amino acids the chromosomes replicated to the end. When cells were allowed to complete DNA replication in the absence of mass synthesis, they found that the different genes were present in equal amounts. This indicated that all chromosomes had finished replicating, and there were no growing points in the middle of the genome.

The earliest report of bidirectional replication came from experiments using transduction to compare the gradient of markers from cultures grown at different growth rates.[104] Different markers might have different absolute efficiencies of transduction, so two cultures were compared, one with a steep gradient of marker replication and one with a shallow gradient. The results were supportive of bidirectional replication from an origin at approximately 66 minutes on the genetic map. Although bidirectional replication was observed, the gradient was not symmetrical, suggesting that there may be different rates of replication of the DNA depending on the direction of replication.

Other experiments in *Salmonella*[105] confirmed the bidirectional replication pattern. But the final proof of bidirectional replication came from a series of papers in which direct biochemical methods were used to analyze the origin and direction of replication.

One of the first and most spectacular demonstrations of bidirectional replication was that of Prescott and Kuempel,[106] who presented autoradiographs demonstrating bidirectional replication. The simple experiment they performed, using the method of Cairns,[107] was to align chromosomes, allow growth to initiation, and then allow a simultaneous

initiation of the chromosomes. By using different specific activities of the
the label, they were able to show that replication proceeded outward
from a particular point in both directions. Shortly thereafter, McKenna
and Masters[108] presented biochemical evidence for bidirectional replication. They incorporated a label sensitive to UV irradiation at the origin;
irradiation caused breaks in the DNA. They reasoned that if replication
were bidirectional, then after irradiation of DNA labeled at the origin
with bromouracil, there would be a halving in molecular weight of the
DNA. They found a specific breaking of DNA, thus confirming the
bidirectional mode of replication.

One of the clearest demonstrations of bidirectional replication came
from the work of Marsh and Worcel.[109] They aligned chromosome replication by raising the temperature in a mutant affected in initiation of
DNA replication, and then lowered the temperature to allow initiation at
all origins. The DNA was labeled and analyzed using restriction enzymes
and electrophoresis of the fragments. Different fragments became labeled at different times, and when the fragments were ordered, the
pattern of labeling demonstrated bidirectional replication.

All of this work led to the eventual observation of bidirectional replication in the electron microscope and to the final demonstration of bidirectional replication in a cell-free enzymatic system.[110] Bidirectional replication appears to be found in animal and bacterial viruses, and even in
eukaryotic DNA replication. Except for a few other organisms, the absolute generality of this mode of replication has been assumed rather than
demonstrated. Bidirectional replication was further shown using minichromosomes containing the replication origin.[111]

C. On the Order of the Genes in the Genome

One very early idea on the arrangement of genes on the genome was that
there was some functional reason for having a particular order, related to
the cell cycle. From consideration of the replication pattern at different
growth rates, it appears that different parts of the genome replicate at
different times in the division cycle. Any gene could replicate at different
times with regard to different cell cycle events—such as the start of
invagination—so no relationship would be expected between the division cycle and the order of genes on the genome.

What reasons explain the particular order of the genome? One of the
most plausible ideas is that the order is determined by the need of the cell
to maximize its ability to synthesize the machinery required for protein
synthesis (Chapter 4). As the interdivision time decreases, there are
multiple forks, and this leads to a greater ratio of origins to termini. If the
number of genes required for ribosome synthesis is limiting, the cell

could maximize its growth rate if the ribosomal genes were close to the origin. Consider a cell growing in minimal medium and suddenly shifted to a rich medium. As noted in Chapter 4, there is a cessation of synthesis of the mRNAs for the compounds added to the medium. After the ribosomes complete their transit of the extant messenger, there is a shift in the fraction of ribosomes and RNA polymerases engaged in the synthesis of particular proteins. The result of the shift is that there is an increase in the proportion of the protein-synthesizing system devoted to the synthesis of the messengers, RNAs, and proteins for the ribosomes, polymerases, and other enzymes involved in macromolecular synthesis. If the gene concentration is limiting, then a faster rate of synthesis of these components could be obtained by increasing the relative concentration of the genes by placing them near the origin.

An engaging subvariant of this theory has been proposed by Snellings and Vermeulen[112] who observed a correlation between the genetic map position of the genes for amino acids and the bulk requirement of that amino acid. The more an amino acid was present in the bulk of the cell, or the more of that amino acid was synthesized, the closer that gene was to the origin. This is consistent with the general idea that the more a function is required at faster growth rates, the more likely it is for the cell to place the function nearer the origin. This ensures that there will be a preferential increase in the synthesis of that function.

D. Is There a Unique Order of Replication?

It is possible to imagine that chromosome replication in bacteria occurs by having one round of replication start at any of a number of places on the genome, and proceed around the genome. All that would be required to make this system tenable would be the presence of one replication point on the genome suppressing initiation of replication points at other places on the genome. The alternative would be to have a unique origin at which replication started. When this question arose, the circularity of the genome was already suspected on the basis of genetic evidence,[113] and it was conceivable that replication could start at many different places along the genome. This conjecture was supported by the well-known idea that different Hfr male strains could donate DNA to a recipient strain by inserting DNA at a number of different genetic positions along the genome.[114]

One of the earliest pieces of evidence against this idea was the classic Meselson–Stahl experiment.[115] That experiment indicated that DNA replicated completely before any material replicated a second time. If replication could start at different points at random, then a region that had just finished replication might be the site of a new origin soon thereafter, and thus there would be a twice-replicated portion of the genome in much less than one generation. In retrospect, it is difficult to know what level of resolution the Meselson–Stahl experiment had in order to eliminate this possibility, but shortly thereafter, many experiments were performed that supported the results obtained with the density transfer measurements.

Nagata[116] studied the replication of different prophages in synchronized cultures and concluded that replication of the DNA proceeded from one end. In *Bacillus subtilis*,[117] a similar conclusion was reached by a transformation experiment. One of the clearest indications that there was a unique origin was provided by Lark, Repko, and Hoffman.[118] In these experiments, cultures were put through a series of treatments that first specifically labeled the origin, and then following growth, the cultures were again treated to see whether, after the same treatment, the same origin was the site of initiation of replication. Briefly, the experiment was to starve cells and allow chromosomes to complete replication. The cells were then initiated in a radioactive medium so the origins were preferentially labeled. The label was removed, and the cells were allowed to grow and randomize their growth pattern. A second round of starvation and initiation, this time in the presence of a density label for DNA, tested whether the original labeled material is preferentially replicated a second time. The result was that there was a preferential replication of the origin label, and thus there was a unique origin.

A unique origin of replication is not an absolute requirement for growth. It is possible to alter the point on the chromosome where replication is initiated by inserting a different origin into a cell that is temperature-sensitive for normal initiation. When this is done, for example by insertion of a resistance plasmid, there is a change in the origin of replication.[119]

XV. THE INITIATION PROBLEM

A. Does Cell Division or Termination Regulate Initiation of DNA Synthesis?

The division cycle is sometimes considered a circularly regulated system whereby each part of the cycle starts or initiates the next phase of the cycle. Consider cells growing with a 90-minute interdivision time as illustrated in Fig. 5–18d. Looking at this division cycle alone, it is possible to imagine that termination of DNA synthesis in the previous cycle, or the cell division leading to the birth of this cell, could start a series of events that lead to the initiation of DNA synthesis 30 minutes into the division cycle. Now consider that we slowly increase the growth rate, and keep the C and D values constant. When the cells reach a 60-minute interdivision time (Fig. 5–1), initiation occurs simultaneously with cell birth. Termination is still a possible point of regulation of DNA replication. As we speed up the growth rate (Fig. 5–1), the initiation occurs before cell division, as shown in the cells with a 50-minute interdivision time. This eliminates cell division as a possible regulator of initiation of DNA synthesis. With further increases in growth rate, initiation moves back to occur at termination in cells with a 40-minute interdivision time. With further increases in growth rate, initiation occurs before that termination, as in the cells with a 35-minute interdivision time (Fig. 5–1). This moving back now eliminates termination as a regulator or trigger of initiation of DNA replication. With even further increases in growth rate the time of initiation moves back past the previous cell division, eliminating even that one as the starting point for a series of events leading to initiation.

The conclusion is that the idea of a particular event in the division cycle (such as cell division or termination) triggering succeeding events is not consistent with the basic patterns of DNA replication. This reasoning led to the generalization that it is the continuous production of some substance between the starts of DNA replication that leads to initiation of DNA replication.

B. Is Initiation Regulated by Positive or Negative Controls?

It is difficult to overestimate the impact the proposal of negative regulation of β-galactosidase has had on thinking about regulation in general. It is now revealed with hindsight that the proposed generality of negative regulation of enzyme synthesis hindered the acceptance of alternative or positive modes of regulation.[120] The dominance of negative regulation has also had a strong impact on ideas regarding the regulation of DNA synthesis initiation.

The observed pattern of enzyme synthesis regulation in response to a given substance has no relationship to positive or negative systems of regulation. Adding lactose to a cell induces an enzyme, and adding tryptophan to a cell represses the synthesis of an enzyme, yet the mode of regulation in both cases is the same, negative regulation. Further, negative regulation is not inherently better than positive regulation.[121] When compared from the viewpoint of their physiological potential for regulation, both positive and negative regulation can give exactly the same kinetics and responses to inducers. A cell chooses a positive or a negative mode of regulation depending solely on the demand for the process. Savageau looked at a wide array of enzymes and observed that negative regulation occurs when the demand for an enzyme is low, and positive regulation occurs when the demand is high. The explanation for this strong correlation is that mutational pressure will exist to select for positive control in a high-demand situation and negative control in a low-demand situation. Demand, in this context, is loosely regarded as a measure of how often, as a fraction of time, the cell requires the product. If the cell has a high demand for the product, the regulation will be positive, and if it has a low demand, the regulation will be negative.[122]

Further, it should be noted that regulation can be defined as positive or negative only when complementation tests are done. A mutant in the regulatory system must be placed in the same cell with the parental type and the phenotype of the hybrid observed. This, of course, has not been done with respect to initiation of DNA synthesis.

The problem is exemplified by the following quote from Halvorson, Carter, and Tauro:[123]

> Let us now consider the Cooper and Helmstetter (1968) [positive] and the Pritchard, Barth and Collins (1969) [negative] models in light of the cell-size hypothesis. In the latter case, given a periodic synthesis of an inhibitor [as postulated by Pritchard, et al.], it is likely that the increase of cell volume would lead (by dilution below a threshold) to periodic DNA synthesis and division. The mathematical predictions of this model have yet to be critically evaluated. The model of Cooper and Helmstetter (1968) on the other hand, is dependent upon the amount (concentration?) of an initiator protein which is produced continuously during the cell cycle. When a critical amount (proportional to *size*) is reached, DNA synthesis commences and subsequently cell division occurs. What is not clear is how once a critical threshold per mass or per volume is achieved, initiation of DNA replication is now discontinuous. Either the hypothetical initiator protein is synthesized periodically or during initiation is consumed. In the latter case the assumption of continuous synthesis could be retained. The continuous synthesis of a stable initiator protein is inconsistent with the observed facts. Apart from the obvious necessity to invoke a discontinuous cellular element, the cause of bacterial cell division is unclear: the situation in eukaryotic organisms has not yet reached the speculative phase.

As is noted in this paragraph, Pritchard et al.[124] have proposed that there is the sudden synthesis of an inhibitor protein during the division cycle, and initiation occurs when cell volume increases so that the inhibitor or negative regulatory molecule is diluted to subinhibitory levels. As seen Chapter 4, there is no evidence for any sudden synthesis of any particular molecule during the division cycle; syntheses occur continuously during the division cycle. Thus, the postulation of periodic synthesis is not borne out by experiment.[125] Further, by focusing on the notion of concentration or amount per cell, Halvorson *et al.* do not consider the possibility, at least on a formal level, of having the ratio of amount of material per origin be the determining factor. By formally postulating this mechanism of the regulation of DNA synthesis, we can have the observed periodic initiation of DNA synthesis independent of a continuous synthesis of initiator. When the origins double, there is a halving of the amount of initiator per origin; this allows the cells to wait for one more doubling time before a new round of replication is initiated.

Both the negative and positive control models are equally satisfactory, but the number of postulations that do not fit the facts is greater for the negative control model.[126] In particular, no temperature-sensitive mutation has ever been found that, when the temperature is raised, leads to a continuous synthesis of DNA. If there were such a temperature-sensitive negative regulator, it would be expected that raising the temperature would allow continuous synthesis of DNA independent of the increase in cell mass.

C. Is Initiator Consumed at Initiation?

Although the idea that initiator is not necessarily consumed at each initiation event is implicit in the above analysis of initiation, the idea still persists that we must have a large decrease in initiator at the time of initiation. The reasoning is that if we did not have this decrease, then there would be continuous initiations when a certain amount of initiator was present. It should be clear from all of this discussion that by describing the mechanism as a titration system, whereby the amount of a certain substance (initiator) is measured against the amount of a different substance (origins), we can have instants of initiation without having the initiator decay, disappear, or inactivated.

XVI. EVENTS LEADING TO INITIATION

A. The Proposal of Bacterial Restriction Points

It was assumed in the discussion of regulation of initiation of DNA synthesis that there was no sequence of events leading up to initiation. Other types of models that propose such events have been proposed (see Chapter 4); let us look at one particular model of this type. Lark and Renger[127] proposed that there were two identifiable events leading to initiation of DNA synthesis. One event, occurring about 10 minutes before initiation of DNA replication, was sensitive to inhibition by 150 μg of chloramphenicol. Another event occurred about 25 minutes before initiation and was sensitive to 25 μg of chloramphenicol.[128] Following this proposal, there were other groups who reported two such points in different bacteria.[129] The effective chloramphenicol concentrations were different in different strains, but in all cases it was concluded that two points, differentiated in time and sensitivity to chloramphenicol, could be identified as occurring before initiation of DNA replication.

This proposal was tested by adding a wide range of chloramphenicol to cells and measuring the residual DNA synthesis. There was a continuous increase in residual DNA synthesis as the chloramphenicol concentration decreased.[130] The range of concentrations in which the DNA synthesis increased coincided with the concentrations that were less and less able to inhibit protein synthesis. It was concluded that the postulated two events were merely due to leakage of protein synthesis as the concentration of chloramphenicol was lowered. Cells that were within 10 minutes of initiation would leak through with 25 μg of chloramphenicol, but cells further from initiation would be able to initiate even with lower concentrations of chloramphenicol. The conclusion of these experiments was that if there were two such events, then there were essentially an infinite number.

To generalize the problem, all of the experiments demonstrating two chloramphenicol-sensitive events depended on what is commonly called an execution-point assay. If there is an execution point 10 minutes before some event actually takes place, then if an inhibitor is added eleven minutes before the event, the event will not occur; if the inhibitor is added 9 minutes before, the event will occur. The execution point is therefore the point at which the final prerequisites occur for an event. But let us say our inhibitor is not perfect. If this is the case, then the conclusion depends on how long we wait to examine whether the event occurred. If we wait long enough, the event will occur, because a small amount of leakage for a long time will lead to the accumulation of the requisite signal, allowing the event to occur. In the experiments of Lark and Renger, the initiation of DNA replication was measured after 90 minutes of incubation.

B. Criterion for Demonstrating Restriction Points

These experiments on chloramphenicol restriction points led to the proposal of a criterion for recognizing such events. If there were discrete events sensitive to particular amounts of an inhibitor, then the inhibition would not be sensitive, over some range, to the precise concentration of an inhibitor. For example, if an event were sensitive to 25 μg of chloramphenicol, we would not expect the event to be resistant to 24 μg of chloramphenicol. The criterion for such restriction points is that there must be a plateau of inhibition when a wide range of inhibitor concentrations is used.[131] If no plateau is found, then leakage could explain the degree of inhibition.

C. Accumulation of Initiator in the Absence of DNA Synthesis

When DNA synthesis is inhibited or decreased by any of a number of different approaches,[132] the accumulation of the ability to initiate DNA synthesis continues unabated. This has been shown for cells where DNA synthesis was slowed by growth in low concentrations of thymine,[133] growth in the absence of thymine,[134] growth in bromouracil instead of thymine,[135] or growth in nalidixic acid, an inhibitor of DNA synthesis. It is interesting to note that when spores are germinated in a medium lacking thymine, they achieve the ability to initiate DNA synthesis when they would normally initiate new rounds of DNA replication in complete medium.[136]

XVII. PLASMID REPLICATION DURING THE DIVISION CYCLE

A special case of the problem of regulation of DNA replication is the question of how small independent pieces of DNA, plasmids, replicate during the division cycle. Are these plasmids models of chromosome replication? Or do they have some other mode of replication that may illuminate the process of chromosomal replication?

A. *A Priori* Ideas on Plasmid Replication during the Division Cycle

Plasmids are pieces of DNA that replicate to provide all cells in a culture with the plasmid. There are two separate problems regarding plasmid maintenance. One is the specific mechanism of plasmid replication during the division cycle; the other is the mechanism by which the DNA is partitioned or segregated to the daughter cells so that each cell has at least one plasmid copy. In a situation with a large number of randomly replicating plasmids in a cell, if these plasmids were apportioned randomly into the progeny daughter cells, the absence of segregants could be explained by the large statistical improbability of the plasmids all appearing in only one of the daughter cells. Consider a bacterial strain where the newborn cell has an average of 30 copies of the plasmid per cell, and a dividing cell has an average of 60 copies per cell. No specific mechanism is required to assure that both daughter cells have at least one copy, because it is unlikely that all 60 copies will wind up on one side of the septum, leaving the other daughter cell without any plasmid DNA.

With low-copy-number plasmids—such as 1 to 2 per cell—the situation is more acute. If a newborn cell has one copy and the dividing cell two copies, then there is a strong possibility that by chance, if the plasmids are visualized as merely floating in the cytoplasm, both plasmids will be in one daughter cell, and the other daughter cell will have none. On a purely random basis, this would occur in half of the divisions, leading to a high degree of instability of the plasmid. Thus the replication and segregation problem is much more important for low-copy-number plasmids. Not only would we expect a more precise replication system, but we would also expect a precise segregation system.

To see the replication problem more clearly, consider that the plasmid in a cell was able to replicate with an equal probability in any part of the division cycle. The problem arises when we consider a plasmid in a cell at the very end of the division cycle. If this plasmid has a purely probabilistic replication-control mechanism, how does the plasmid ensure that it has replicated at least once before cell division? If the mechanism is purely probabilistic, as the cell approaches division, there is no greater tendency for the plasmid to replicate than at any other time during the division cycle. We would need an additional rule to make sure that plasmid-free cells were not formed. The rule would be that the cell would not divide until plasmid replication had taken place. This mechanism would suggest that a plasmid might affect the cell division parameters of the cell.

An alternative possibility (which has been demonstrated but which will not be discussed here) is that replication and segregation may not be precise, but the absence of segregants is due to the death of cells that have lost a plasmid. If cells that had a plasmid die if they lose the plasmid, this would give the appearance of stability even though there was a significant production of plasmid-free cells.

Considering the replication of plasmids with high copy numbers, it is possible for replication to occur throughout the division cycle with a continuous increase in plasmid number until division. If we liken a cell to a flask of nutrients for plasmid growth, then a lower starting concentration of plasmids would grow more rapidly than a higher concentration of plasmids. This would allow a steady-state concentration of plasmids per cell to be reached.

B. Plasmid Replication during the Division Cycle

1. F*lac*

The analysis of F*lac* replication during the division cycle has one of the most unusual histories in science. All manner of results have been obtained, all possible types of regulation appear to have been reported, and although it has been studied longer than any other plasmid, its mode of replication during the division cycle is still uncertain. The changes that have occurred can be seen most clearly by looking at the field in its chronological development.

The first work on F*lac* replication used filtration to produce synchronized cultures (see Chapter 3 for a discussion of this method). The cells were induced for β-galactosidase at different times.[137] A step in inducibility was observed, suggesting that there was a specific time when the F*lac* plasmid replicated. Another enzyme, serine dehydratase, used as a control, did not give steps. The enzyme-inducibility method was then repeated using cells synchronized by sucrose gradients, and it was also determined that there was a specific time when the plasmid replicated.[138]

It has been emphasized in Chapter 4 that there are no changes in the activities or rate of biosynthesis of enzymes during the division cycle. This idea of a doubling in the rate of synthesis of an enzyme appears to contradict this proposal. The main distinction to be made is that in continuous, steady-state growth, with no sudden changes in conditions, we have the exponential, invariant synthesis of enzymes during the division cycle (Chapter 4). In the experiments on F*lac* replication, the steady-state is disturbed when a short pulse of inducer is added to the growing cells. In this case there is a difference depending on the concentration of genetic material for that enzyme in the cell. To understand this difference, consider a cell growing in a steady state. When the cell is stimulated to produce an enzyme by the sudden addition of an inducer, it now appears that the DNA is limiting, as the cell attempts to make as much enzyme as possible in the presence of the inducer. A cell with twice as much genetic material for a particular enzyme will make twice the amount of enzyme during a pulse of the inducer. The capacity of a cell to synthesize an enzyme when suddenly induced does not affect the arguments for exponential cytoplasm growth made in Chapter 4.

Zeuthen and Pato[139] used the backwards membrane-elution method to measure the replication of F*lac* and its relationship to chromosome replication. They added an inducer of β-galactosidase and a radioactive label for DNA to a growing culture, then placed the cells on a membrane filter and eluted the newborn cells. They found that there was a doubling in the enzyme activity in the middle of the division cycle at all growth rates studied. They studied cultures growing with doubling times from 24 minutes to 110 minutes, so this result implied that there is no relationship of the time of plasmid replication to the time of initiation of chromosome replication.

This same method was repeated using a shift-up rather than different growth rates to vary the chromosome pattern during the division cycle. The logic of this method requires some discussion. When slow-growing cells are shifted up to a richer medium, there is a continuous change in the pattern of DNA replication during the division cycle. The pattern may be summarized by stating that at all times during the transition from slow growth to fast growth, the DNA patterns during the division cycle are the same as the DNA replication patterns that would exist at all growth rates between the two extreme growth rates. At any time during the transition, there exists a DNA replication pattern that is the same as any given DNA replication pattern at any particular growth rate. This means that we may select a particular DNA replication pattern by choosing a particular time during the shift-up rather than attempting to adjust the growth rate by changing the medium. Between cells growing with interdivision times of 40 and 25 minutes, in both of which the initiation of DNA replication occurs in the middle of the division cycle, there exists a doubling time with DNA replication initiating at cell division as in a cell with a 30-minute interdivision time (Fig. 5–1). The shift-up simplifies matters by allowing choice of particular times during the shift-up to perform the pulse-induction experiment. This is simpler than searching for the particular conditions that give a particular growth rate. When the shift-up was performed, it was found that there was a coincidence between the time of initiation and the time of plasmid replication.[140] This was seen as a moving back of the time of initiation of plasmid replication, so that at some time plasmid replication occurred early in the division cycle, then at the time of division, and then late in the division cycle. This is the same move-back as is observed in the initiation of chromosomal DNA replication (see Chapter 5). Thus, replication coincided with initiation rather than being consistently midcycle.[141]

Another approach to determining when the plasmid replicated during the division cycle was to measure the relative amount of the plasmid compared to the chromosome at different growth rates. Collins and Pritchard[142] used DNA-DNA hybridization to measure the amount of plasmid present at different growth rates. They considered the predictions made according to three different models: first, initiation of plasmid replication occurred at the initiation of chromosome replication; second, plasmid replication occurred at a fixed cell age, and third, plasmid replication occurred coincident with termination of chromosome replication. Their results fit the termination model best. The possibility that plasmid replication occurred at a fixed cell age could not be eliminated by the data.

There are a number of problems with this method. The main one is that accurate determinations of a very small amount of DNA must be performed. Even if these measurements are accurate, a number of assumptions must be made in order to properly calculate the time of plasmid replication. This method is clearly an integral method of analysis rather than a differential method. The results are the result of the integration of the DNA contents over the entire division cycle. This minimizes the accuracy of the method (Chapter 3).

More work using the membrane-elution method, again with β-

galactosidase inducibility as a marker of plasmid replication, led to the proposal that replication occurred at a constant cell mass.[143] This result was at least consistent with the same proposal of mass regulation for the normal chromosome. Further work using quantitative hybridization techniques[144] led to a new proposal, that plasmid replication was not coupled to any stage of the replication cycle, the host chromosome, or to division. The general conclusion was that the F*lac* plasmid replicated after initiation.

A new type of result, using the membrane-elution method, suggested that there was a particular time when replication occurred, but the time of replication was the same as the replication of the chromosomal *lac* gene.[145] This result held primarily for the more rapid growth rates, with a different pattern appearing at slow growth rates.

Recently it has been proposed that there is no special time during the division cycle when the F*lac* plasmid replicates. This proposal suggests that plasmids replicate randomly during the division cycle. One experiment used a zonal rotor to produce synchronized cells and measurements of enzyme inducibility during the division cycle; the conclusion was that F*lac* replicated randomly.[146] A control experiment indicated that the normal chromosome *lac* gene replicated at a particular time during the cycle. This work was supported by measurements of the randomness of plasmid replication without respect to the division cycle. Using density shifts, it was shown that the F*lac* replicates randomly.[147] Additional support for a precise time of replication came from work again on the backwards membrane-elution method where it was shown that the plasmid replicated at the same time as the *lac* gene.[148]

The latest result uses the membrane-elution method to determine when a number of plasmids replicate. Using autoradiography, it was concluded that the plasmid replicated randomly during the division cycle. A plasmid that was known to replicate randomly and one known to replicate nonrandomly were included in the same cell as a negative and a positive control. Did F*lac* replicate like the nonrandom plasmid or like the random plasmid? Quantitative analysis of the autoradiograms indicated that F*lac* replicated randomly. Helmstetter and Leonard found that the F*lac* was synthesized exponentially during the division cycle, and there was no particular time of replication during the cycle. In a cell containing both an F*lac* plasmid and a minichromosome (see below), there was a periodic replication of the minichromosome but a constant replication of the F*lac* plasmid.[149] Only a rapidly growing cell was used.[150]

This short history illustrates the difficulties in determining even the simplest pattern of plasmid replication during the division cycle. At the moment, the final answer is not in. I believe that the ultimate result may be that F*lac* replicates at a particular time during the division cycle. How this time is chosen, and how this time relates to the initiation of the chromosome, is yet to be determined.

2. Resistance Plasmids

Although plasmid R1, an antibiotic resistance plasmid, is a low-copy-number plasmid, it has been reported to replicate randomly during the division cycle.[151] This random replication was measured using density-shift experiments. After one generation time, there was some twice-replicated plasmid DNA, demonstrating that the plasmid replicated randomly during growth. This is a low-copy-number plasmid, and the result may be taken as experimental support that low-copy-number plasmids, in spite of arguments to the contrary, replicate randomly during the division cycle. A recent result by Bezanson[152] suggesting that the plasmid replicates nonrandomly during the division cycle has cast doubt upon this conclusion.

3. Minichromosomes

Minichromosomes are pieces of DNA that carry the normal origin of replication. This normal origin of replication enables the DNA to replicate in the same manner (i.e., using the same biochemical elements as determined in a cell-free system) as the chromosome. The question arises as to when during the division cycle the minichromosome replicates. Using the membrane-elution method, it was found that minichromosomes replicate in a cell-cycle–specific manner.[153] They do not replicate at random, even though there is a very high number of minichromosomes per cell. More important, they replicate at the same time as the normal chromosome initiates replication (Fig. 5–19).[154]

4. Colicinogenic Plasmids

As a control for other experiments, the colicinogenic plasmid was also studied as a function of its replication during the division cycle. Replication appeared random.[155] This plasmid is a high-copy-number plasmid, so this result fits the logical analysis given above.

5. P1 Prophage

In the prophage state, phage P1 exists as an independent circle of DNA. It is present in a low copy number and is stable. Analysis of the pattern of P1 replication during the division cycle indicated that there was a unique time during the division cycle when replication took place.[156]

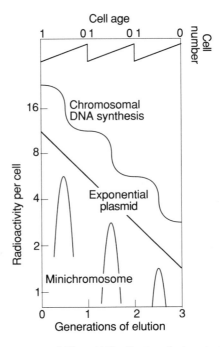

Figure 5.19   Minichromosome and Plasmid Replication during the Division Cycle Using the Membrane-Elution Method. When cells labeled with thymidine are analyzed for plasmid replication by the membrane-elution method, minichromosomes are replicated at discrete times during the division cycle, while *Flac* plasmids appear to be replicated continuously and exponentially during the division cycle. The minichromosome curves are the rates of minichromosome replication during the division cycle. Since replication occurs only during a short part of the division cycle, oscillations occur in the rate of DNA synthesis in the minichromosome band on agar gels.

6. Summary of Plasmid Replication during the Division Cycle

Considering the amount of work that has gone into understanding plasmid replication and plasmid biology, hardly anything is known about how plasmids replicate during the division cycle. This is the key to understanding plasmid replication. Most studies of the regulation of plasmids consider the plasmids isolated chemicals in an amorphous medium consisting of repressors and initiators of replication.[157] The replication of the plasmid is controlled by the interaction of various soluble elements. But none of these studies has considered how this replication is related to the division cycle. For those plasmids with a continuous replication pattern during the division cycle, the chemical model is sufficient. For those with a cycle-specific pattern of replication, all of the chemical modeling must be related to producing a cell-cycle–specific replication pattern.

XVIII. THE INITIATION AND TERMINATION OF DNA SYNTHESIS

This discussion of DNA synthesis during the division cycle of bacteria has not dealt with the precise details of DNA replication, to mention nothing of the details of initiation of replication and termination of replication. It was assumed that such events can occur, and that the cell has the requisite material to initiate and terminate DNA replication. These topics have been reviewed and analyzed, and here mention will be made only of some relevant results and some key references.

A. Initiation of DNA Replication

The initiation of DNA replication has been well reviewed.[158] The origin of replication was identified by a number of different groups,[159] using the general strategy of cloning the origin on various vectors and screening for one that had an autonomous replication pattern. Assiduous work led to the eventual sequencing of the origin, now called *oriC*.[160]

The process of initiation of DNA replication at the chromosomal origin is complex. A number of elements interact to start replication; chief among them is the origin itself. An important protein for initiation of DNA replication is the *dnaA* protein. This protein is specified by the *dnaA* gene and has been proposed by many to regulate the frequency of initiation. It is required for initiation of DNA replication *in vivo* and *in vitro* from the biological origin of replication, *oriC*.[161] It was originally identified as a mutation that affected the initiation process.[162] More interesting, defects in this protein can lead to overinitiation.[163] Support comes from evidence that its synthesis is autoregulated.[164]

B. Replication of DNA

The replication of DNA, once initiated, has been well reviewed; the books by Kornberg[165] give most of the history and details of DNA replication. Other reviews on DNA replication have added to the body of information.[166]

C. Termination of DNA Replication

The termination reaction of DNA is less well understood than the initiation reactions, but some progress has been made in recent years.[167] Replication slows down in specific regions of the chromosome, and these places are related to the act of termination.[168] There are also specific protein factors that interact with these terminator sequences.[169]

XIX. DNA REPLICATION DURING THE DIVISION CYCLE OF BACTERIA

The enormous amount of effort expended on DNA replication during the division cycle has given us a very clear view of how DNA replicates. The pattern is simple, can be understood in molecular terms, and leads to an understanding of how the cell makes just enough DNA to give two genomes to each of its daughter cells. The utility of our understanding of DNA replication in the gram-negative, rod-shaped bacteria does not end with these bacteria. There are many applications, and it is hoped that in future years many more applications will be discovered for this wealth of information.

NOTES

1. Helmstetter, 1967; Helmstetter and Cooper, 1968; Cooper and Helmstetter, 1968.
2. The reader is reminded that there are actually two replication points going to six replication points as replication is bidirectional. This precise distinction is unncessary in considering the relative rate of DNA synthesis during the division cycle.
3. This decrease in the rate of synthesis does not mean DNA synthesis actually decreases with growth rate. At division, considering the rate of DNA synthesis from the mother to the two daughter cells, the rate of DNA synthesis actually increases by a factor of three.
4. Koch, 1980.
5. Lark and Maaløe, 1954, 1956.
6. Howard and Pelc, 1951a,b.
7. Schaechter, Bentzon, and Maaløe, 1959. This experiment indicated that over 80% of the cells in the culture were synthesizing DNA.
8. Pachler, Koch, and Schaechter, 1965; see also McFall and Stent (1959) who used radioactive phosphate incorporation to demonstrate that over 70% of the cells in a culture were synthesizing DNA at any time.
9. Abbo and Pardee, 1960.
10. This problem does not apply to the determination of the exponential rate of cytoplasm synthesis (see Chapter 4), as the membrane-elution method has been shown to be adequate to resolve different modes of biosynthesis during the division cycle.
11. Van Tubergen and Setlow, 1961.

12. Forro and Wertheimer, 1960; Forro, 1965.

13. The replication of DNA is bidirectional, so the finding of only two or four segregating units in microcolonies means that there is a physical connection between the strands replicating in opposite directions. This must mean that there is a physical continuity, unbroken during subsequent rounds of replication, between the clockwise and counterclockwise strands across the origin.

14. Bonhoeffer and Gierer, 1963.

15. Meselson and Stahl, 1958.

16. Lark and Lark, 1965; Lark, 1966.

17. Koch and Pachler, 1967, unpublished.

18. Clark and Maaløe, 1967.

19. Helmstetter, 1967.

20. Helmstetter and Cooper, 1968; Cooper and Helmstetter, 1968.

21. Pierucci, 1972.

22. Schaechter, Maaløe, and Kjeldgaard, 1958. See also Shehata and Marr (1975) for a confirmation that cell size does not change with temperature.

23. Cooper and Helmstetter, 1968.

24. Cooper and Helmstetter, 1968.

25. Cooper and Helmstetter, 1968.

26. Cairns, 1963b; Condemine and Smith, 1990.

27. Cooper and Helmstetter, 1968.

28. Cooper and Ruettinger, 1973.

29. Skarstad, Steen, and Boye, 1985.

30. Bird, Louarn, Martuscelli, and Caro, 1972.

31. The case for exponential synthesis of mass during the division cycle was presented in Chapter 4. The analysis presented here is independent of exponential growth. All that is required is that the pattern of mass synthesis be the same at all growth rates. For example, if mass synthesis were linear at all growth rates, the arguments would be the same. For a constant pattern of size increase during the division cycle, the size of the cell at any particular time is a valid predictor of the size of the cell at birth or division.

32. Donachie, 1968.

33. Donachie, 1968; Pritchard, Barth, and Collins, 1969.

34. It is important to realize that I is not the period from cell division to initiation. The period between birth and initiation is called the B period. The I period is equal to the interdivision time of a culture growing in steady state. The beginning of the $I+C+D$ sequence may begin one or more generations before the division cycle during which initiation occurs, or the division cycle that ends with the division.

35. This point is not often appreciated, and it has been proposed that after mass synthesis resumes, there is a synchronous initiation of DNA replication. An example of this is present in a recent review (Smith and Condemine, 1990).

36. Throughout this discussion of DNA synthesis and cell division, it has been tacitly assumed that the termination of DNA synthesis may be visualized as *causing* or *initiating* or *producing* a cell division. It must be emphasized that there is no evidence that termination is the proximal cause of cell division, and in fact

there are suggestions that termination is unrelated to the initiation of cell division. The discussion is made easier, however, by analyzing the patterns of DNA synthesis, mass synthesis, and cell division as though there were some causal relationship between DNA synthesis and cell division.

37. Maaløe and Hanawalt, 1961; Hanawalt, Maaløe, Cummings, and Schaechter, 1961.

38. Churchward and Bremer, 1977; Churchward, Bremer, and Young, 1982; Churchward, Estiva, and Bremer, 1981; Woldringh, de Jong, van den Berg, and Koppes, 1977; Zaritsky and Pritchard, 1971.

39. Pritchard and Zaritsky, 1970; Spratt and Rowbury, 1971.

40. Pritchard and Lark, 1964.

41. Pierucci, 1969.

42. Kjeldgaard, Maaløe, and Schaechter, 1958.

43. Cooper, 1969.

44. An extremely illuminating analysis of the shift-up and the growth of the cell in balanced growth before and after the shift-up is given by Bleecken (1969a, 1969d, 1971). This work was published soon after the patterns of DNA synthesis were published, and it is one of the earliest papers on this subject to come from the German Democratic Republic.

45. A complete formulation of this concept is presented in Chapter 15. Although the formal presentation there is given in terms of eukaryotic cells, it applies directly to prokaryotes as well.

46. Margolis and Cooper, 1971.

47. Cooper, 1969.

48. Kjeldgaard, Maaløe, and Schaechter, 1958.

49. Cooper, 1969.

50. Cooper, 1969.

51. Robin, Joseleau-Petit, and D'Ari, 1990; Kepes and Kepes, 1985; Sloan and Urban, 1976; Kepes and D'Ari, 1987.

52. Cooper, 1969.

53. Cooper, 1969.

54. An interesting and speculative view of the shift-down has been presented by Bleecken (1969b).

55. Bremer and Churchward, 1977a,b.

56. If this were not taken into account, there would have been a larger increment in the residual DNA synthesis, suggesting the conclusion that the value of C in the exponentially growing culture was greater than the correct value. This is because the value of C, as a function of τ, gives a greater increase in DNA, as the value of C is larger relative to the interdivision or doubling time of a culture.

57. Zaritsky, 1975b.

58. Zaritsky and Pritchard, 1971.

59. Beacham, Beacham, Zaritsky, and Pritchard, 1971.

60. Cairns, 1963a.

61. Cairns, 1963b.

62. Chandler, Bird, and Caro, 1975; Lane and Denhardt, 1975; Chandler, Funderburgh, and Caro, 1975.

63. Churchward and Bremer, 1977.

64. Bremer and Churchward, 1977a.

65. Gudas and Pardee, 1974.

66. Louarn, Funderburgh, and Bird, 1974.

67. Hanks and Masters, 1987.

68. Manor, Deutscher, and Littauer, 1971.

69. Pritchard and Zaritsky, 1970; Zaritsky, 1975b.

70. Lane and Denhardt, 1974, 1975.

71. Pato, 1975.

72. Cooper and Helmstetter, 1968.

73. Kjeldgaard, Maaløe, and Schaechter, 1958.

74. Woldringh, 1976.

75. Brock, 1988. This article is a superb example of the writing of the history of modern science. It shows the development of the idea of the bacterial nucleoid and how it changed as different techniques were employed to study and observe the nucleoid.

76. Helmstetter and Pierucci, 1968.

77. Cooper, 1969.

78. Kubitschek, 1974.

79. Kubitschek, 1974.

80. Marunouchi and Messer, 1973.

81. Lark, 1966; Helmstetter, 1967; Kubitschek, Bendigkeit, and Loken, 1967; Clark, 1968; Bird and Lark, 1968.

82. Helmstetter, 1967.

83. The increase in DNA synthesis in the oldest cells is seen, experimentally, as a decrease in the amount of thymidine label during a membrane-elution experiment. This is because the cells are eluted from the membrane in reverse order to their age at the time of labeling.

84. Helmstetter and Pierucci, 1976.

85. Kubitschek and Freedman, 1971.

86. Kubitschek and Freedman, 1971.

87. Painter, 1974.

88. Chandler, Bird, and Caro, 1975.

89. Helmstetter, 1967.

90. Gudas and Pardee, 1974.

91. Koppes, Woldringh, and Nanninga, 1978.

92. Kubitschek and Newman, 1978.

93. Kubitschek, 1970b.

94. Pato and Glaser, 1968.

95. Helmstetter, 1968.

96. Ward and Glaser, 1969c.

97. Cerda-Olmeda, Hanawalt, and Guerola, 1968.

98. Wolf, Newman, and Glaser, 1968.

99. Pritchard and Lark, 1964.

100. Louarn, Funderburgh, and Bird, 1974.

101. Bird, Louarn, Martuscelli, and Caro, 1972.

102. Zaritsky and Pritchard, 1971.

103. Maaløe and Hanawalt, 1961; Lark, Repko, and Hoffman, 1963.

104. Masters and Broda, 1971.

105. Nishioka and Eisenstark, 1970; Fujisawa and Eisenstark, 1973.

106. Prescott and Kuempel, 1972; Rodriguez, Dalbey, and Davern, 1973. See also Rodriguez and Davern (1976), who showed that bidirectional replication was not altered by growing cells at very slow growth rates. This rules out the possibility that slower growth rates are obtained by changing the topology of replication from directional to unidirectional.

107. Cairns, 1963a,b.

108. McKenna and Masters, 1972.

109. Marsh and Worcel, 1977.

110. Kaguni, Fuller, and Kornberg, 1982.

111. Meijer and Messer, 1980.

112. Snellings and Vermeulen, 1982.

113. Jacob and Wollman, 1961.

114. Hayes, 1957; Lederberg, Cavalli, and Lederberg, 1952; Jacob and Wollman, 1961.

115. Meselson and Stahl, 1958.

116. Nagata, 1962, 1963.

117. Yoshikawa and Sueoka, 1963a,b.

118. Lark, Repko, and Hoffman, 1963.

119. Bird, Chandler and Caro, 1976.

120. Beckwith, 1987.

121. Savageau, 1989.

122. The argument regarding the choice between positive and negative regulation is as follows. If a process is used by a cell almost all the time, and a negative control is removed or inactivated almost all the time, then mutations inactivating that control system will not be strongly selected against. For most of the lifetime of the cell there will be no distinction between the mutant and the wild-type, parental cell. There will be no strong selective pressure to remove cells that have lost the control system. The same argument can be made for positive control of a rarely used system. If the positive control system is lost, then there will be little selection against the mutated cell until that rare time when the system is required. However, with negative controls for functions with low demand, and positive controls for functions with high demand, mutants with inactivated control systems will be selected against almost immediately. In the case of the loss of negative control, the cell will make an unneeded enzyme and be less efficient than the parental cell most of the time. In the case of positive control, the loss of the positive control would mean that the cell will not be able to make a molecule required most all the time. This explanation follows demonstrations by Savageau (1989) that there is no functional, physiological difference between what is possible by positive or negative control. The reason a cell chooses positive

or negative control to regulate the synthesis of a particular enzyme depends upon the demand for that particular function.

123. Halvorson, Carter, and Tauro, 1971.

124. Pritchard, Barth, and Collins, 1969.

125. A specific negative regulatory model was proposed by Rosenberg, Cavalieri, and Ungers, 1969.

126. For an incisive analysis of one negative control model, see Margalit and Grover, 1987.

127. Lark and Renger, 1969.

128. It should be explicitly noted that the later point (the high chloramphenicol-sensitive point) is proposed to be insensitive to the low concentration of chloramphenicol, but the converse is not true, and the low concentration sensitive point is sensitive to high concentrations as well.

129. Ward and Glaser, 1969a, 1970; Messer, 1972.

130. Cooper and Wuesthoff, 1971. Lark (1973) criticized these experiments by proposing that DNA replication does not proceed to the end of the chromosome in some concentrations of chloramphenicol.

131. Cooper, 1974.

132. Maaløe and Rasmussen, 1963.

133. Pritchard and Lark, 1964.

134. Yoshikawa and Haas, 1968; Pierucci, 1969.

135. Boyle, Goss, and Cook, 1967; Hewitt and Billen, 1965; Doudney, 1965.

136. Yoshikawa and Haas, 1968.

137. Nishi and Horiuchi, 1966.

138. Donachie and Masters, 1966.

139. Zeuthen and Pato, 1971.

140. Cooper, 1972.

141. Cooper, 1972.

142. Collins and Pritchard, 1973.

143. Davis and Helmstetter, 1973.

144. Pritchard, Chandler, and Collins, 1975.

145. Finklestein and Helmstetter, 1977.

146. Andresdottir and Masters, 1978.

147. Gustafsson, Nordstrom, and Perram, 1978.

148. Steinberg and Helmstetter, 1981.

149. Leonard and Helmstetter, 1988.

150. Keasling has observed a periodic, cycle-specific replication of Flac plasmid during the division cycle. This was seen at all growth rates, but the result was clearest at slower growth rates.

151. Gustafsson and Nordstrom, 1975; Gustafsson, Nordstrom, and Perram, 1978; Koppes and Nordstrom, 1985.

152. Bezanson, 1980.

153. Leonard and Helmstetter, 1986.

154. Helmstetter and Leonard, 1987a.

155. Leonard and Helmstetter, 1988.

156. Prentki, Chandler, and Caro, 1977.

157. For example, see Womble and Rownd, 1986a,b, 1987.

158. Kolter and Helinski, 1979; Messer, 1987; Bramhill and Kornberg, 1988.

159. Hirota, Yasuda, Yamada, Nishimura, Sugimoto, Sugisaki, Oka, and Takanami, 1979; Messer, Bergmans, Meijer, Womack, Hansen, and von Meyenburg, 1978; Yasuda and Hirota, 1977; von Meyenburg, Hansen, Riise, Bergmans, Meijer, and Messer, 1979.

160. Meijer, Beck, Hansen, Bergmans, Messer, von Meyenburg, and Schaller, 1979; Sugimoto, Oka, Sugisaki, Takanami, Nishimura, Yasuda, and Hirota, 1979. The minimal origin was determined by deletion analysis to be a sequence of 245 base pairs by Oka, Sugimoto, Takanami, and Hirota, 1980.

161. Atlung, Rasmussen, Clausen, and Hansen, 1985; Carl, 1970; Chakraborty, Yoshinaga, Lother, and Messer, 1982; Fuller, Funnell, and Kornberg, 1984; Hansen and Rasmussen, 1977; Fuller and Kornberg, 1983; von Meyenburg, Hansen, Atlung, Boe, Clausen, van Deurs, Hansen, Jorgensen, Jorgensen, Koppes, Michelsen, Nielsen, Pedersen, Rasmussen, Riise, and Skovgaard, 1985; Løbner-Olesen, Skarstad, Hansen, von Meyenburg, and Boye, 1989; Skarstad, Løbner-Olesen, Atlung, von Meyenburg, and Boye, 1989.

162. Carl, 1970; Frey, Chandler, and Caro, 1981; Hirota, Mordoh, and Jacob, 1970; Hirota, Ryter, and Jacob, 1968; Schaus, O'Day, Peters, and Wright, 1981; von Meyenburg, Hansen, Riise, Bergmans, Meijer, and Messer, 1979.

163. Frey, Chandler, and Caro, 1984; Kellenberger-Gujer, Podhajska, and Caro, 1978.

164. Atlung, Clausen, and Hansen, 1985; Braun, O'Day, and Wright, 1985.

165. Kornberg, 1974, 1980, 1982, 1983, 1984, 1988.

166. McMacken, Silver, and Georgopoulos, 1987; Marians, 1984; Wickner, 1978.

167. Kuempel, Pelletier, and Hill, 1989.

168. De Massey, Bejar, Louarn, Louarn, and Bouche, 1987; Hill, Henson, and Kuempel, 1987.

169. Kobayashi, Hidaka, and Horiuchi, 1989.

6

Synthesis of the Cell Surface during the Division Cycle

The cell-cycle pattern of surface synthesis is slightly more complicated than DNA or cytoplasm synthesis. This is because of the particular rod shape of the cells that have been the center of our discussion. As the biochemistry of the bacterial cell surface is not so well known as that of other cell components, a review of the structure and biochemistry of wall synthesis relevant to the problem of synthesis during the division cycle will first be presented.

I. THE STRUCTURE OF THE CELL SURFACE OF GRAM-NEGATIVE BACTERIA

The gram-negative bacterial cell[1] is covered with a three-layered coat consisting of an inner membrane adjacent to the cell cytoplasm, a peptidoglycan or murein[2] layer encompassing the inner membrane, and an outer membrane layer enclosing the cell (Fig. 6–1). The surface is usually described as composed of a cylindrical side wall capped by two hemispherical poles. During the early part of the division cycle, when cells are not invaginating, cells grow in the cylindrical wall area. In the later part of the division cycle, cells invaginate and produce two new poles in the middle of the parental cell.

The problem to be addressed in this chapter is when, where, how, and how much cell surface is made during the division cycle. What is the rate of synthesis of each of the wall components during the division cycle? Where is the wall inserted? And how are these rates and locations determined? Further, how are these rates of biosyntheses related to the observed regularity of division during the cell cycle? Most of the work on the biosynthesis of the cell surface is specifically related to peptidoglycan synthesis, as peptidoglycan appears to be the most stable portion of the cell surface. Peptidoglycan may also be the structure that is involved with the determination of cell shape.

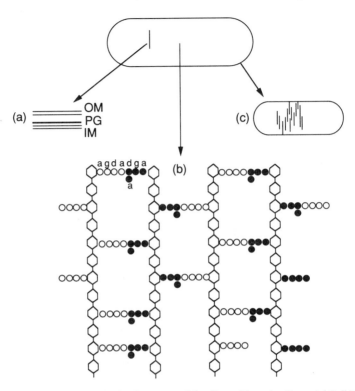

Figure 6.1 Cellular and Molecular Structure of the Gram-Negative Bacterial Cell Wall. The rod-shaped, gram-negative cell wall is composed (a) of a peptidoglycan (PG) layer sand-wiched between an outer membrane (OM) and an inner membrane (IM). At a molecular level (b), the peptidoglycan consists of chains of sugar residues cross-linked by amino acid chains. The amino acids are labeled, indicating L-alanine (a), D-glutamic acid (g), diaminopimelic acid (d), and D-alamine (a). The sugar chains are arranged perpendicular to the long axis of the cell as indicated in (c). There is no information available on the orientation of strands in the spherical polar regions. As the cross-links are drawn, the open circles are the donors to the closed circles.

A. Peptidoglycan Structure

1. Composition and Cross-Linking of the Peptidoglycan Subunits

The peptidoglycan cross-linked network is made up of chains of subunits composed of two sugars and four amino acids.[3] The sugars are arranged in chains, and the amino acids are engaged in the cross-linking between glycan chains. After insertion of subunits, the linear strand is composed of alternating N-acetylglucosamine and N-acetylmuramic acid subunits.[4] The tetrapeptide side chain is attached to the muramic acid and may be either free or involved in cross-linking between chains.

The amino acids of the side chain are L-alanine, D-glutamic acid, *meso*-diaminopimelic acid and D-alanine. If a cross-link is formed between strands, there is a peptide bond between the carboxyl group of the D-alanine in one chain and the ε-amino group of the diaminopimelic acid in an adjacent strand. By a combination of cross-linking and strand extension, a large network is produced that completely encloses the cell. All of the peptidoglycan material is covalently attached to other peptidoglycan material, so the peptidoglycan sacculus may be visualized as a single macromolecule (Fig. 6–1). The problem is to retain this macromolecular structure, while allowing the cell to grow as new material is inserted into the peptidoglycan.

Because the peptidoglycan is a single macromolecule, it does well at its main function, holding the cell together against the large internal turgor pressures found within bacterial cells. (The internal pressure has been estimated to be 75 to 90 lbs/in^2; this is approximately the pressure found inside a high-pressure bicycle tire.) Without the presence of the strong peptidoglycan layer, the cell would burst under the internal pressure. This conclusion is supported by the observation that enzymatic digestion of the peptidoglycan leads to cell lysis unless the cells are suspended in a hypertonic medium (high concentrations of sucrose or glycerol). Of course, the fact that the peptidoglycan is a single covalently linked network leads also to the central problem of wall growth: how does the cell wall grow without weakening the essential structural features of the cell wall?

2. Amount of Peptidoglycan on a Cell

What is the arrangement of glycan strands on the surface of a bacterial cell? One pertinent question relates to the thickness of the peptidoglycan. How many layers of the peptidoglycan are there on a cell? Determining the amount of peptidoglycan per cell can reveal how this amount of peptidoglycan covers the cell. Early work using chemical determinations of the amount of diaminopimelic acid per cell[5] gave a value of 2.7×10^6 diaminopimelic acid residues per cell. The surface of the cells analyzed had an average surface area of 30 to 40 μm^2 per sacculus. This gives a calculated surface area per diaminopimelic acid residue—the diaminopimelic acid moiety indicating the *unit tetrapeptide* of the peptidoglycan—of 12 nm^2 per subunit. In actuality, the subunit cannot cover the area (see below).

More refined chemical measurements of the diaminopimelic acid content of cells using amino acid analysis revealed that there were 3.1×10^6 diaminopimelic acid residues per cell for cells growing slowly in minimal medium and 5.6×10^6 residues per cell for cells growing rapidly in rich medium. The surface area of rapidly growing cells is larger than that of slower growing cells (Chapter 1), and these measurements give a slightly more reasonable amount of diaminopimelic acid per cell-surface area. More refined measurements of the diaminopimelic acid per cell, combined with surface measurements on the same cells,[6] gave a value of 4 to 5×10^6 subunits per cell, with a value of 7 nm^2 per diaminopimelic acid residue. Interpretation of these values requires consideration of the dimensions of a pentapeptide subunit. Model-building and x-ray–scattering data reveal that the length of a repeating disaccharide subunit is 0.98 nm, and the average separation of glycan chains is about 1.9 nm. Other values give 2.5 nm between chains. With the above data, the area per disaccharide unit is 2 to 2.5 nm^2. With 4×10^6 units per cell, this is enough peptidoglycan to cover about 9 to 12 μm^2. The average surface area of these smaller cells is 7 μm^2; this is too much for a monolayer, too little for a trilayer, but fits a bilayer.

Given the uncertainty about the amount of peptidoglycan on the bacterial cell surface, we can only attempt to fit the data to a reasonable model of surface structure. One possible way of looking at surface structure is presented in Fig. 6–2. In this three-dimensional representation of the cell there is a single load-bearing monolayer of peptidoglycan with additional, non-load-bearing layers beneath. The strands below the surface are only loosely connected by cross-links between the glycan strands. This view of the load-bearing structure of the cell incorporates the idea of *make-before-break*. The taut bonds of the peptidoglycan are eventually going to be severed, allowing new material to be inserted. By having the new material in place before the bond cutting, new surface growth can take place without relinquishing the structural integrity of the peptidoglycan layer. As shown in Fig. 6–3, there may be more than one layer of strands below the load-bearing layer. When the cross-links in the load-bearing layer are cut, a large increase in cell surface can be produced without any break in the structural continuity of the cell surface.

This view of the cell surface allows a variable amount of cell peptidoglycan depending on how many strands are located below the load-bearing layer. It is not necessary, as depicted in Figs. 6–2 and 6–3, to have every interstrand region filled with non-load-bearing strands. It may be that only a fraction of the load-bearing strands have these sublayer strands in place. Depending on growth conditions, for example, the density of peptidoglycan per unit of surface area could vary, with only the restriction that there must be at least the minimal load-bearing layer of peptidoglycan. If some but not all interstrand regions have loosely connected strands below, the cell must have a mechanism for ascertaining that a particular strand has a strand in place before the load-

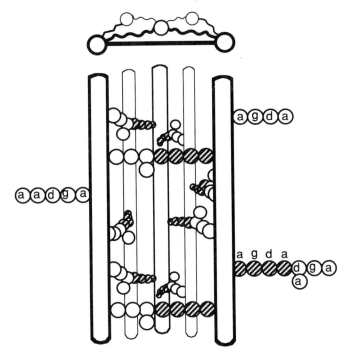

Figure 6.2   Three-Dimensional Representation of Peptidoglycan Structure. This is an idealized representation of peptidoglycan structure as seen from the outer surface. The thick
bars represent chains of sugars at the outside of the peptidoglycan layer. The thinner bars
represent chains below the outer layer. The stretched chains of circles represent amino
acids cross-linking the glycan chains. The chains below the stretched surface of the cell rise
to the outer layer when the taut layers of the peptidoglycan are hydrolyzed. Above is a
cross-sectional view through the glycan chains illustrating the taut outer layer and the more
loosely inserted inner material.

bearing cross-links are severed. If a cell had such a protective mechanism, no further severing of cross-links would occur in the rightmost
diagram of Fig. 6–3 until new non-load-bearing strands and cross-links
were inserted between the load-bearing glycan strands.

3. Heterogeneity of Peptidoglycan Structure

This simplified picture of peptidoglycan structure has been made more
complicated but more interesting by the finding of a large variety of
fragments in peptidoglycan digests. By the use of high-performance
chromatography, it is possible to observe up to 80 different fragments
from the cell-wall peptidoglycan.[7] In addition to the monomers and

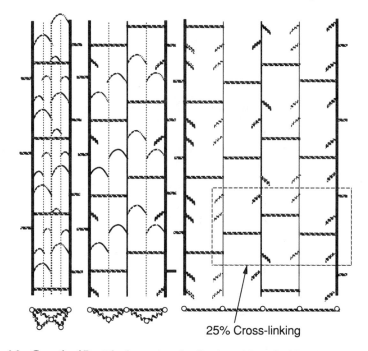

25% Cross-linking

Figure 6.3   Growth of Peptidoglycan Area by Cutting of Stretched Bonds. A taut peptidoglycan layer is shown at the left with two layers of inserted peptidoglycan below. As the
stretched bands are cut, the diagrams to the right illustrate the growth in surface area.
When the fully stretched peptidoglycan is produced, the free amino acid chains can be
acceptors for further new chain insertion. A unit area of peptidoglycan illustrating the
degree of cross-linking is shown in the dotted box. One-quarter of the possible bonds are
involved in cross-linking, thus the degree of cross-linking is 25%. This calculation is made
by noting that each side chain contains the material for one complete cross-link, a donor and
an acceptor. There are a total of 16 side chains within the box, and there are four complete
cross-links, so 25% of the possible cross-links are made.

dimers[8] containing tetrapeptides, a large number of fragments are
formed by permutations involving different numbers of amino acids
(tripeptides, tetrapeptides, pentapeptides), different cross-links (DAP-
DAP in addition to the D-ala-DAP link), and different sugar structures,
depending on whether the sugars were located internal to a glycan chain
or at the end of a glycan chain. The covalent attachment of lipoprotein to
a significant fraction of the peptidoglycan produces additional variability. The question arises as to whether any of these minor fragments have
any function within the cell. It could be imagined that these minor
components of the peptidoglycan are regulatory signals that allow the
cell to start constriction at a particular time and at a particular place.

B. Arrangement of Peptidoglycan Strands on the Cell Surface

Electron-microscopic analysis of partially digested cell walls,[9] and the results of controlled sonication studies,[10] suggest that the strands of the cylindrical side wall are arranged primarily in a hoop direction; i.e., the strands encircle the cylindrical side wall as hoops encircle a barrel. When the walls were observed after fragmentation, the strands tended to go preferentially in one direction perpendicular to the long axis of the cell. Although there may not be a perfect alignment, as has been argued by Koch,[11] a hoop arrangement is the theoretically best way to imagine placing peptidoglycan strands to allow lengthwise extension of the growing cell (Fig. 6-1). New strands are inserted between pre-existing glycan chains, and the cell grows primarily in the lengthwise direction between divisions. If the glycan chains were placed in the axial direction—i.e., in the lengthwise direction of the cell, with the cross-links in the hoop direction—then insertion of new strands between pre-existing glycan chains would lead to an increase in cell circumference. It is clear, from even the earliest observations of cells in the light microscope,[12] that rod-shaped cells grow primarily lengthwise. This is strong *a priori* support for the weaker experimental evidence that the strands are arranged primarily as hoops around the cell circumference.

Biochemical analysis indicates that glycan strands are relatively short and cannot extend around the entire circumference of the cell.[13] This means that the peptidoglycan in the hoop orientation is actually made up of short strands that collectively go completely around the cell. The absence of long-range order in the arrangement of peptidoglycan strands around the circumference of the cell cylinder should not obscure the conclusion that the insertion of new glycan strands between resident glycan strands leads to the growth of the cell in the axial or lengthwise direction.

II. BIOCHEMISTRY OF PEPTIDOGLYCAN SYNTHESIS

A. Synthesis of the Pentapeptide Precursor

Although the subunit of the peptidoglycan was described as a tetrapeptide, the building block for peptidoglycan is actually a pentapeptide. The biosynthesis of the pentapeptide precursor in the cytoplasm has been described.[14] The final product of a series of biosynthetic steps is transferred to a membrane-bound lipid carrier that transports the subunit out of the cytoplasm for insertion into the extant peptidoglycan. There is no evidence that the synthesis or pool of this precursor varies during the division cycle. It is assumed that the pentapeptide precursor is made in an amount that reflects the rate of synthesis of cell wall at any time during the division cycle.

B. Transfer of the Pentapeptide to the Growing Chain

There are a number of transglycosylases that appear to be involved in the transfer of the pentapeptide to the peptidoglycan.[15] Chains are extended by the transfer of the disaccharide-pentapeptide to the end of a chain. No information is available as to how new chains are started. New chains presumably start in some random manner, and extend in a progressive manner until chain extension stops randomly. Stopping occurs relatively frequently, at least compared to the progressive synthesis of protein, DNA, or RNA, therefore producing chains that are short, between 6 and 50 subunits long.

C. Cross-Linking of Peptidoglycan Chains

After, or perhaps simultaneous with, glycan chain extension, there is a cross-linking of the newly inserted chain to an adjacent resident chain. This occurs by the formation of a peptide link between the *donor*[16] penta-peptide and an ε-amino group of an adjacent *acceptor* diaminopimelic acid residue. The energy for the cross-linking appears to be located in the peptide bond between the fourth and fifth amino acids of the donor chain. At the time of cross-linking, the D-ala-D-ala bond in the donor pentapeptide is broken, and the donor chain is converted to a tetra-peptide.

In the cell there are carboxypeptidases that remove the final D-alanine from the pentapeptide, whether or not cross-linking has occurred.[17] Shortly after a pentapeptide is inserted into the surface layer, the penta-peptide is either cross-linked or is shortened by carboxypeptidases to a tetrapeptide subunit. In either case, the ability to serve as a donor in cross-linking is essentially lost.[18] The tetrapeptide, however, can still serve as an acceptor in cross-linking when new strands are inserted adjacent to the tetrapeptide.

III. THE RELATIONSHIP OF MASS SYNTHESIS TO PEPTIDOGLYCAN SYNTHESIS

The important point which must now be considered is the regulation of cell-surface synthesis. It appears that the cell does not make too much or too little peptidoglycan. Morphological studies reveal a uniform peptido-glycan layer with no buckling due to an excess, and the cell integrity is maintained, indicating that there is no deficiency in peptidoglycan. More important, if there were no breakage of the pre-existing cross-links, the surface would not grow in area but would merely get thicker. Of course, the cutting of the cross-links must be strictly matched to the insertion of new peptidoglycan and the growth of the cell.

As a starting point in analyzing the relationship of peptidoglycan synthesis to the synthetic processes occurring in the rest of the cell, consider that Fig. 6–2 is a schematic diagram of the murein of a cell. The stretched linkers are at the outside of the cell. Before the stretched cross-links are broken, new cross-links are in place, connecting the glycan strands destined to separate. Koch[19] proposed that the stretching of peptidoglycan leads to bends in the bond angles in the cross-links, lowering the energy of activation of the cutting reaction. The energy of activation for the hydrolysis of a peptide bond is approximately 10–20 kcal/mol. If a stress is applied to this bond—by stretching it, for example—the calculated decrease in the energy of activation is approximately 4 kcal/mol. This leads to an increase in the rate of hydrolysis by a factor of about 4×10^{10}, comparing stressed to unstressed peptidoglycan.[20] Now consider that over a short time there is a small increase in the mass of the cytoplasm leading to a small increase in the turgor pressure over the entire cell surface. The pressure increase occurs because over this short time, the mass increases by taking in nutrients, synthesizing macromolecules, and performing other activities, while the cell surface does not grow and the cell does not increase in volume. This increase in turgor pressure leads to an increased stress on the load-bearing bonds all over the surface. As the bonds stretch, there is a steady lowering of the energy of activation for cutting the stretched cross-links. Somewhere in the cell, the energy of activation is low enough to allow an enzyme to hydrolyze an existing load-bearing cross-link. Cutting of load-bearing bonds leads to a separation of the strands previously held together by the cross-link. We shall assume that the cut does not remain localized but continues down the strands in a zipper-like fashion. This allows the insertion of the strand that was below the surface into the load-bearing layer. The zipper-like movement of the hydrolytic activity may come about because a single cut leads to a large increase in the stress on the adjacent bonds; thus the cutting action would proceed processively between two adjacent glycan strands. As the strands separate, there is a slight increase in the total volume of the cell (Fig. 6–3). This increase in volume relieves the stress throughout the cell surface. But as the mass increases continuously (see Chapter 4), this respite for the cell wall does not last. Again there is a round of mass increase, internal pressure increase, bond stretching, cutting, strand separation, and finally cell volume increase. As the cutting of the bonds is distributed randomly over the surface, there is a diffuse intercalation of new strands between the old strands. The volume of the cell just accommodates the volume of the mass through this mass increase and surface growth. If mass synthesis is inhibited, there is a cessation of cell-surface growth.

A simple way to consider the cell is to imagine a balloon made of a special material. As more air is put into the balloon, the volume and the surface area increase, with new material entering the balloon's surface. The new material enters the balloon at such a rate as to keep the internal pressure of the cell constant. The volume of the balloon increases to accommodate the increased air placed in the balloon. Similarly, in the case of the cell, the stress on the cell surface is constant, the internal pressure is constant, and as the volume expands to just enclose the cell mass, the cell density is constant. This view of the growth of the cell surface has been termed the surface stress model by Koch.[21]

The load-bearing layer of peptidoglycan is a monolayer of material surrounding the cell. When the cell grows, there must be a cleavage of this load-bearing layer. If the layer were merely cut, with no previous preparation, then the cell would lyse and the cell cytoplasm would burst out of the cell. But the cell presumably has a rule that makes sure the cell does not cut a pre-existing load-bearing bond before there is a bond in place. As an analogy, consider a high pressure bicycle tire. The pressure inside this rubber tube is of the same order of magnitude as that inside a bacterial cell. One way to make the tire longer is to cut the tube perpendicular to the circular axis, enlarge the tube, place a new piece of rubber between the exposed ends, and seal the new edges to the older rubber. It should be immediately obvious that as soon as the tire is cut the air would leak out, and the tire would be deflated. An alternative is to go inside the inflated tire, seal in place a patch of approximately the size we wish the tire to lengthen, then go outside the tire and cut the outer rubber. The tire will grow, the patch will expand, and the air will be retained in the tire. This is because the patch was inserted before cutting the tire. The surface-stress model postulates that there must always be an intact cell wall with new strands in place before cutting any part of the cell wall under stress. This is what is meant by *make-before-break*; strands to be inserted are in place before any cutting of load-bearing bonds. This view of the internal pressure of the cell—approximately 5 to 7 atmospheres—means that the internal pressure is not only a problematic force that the cell must contend with, but is actually a positive force that allows the cell to grow. It is the turgor pressure, producing the stress on the cell surface, that leads to the continuous increase in cell surface and cell volume during the division cycle.

With this view of the regulation of peptidoglycan synthesis, we can now look at the particular pattern of cell-surface synthesis in rod-shaped, gram-negative bacteria. But before describing this pattern, it is important to see the historical context from which our understanding has emerged.

IV. EARLY STUDIES ON THE PATTERN OF CELL-SURFACE SYNTHESIS DURING THE DIVISION CYCLE

The history of the analysis of cell-surface synthesis during the division cycle is different from the analysis of DNA synthesis during the division cycle. The pattern of DNA synthesis was discovered relatively early (in 1967 through 1968), and the large amount of work on DNA synthesis that followed was placed within a single framework (described fully in Chapter 5). In contrast, for many years there was no consensus on the synthesis of cell surface during the division cycle. There were many different views, many different interpretations, many different experimental results, and there was no common thread to the analysis. Only within the last few years has there emerged a solid foundation for understanding cell-wall synthesis during the division cycle. Some of the early work on cell-surface synthesis is obsolete. Nevertheless, it is important to review these early studies if only to see what was correct and what must be reinterpreted.

A. Early Studies on the Location of Cell Surface Synthesis

The first two decades of the study of bacterial cell-wall growth were dominated by the idea of growth zones. A growth zone is a localized area of cell-surface growth. Regulating the number of these zones could help explain the rate of surface synthesis. There are three sources for the idea of zones. One is the early recognition that in *Streptococcus*, the cell wall is a rigid structure that grows at one edge (see Chapter 12). Old cell wall is not metabolized, and new material is not inserted within the old wall material. This zonal growth pattern served as a model for zonal growth in gram-negative bacteria. The second source, and one that may be more important, is the analogy to DNA synthesis. In Chapter 5 it was concluded that DNA replication is regulated by the insertion, at appropriate times, of new replication points at the origins of DNA. This idea, applied directly to cell-wall synthesis, suggests that there are a particular number of growth zones, and at some time or times during the cycle, new growth zones would be inserted. At the time of insertion or activation of these zones, there would be a change in the rate of cell-wall synthesis. At the simplest level, if a cell had one zone, and another zone were activated, then there would be a sudden doubling in the rate of surface synthesis. The third element leading to the idea of zonal growth was the proposal of the replicon model to explain DNA segregation. In the absence of a visible mitotic apparatus, it was proposed that the regular segregation of DNA at division could be explained by the binding of DNA strands to the cell surface with cell-wall growth taking place between the bound DNA strands.[22] The wall growth between the surface-bound DNA strands could lead to their separation and sequestration in the two new daughter cells. It appeared that zonal growth (Fig. 6–4), particularly in the center of the cell, was an important requirement for DNA segregation.

Zonal growth of the cell wall may be contrasted with diffuse growth, where the entire surface of the cell is available as sites of new cell-wall growth (Fig. 6–4). Different types of zonal growth have been postulated, from the proposal of a conserved unit cell, to the growth of one or a few zones over the cell surface.

1. Experimental Support for Zones of Wall Synthesis

When cells were labeled for a short time with diaminopimelic acid and analyzed for the location of grains by autoradiography using the electron

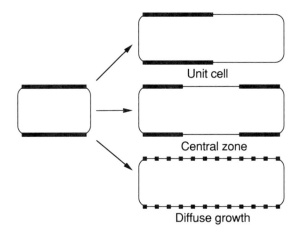

Unit cell

Central zone

Diffuse growth

Figure 6.4 Zonal and Diffuse Cell-Wall Synthesis. A rod-shaped cell at the left grows by elongation of the cylindrical side wall. If extension takes place at only one side of the cell, the unit-cell mode of growth occurs; a unit cell remains conserved while one cell is completely new. If there is a zonal growth in the center, then new material appears in the center, and there is a conservation of pre-existing material as a unit within each of the daughter cells. With diffuse growth, there is a continuous dispersion of the pre-existing material with new material appearing between older material. Different degrees of diffuse growth can be considered, from intercalation of new strands between all pre-existing peptidoglycan strands, to a less complete intercalation model with the preservation of zones of peptidoglycan that are relatively resistant to the insertion of new strands.

microscope, cells of all sizes had a preferential location of zones of incorporation in a relatively narrow band at the center of the cell.[23] This result was interpreted as indicating a preferential zone of growth in the center of the cell. New zones would appear in the new daughter cells at some later time. Immunofluorescent analysis also supported the insertion of discrete zones of synthesis in the side wall of rod-shaped cells.[24] When pulse-chase experiments were performed, the evidence for zones was ambiguous,[25] and it was concluded that there was a randomization of the material in the initial zone. At the same time that autoradiography was suggesting a zonal growth mechanism, kinetic measurements of cell-wall synthesis during the division cycle indicated that in the middle of the cycle, there was a sudden doubling in the rate of peptidoglycan synthesis.[26] This result was compatible with the idea of zonal growth.

The main controversy to emerge from these studies was not whether there were zones, but when zones were activated. One model proposed that growth zones were produced at particular times during the cell cycle and the zones grow at rates proportional to the growth rate.[27] Pierucci[28] proposed that new growth zones, with a finite life span, were activated at the initiation of new rounds of chromosome replication. Pritchard[29] proposed that cell-wall synthesis was determined by an unregulated gene located near the terminus of the chromosome. A doubling in the rate of surface synthesis was predicted to occur when this gene replicated; the rate of envelope synthesis at each of these zones was assumed to be constant. Another model was proposed by Donachie, Begg, and Vicente[30] wherein cells abruptly increased their rate of elongation at a critical length; this length was proposed to be twice the minimal cell length.

2. The Unit-Cell Model

The unit-cell model has been proposed as a particular zonal growth pattern.[31] As illustrated in Fig. 6–4, the unit-cell model proposes that the cells grow only from one pole, producing one daughter cell with completely new peptidoglycan in the side wall and one daughter with completely old peptidoglycan in the side wall. The experiments demonstrating the unit cell were observations in the microscope of cells growing in only one direction. The preferential attachment of one cell pole to the substrate and the free movement of the other pole could not be eliminated; this preferential attachment could produce the appearance of growth in one direction. This proposal was supported by additional data[32] indicating that phage attachment sites are inserted asymmetrically on the cell surface. Verwer and Nanninga[33] decisively eliminated the unit-cell model (at least for peptidoglycan) by analyzing the distribution of radioactive diaminopimelic acid on each of the two cell halves of dividing cells. Statistical considerations alone suggest that one of the two halves would show more grains than the other, so they determined whether the lower number of grains in one half of the cell was attributable solely to statistical variation or whether there was a bias to fit an asymmetrical pattern of peptidoglycan synthesis. Their results showed that synthesis of cell wall in the two cell halves was the same; this result eliminated the unit-cell model.

3. Evidence against Zonal Growth

The earliest experiments on cell-wall growth using fluorescent antibodies to label the cell surface indicated that the wall grew diffusely with no conserved areas.[34] Additional evidence against zonal growth was based on autoradiographic evidence that new cell-wall material can be inserted over the entire surface of the cell.[35] Similar results were obtained for the matrix protein attached to the peptidoglycan.[36] More refined autoradiographic evidence was then presented, which showed that there were no apparent zones of synthesis before invagination.[37] Biochemical support came from studies of the acceptor-donor radioactivity ratio, a technique that measures the pattern of strand insertion into the cell wall. These studies indicated that there were no conserved areas of the cell wall, and new material was inserted between any two strands.[38]

B. Early Studies on Rate of Peptidoglycan Synthesis during the Division Cycle

Measurements of the rate of peptidoglycan synthesis during the division cycle have supported the zonal growth model. For example, the rate of diaminopimelic acid incorporation increased shortly before the end of the division cycle.[39] Other studies indicated that the rate of peptidoglycan synthesis accelerated toward the end of the division cycle.[40]

A different approach to the problem was to take advantage of the measurement of cell growth during the division cycle. The cell surface is relatively rigid, so we can measure the lengths and widths of cells during the division cycle and derive the pattern of cell-surface growth. When this was done for synchronized cells, two different results were obtained for two different strains.[41] Later analyses suggested that the original data were compatible with exponential length growth during the division cycle.[42] Observations of single cells suggested that growth was also close to exponential, or at least was continuous.[43]

Another approach to the growth problem is the static Collins–Richmond method (Chapter 3) for measuring the pattern of cell growth during the division cycle. Length is the easiest parameter to measure, so most of the work has centered on the determination of length as a function of cell age. This method, as described in Chapter 14, is not so useful as originally believed. For example, one set of analyses, on very good data, was unable to distinguish among a number of different models of cell growth during the division cycle.[44]

C. Summary of Early Work on the Rate of Cell-Surface Synthesis

Two fundamental ideas emerge from the early work on surface growth during the division cycle. The most important idea on the regulation of cell-surface growth was that triggers and mechanisms regulated the synthesis of peptidoglycan independent of other synthetic processes occurring in the cell.[45] The second idea to emerge was that the current methodology could not solve the cell-surface growth problem. For every question there were at least two answers. Conflicting data abounded. Results were interpreted without considering the growth of the entire cell.

V. RATE AND TOPOGRAPHY OF PEPTIDOGLYCAN SYNTHESIS DURING THE DIVISION CYCLE

A pattern for cell-wall synthesis that considers cell-wall synthesis within the context of total cell growth can accommodate and explain almost all of the data on cell-surface growth. This proposed pattern is interesting because it is derived from *a priori* considerations as well as experimental results.

According to the surface-stress model, the relationship of mass synthesis to the growth of cell surface implies that cell surface is made to perfectly enclose, without excess or deficit, the cytoplasm synthesized by the cell. Cell cytoplasm increases continuously during the division cycle (Chapter 4), and therefore cell surface is made continuously. We now come to the question of the actual rate of surface synthesis during the division cycle. What is the precise pattern of synthesis for cell wall, or peptidoglycan in particular, during the division cycle?

Consider an imaginary cell in which the cytoplasm is enclosed in a tube that is open at each end. Assume that the cytoplasm remains within the bounds of the tube. The cytoplasm in the newborn cell is encased in the cylinder of cell surface made up of membrane and peptidoglycan. As the cytoplasm increases exponentially, tube length increases to exactly enclose the newly synthesized cytoplasm. Cell surface increases exponentially in the same manner as the cytoplasm. For each amount of cytoplasm, the cell length is directly proportional to the amount of cytoplasm present, and thus cell surface in this particular (and imaginary) cell increases exponentially. When the cell cytoplasm doubles, the tube divides into two new cells, and the cycle repeats. In this imaginary cell, the cytoplasm increases exponentially, the internal volume of the cell increases exponentially, and the surface area increases exponentially. The density of the cell, i.e., the total weight per cell volume, is constant during the division cycle. Further, because both the surface and the cytoplasm increase exponentially, the ratio of the rate of cytoplasm synthesis to the rate of cell-surface synthesis is constant throughout the division cycle. But a real rod-shaped cell does have ends, and therefore the pattern of cell-surface synthesis during the division cycle is not exponential. If the cell surface were synthesized exponentially, the cell volume could not increase exponentially and there would have to be a change in cell density. Cell density, however, is constant during the division cycle (see Chapter 7).[46]

A proposal for cell-surface synthesis that allows an exponential increase in cell volume, and therefore a constant cell density, is presented in Fig. 6–5. Before invagination the cell grows only in the cylindrical side wall. After invagination, the cell grows in the pole area and the side wall. Any volume increase required by cell-cytoplasm increase that is not accommodated by pole growth, is accommodated by an increase in side wall. The cell is considered a pressure vessel,[47] and when the pressure in the cell increases, there is a corresponding increase in cell-surface area. The pole is preferentially synthesized by mechanisms that are not yet known. Because there is a continuously changing volume in the growing poles, there is a varying amount of growth in the cylindrical wall to just accommodate the exponentially increasing cell mass.[48]

The resulting pattern of synthesis is approximately exponential (Fig. 6–5b). The formula describing surface synthesis during the division cycle is complex, including terms for the shape of the newborn cell, the cell age at which invagination starts, the pattern of pole synthesis, and the age of the cell. It is simpler to understand the pattern of surface synthesis by considering and measuring the ratio of the rate of surface synthesis to the rate of cytoplasm synthesis. Before invagination, cell-surface growth occurs only by cylindrical extension. As the radius of an individual cell is constant during the division cycle, the rate of surface synthesis before invagination is directly proportional to the rate of cytoplasm synthesis. That this is so can be seen by imagining all synthesis in the side wall occurring in a narrow coin-shaped disc as indicated by the shaded areas on the cells between ages 0.0 and 0.5. Each of these coin-shaped volumes has an edge and a volume. The volume of two coins is twice that of one coin, and the edge area of two coins is also twice that of one coin. The thickness of the shaded area is a measure of the rate of both area and

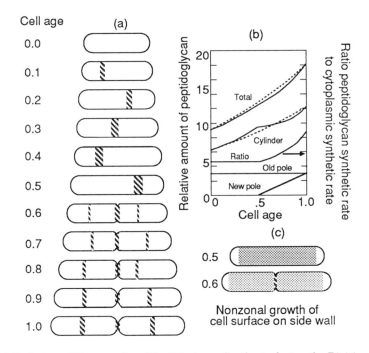

Figure 6.5 Rate and Topography of Peptidoglycan Synthesis during the Division Cycle. The newborn cell at the left (a) is drawn with a cylinder length of 2.0 and a radius of 0.5. Before invagination the cell grows only by cylinder extension. Each cell in (a) is drawn to scale, with the volume of the cells increasing exponentially during the division cycle. The shaded regions of the cell indicate the amount and location of wall growth (whether in pole or side wall), each tenth of a division cycle. The width of the shaded area is drawn to scale. Cell-surface growth actually occurs throughout the side wall (c) and not in a narrow contiguous zone. Before invagination the ratio of the rate-of-surface-increase to the rate-of-volume-increase is constant. When pole synthesis starts, at age 0.5 in this example, there is an increase in the ratio of the rate-of-surface-increase to the rate-of-volume-increase. Any volume not accommodated by pole growth is accommodated by cylinder growth. At the start of pole growth there is a reduction in the rate of surface growth in the cylinder. As the pole continues to grow, there is a decrease in the volume accommodated by the pole, and an increase in the rate of growth in the side wall. This is schematically illustrated by the thinner sector in the expanding side wall immediately after the start of constriction. As the new pole increases, in increments of equal area between the indicated ages, the volume accommodated by the new poles is continuously decreasing. Therefore, the growth rate in the cylindrical portion increases continuously during the constriction period. At the end of the division cycle, the rate of synthesis in the cylinder is the same as the rate for a newborn cell. There is no sharp change in the rate of cylinder elongation at the instant of division. At the upper right (b) is a plot of the expected pattern of accumulation of peptidoglycan or cell surface during the division cycle. The total accumulation of peptidoglycan is the sum of the individual accumulations of new pole, old pole, and cylindrical side wall. The dotted line represents the expected pattern for exponential synthesis. The ratio of the rate-of-surface to the rate-of-cytoplasm synthesis is indicated. At the lower right (c) an explicit illustration of the dispersive, nonzonal growth of the side-wall is illustrated for cells of ages 0.5 and 0.6. There is a decrease in the density of incorporation of new cell wall material after invagination starts, as indicated by the side-wall shading.

volume increase in the growing cell, so the ratio of the rate of surface synthesis to the rate of volume or mass increase is constant before invagination. After invagination starts, there is an increase in the rate of surface synthesis relative to the rate of cytoplasm synthesis. This is due to a higher ratio of surface to volume in a sphere compared to a cylinder. During pole formation, the relative increase of surface is greater than the increase in volume. As drawn in Fig. 6–5a, the surface of the pole increases with equal areas between each cell age. The volume accommodated by the pole decreases as the growing pole edge goes out from the cylinder, so the ratio of surface to volume must increase.

When invagination starts, the growth of the pole accommodates some of the volume increase required by the increase in cell mass or cytoplasm. Whatever mass is not accommodated by pole growth is accommodated by side-wall growth. This model predicts that when invagination starts there will be a decrease in the rate of side-wall synthesis. By having this pressure relief system, the volume of the cell increases perfectly exponentially to accommodate the exponential increase in cell mass (see Chapter 4). This view of cell-surface synthesis leads to the powerful *ratio-of-rates method*, in which the ratio of the rate of surface synthesis to the rate of cytoplasm synthesis is measured, rather than the rate of surface synthesis alone. This ratio is predicted to be constant during the first part of the division cycle, when there is only side-wall growth, and to increase during invagination (Fig. 6–5b). This pattern of cell-surface synthesis has been observed in *Salmonella typhimurium*[49] and *Escherichia coli*.[50] One consequence of this model of surface synthesis is that at no time is surface synthesis exponential, because the rate of surface synthesis is not proportional to the amount of surface present over any period during the division cycle.

A. Quantitative Analysis of Wall Growth during the Division Cycle

What is the formula that describes the pattern of surface synthesis during the division cycle? Historically, much of the effort on synthesis of the bacterial cell surface has been spent on the search for a particular formula—linear, bilinear, or exponential—and performing experiments that would prove one or another formula correct. Some of these models were of interest because they predicted that there would be a variation in density during the division cycle; these variations could be clues to the control of cell-cycle events such as the initiation of DNA synthesis. A qualitative description for cell-wall synthesis was presented above with a central assumption that the cell density during the division cycle is constant. Let us look at this model in a more rigorous and quantitative way.

A general formula for the amount of surface synthesis present on a cell can be derived for the proposal illustrated in Fig. 6–5. The assumptions made for the calculation are that (1) the rate of mass and volume increase is exponential during the division cycle; (2) cell density is constant during the division cycle (which means that cell volume increases exponentially); (3) constriction starts at a particular age during the division cycle; (4) the cell can be approximated by a cylinder capped with two hemispheres; (5) the cell grows with a constant width (diameter) during the division cycle; (6) the new pole grows at a constant rate of area increase after the start of constriction (although this particular assumption can be relaxed to give a more general equation); and (7) any volume increase in the cell not accommodated by the increase in new pole volume is accommodated by an increase in the cylindrical wall of the cell.

Given these assumptions, the surface area at any time during the division cycle is given by[51]

$$A_{\alpha_{tot}} = \frac{8}{3}(2^{\alpha})\pi r^2 + 2(2^{\alpha})\pi r L_{0_{cy}} + \frac{4}{3}\pi r^2 + \frac{4}{3}\pi \frac{h_{\alpha}^3}{r}$$

In this formula r is the radius of the cell, L the length of the cylindrical portion of the cell, and h the height of the growing pole measured from the end of the cylindrical portion of the cell. Inspection of the equation of surface growth during the division cycle clearly indicates that the formula is not exponential. The precise pattern is very close to exponential (Fig. 6–5b), and by direct, standard methods (synchronization, single labels, etc.) this pattern would be very difficult to distinguish from exponential.

As analyzed in Chapter 4, the mass at any time during the division cycle is given by

$$M_{\alpha_{tot}} = 2^{\alpha}M_0$$

If we now look at the rate of surface and mass synthesis during the division cycle, i.e., first take the differential of each of these formulae describing the amount of surface and mass present at any time during the division cycle, and then the ratio, we get the ratio of the rate of surface to the rate of mass synthesis during the division cycle:[52]

$$\lambda\left(\frac{dA}{dM}\right)_{\alpha} = 1 + \kappa\left(\frac{h_{\alpha}^2}{2^{\alpha}}\right)$$

Before invagination, when h, the height of the new pole, is zero, the ratio of the rate of surface to the rate of mass synthesis is constant. After invagination this ratio increases, as the second term is a positive value.

The exact pattern of increase depends on the relationship of the height of the new pole at particular ages during the division cycle.

A general formula, independent of the pattern of pole synthesis or cell density, has been derived by Keasling[53] and is given by

$$\frac{dA}{dM} = \left(\frac{1}{\rho} + \frac{1}{\gamma} \cdot \frac{h(\alpha)^2}{2^\alpha} \cdot \frac{dh(\alpha)}{d\alpha}\right) \cdot \frac{1}{\beta}$$

β is the ratio of the cylinder surface area to volume, ρ is the density of the cytoplasm, γ is the adjusted initial cell mass, and $h(\alpha)$ is the height of the new pole as a function of the cell age. When pole synthesis is linear, the rate of pole synthesis is constant [i.e., $dh(\alpha)/d\alpha = K$], and the general formula reduces to the formula given above.

The precise rate of surface synthesis and its measurement is less important than the unanticipated conclusion, presented here, that there is no description of the rate of peptidoglycan synthesis in terms of a simple mathematical formulation. Peptidoglycan synthesis is neither exponential nor linear, but is a complex pattern that is easy to describe. Before the start of constriction, when cylindrical extension is the only means of cell growth, the rate of peptidoglycan synthesis *appears* exponential because the differential rate of wall increase is similar to the differential rate of mass and volume increase. But it is incorrect to say that synthesis is exponential. Exponential synthesis means that the rate of synthesis is exponential and that the total amount of material increases exponentially. At no time during the division cycle is this the case for cell surface.

Woldringh and his colleagues have analyzed the pattern of cell-surface growth during the division cycle. They noted that when the cell grows before invagination there is an increase in the amount of mass per cell surface compared to that present at the start of the division cycle. Therefore, during invagination there must be a preferential accumulation of cell surface compared to mass in order to have the correct ratio of surface to mass at division.[54] This concept, which stimulated the proposal of the jump-invagination model[55] is rigorously explained by the equations presented here. If we take the ratio of the amount of surface to the amount of mass at any time during the division cycle, rather than the ratio of the rates of synthesis, there is an initial decrease in the ratio followed by an increase in the ratio in the latter portion of the division cycle. Thus, the proposed pattern of wall synthesis accounts for the ratio of the surface to mass during the division cycle, as well as the ratio of the rates of synthesis of surface and mass during the division cycle.

B. Experimental Support for the Pressure Model of Cell-Surface Synthesis

Given the constant density of the cell, and the regulation of surface synthesis by mass synthesis, the predicted pattern is *a priori* correct. In addition, there is a set of experiments that strongly supports the proposal.

1. Ratio of Peptidoglycan Synthesis to Protein Synthesis during the Division Cycle Using Membrane Elution

A simple proof of the proposed pattern of wall and pole synthesis could come from measurements of the rate of peptidoglycan synthesis during the division cycle. The problem, however, is that the predicted rate is so similar to exponential that it is difficult to perform the requisite measurements and demonstrate the precise nonexponential pattern. The approach that succeeded analyzed the prediction that the ratio of the rates of mass and surface synthesis would vary in a predictable way during the division cycle. The *ratio-of-rates* method uses two radioactive labels, one for cell surface (diaminopimelic acid) and one for cytoplasm (e.g., leucine), and determines the ratio of the rates of synthesis during the division cycle. The experiment is simple. Exponentially growing cells are pulse-labeled with both diaminopimelic acid and leucine. The labeled cells are placed on the membrane-elution apparatus, the newborn cells are eluted, and the radioactivity in the eluted cells is determined for the two labels. The ratio of the rates expected according to the proposed model is shown in Fig. 6–6. A decrease in the ratio is expected at the beginning of elution, followed by a plateau in the ratio as the cells that were not constricting (the youngest cells at the time of labeling) now give newborn cells to the eluate. A second peak in the ratio follows in the next generation of elution. In the third and subsequent generations of elution, there are no further peaks. The reason that there are only two peaks is illustrated in Fig. 6–7. The excess surface synthesized, relative to cell mass, is present only at the newly forming internal poles, so this elevated ratio occurs only in the first and second generations of elution. In later generations, only material labeled in the side wall is eluted, and therefore the ratio remains constant.

The experimental results are fully consistent with the predictions.[56] Two peaks separated by the requisite plateaus were found. The data, however, are not precise enough to prove that pole growth is linear. The important point regarding the ratio-of-rates method is that extremely slight deviations from exponentiality are experimentally detectable.

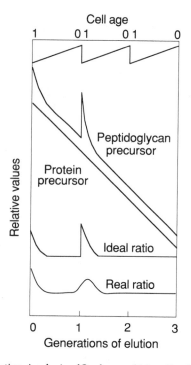

Figure 6.6 Membrane-Elution Analysis of Surface and Mass Synthesis during the Division Cycle. Cells labeled with precursors of peptidoglycan and protein are analyzed by the membrane-elution method. The cells coming off the membrane represent the reverse order of cell ages, so the increase in the ratio of peptidoglycan synthesis to protein synthesis appears as a decrease. In the second generation of elution, there is another peak in the ratio. In practice, the peaks in the peptidoglycan graph alone are difficult to see, but the ratio analysis illustrates them clearly.

2. Cell Density

A number of models of cell growth have used variations in cell density to regulate cell division.[57] In contrast to those theoretical proposals, many experiments indicate that the density of *Escherichia coli* (and by extension to a homologous organism, the density of *Salmonella typhimurium*) is invariant during the division cycle.[58] The relationship of the constant density to the pattern of wall synthesis during the division cycle can be looked at in two complementary ways. The proposed pattern of the synthesis of cell surface and its relationship to cell volume and mass means that the cell density is invariant during the division cycle. Thus, the stress model proposed here is consistent with constant cell density. Alternatively, for the assumptions made in the quantitative analysis, any

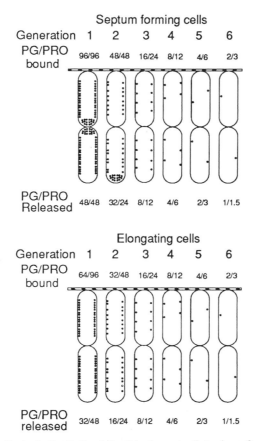

Figure 6.7 Why Peaks in the Ratio of Peptidoglycan-to-Cytoplasm Synthesis Occur Only Twice. A cell in the process of forming poles and labeled with similar amounts of peptidoglycan and protein label (taken, for example, as 96 units in each case), is allowed to be eluted from a membrane-elution apparatus. The leucine eluted in subsequent generations is halved at each generation (48, 24, 12, 6, 3, 1.5, etc.). The behavior of the peptidoglycan is different from protein, as illustrated by the dots on the periphery of each cell. There are 96 dots on the first cell undergoing constriction. The poles are assumed to have 32 dots, and the walls have 64 dots, with the dots proportional to the radioactivity present. At the first division, 48 dots (or units of peptidoglycan) are released. At the next division, there is an unequal partition of the peptidoglycan, as 32 units of peptidoglycan are released, and 16 units remain in the membrane bound cell. In the third generation, only lateral wall material is present, so only eight dots are released in the third, four in the fourth, two in the fifth, and one in the sixth generation of elution. As the leucine has an exponential pattern (see Chapter 4), the ratio of radioactivity in peptidoglycan to that in protein differs in each generation. There is an increased ratio at the start of the first and second generations of elution (when the cells that were invaginating are presenting newborn cells to the eluate), and a constant ratio afterward. The constant ratio in the third and later generations is an indication that there is both random segregation of the side-wall material in each generation, and that there is no measurable turnover and release of side-wall material to the medium.

model that is not equivalent to the proposal made here must predict a variation in cell density during the division cycle. Given a constant density, and the rod-shaped organism, it is necessary that the organism grow as described in the model in Fig. 6–5.[59]

3. Segregation of Peptidoglycan

Van Tubergen and Setlow[60] were the first to look at the segregation of peptidoglycan. They noted that there was a uniform segregation pattern indicating a large number of randomly segregated subunits. This was supported by the theoretical arguments of Koch[61] and by other experiments demonstrating a random dispersion of peptidoglycan.[62]

The membrane-elution apparatus gives a very simple approach to the segregation problem. The results from long-term elution of diaminopimelic acid and leucine from the membrane-elution apparatus indicate that cell-wall material is subdivided randomly for at least six generations. The results are consistent with a halving of the lateral wall at each division and a random distribution of material at cell division. Any unit-cell or zonal-growth model is not compatible with the experimental results.

4. Rate of Cell-Length Growth

From measurements of the cell-length distribution of exponentially growing cells it was concluded that cell growth is exponential.[63] The length of the cell, according to the model proposed here, increases almost exponentially; the length of the cell increases in the same manner as the total surface area (Fig. 6–5b). Given the uncertainties in length measurements of cells, and in particular, measurements of living cells growing synchronously, measurements of the pattern of length or surface increase during the division cycle are consistent with the proposal made here.

5. Location of Newly Synthesized Peptidoglycan

Woldringh, Huls, Pas, Brakenhoff, and Nanninga[64] noted that the incorporation of diaminopimelic acid into the lateral walls of cells with constrictions was significantly less than incorporation into the lateral walls of cells without constrictions. This unexpected observation is predicted by model proposed here (Fig. 6–5a,b). At the start of invagination, lateral-wall synthesis decreases. Woldringh *et al.* suggest that this increase in synthesis at the pole occurs ". . . at the expense of the activity [i.e., synthetic or growth activity] in the lateral wall." What is the mechanism of this redistribution of synthesis between the pole and the side wall? If there were a limited supply of enzymatic machinery or wall precursors available for peptidoglycan synthesis, when pole synthesis started there would be a reduction, by competition, in the rate of lateral-wall synthesis. An alternative view, presented in the model proposed here (Fig. 6–5a,b), suggests that there is no limitation in the amount of precursor or the enzymatic machinery for cell-wall synthesis. When pole synthesis starts, the increase in cell volume by pole growth relieves the stress in the cylinder area. Because of this reduction in stress, the rate of insertion of peptidoglycan in the lateral wall is reduced. Within the terms of the surface-stress model, no proposal is made for any specific mechanism for changing the rate of wall synthesis other than the passive one of altering the amount of stress that a particular part of the cell receives during the division cycle.

An alternative proposal was made by Woldringh, Huls, Pas, Brakenhoff, and Nanninga[65] regarding the pattern of pole synthesis; they assumed an exponentially increasing pole area during constriction. The model proposed that the synthesis of pole area begins slowly and increases when the pole area yet to be synthesized is continuously decreasing. Not discussed in the model of Woldringh et al. is how new cell-wall synthesis is partitioned between the new pole and the lateral wall during the constriction period.[66] Another possible model that should be considered proposes that pole area is synthesized in proportion to the circumference at the leading edge of the pole growing area. Synthesis would start out at a high rate at the start of constriction and decrease as the diameter of the growing area at the pole decreased. The model proposed in Section V above (Fig. 6–5), pole growing with equal areas per unit time, is somewhere between the exponential and the leading-edge models. At this time, no experimental distinction between the different modes of pole growth can be made.

6. Leading-Edge Model of Pole Growth

How is pole synthesized during the period of constriction? This is a very difficult problem. The different patterns that might be suggested for pole synthesis do not make strikingly different predictions for the pattern of wall synthesis. In addition, experimental measurements of the rate of pole synthesis separate from side-wall synthesis are very difficult to perform and even more difficult to interpret. Wientjes and Nanninga[67] have looked at the incorporation of radioactive diaminopimelic acid into cells with slight, medium, and deep constrictions. They found that the amount of radioactivity in the pole area appeared constant, and the width of the peak of incorporation was also constant. They interpreted their results as indicating that the rate of pole synthesis was constant and that synthesis occurred only in a narrow *leading edge* at the junction of the two daughter cells. This proposal is the same as that for the case in which pole is synthesized linearly during invagination (Fig. 6–5).

7. Variation or Constancy of Diameter during the Division Cycle

One of the assumptions of the pressure model (Fig. 6–5) is that cell diameter, within a division cycle, is constant. The evidence for this proposition is ambiguous; some experiments support a constant width, and others, a decreasing width during the division cycle. This will be discussed in detail, but the conclusion to be made is that the diameter of an individual cell does not vary systematically during the division cycle.

8. Stability and Turnover of Peptidoglycan

There have been a number of reports of turnover of peptidoglycan in *Escherichia coli*. There is a significant release of labeled cell-wall material to the external medium.[68] Cells can take up some of these excreted fragments, so the amount released to the medium is considered a minimal measure of the turnover of peptidoglycan. Throughout our prior discussion of peptidoglycan synthesis, no mention was made of turnover. It is suggested here that this turnover may be a special case, as studies using the membrane-elution method reveal that the peptidoglycan is stable.

The standard method for measuring turnover is to label cells, remove them from the label, grow them in unlabeled medium, and measure how much of the label is either excreted into the medium or found in the soluble pool of the cell. How do we measure the turnover of peptidoglycan using the membrane-elution method? Cells are labeled in both the cell wall and cell protein. The cells are filtered onto a membrane, the membrane inverted, and the bound cells are allowed to grow. Consider that there are 100 counts in each of these components. The protein would be released from the membrane-elution apparatus over each generation with the amount decreasing by half each generation. Thus, there would be 50, 25, 12.5, 6.25, etc. cpm released each generation. If there were turnover of peptidoglycan, the first generation would be expected to give 50 cpm by division, with an additional 5 cpm if there were 10% turnover from the bound sister cell. Thus 55 counts would be released to the medium in the first generation. In the next generation, half of the remaining 45 cpm would be released, plus 10% of the remainder, which would be approximately 25 cpm; in the next generation, there would be 11 cpm, and so on. The ratio of diaminopimelic acid to leucine released from the membrane would decrease (55:50, 25:25, 11:12.5, etc.) over time. A decrease is not seen in the membrane-elution experiments and the conclusion is that peptidoglycan, in *Salmonella*, is at least as stable as protein.[69] The same stability of peptidoglycan also holds true for *Escherichia coli* B/r.[70]

The degradation of peptidoglycan in *Escherichia coli* has been the center of a general model of peptidoglycan synthesis that proposes that a large component of wall biosynthesis is due to an inside-to-outside movement of peptidoglycan as in *Bacillus subtilis*.[71] It has been argued[72] that there is turnover even in *Salmonella*, but the release of degradation products to the medium is prevented by the outer membrane, and the fragments are reutilized by further growth. Thus, the absence of released material is not necessarily an impediment to the inside-to-outside model of growth.

The difference between *Escherichia coli* and *Salmonella typhimurium* with regard to turnover may be attributable to the fact that almost all the experiments on turnover in *Escherichia coli* were carried out with a diaminopimelic acid auxotroph, strain W7. *Escherichia coli* is rather impermeable to diaminopimelic acid,[73] and large amounts of diaminopimelic acid must be present in the medium to ensure growth. It is possible that the physiology of the organism is disturbed by diaminopimelic limitation. This is supported by Driehuis and Wouters,[74] who noted that limiting diaminopimelic acid led to the production of abnormal peptidoglycan.

An inside-to-outside mode of peptidoglycan growth, similar to that for *Bacillus subtilis*, has been proposed for *Escherichia coli*.[75] This proposal is based on the observed recycling of murein, the calculated amount of peptidoglycan per cell, and the existence of trimeric and tetrameric fragments consistent with a multilayered peptidoglycan structure. The insertion into cell wall of new peptidoglycan strands in an unstressed configuration before their movement into the load-bearing layer of the peptidoglycan can explain all of these observations. The recycling of peptidoglycan may be a strain specific result, as there is no apparent release or recycling of peptidoglycan in *Salmonella typhimurium*[76] or in *Escherichia coli* B/r.[77] Koch[78] has presented the arguments against such an inside-to-outside mechanism of peptidoglycan growth, but at this time the inside-to-outside mode of surface growth cannot be excluded.

VI. CONTROL MECHANISMS FOR WALL SYNTHESIS

The occurrence of a discrete event, such as invagination during the division cycle, immediately inaugurates the search for a specific mechanism for such an event. Some of the proposals that have been made for specific regulation of cell wall synthesis will be briefly reviewed.

A. Stringent Regulation of Wall Synthesis

It has been proposed that the synthesis of cell wall is under the control of the stringent-relaxed system for the regulation of RNA synthesis.[79] Conditions that lead to increased RNA synthesis (mutation to relaxed phenotype, addition of chloramphenicol to amino acid-starved cells), also lead to an increase in peptidoglycan synthesis. A simple explanation of this finding is that the increase in RNA leads to the increase in cell-wall synthesis. The apparent regulation of peptidoglycan through the stringent response may be nothing more than the response of the cell surface to an increase in cytoplasm due to increased RNA synthesis. In order to demonstrate a direct stringent control mechanism, it is necessary to distinguish cytoplasm synthesis as the immediate cause of surface increase from direct regulation by the stringent control mechanism.

B. Regulation of Surface Synthesis by Cytoplasmic Signals

There have been a number of proposals stating that specific signals at particular times during the division cycle affect cell shape and surface synthesis. For example, a sudden change in the concentration of the cytoplasmic calcium concentration has been proposed as a signal for a number of cell-cycle events.[80] Cyclic nucleotides have also been suggested as being involved in cell-cycle regulation.[81] The evidence for variations, during the division cycle, of concentrations or amounts of various substances in the cytoplasm have been considered extensively in Chapter 4. The conclusion reached there is that there is no compelling evidence for variation of any substance in the cytoplasm as a function of the division cycle.

C. Regulation of Wall Synthesis by Peptidoglycan Structure

The structure of the peptidoglycan in different parts of the cell has been proposed as a trigger of invagination. This triggering may be due to a change in the chemistry of cell-wall synthesis. For example, the peptidoglycan structure in the pole may be slightly different from side-wall structure; by having different types of subunits prepared or different cross-links made, invagination may be initiated. Mutant strains that grow as spheres show a rise in total cross-links and a decrease in the average length of the glycan strands.[82] But other work shows that there are no major differences between cells growing as either long rods or spheres; such differences were due to different growth rates.[83] Chemical analysis of poles from minicells compared to whole cells did not reveal any difference in composition between poles and side walls.[84] More direct evidence comes from studies of cells labeled during the cell cycle. No changes in the composition of peptidoglycan synthesized during cell elongation or septum formation could be detected.[85] Although it is not clear that the decisive experiment has been performed, no available evidence indicates that peptidoglycan composition changes during invagination.

D. Heat Shock and Cell Division

When cells are exposed to a sudden increase in temperature, there is a period during which a small number of diverse proteins are preferentially synthesized.[86] It has been suggested that the heat-shock response, or the heat-shock proteins, may be involved in cell division.[87] This suggestion comes from the observation of altered patterns of cell division following a temperature shift in cells genetically altered in their heat-shock response. The problem is that slight changes in temperature affect division of cells irrespective of their genetic background. It is very difficult to make a clear distinction between a pleiotropic effect of the heat-shock response, and a defined relationship between heat shock and cell division.

E. Cell Division Related to Pattern of Cytoplasmic Protein Synthesis

It is widely thought that the protein determined by the *ftsZ* gene may play a key role in triggering cell division.[88] Evidence has been presented that the protein determined by *ftsZ* is synthesized linearly during the division cycle.[89] The interesting aspect of this report is that it was thought that the protein might have been synthesized at a particular time during the division cycle. When it was found that the *ftsZ* gene product was synthesized throughout the division cycle, the experimental measurements were plotted on rectangular graph paper to demonstrate that there were periods of linear synthesis of the protein. As noted in Chapter 3, the method used to synchronize cells, the phosphate-starvation-entrainment method, may produce artifacts. A large number of starvations may produce abnormal and altered cells. As noted before, it is difficult to distinguish exponential synthesis from linear synthesis by total measurement, so a rate measurement is necessary (Chapters 3 and 4). Other enzymes related to peptidoglycan synthesis have been reported to vary during the division cycle. For example, a carboxypeptidase has been reported to have its maximal activity at the time of division.[90]

F. Control of the Cell Cycle by Phospholipid Flip-Out

A general model for the regulation of the division cycle envisions a once-per-cycle *flip-out* of phospholipids from the inner membrane to the outer membrane.[91] It is conjectured that there is an increase in the phospholipid density in the inner membrane during the division cycle, and at a critical density, there is a rapid movement of lipid to the outer monolayer. When this occurs, a variety of events can be triggered. The DNA initiation complex seems to be bound to lipid, so a change in membrane fluidity at a particular time in the division cycle may allow an initiation of DNA replication. The phospholipid flip-out may control the cell cycle, it is further conjectured, by affecting the calcium concentration in the cell. But the phospholipid flip-out model is speculative. I have suggested that the surface density of the membrane remains constant as membrane is synthesized in direct proportion to the cell-surface area. No change in density occurs during the division cycle.

VII. MATURING OF PEPTIDOGLYCAN

A. Change in Acceptors and Donors

If new strands of peptidoglycan are inserted between older strands as single strands, and if cross-links are formed only from pentapeptides present on the newly inserted strands, then we would expect all donor links to come from the new strand. This means that the resident strand is always an acceptor. If a cell is labeled for a short time with diaminopimelic acid, all of the radioactive diaminopimelic acid will be in the newly inserted strand. The nonradioactive resident strand will accept the cross-link from the adjacent D-alanine, so the two types of diaminopimelic acids in a cross-link can be chemically distinguished.[92] After a short labeling period, there will be no label in the acceptor portion of dimers, and all of the label will be in the donor moiety. The results of measurements of acceptors and donors can be summed up in the acceptor-donor radioactivity ratio, the ADRR.[93] An ADRR value of 0.0 means there are no acceptors and that all of the material is present in the donor form. Measurements of initial ADRR values and their change during extended growth indicated that (1) the peptidoglycan strands entered the cell wall as pairs, with two new strands being inserted adjacent to each other; and (2) after 8 minutes the strands encircled the cell and returned to their original point of insertion. This was based on the finding[94] that the initial ADRR was not zero, and that the value was constant for 8 minutes before it began to increase. Other investigators[95] have found lower initial ADRRs indicating that single strands are inserted during peptidoglycan synthesis. The insertion of single strands is indicated in Figs. 6–2 and 6–3. The precise kinetics of the increase is a matter for further work, but the final value of the ADRR is an indication of the degree to which strands can be separated and new strands placed between them.

B. Increase in Cross-Linking

The peptidoglycan also matures by increasing the cross-linking frequency with time after synthesis.[96] The observation that cross-linking increases after a pulse-label is unexpected and paradoxical. Consider a

strand with 1000 pentapeptide subunits inserted during a labeling period. Assume that the frequency of cross-linking is such that 25% of these
inserted peptides will form cross-links with adjacent strands. These 250
cross-links are distributed equally to each side, 125 to the adjacent left
strand and 125 to the adjacent right strand. At some later time, a new
strand is inserted between the radioactive strand and the original unlabeled material. This newer strand acts the same way as the original
labeled strand; it forms 250 cross-links with its adjacent strands. These
cross-links are also equally distributed with 125 cross-links to the unlabeled strand and 125 to the labeled strand. Thus, 125 cross-links were
broken (from one side of the labeled strand) when the second strand was
inserted, but 125 cross-links were re-formed. No change in the degree of
cross-linking is expected. Although the actual subunits linked may or
may not be the same, the total cross-linking fraction of the original
material does not change.

How can we explain the observed increase in the total amount of
dimers? I have developed the following explanation for the increase in
cross-linking in peptidoglycan. Consider that the number of cross-links is
not precisely the same for each strand but is statistically distributed about
some mean value. For two strands inserted in different parts of the cell,
one strand may have 300 cross-links formed from 1000 inserted subunits,
and the other strand may have 200 cross-links. The average would be the
same as described above, 250 cross-links per strand. But now assume that
the next strands inserted preferentially replace the links between the low
cross-linked strands; there is a preference that the 200 cross-links will be
replaced by 250 cross-links the next time a new strand is inserted, so that
strands with 300 cross-links are relatively stable. The strand with 200
cross-links has a below-average cross-linking density, so it is expected
that the newly inserted material will have a higher cross-linking density.
Thus there will be a steady increase in the cross-linking fraction as the
low cross-linked material is replaced by higher cross-linked material.
This proposal is illustrated in Fig. 6–8.

The way to think about this mechanism for increasing cross-linking is
to consider the different stresses on strands connected by either 200 or
300 cross-links. The internal pressure of the cell makes the turgor pressure the same all over the cell surface, so it is expected that the 300 links
will have less stress per cross-link than the strands with 200 links between them. For a given length of strands, the higher number of links
means that there is less stress per link. The surface-stress model predicts
that the replacement will preferentially occur between the strands that
are poorly cross-linked. The average replacement strand is above the
cross-linking value, so there will be a steady drift to higher cross-linking

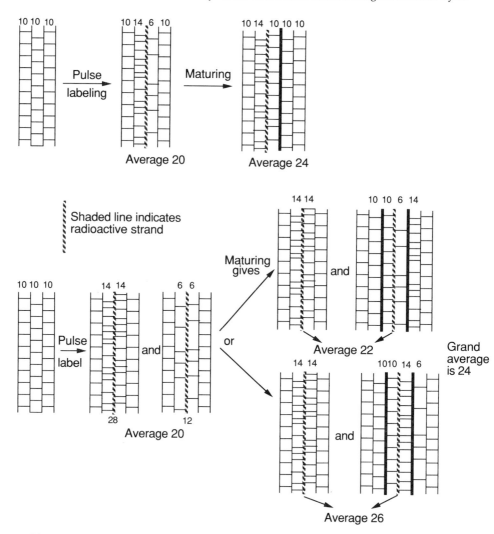

Figure 6.8 Increase in Cross-Linking by Natural Selection. In the upper figure a single strand is inserted to give an average of 20 cross-links. The cross-links on the left and right are not equal. When another, unlabeled, strand is inserted, it will preferentially replace the less dense cross-links (the 6 cross-links), and there will be an increase in cross-linking. In the lower diagram it is shown that the inequality of strand cross-linking at the time a labeled strand is inserted, does not have to be restricted to one strand. Over the population, different strands may have cross-linking densities above or below the mean. When new strands are inserted during further growth, the lower density cross-links are preferentially replaced by new strands. Although sometimes the replacement does not change the cross-linking values, over the entire population there will be the same total increase in cross-linking.

values (Fig. 6–8). By this mechanism the cell will continue to strengthen the peptidoglycan by a drift to more and more cross-links.

VIII. REGULATION OF CONSTRICTION IN ROD-SHAPED CELLS

It is not known how cells initiate constriction, how the constriction is timed, or how constriction proceeds. Nevertheless, there are some proposals regarding the regulation of invagination.

A. Nucleoid-Occlusion Model

One possible way in which DNA replication may affect cell division is by exerting a negative effect on the synthesis of the septum by the mere presence of DNA in the middle of the cell. This idea was first articulated by Helmstetter and his colleagues.[97] They proposed that if chromosome replication were not completed, the chromosomes would not segregate, and division would not take place. This proposal of DNA's interfering with division has been extended by the nucleoid-occlusion model. The nucleoid-occlusion model also proposes that DNA itself exerts a negative effect on septum formation.[98] A nucleoid that has not finished replicating is located in the middle of the cell, so it is proposed that the centrally located nucleoid prevents septum formation from starting. Upon termination of DNA replication, the nucleoids move apart (by mechanisms unrelated to the cell surface); this separation frees the center of the cell for invagination. The mechanism for nucleoid separation involves the idea that ribosome-assembly compartments develop around the duplicated gene clusters coding for ribosomal RNA and protein. This accumulation of newly synthesized ribosomes causes the gene clusters, together with the interposed origin, to drift apart.[99] The central proposition of the nucleoid-occlusion model is that there is a negative nucleoid effect ". . . and a positive compensating termination signal."[100] Both the negative and positive effects originate from the actively transcribed and replicating nucleoid and produce a signal that influences the peptidoglycan synthesizing system in the plasma membrane. This balance of forces determines when and where invagination will occur. A formal description of the nucleoid occlusion model has been presented.[101]

B. The Periseptal Annulus

The periseptal annuli are two concentric rings found beside a newly forming septum.[102] The annuli are zones of adhesion between the inner membrane and the peptidoglycan. When cells are plasmolyzed by placement in a hypertonic medium (e.g., a high sucrose concentration), we can observe these zones of adhesion that delimit the plasmolysis bays. Besides their association with the septum, the periseptal annuli can appear at sites of future septum formation before any septum is visible.[103] In the longest cells of a population, periseptal annuli were observed at one-quarter and three-quarters of the cell length, presumably where the septum will be forming in the future daughter cells. A second type of support for the involvement of periseptal annuli in the invagination process comes from the study of temperature-sensitive mutants. At elevated temperatures, as filaments formed, the annuli were present with normal spacing, and the sites were separated by a distance equivalent to a unit cell. When the cells were allowed to defilament by lowering the temperature, the new septa formed at the sites where the annular structures were found at the elevated temperatures.

How does the cell localize the annuli at the precise midpoint of the future daughter cells? It appears that new annuli come from pre-existing annuli. They first appear close to the older annuli; then proceed away from the center of the cell (possibly by membrane synthesis between the septum and the periseptal annulus), coming to rest at the midpoint between the pole and the septum.

The periseptal-annulus proposal suggests that sites for invagination are chosen at some time earlier than the division cycle that is currently under observation. The formation of periseptal annuli is analogous to the premature initiation of DNA within division cycles before the division cycle in which DNA synthesis is terminated. After the annulus is produced, there are two possible mechanisms by which the annulus moves to, and rests at, its proper location. One involves preferential membrane growth between the midpoint of a cell and a periseptal annulus. Upon reaching the midpoint of the cell, a balanced membrane synthesis on both sides would keep the periseptal annulus at the midpoint. Alternatively, a new generation of annuli is produced that moves away and keeps the annulus at the midpoint of the future cells.

What is the trigger for the formation of periseptal annuli? One possibility is that the cell titrates something in a manner similar to the formal titration related to the initiation of DNA synthesis. For example, if the cell titrated the amount of surface area per cell pole, then when the area per two poles was above a certain value, a new set of periseptal annuli would be initiated. At the instant new poles were initiated, there would be a decrease in the ratio, and thus a stable situation. Another way of looking at this model is presented in Fig. 6–15 where the constant-area model is illustrated. One prediction of this model is that new periseptal annuli would appear earlier and earlier at faster growth rates. This prediction is in accord with the limited data available.

C. Variable-T Model for Pole Synthesis

The surface tension (T) of an area or surface is the work required to expand the area by a unit amount. If the surface tension is high, then it takes a lot of work or energy to add to that surface. A low–surface tension means that it takes less energy to make a unit amount of cell surface. For a given amount of energy, a low–surface-tension material will increase in area more than a high–surface-tension area. This formal concept of surface tension offers a way of verbalizing the pole-formation problem, and may even present ideas as to the mechanism of pole formation.

In the basic pattern of pole synthesis (Fig. 6–5), when cells are invaginating, the increase in surface area per increase in cell volume is greater in the pole area than in the cylindrical-wall area. The surface tension of pole synthesis is less than the surface tension of side wall synthesis. No mechanistic explanation for this difference in surface tension is available. Koch and Burdett[104] have analyzed the shape of the poles of gram-negative bacteria and have proposed the variable-T model for pole growth. They proposed that very slight decreases in the surface tension of wall synthesis in the middle of the cell, and a continuous change in this value with continued constriction, can explain the shape of the bacterial pole.

D. When Does Invagination Occur?

It is generally believed that invagination occurs sometime in the middle of the division cycle. Woldringh[105] presented evidence to link the start of invagination with the termination of DNA replication or the start of the D period. This experiment is evidence for the idea that termination of DNA synthesis is related to, or possibly triggers, invagination. Recently this proposal has been extended into a more specific regulatory mechanism whereby the DNA or nucleoid itself prevents invagination by its very presence in the center of the cell. This nucleoid-occlusion model[106] proposes that the charge on the DNA or some other property of DNA has a negative effect on invagination. When the nucleoids separate there is a removal of the negative effect, and the cell invaginates. Pictures in the literature, however, show cells invaginating even in the presence of a definite nucleoid stretching along the inside of the cell.[107]

If cells do not invaginate until the middle of the division cycle, then newborn cells should not show any signs of invagination. Cells eluted from a membrane-elution apparatus have been looked at in the electron microscope, and a significant amount of invagination was observed.[108] It was proposed that the newborn cells were contaminated with a significant number of cells that were randomly eluted from the membrane. By calculating the degree of contamination (assuming that true newborn cells were not invaginated), it was proposed that up to 57% of the eluted cells were randomly eluted and not newborn (see Chapter 3 for a full discussion of this point). Of course, if this were the case, then we would expect an ever-decreasing number of cells eluted from the membrane at each generation of elution. A decrease in cell number is not usually observed, so it may be valid to conclude that newborn cells can be invaginated. Any models that related termination of DNA replication and the start of invagination would be eliminated if newborn cells could invaginate.

IX. MEMBRANE SYNTHESIS DURING THE DIVISION CYCLE

Less is known about membrane synthesis during the division cycle than about peptidoglycan synthesis. The model presented in this chapter predicts that the membrane would increase in the same manner as the peptidoglycan. This prediction follows from the aggregation argument made in Chapter 2.

A. Lipid Synthesis during the Division Cycle

If the lipid of the bacterial cell is in close association with the peptidoglycan, and increases along with the peptidoglycan, then we would expect a pattern of synthesis similar to that of peptidoglycan. To observe such a pattern, the data must show very slight deviations from exponential. Although the conclusive experiments do not yet appear to have been done, there have been a number of studies of membrane synthesis during the division cycle; none of the results fits this expectation.

There have been a number of studies on lipid or membrane biosynthesis during the division cycle. Ohki[109] concluded that cytochrome b_1 (localized predominantly in the cytoplasmic membrane) and L-α-glycerolphosphate transport increased in steps during the division cycle. Further, the turnover of phospholipid was proposed to vary in steps during the division cycle. Carty and Ingram[110] observed abrupt increases in lipid synthesis coincident with the initiation of cross walls. Similar transient increases in glycerol incorporation in synchronized cultures of both *Escherichia coli* and *Bacillus megaterium* were seen by Daniels.[111] James and Gudas described a cell-cycle–specific incorporation of lipoprotein into the outer membrane.[112] In contrast, a continuous and nonstepwise increase in membrane components was reported by two groups.[113] More recently, Joseleau-Petit, Kepes, and Kepes[114] have proposed a bilinear pattern for membrane synthesis during the division cycle. Doublings in phospholipid synthesis were also reported by others.[115]

Although the discussion in this chapter of the experimental determination of the rate of surface synthesis during the division cycle has dealt primarily with peptidoglycan, it applies equally to membranes and other surface-associated elements. This is because, as noted above, the cell membrane grows in response to the increase in peptidoglycan surface, and coats the peptidoglycan without stretching or buckling. Therefore, the area of the membrane should increase in the same way as does peptidoglycan. The observed doubling in the rate of phospholipid synthesis (DROPS) during the division cycle[116] may be attributable to artifacts of the synchronization procedure.[117] Pierucci's[118] data on phospholipid synthesis, obtained with the membrane-elution method, are consistent with the rate of phospholipid synthesis being similar to peptidoglycan synthesis. Experiments using the membrane-elution method and the ratio-of-rates analysis fully support the suggestion that the rate of membrane synthesis is similar to that of peptidoglycan.[119]

B. Membrane Protein Synthesis during the Division Cycle

In addition to lipid, the membranes of the cell contain specific membrane proteins. Measurements of the amount of protein per unit surface area for cells growing at different rates indicate that the density of proteins is constant.[120] This means that the protein composition in the membrane does not vary as the cell surface varies over a factor of two. If membrane proteins were inserted into the membrane in direct proportion to the increase in cell surface area, then we would expect that the pattern of synthesis of a membrane protein would be similar to that observed for peptidoglycan synthesis. Measurement of protein synthesis during the division cycle indicated that the bulk membrane protein was synthesized at a constant rate throughout the cycle with an abrupt doubling in rate approximately 10–15 minutes before division.[121] This result is reminiscent of the reported increase in peptidoglycan synthesis (relative to total mass synthesis) at the time invagination starts.

X. THE SHAPE OF ROD-SHAPED, GRAM-NEGATIVE BACTERIA

"What is the shape of a rod-shaped cell?" It sounds like a facetious question from the 1940s' "It Pays to be Ignorant" radio show. But the question is a serious one, and the answer has important implications for the regulation of cell growth and division. The actual shape of a bacterial cell is not obvious to us. We see the shape of the cell, and know that a rod-shaped cell is rod-shaped, but no notice is taken of the particular dimensions of the cell. Why does a bacterial cell have the shape it has? Why isn't it longer or shorter, or thicker or thinner? The question asked here is not whether *Escherichia coli* is precisely rod-shaped, or is some other similar shape, but rather, why does the cell have the particular length-to-width ratio, and how is this shape maintained and determined? The shape of a bacterial cell is defined by the ratio of cell length to width (Fig. 6–9).

Cells growing rapidly in rich medium are significantly larger than cells growing more slowly in minimal medium.[122] How does shape vary as size varies? How does shape vary within the cells of a culture growing in a steady state at a particular growth rate and with a particular average size? The answer to the first question is that the shape of the cell is constant and independent of the growth rate and cell size. The answer to the second question is that the cell shape varies so that the volume of the cell is relatively constant at different points in the division cycle. This proposal implies an inverse relationship between the length and width of cells in a culture. The combination of these ideas—a constant shape at

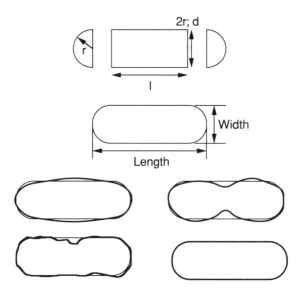

Shape factor = length/width

Figure 6.9 Definition of Cell Shape. Rod-shaped cells are considered to be composed of a cylindrical central area capped by two hemispheres. The total length of the cell is the length of the cylinder plus two radii. Examples of actual cell shapes are shown superimposed on an ideal cell shape. Irrespective of the actual shape of a cell—whether the cell is of ideal shape, invaginating or bulging, or even amorphous—we can approach the problem of shape by considering the ratio of cell length to width. If a consistent width-to-length measurement is used, the relative shape of cells can still be determined.

different growth rates, and a slightly variable shape for the cells within a culture—will explain a large amount of data.

A. Constant Shape of Cells at Different Growth Rates

Cell shape is constant. The meaning of a constant shape for cells growing at different growth rates is illustrated in Fig. 6–10. Cells growing at faster rates in richer medium are larger than slower-growing cells. The size of the cells may be calculated by the formula:

$$M_{avg} = K' \cdot 2^{(C + D)/\tau}$$

This means that, for a constant value of $C + D = 60$, the mass for cells with 60-, 30-, and 20-minute doubling times varies as 1:2:4. Because cell density is constant at different growth rates and during the division cycle,[123] the volume of a cell may be equal to its mass. The cells in Fig. 6–10 are drawn with their linear dimensions varying as the cube root of the relative volumes. Thus the volume of these cells varies as 1:2:4. In Fig. 6–10, only the newborn cells and the dividing cells are drawn. Actually, there is no unique shape for growing cells, as shape varies during the division cycle. If the pattern of growth between birth and division is the same at all growth rates, then any one particular cell age may serve as an indicator of the shape of the cell. In much of the discussion that follows, the shape of the cell will be defined as the shape of the newborn cell, although any particular cell age may serve as an indicator of cell shape.

The data supporting a constant shape at different growth rates are not perfect. The difficulties in measuring the diameter of a cell without artifacts are significant.[124] The fact that there is a tendency for width to increase with growth rate in the same manner as length (see below) is one of the best arguments for the existence of a constant shape. If cell shape were not constant, then the ideal mode of growth might be a constant diameter at all growth rates. Because the cell grows during the division cycle primarily by elongation at a constant width,[125] the cell could continue in this manner without a change in diameter; large cells would appear long and thin and small cells would appear short and wide. When cells growing in minimal medium are shifted to a richer medium, a shift-up,[126] the cells grow larger because there is a delay before the rate of cell division increases, although the rate of mass increase changes rapidly to the new rate. If width stayed constant at all growth rates, the cells would merely increase in length and achieve the new steady-state size with a major change in cell shape. If the cell increases both its length and width when cell size increases, we may ask whether there is a direct proportionality between the two, or whether there is some other numerical factor relating the increase in width and length.

Figure 6.10 Constant Shape of Rod-Shaped Cells of Varying Size. Newborn and dividing cells for three different growth rates are illustrated. Cells are drawn to scale with volumes varying in proportion as $1:2:4$. These are cells growing with 60-, 30-, and 20-minute interdivision times respectively. For all of the newborn cells at the left or dividing cells at the right, the ratio of length to width is constant. Cell shape is constant if cell length and cell diameter increase or decrease by the same factor as the size of the cell varies.

Volume 1.0

Volume 2.0

Volume 4.0

The earliest proposal of a constant shape was that of Zaritsky,[127] who reanalyzed the classical size and diameter measurements of Schaechter, Maaløe, and Kjeldgaard[128] on *Salmonella typhimurium*. At growth rates of 2.73, 1.85, 1.10, and 0.61 doublings per hour, the diameters of the cells were 1.43, 1.22, 0.93, and 0.87 μm. There is a constant ratio between the diameter values cubed and the cell-size measurement, indicating that cell shape is constant.

Experimental evidence in favor of a constant shape consists of reports in which both the width and length of cells have been measured. A number of these measurements are summarized in Table 6-1.[129] It appears that cell shape is constant when cell size varies. Woldringh *et al.*[130] determined the shape factor for *Escherichia coli* B/r growing at 17 different rates or sizes and found that the shape factor was constant. Recently, Woldringh *et al.*[131] plotted the length and width values for a number of

Table 6-1 Dimensions of Cells Growing at Different Rates

Doubling Time (minutes)	2R	L	L/2R
Strain B/rH, 37°, electron-microscope analysis			
22.5	0.96	2.64	2.75
60	0.73	2.00	2.74
109	0.57	1.51	2.649
125	0.61	1.66	2.721
194	0.53	1.68	3.17
Strain B/rH, 37°, electron-microscope analysis			
19	1.1	3.81	3.464
48	1.7	3.06	4.371
78	0.51	2.84	5.569
80	0.57	2.94	5.158
119	0.52	2.66	5.115
Strain B/rH, 37°, light-microscope analysis			
19	0.9	4.26	4.733
48	0.74	3.37	4.554
119	0.57	2.58	4.526
Strain B/rH, 22°, light-microscope analysis			
63	0.86	3.69	4.291
164	0.83	3.08	3.711
551	0.66	2.32	3.515

different growth rates; the lines were parallel on semilogarithmic paper, indicating that cell shape is constant.

Donachie et al.[132] proposed that the cell length of cultures growing at different growth rates increased in proportion to the one-third power of the growth rate. They noted that the mass or volume increased in proportion to the power of the growth rate; this means that the linear dimension of cells produced a constant cell shape. As summarized by Donachie and Robinson,[133] ". . . these two relationships suggest that the ratio of mean cell length to mean cell width must be very similar in populations of cells growing at any of these rates; i.e., although the absolute size (volume, mass) varies greatly with growth rate, the ratio of average length to average width does not change."

A subtle physiological argument in support of the constant shape of bacteria growing at different growth rates comes from the studies of the segregation patterns of chromosomes and minichromosomes. Helmstetter and Leonard[134] explained the observed variable nonrandom segregation pattern of DNA at different growth rates in terms of the fraction of the cell surface that is pole area. The observed degree of nonrandom segregation in cells growing at different rates[135] is constant when analyzed by the membrane-elution method.[136] This result is consistent with constant cell shape at different growth rates.[137] (See Chapter 9 for a detailed analysis of this argument.)

Finally, the observed overshoot in cell length, and the return to the final steady-state length following a shift-up,[138] is most easily explained by the cell's maintaining a constant shape at different growth rates.[139] The overshoot is explained by the slow, steady increase in the radius of the cell to the new, larger radius, and the accommodation of the increased cell size during the transition by a preferential increase in cell length.

B. Energetic Argument for Constant Shape

The minimal surface for a given volume is a sphere. Energy-minimizing arguments (or surface-minimizing arguments) could be applied if Escherichia coli were a sphere. But the cell grows as a rod, so energetic arguments support a constant shape. If shape were not constant, then when there is a change in cell size, either one or the other size will have a less-than-optimal shape based on energy- or surface-minimization considerations. For example, if a cell of a given shape were shifted to a faster growth rate, and grew to a larger size solely by length extension, the surface area of the cell at the larger size would be greater than the surface area if the cell maintained a constant shape. Given that cells grow as rods, and that cells can have a nearly constant shape, we may suspect that such energy considerations may have led to the evolution and optimization of a constant shape.

C. Exceptions to Constant Shape

Not all experimental evidence supports constant shape. One set of measurements on *Salmonella typhimurium* that varied in growth rate from 0.31 doublings per hour to 2.40 doublings per hour had the length increase 1.53-fold while the diameter increased only 1.28-fold.[140] This would support a variation in cell shape as the growth rate varied. A more detailed analysis of the data also indicated that the cell mass (as determined by absorbance per cell) increased 4.61-fold while the cell volume (determined by electron-microscopic measurements of the cell dimensions) increased only 3.21-fold. The difference in the increase in cell mass and cell volume would predict a change in cell density with growth rate. As we shall see in Chapter 7, such a variation in density is not observed. This analysis does, however, suggest that there is enough error in the measurements of cell diameter and cell length to make the shapes of these cells relatively constant as growth rate varies.

Cell-shape constancy is not an absolute requirement, and the cell shape can vary when growth conditions are altered. When cells were grown in limiting thymine, the cell shape was altered with mass increasing primarily due to an increase in cell width.[141] These observations suggested that cell-length extension was determined, in some way, by the completion of chromosome replication. Examination of the shape of *rep⁻* cells that have a slightly longer replication time (see Chapter 5), indicated that cell width did not change with a change in DNA replication rate.[142] This left open the possibility that low thymine concentrations may change cell shape by a direct effect on the biosynthesis of cell wall. Slightly different results were obtained by others,[143] although a change in shape was also observed during growth in low thymine concentrations. Thus cell shape is not an absolutely constant factor, but may be varied. How specific starvation or limitation regimens affect cell shape is not understood.

Shape is affected by temperature. At lower temperatures, the cells become squatter. Length decreases more than width, and so the length-to-width ratio decreases.[144] As we shall see in Chapter 9, this result is predicted by the pattern of segregation at different temperatures.

D. Cell-Width Variability within a Culture

Nothing is perfect. Even the most precisely made instrument has some variability. This is usually described by the statistical variation, about some mean value, of the instrument's ability to perform some task. This absence of perfection is even more noticeable in biology. When a cell is described as having a particular width, it is not meant that all cells in a growing culture have precisely and exactly the same width, but that there is an average or mean width, with a relatively narrow variation about the mean. While all cells in a culture could have a constant and identical width, and the observed variability might be owing solely to experimental error, the variability discussed now is the actual variability within the cells in a culture.

A growing bacterial culture contains cells of different widths. Although the observed spread in cell width is narrow, the observed variation in width is greater than experimental error.[145]Consider a series of bacterial cultures each derived from single cells, with the initial cells having a range of cell widths. The final width distribution is the same in all of these different cultures (otherwise there would be a genetic inheritance of cell width, which is not observed), so cell width must vary during the growth of the culture. This homeostasis of cell shape is due to both narrow and wide cells producing cultures with both narrow and wide cells. I propose that width changes slowly over a time equivalent to many cell interdivision times (Fig. 6–11). There is a tendency for cells to adjust their width toward the mean; wide cells tend to narrow, and narrow cells tend to widen. Also, cell width varies over times that are long compared to a single cell-division cycle (perhaps over 10 or more generation times), so during a single division cycle, cell diameter appears constant.[146] At a minimum, there is no systematic variation in diameter, as diameter can increase or decrease during any individual division cycle. In general, cells that are wider than normal would more likely become thinner, but this is compensated for by thinner cells that would be getting wider. A constant width for an individual cell during the division cycle is also implied by the constrained-hoop model of cell-wall synthesis.[147]

E. Cell Length and Width in a Culture

There is a great deal of evidence that the cell initiates DNA synthesis as though the size per origin of DNA at initiation were constant at different growth rates (Chapter 5).[148] Irrespective of the mechanism of this control of initiation of DNA synthesis, this idea will now be applied to the size of cells within a culture. Assume that cell size at initiation of DNA synthesis is invariant, that the time for a round of replication (the C period) and the segregation period (the D period) are also constant and invariant, and that the rate of mass synthesis is constant, and the same, in all cells within a culture. (None of these assumptions alters the arguments below; any variability, which undoubtedly does exist owing to the imperfection of regulation by size, will make only slight changes in the calculations.)

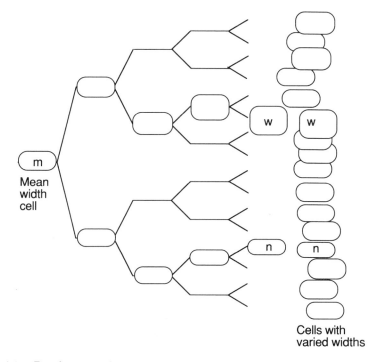

Figure 6.11 Development of Variable Widths during Bacterial Growth. The development of a narrow (n) or a wide (w) cell from a cell of mean width (m) occurs over many interdivision times. At the end of a growth period, there are cells of all widths, indicated by the column of cells at the right. Cell width is relatively constant over any particular interdivision time, but over many interdivision times, there will be a divergence in width. The further a cell is from the mean width, the more likely it is to produce progeny that are closer to the mean width. Within any one interdivision period there is essentially no observable change in cell width, so this proposal is consistent with a constant cell width during the division cycle.

These assumptions mean that cells in a culture will be born with approximately the same size or cell mass. Dividing cells will all have the same size, twice that of the newborn cells. Termination of DNA replication would also be expected to occur at a relatively constant cell mass. As there is evidence for a strong correlation between termination of DNA synthesis and the initiation of invagination,[149] there would be a constant cell mass at the start of invagination. Thus the cell mass or volume at birth is relatively constant for cells within a culture, irrespective of the variation of cell width or cell length. The implications of this conclusion will be analyzed and applied to a number of experimental observations in the literature.

Given a constant cell volume at birth and division, the expected pattern of cell width and cell length in a culture is illustrated in Fig. 6–12. Cells at birth, which are drawn with a constant volume, have an inverse relationship between cell length and width. Assume that the cells in a culture have a mean diameter of 1.0 and that the diameter varies between 0.8 and 1.2 units. Presumably not all diameters are equally represented; there is a distribution about the mean, with the frequency of cells decreasing as the diameter departs from the mean. For the assumptions of constant birth size and division size, the dimensions of cells in such a culture can be calculated for different widths. All of the newborn cells, regardless of their diameter, have a volume of 2.094 cubic units; dividing cells have a volume of 4.188 cubic units. For the diameters illustrated in Fig. 6–12, the length of the narrowest newborn cell is almost the same as that of the widest dividing cell. The length distribution of this culture varies from 2.25 units to 9.90 units. It is generally observed that the cell-length distribution of growing cultures varies over a factor of almost four.[150]

F. Apparent Variation of Cell Diameter during the Division Cycle

Trueba and Woldringh[151] reported an inverse correlation between length and width for the cells in a growing population. The shortest cells in a population were widest, while the longer cells were thinner. They concluded that cell width decreased during the division cycle. This conclusion was based on the assumption that the length of the cell is an accurate

---

Figure 6.12 Length and Width Distribution in a Cell Population. The cells of a culture are drawn here with nine different diameters. The diameter is held constant during the growth of an individual cell, with the volume increase during the division cycle expressed solely by an increase in cell length. All newborn cells are assigned exactly the same cell volume. The newborn cell with a mean diameter of 1.0 has a cylindrical wall length of 2.0, giving a total length of 3.0. The volume of this cell is 2.094 cubic units. All of the other newborn cells (shown at the left) are drawn with this volume. The cell length varies depending on the diameter; thinner newborn cells are longer, and thicker newborn cells are shorter. For example, for the cell with a diameter of 1.2, the two polar hemispheres have a total volume of 0.9084 leaving a volume of 1.1896 in the cylindrical part. Solving for the height of the cylinder with a diameter of 1.2 gives a cylindrical length of 1.0518. Adding the two hemispheres of radius 0.6 gives a total length of 2.25. The dividing cells are drawn twice the newborn length. The thickness of the lines depicting the cells indicates their presumed frequency in the culture; the mean cell width (thickest line) is most common, with the other cells occurring at decreasing frequencies as the width deviates from the mean. The precise distribution of cell widths was not known. Inset: Graph of width as a function of length for cells in a culture. For each cell length, the average width is calculated. This calculation does not take into account the distribution of cell widths in the culture (with more cells having mean widths than being either very wide or very narrow). Also, the calculation does not take into account the age distribution of the cells of each width.

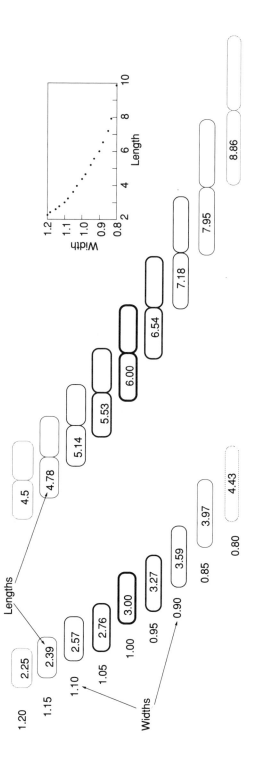

measure of cell age. Applying the variable-width proposal to a growing culture presents an alternative explanation of these experimental measurements.

The relationship of length to width, determined from the cells in Fig. 6–12, is graphically illustrated in the inset to Fig. 6–12. There is an inverse correlation between cell length and cell width similar to the experimental observations of Trueba and Woldringh.[152] The negative slope of the points in the inset to Fig. 6–12 is independent of the frequency of cells of different widths in the culture. There does not have to be a systematic variation in cell width during the division cycle to obtain the observed inverse correlation of length and width in a cell population.

There is conflicting evidence on variability in cell diameter during the division cycle. For example, Aldea et al.[153] reported that cell width in *Salmonella typhimurium* is constant during the division cycle. These results are not necessarily in conflict with those of Trueba and Woldringh.[154] It may be that the width variation in the cells used by Aldea et al. was much narrower, and thus was essentially undetectable by the methods they used. It is proposed here that the basic shape law is that of an inverse length–width distribution, but that this relationship can be experimentally demonstrated only under favorable experimental conditions. One of these conditions might be a relatively large variation of cell width within a culture for a particular bacterial strain. The results of Trueba and Woldringh may be a good example of this width variation.

G. Width Variability and Peptidoglycan Synthesis during the Division Cycle

Wientjes and Nanninga[155] observed that at intermediate cell lengths, in both septate and nonseptate cells, there was a higher rate of incorporation of diaminopimelic acid (a precursor of cell-wall synthesis) in septate cells than in nonseptate cells. They interpreted this observation as indicating a discrete increase in the rate of peptidoglycan synthesis at the start of invagination. Woldringh et al.[156] had previously proposed this jump in peptidoglycan synthesis at the start of invagination. A sudden jump in peptidoglycan synthesis is incompatible with the proposal that there is a smooth, continuous increase in the rate of peptidoglycan synthesis during the division cycle.[157]

An alternative explanation for the result of Wientjes and Nanninga[158] is given in Fig. 6–13. Cells are born with a constant size and divide with a constant size. Assume that cells of all widths initiate septation at a cell volume or mass equivalent to the middle-aged cell (age 0.5) of the median width (width 1.0). If a cell increases its mass exponentially, i.e., mass increases in proportion to the mass present,[159] a cell with a mean width of 1.0 will have a length of 4.1046 at age 0.5. Narrower cells will initiate septation when longer and wider cells will initiate when shorter. For each width, cells of the length 4.1046 are added to the diagram (Fig. 6–13). At this cell length, the widest cell is well along in septation (as indicated by the vertical septum), as it has reached the mass for the initiation of septation at a short length. Conversely, the narrow cells of the same length as the middle-aged mean-width cell have not yet achieved the mass required for initiation of invagination. The expected rate of peptido-glycan synthesis in these cells can be calculated for a 1% increase in cell mass or volume (Fig. 6–13, inset). This is equivalent to labeling the cells for a period of time so that all cells have a 1% increase in cell mass. The wide cells make more mass and increase their absolute surface more than narrow cells. The relative rate of peptidoglycan synthesis (rate of surface synthesis per rate of mass synthesis) is greater in the septate cells than in nonseptate cells of the same length. Variable cell width, with a relatively constant cell mass or volume at the initiation of septation, can explain the observed higher rate of peptidoglycan synthesis in septating cells without invoking any jump in peptidoglycan synthesis at the start of invagination. This analysis is consistent with the proposed smooth increase in the rate of peptidoglycan synthesis during the division cycle.[160]

H. Width Changes after EDTA Treatment or Amino Acid Starvation

Grossman, Ron, and Woldringh[161] reported the average cell width decreased after starvation for amino acids. A similar result has been reported after treatment with ethylenediaminetetraacetate (EDTA).[162] Both of these conclusions are based on width measurements before and after the treatment; a decrease in cell width during treatment was not observed since the requisite width measurements can only be obtained by electron-microscopy. One possible explanation for these observations is that there may be a preferential division of narrow cells following various treatments. If narrow cells, for reasons not yet known but perhaps owing to the smaller area connecting two putative daughter cells, were to preferentially divide during treatment, the numerical average for the width would decrease, because there would be more narrow cells in the resulting cell population. While this explanation fits much of the data, some experiments cannot be explained this way. For example, when newborn cells (produced by membrane-elution) were starved for an amino acid, there was a decrease in average width without any cell division.[163]

I. Central Zone Surface Synthesis during the Division Cycle

In one of the earliest studies on cell-surface synthesis in *Escherichia coli*, Ryter, Hirota, and Schwarz[164] observed that cells of all length classes had a central *zone* of increased peptidoglycan synthesis. Peptidoglycan

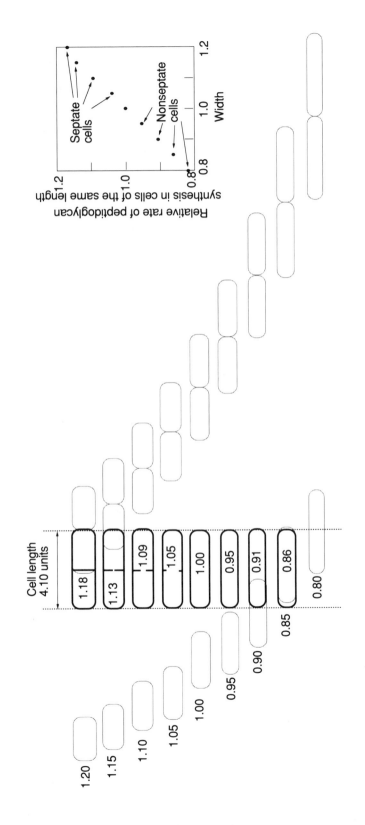

synthesis was measured by tritiated diaminopimelic acid incorporation into growing cells and subsequent autoradiography. The pattern of incorporation observed was interpreted as a preferential synthesis in the central portion of the cell. This result was important as it supported the idea that growth in the center of the cell could account for the segregation of DNA; the DNA would presumably be attached to the surface of the cell astride the central growth zone. Recent data have contradicted the initial proposal, as it has been shown that in nonconstricting cells, there is no central zone of peptidoglycan synthesis.[165] Nevertheless, the variable-width proposal can explain these earlier results.[166] Only four length groups were analyzed in the original experiments. If these were chosen to have an equal number of cells in each group, then a large number of cells would be old because they were wide in the short age class; i.e., in the class of short cells, older cells would be invaginating because these cells are wider than average. According to Fig. 6–13, it would be expected that some of these short, wide cells would be constricting. Thus, even the smallest size group would have a class of constricting cells. These cells could give the observed central zone of diaminopimelic acid incorporation. The variable-width proposal can thus explain the original zonal-synthesis observations. The importance of this reanalysis is attributable less to its application as a clarification of the zonal-diffuse surface-growth controversy, than to the small but additional support for the proposal of an inverse length-width relationship.

J. Length Distribution of Newborn Cells

When the length distributions of newborn cells, obtained by membrane-elution, are compared to the length distribution of the total culture, the two distributions show an unexpected, but decided, overlap.[167] The

---

Figure 6.13 Peptidoglycan Synthesis in Septate and Nonseptate Cells of the Same Length. The cell pattern of Fig. 6.12 is redrawn with the cells of length 4.10 indicated between the vertical lines. This length is equal to the length of a mid-cycle (age 0.5) cell of median width (diameter 1.0). Assume that this mean-width cell initiated invagination in mid-cycle, i.e., at age 0.5. Further assume that cells of any width start invagination at a volume equal to the mid-cycle cell of width 1.0. The wider cells will have initiated invagination before reaching length 4.10. Narrow cells will not yet have started invagination at this length because the narrower cells will have a smaller size than that required for initiation of invagination. The value written on the cells between the vertical lines are the calculated rates of peptidoglycan synthesis for those cells determined by calculating the absolute increase in cell surface for a 1% increase in cell mass. The cell-surface area before and after such an increase was calculated, the difference determined and presented as values normalized to 1.0 for cells of mean cell width. Inset: Peptidoglycan synthesis as a function of cell width for cells of the same length. The relative rates of peptidoglycan synthesis as a function of cell width at a constant cell length are plotted.

variable-width hypothesis predicts that there will be a great deal of variation in the length of newborn cells, with some of the thinner cells being quite long (Fig. 6–12). This variation can account, at least in part, for the observed overlap in the length distributions of newborn cells and the total population. The observation that the size (volume) of cells measured electronically with a Coulter Counter has a narrower distribution of newborn sizes, compared to the size distribution of the entire bacterial population, is consistent with the proposal of a relatively constant size at birth and the variable-width model. (In Chapter 14 the length-growth law will be derived; see Fig. 14–2.)

K. Molecular Mechanisms of Shape Determination

Previous attempts at understanding cell shape have relied primarily on mechanisms that might determine cell length or cell width as individual components of cell shape. For example, the minimal-cell-length model of Donachie, Begg, and Vicente[168] deals with length determination, and the nucleoid-occlusion model of Woldringh et al.[169] deals with width determination. The nucleoid-occlusion model proposes that a large nucleus determines, in some way, a wider cell. Experiments with a thermosensitive dnaA mutant grown at intermediate temperatures did not reveal any change in cell width, although there was a significant change in the concentration of DNA per cell.[170] The view presented here is that the cell does not determine width or length alone, but measures or determines, in some way, the ratio of length to width.

How is this apparently constant shape determined? As seen in Figs. 6–10 and 6–12, there is no ideal length or width that the cell uses to determine its shape. Rather, the cell adjusts its length and width to accommodate, about some range of values, the volume or mass of a cell at a particular growth rate. No definite answer can be given at this time, but it is possible to speculate on possible molecular mechanisms for shape determination.

One possibility is that cell shape is determined by the chemical or physical properties of the peptidoglycan.[171] It may be that the peptidoglycan has properties that lead to invagination when the length of the side wall, in relation to the width, is too great; this could give a constant shape. If, however, the regulation of the shape is not inherent in the peptidoglycan, then we must look to some determinant of cell shape that is external to the peptidoglycan.

One possible way that membrane may allow the cell to adjust its shape is illustrated in Fig. 6–14. In this idealized view, the peptidoglycan tries to grow in length, while the outer membrane (or alternatively, the inner membrane) tries to assume a spherical shape. When the tube of peptido-

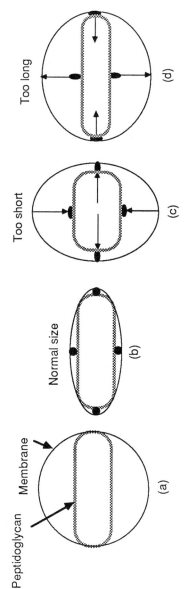

Peptidoglycan   Membrane

(a)   Normal size   Too short   Too long
           (b)   (c)   (d)

Figure 6.14 Membrane-Peptidoglycan Interaction for Shape Maintenance. In (a) the peptido-glycan shell, and the circle represents the shape the membrane would achieve if it were not constrained by the peptidoglycan. The membrane is attached to the peptidoglycan by linkers, primarily the lipoprotein covalently linked to the murein, so the membrane is actually found approximating the peptidoglycan (b). Because of this dissonance of shapes, the membrane is pulling outward on the side wall and pushing inward on the ends of the peptidoglycan. For the normal-shaped cell, there has been an accommodation of these two shapes to produce the observed cell shape. When the peptidoglycan shape deviates from the mean, becoming either too wide (and therefore being too short), as in (c) or too narrow (and therefore being too long), as in (d), there is a stress created that leads to the restoration of the mean shape. When the cell is too long (and too narrow), the stress on peptidoglycan in the longitudinal direction is weakened because of the inward pressure of the membrane on the outer ends of the cell. This leads to molecular rearrangements producing the volume growth in the hoop direction; this leads to a wider cell. Each increase in cell mass is accommodated more in the hoop direction than in the axial direction. When the cell is too short (and too wide) as in (c), the inverse happens; there is more pulling outward stress on the ends of the cell. This leads to growth in the side wall and an increase in length relative to an increase in width. Thus, with a particular balance of forces there would be a restoration of the shape determined by the stress relationships of the peptidoglycan and the membrane.

glycan is too long (or too short) the pressure (or pull) from the membrane systems either pushes (or pulls) the ends of the peptidoglycan. This stress leads to preferential incorporation of new peptidoglycan material either in the hoop or axial direction.[172] In this way, the cell grows so that it increases (or decreases) preferentially in width as the membrane pulls out on the peptidoglycan and pushes in on the sides of the peptidoglycan. Cells that are too short grow preferentially in length.

XI. THE VOLUME, SURFACE AREA, AND DIMENSIONS OF CELLS AT DIFFERENT GROWTH RATES

It is possible to calculate the cell's surface area at different growth rates from the assumption that the shape of the cell is constant. Theoretically, the cell mass or volume increases proportional to $2^{(C + D)/\tau}$, with τ equal to the doubling time.[173] Therefore, surface area increases proportional to $2^{2(C + D)/3\tau}$, and the radius and length increase proportional to $2^{(C + D)/3\tau}$. This is illustrated in Fig. 6–15. Measurements of cell length over a range of 0.6 to 3.0 doublings per hour fit this prediction.[174] This is additional support for the proposal that the cell maintains a constant shape at different growth rates. For each growth rate, and for a given volume and a constant shape, the length, radius, and area of a cell can be derived.

XII. THE CONSTRAINED-HOOP MODEL

The cell wall is extended in a passive manner in response to the increase in cell mass; the cell is visualized as a pressure vessel. For each increase in cell mass, wall growth accommodates the increase by increasing the cell volume.[175] The turgor pressure on the cell wall is constant. When cell poles grow over the latter portion of the division cycle with constant increases in pole surface area, there is an increase in the ratio of the rate of surface synthesis to the rate of mass synthesis.[176] Let us now consider what happens when there is a shift-up in growth rate. When *Escherichia coli* or *Salmonella typhimurium* are grown in minimal medium, and are shifted-up to a richer medium, the rate of mass synthesis increases almost immediately to the new growth rate.[177] There is a delay (rate maintenance) before the rate of increase in cell number changes to the new and faster rate. The cells then grow in a steady state at the higher growth rate with a larger mass. Woldringh, Grover, Rosenberger, and Zaritsky[178] reported the surprising result that the cell length overshoots

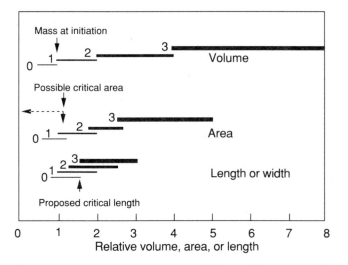

Figure 6.15   Volume, Surface Area, and Length of Cells at Different Growth Rates. For cells at four different growth rates, 3, 2, 1, and 0 doublings per hour, the volumes expected for the cells assuming a constant C and D period is presented. The volumes of the cells vary in the ratio of 1:2:4:8. By taking the two-thirds root and the one-third root for each of these beginning and ending values, we can calculate the expected area and length of these cells if we assume a constant shape. The values for the area and the length are normalized, so that the values for the 1-doubling-per-hour cells now increase from value 1 to 2. The other values are adjusted accordingly. There is one length that occurs at all growth rates; that is, the largest cell from the smallest set of cells is the same as the smallest cell from the largest cell culture. Also noted for the lines associated with different cell areas is the possibility that initiation of periseptal annuli can occur at some particular and constant surface area. If this were the case, there would be an initiation of periseptal annuli before the cell division in which it acted, as the growth rates became faster and faster. This is analogous to the premature initiation of DNA synthesis in division cycles before the division cycle in which termination of DNA synthesis occurred.

the final cell length during a shift-up. The average cell length increases after a shift-up to a value higher than the final cell length, and then decreases to the final cell length.

The cell grows during steady states by increasing in length with a constant width or diameter. The mechanism for length extension is becoming easier to visualize,[179] but changes in diameter accompanying larger cell size are more difficult to envision. Koch[180] outlined the problems involved in this change. An exponentially growing, constant-diameter cell can be maintained just by placing the newly inserted strands in a one-to-one correspondence with the pre-existing material encircling the cell perpendicular to the long axis. As more hoops of peptidoglycan, made up of relatively short lengths of glycan chain, are inserted into the cell wall between the resident hoops, the cell grows in length.[181] When the diameter increases, there must be a greater length of new peptidoglycan in the hoop direction compared to that in the old peptidoglycan. The cell slowly accommodates its diameter to the new cell size. During a shift-up, the rate of mass synthesis changes immediately, and requires a concomitant immediate increase in the cell volume to maintain a constant density[182] and turgor pressure.[183] The only way for a cell to grow is for it to preferentially increase in length. The cell is like a balloon that is constrained in a tube of constant diameter. If air is inserted while the balloon is prevented from expanding in circumference, the balloon must extend its length. So it is with bacteria. The relatively slow increase in the radius of the cell leads to an overshoot in the length of the cell.

The increase in length and area of a cell during a shift-up with a linear increase in the radius over a period longer than 60 minutes is illustrated in Fig. 6–16. The results are calculated assuming that the time required for the radius to increase is 200 minutes. The linear increase in radius is accompanied by an overshoot in cell length to a value greater than the final steady-state value. The observations[184] on the overshoot in cell length during a shift-up are consistent with this proposal.

The expected area with a constrained hoop during a shift-up is illustrated in Fig. 6–17. The surface area increases faster than the area of cells growing with a constant shape. There is an asymptotic approach to the final value by 200 minutes. The only shift-up experiment in which peptidoglycan synthesis was measured[185] gave results that are consistent with the predictions made here.

XIII. IS THERE A MINIMAL OR CRITICAL CELL LENGTH?

The achievement of a minimal or critical cell length has been proposed as a key regulatory point in the division cycle of rod-shaped bacteria.[186] The proposal of the model illustrated in Fig. 6–5 implies that cell length is neither a critical nor a regulatory factor. Rather, cell length follows from the constant shape of the cell and the cell size at different growth rates.

The proposal of a minimal or critical cell length derives, in part, from the observation that over the range of growth rates considered here (infinitely long interdivision times down to 20-minute interdivision times), every growth rate contains some cell of a particular length (Fig. 6–15). A simple calculation in which the length of the cell is proportional

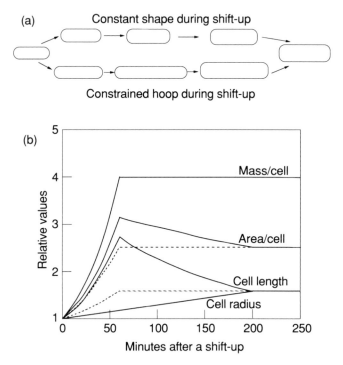

Figure 6.16 The Constrained-Hoop Model. When a cell is shifted to a richer medium, the average mass per cell increases. If the cell grows with constant shape before the new steady state is reached—with both length and width increasing proportionately—the growth is as described in the upper branch of (a). If, however, as proposed by the constrained-hoop model the radius does not increase rapidly, then the length extends to accommodate the new mass. This leads to an overshoot in cell length before a new steady-state length is reached. In (b) a plot of the changes in mass, area, and cell dimensions is presented. When a cell is shifted to a richer medium, leading to a fourfold increase in cell size, there is a corresponding change in the surface area and the cell dimensions. The dashed line indicates the changes expected with a constant shape. If there is a constrained hoop, and the radius or width of the cell is assumed to increase linearly over a period of 200 minutes (while the steady-state mass is achieved by 60 minutes), then there is an overshoot in cell length.

to the cube root of its volume (which follows from the constant-shape hypothesis) indicates that we would expect the longest cell of extremely slow-growing cells to have a length of 1.59, while the shortest cell from a rapidly growing culture would have a length of 1.58 (Fig. 6–15). [These values are normalized to cells of volume 1.0 (60-minute cells) increasing in length from 1.0 to 2.0.] At all growth rates there exists a common length. Width variability means that some cell interdivision length patterns, for example, that from the slow-growing wide cells and that for the

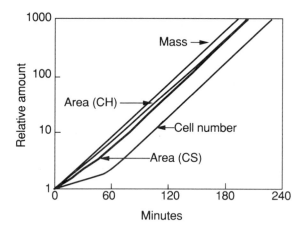

Figure 6.17 Surface Synthesis during a Shift-Up. During a shift-up the mass and cell number increase as indicated, but the area of the cell increases differently depending on whether there is a constant shape (CS) or a constrained hoop (CH). The final steady-state value is the same in both cases, but the pattern of wall synthesis indicates that the constrained hoop leads to a slightly more rapid increase in wall area compared to the constant-shape model.

fast-growing narrow cells, would not contain a common cell length; there would be no overlap in the cell-length distributions. This means that a particular cell length cannot trigger any particular event within the cell cycle.

The critical-length hypothesis, as originally proposed, is an empirical result with no mechanistic relationships. This is clearly stated by Donachie and Robinson:[187] "The empirical observation that cells become committed to division at an approximately constant length does not imply or require that they 'measure' their own length in some way, but only that a growth-dependent process required for division be completed at a stage in the cell cycle which is correlated with cell length."

XIV. GENETIC ANALYSIS OF THE DIVISION CYCLE

An enormous amount of effort has been expended to study the division cycle using the tools of genetics. Mutants have been isolated in various processes, the locations of these mutations have been mapped, and the regions sequenced and analyzed by the entire armamentarium of modern molecular genetics. These genetic approaches have made contributions, as a group, to the study of the processes of growth and division. I will discuss how successful has the genetic approach been, where has it been unsuccessful, and what might this tell us about the analysis of other complex biological systems.

A. *A Priori* Considerations of Genetic Analysis

The archetype of a successful genetic analysis is the understanding of a simple biochemical pathway, such as one involved in the synthesis of an amino acid. To study this pathway, mutants that require an amino acid are isolated using standard mutational and enrichment techniques. The location on a genetic map of these mutations is then determined by various genetic techniques. At the beginning of the study, various functional tests are used to classify the mutants. Complementation tests put the different mutants with the same original phenotype into different functional categories. Various molecules accumulate in the mutants, and can be isolated and their structure determined to aid in the elucidation of the biochemical steps in the pathway. Various enzymes that may belong to the pathway can be assayed to see if the enzymes are altered in the mutant strain. A particular enzyme may be altered in its sensitivity to elevated temperatures in a temperature-sensitive mutant. When the biochemical elements—i.e., the enzymes, precursors, and intermediates—of a pathway have been elucidated using genetic techniques, the regulation of the pathway can be analyzed using what have now become standard genetic manipulations. Mutants with altered regulation can be isolated; these can be placed in cells with either the parental or mutant sequences and the regulatory circuits analyzed. The genetic variants can be used to study the mechanism of action of a particular enzyme.

This genetic approach has been applied, with great success, to an enormous number of pathways and enzymes in bacteria. This is the paradigm of genetic success. But this model, applied to phenomena only superficially related to a biochemical pathway, for example the regulation of division, gives a more mixed and modest picture.

B. Mutations in Simple and Complex Processes

One difference between the processes related to classical biochemical pathways and those related to the synthesis of complex structures and the division processes, is the nature of the mutants that can be isolated. Mutants in the synthesis of an amino acid can be isolated by growing cells in the presence or absence of an amino acid. Further, the study of the mutant is aided by the ability to present or remove this amino acid from the medium. Such aspects as cytoplasm, DNA, cell surface, or cell division cannot be added to the medium. A mutant in DNA polymerase

Table 6.2 Classification of Morphogenes (according to Donachie, Begg, and Sullivan)

Map Location	Gene Symbol	Function
	Class I, Net peptidoglycan synthesis	
2	mraA	D-alanine carboxypeptidase
2	mraB	D-alanine requirement
2	murE	meso-diaminopimelate adding enzyme
2	murF	D-alanyl:D-alanine adding enzyme
2	murG	PG biosynthesis
2	ddl	D-alanine-D-alanine ligase
2	murC	L-alanine adding enzyme
4	mrcB	PBP1B
69	dacB	PBP4, D-alanine carboxypeptidase
75	mrcA	PBP1A, transpeptidase, transglycosylase
90	mrbA	UDF-N-acetylglucosaminyl-3-enolpyruvate reductase
90	mrbB	PG biosynthesis; D-alanine required PG biosynthesis
	Class II, Outer membrane morphogenes	
22	ompA	outer membrane protein 3a (II*;G;d)
36	lpp	murein lipoprotein
75	envZ	regulation of outer membrane proteins
	Class III, Cell elongation	
15	dacA	PBP5; D-alanine carboxypeptidase
15	rodA	cylindrical sacculus growth
15	pbpA	PBP2; transpeptidase, transglycosylase
71	envB	cell shape
74	crp	cAMP-binding protein (CAP)
85	cya	adenyl cyclase, cell shape
	Class IV, DNA replication and division	
10	lon	inactivation of sfiA protein, DNA-binding, ATP-dependent proteins La
22	sfiA	inhibitor for ftsZ protein
	Class V, Nucleoid separation and septum formation	
82	pcsA	chromosome segregation
83	gyrB	chromosome segregation and septum placement; DNA gyrase B

(continued)

Table 6.2 (*Continued*)

Map Location	Gene Symbol	Function
		Class VI, Septum initiation
0	*dna*K	septation (heat-shock protein)
2	*fts*I	PBPe, transpeptidase, transglycosylase
2	*fts*Q	septation
2	*fts*Z	septation
2	*sec*A	secretion of envelope proteins
2	*azi*	septation
30	*fts*G	septation but also cell viability
49	*fts*B	septation but also cell viability
		Class VII, Septum formation
0	*dna*K	septation
2	*fts*I	transpeptidase, transglycosylase
2	*fts*Q	septation
2	*fts*A	septation
2	*fts*Z	septation
2	*sec*A	secretion envelope proteins
4	*sef*A	septation
74	*fic*	cAMP-sensitive septation step
76	*fts*E	septation
76	*fam*	murein lipoprotein
76	*fts*S	septation
81	*env*C	septation
82–83	*fts*F	septation
86	*fts*D	septation
86	*fcs*A	septation
90	*fts*124	septation
		Class VIII, Septum (cell) separation
2	*env*A	division, control of N-acetylmuramyl-L-alanine amidase
		Class IX, Inactivation of septum-forming sites
10	*min*A	inactivation of septal sites
26	*min*B	inactivation of septal sites

cannot grow on a plate filled with DNA polymerase. A mutant with impaired cell division, in which septa are not formed, cannot be grown on a plate enriched with isolated septa. In order to study such processes using genetics, the concept of conditional lethal mutations has been developed. Mutants are isolated that can grow in one particular condition and that cannot grow in another particular condition. Temperature sensitivity has been the most popular approach. A number of temperature-sensitive mutants in various aspects of growth and division of the cell have been isolated.[188]

The genetic processes described above for biochemical pathways are then applied to a collection of mutants with a particular phenotype. The mutants are mapped, classified according to phenotype, cloned, sequenced, and placed in combination with other parental and mutant genes. This is one way to understand the regulatory circuits involved in growth and division.

C. Classification of Mutants Affecting the Cell Surface and Division

We may look at the mutants affected in cell division in two separate ways. First, we may classify them as mutants affected in the synthesis of the cell surface. Second, we may concentrate on cell growth, elongation of the rod-shaped cell, and formation of the septum. In practice, there is no easy distinction between these approaches. Donachie, Begg, and Sullivan[189] have listed and classified a large number of mutants affected in cell-surface growth and in cell division. Nine categories of mutants are identified, and in Table 6–2 they are listed in order of their presumed use in the process of cell division. All of the *morphogenes* that have been described are placed into one of these categories. The categories proceed in a linear order from what is perceived to be the initial events of surface synthesis to the final end result of surface synthesis, the two daughter cells. The best information available is for the mutants of Class I, those involved in the biosynthesis of peptidoglycan. This process is similar to any of the other biochemical processes going on in the cell; there is a pathway for the synthesis of precursors, and there are enzymes and various functions that lead to the incorporation of these precursors into the cell wall. This area of genetics has been quite successful.

Once we leave this area of biochemistry, the results are much more problematic. Sometimes the enzyme affected by a particular mutation is known, and the mutation gives a particular phenotype. But there is no understanding of how the particular phenotype is derived from the particular mutation. This problem is highlighted when looking at mutants in Classes V, VI, and VII, which are affected in the formation of the septum. These cells are filamentous at the nonpermissive temperature. It is relatively easy to enrich for filamentous mutants. Such cells are by definition larger than the nonmutant population, so all that one has to do is to pass the cells containing filaments through a filter with pores that allow the normal cells to pass through but that retains the larger cells. Nitrocellulose filters with an 8-micron pore size are suitable for this task,[190] and a large number of mutants have been isolated in this way. But even though these mutants have been mapped, sequenced, studied, mixed with other mutants, and manipulated in every conceivable way, as far as I am aware, none of these mutants has been explained in molecular terms.

This general observation leads to the following speculation. It may be that the paradigm for the analysis of genetic pathways does not pertain when these ideas and rules are applied to higher morphogenetic processes. For example, if it took a miniscule change in the biochemistry of pole synthesis to prevent septum formation, then an extremely minor genetic change could produce very large morphological changes. We may not be able to see changes in enzymes, even though these enzymes may be involved in the processes of septum formation. I propose that we must look at the genetics of these higher processes in a new way, looking for new paradigms that will enable us to understand the processes so different from the simple biochemical pathways.

XV. ON LAWS, CRITICAL TESTS, PREDICTIONS, AND EXCEPTIONS

In this chapter, a number of proposals regarding the rate and pattern of surface synthesis, and the determination of cell shape have been made. Many of these proposals have a speculative component, and some do not fit all of the available data. Although these proposals explain many different observations in the literature, I have pointed out a number of places where the experimental data are not in accord with the predictions. What can we make of these exceptions? The idea of a critical test of models is well known. Each of these exceptions, in a sense, may be viewed as a critical test of the model, and in view of the existence of these exceptions, we may say that the proposals made here may be incorrect. Francis Crick described an experience in his recent autobiography:[191]

> This failure on the part of my colleagues to discover the alpha-helix made a deep impression on Jim Watson and me. Because of it I argued that it was important not to place too much reliance on any single piece of experimental evidence. It might turn out to be misleading, as the 5.1Å reflection

undoubtedly was. Jim was a little more brash, stating that no *good* model ever accounted for *all* the facts, since some data was bound to be misleading if not plain wrong. A theory that *did* fit all the data would have been "carpentered" to do this and would thus be open to suspicion.

The problem of fitting a model to the data is particularly acute when considering the proposal for shape determination and maintenance. In this age of the proliferation of scientific literature, there are many experiments to look at when making a general proposal. It is common that not every piece of data fits every proposal. I have suggested that there is a general law of shape determination. Each strain of *Escherichia coli* has evolved some optimal shape, and it is this shape that we must understand. Despite exceptions to this general proposal, the large number of good experimental results implies that the proposal of a constant-shape, variable-width, size-determined cell cycle is substantially correct.

XVI. ICON FOR CELL-WALL SYNTHESIS DURING THE DIVISION CYCLE

In previous chapters, we have used an icon to summarize the different patterns of synthesis. The continuous synthesis of cytoplasm was indicated by a circle. DNA synthesis was indicated by a line outward from the circle. We now add cell surface to this icon. Mass synthesis and accumulation is the driving force for wall synthesis. First only the side wall is made, and then synthesis is partitioned between the pole and the side wall. At division we come to the end of this process. The icon for wall synthesis is shown in Fig. 6–18.

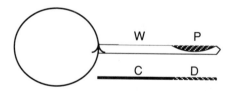

Figure 6.18 Icon for Surface Synthesis. Mass synthesis (the circle) drives wall synthesis and then both wall and pole synthesis. It is not known whether there is a relationship between the start of pole synthesis and the initiation of the D period.

NOTES

1. The discussion in this chapter is restricted to the Gram-negative, rod-shaped bacteria. Chapters 11, 12, and 13 will be devoted to a discussion of the the division cycle and surface growth of other prokaryotes.

2. The word murein was originally coined by Weidel to describe the wall-like properties of the bacterial peptidoglycan (Weidel and Pelzer, 1964). Murein is a word analogous to protein and nuclein, and it does not imply any structural connotations. The term murein has not been used very often, and the term peptidoglycan has dominated the literature.

3. Park, 1987a,b.

4. Muramic acid is N-acetylglucosamine with a D-lactic acid ether substituted at C-3.

5. Braun, Gnirke, Henning, and Rehn, 1973.

6. This analysis was originally done by F. B. Wientjes.

7. Glauner, 1986, 1988; Glauner, Höltje, and Schwarz, 1988; Glauner and Schwarz, 1988; Schwarz and Glauner, 1988.

8. A monomer is a single subunit that is not cross-linked. It contains two sugar residues. A dimer is made up of two cross-linked monomers and thus has four sugars.

9. Verwer, Nanninga, Keck, and Schwarz, 1978.

10. Verwer, Beachy, Keck, Stoub, and Poldermans, 1980.

11. Koch, 1988a.

12. Schaechter, Williamson, Hood, and Koch, 1962.

13. Höltje and Glauner, 1990.

14. Park, 1987a.

15. For review, see Park, 1987a; Matsuhashi, Wachi, and Ishino, 1990.

16. A donor is the peptide chain with the D-alanine participating in the cross-linking, while the acceptor is the chain that has the diaminopimelic acid participating in the cross-linking.

17. De Pedro and Schwarz, 1981.

18. This is not absolutely correct as recent work has revealed many other, albeit minor, links (such as DAP-DAP), which may be brought about by the energy from the bonds between the third and fourth amino acids. See Glauner (1988) and Glauner, Höltje, and Schwarz (1988).

19. Koch, 1983a.

20. Koch, 1983a.

21. Koch, 1983a.

22. Jacob, Brenner, and Cuzin, 1963.

23. Ryter, Hirota, and Schwarz, 1973.

24. Chung, Hawirko, and Isaac, 1964.

25. Schwarz, Ryter, Rambach, Hellio, and Hirota, 1975.

26. Hoffman, Messer, and Schwarz, 1972; Olijhoek, Klencke, Pas, Nanninga, and Schwarz, 1982.

27. Grover, Woldringh, Zaritsky, and Rosenberger, 1977; Grover, Zaritsky, Woldringh, and Rosenberger, 1980; Rosenberger, Grover, Zaritsky, and Woldringh, 1978a,b; Zaritsky, 1975a.

28. Pierucci, 1978.

29. Pritchard, 1974.

30. Donachie and Begg, 1970; Donachie, Begg, and Vicente, 1976.

31. Donachie and Begg, 1970.

32. Begg and Donachie, 1973, 1977.

33. Verwer and Nanninga, 1980.

34. Cole, 1964; May, 1963; Beachey and Cole, 1966.

35. Burman, Raichler, and Park, 1983a,b.

36. Begg, 1978.

37. Woldringh, Huls, Pas, Brakenhoff, and Nanninga, 1987.

38. Burman and Park, 1984.

39. Hoffman, Messer, and Schwarz, 1972.

40. Ryter, Hirota, and Schwarz, 1973; Schwarz, Ryter, Rambach, Hellio, and Hirota, 1975; Koppes, Overbeeke, and Nanninga, 1978; Olijhoek, Klencke, Pas, Nanninga, and Schwarz, 1982.

41. Meyer, De Jong, Demets, Nanninga, and Woldringh, 1979.

42. Koppes, Meyer, Oonk, De Jong, and Nanninga, 1980.

43. Schaechter, Williamson, Hood, and Koch, 1962.

44. Koppes, Woldringh, and Grover, 1987; Grover, Woldringh, and Koppes, 1987.

45. For review, see Sargent, 1978.

46. Kubitschek, 1987.

47. Koch, 1983a.

48. Cooper, 1988b; Cooper and Hsieh, 1988.

49. Cooper, 1988b.

50. Cooper and Hsieh, 1988.

51. Cooper, 1988b.

52. The insertion of the λ modifying the previous formulation (Cooper, 1988b) makes this formula dimensionally consistent.

53. Keasling, 1990, unpublished results.

54. Woldringh, Huls, Nanninga, Pas, Taschner, and Wientjes, 1988; Woldringh, Valkenburg, Pas, Taschner, Huls, and Wientjes, 1985.

55. Woldringh, Huls, Pas, Brakenhoff, and Nanninga, 1987.

56. Cooper, 1988b; Cooper and Hsieh, 1988. The initial results were obtained with *Salmonella* as this organism is fully 30 to 50 times more efficient at incorporating diaminopimelic acid than *Escherichia coli* when short pulses of label are used. Short-term labeling experiments with *Escherichia coli* are therefore impracticable. In later work, once the concept of the double-label experiment was understood, then N-acetylglucosamine could be used to demonstrate that the pattern of wall synthesis in *Escherichia coli* was similar to that in *Salmonella*. When longer incubations are used, the relative incorporation of *Salmonella typhimurium* is only 4 to 8 times more efficient than that of *Escherichia coli*. But the determination of the rate of synthesis of peptidoglycan requires short labeling times.

57. Pritchard, 1974.

58. Koch and Blumberg, 1976; Kubitschek, 1986; Kubitschek, Baldwin, and Graetzer, 1983; Kubitschek, Baldwin, Schroeter, and Graetzer, 1984; Martinez-Salas, Martin, and Vicente, 1981.

59. All that can be varied is the precise pattern of pole formation or pole growth.

60. Van Tubergen and Setlow, 1961.

61. Koch, 1983a.

62. Burman, Raichler, and Park, 1983a,b; Lin, Hirota, and Jacob, 1971.

63. Koppes, Woldringh, and Nanninga, 1978; Kubitschek and Woldringh, 1983.

64. Woldringh, Huls, Pas, Brakenhoff, and Nanninga, 1987.

65. Woldringh, Huls, Pas, Brakenhoff, and Nanninga, 1987. Although the results as summarized in this paper show that there is a jump in the rate of peptidoglycan synthesis in invaginating cells, the original data imply that a distinction cannot be made between a jump in the rate of peptidoglycan synthesis and a smooth increase in the rate of peptidoglycan synthesis.

66. Although the mathematical analysis starts off by noting that cell volume increases exponentially, at no point is this fact brought into the derivation of the rate of cell-wall synthesis. Further, analysis of the final formulas for the rate of peptidoglycan synthesis indicates that they do not fit the pattern derived from a constant cell density. This means that their mathematical analysis does not give a constant cell density during the division cycle.

67. Wientjes and Nanninga, 1989.

68. Goodell and Schwarz, 1985; Goodell, 1985; Chaloupka and Strnadova, 1972. A general review of turnover in a wide number of bacteria has been published (Doyle, Chaloupka, and Vinter, 1988).

69. Cooper, 1988b. It should be noted that Goodell and Asmus (1988) suggest that there is some turnover in the peptidoglycan of *Salmonella typhimurium*. This experiment was done with a strain that did not take up excreted peptides from the medium. My analysis of their Fig. 3 indicates that there is no turnover of peptidoglycan. There is no significant decrease in the radioactivity in the labeled cells over 2 hours, and any slight decrease should be corrected for the loss of some other stable material such as DNA or protein.

70. Cooper and Hsieh, 1988.

71. Höltje and Glauner, 1990.

72. Höltje, personal communication.

73. Cooper and Metzger, 1986.

74. Driehuis and Wouters, 1985.

75. Schwarz and Glauner, 1988; Glauner, Höltje, and Schwarz, 1988.

76. Cooper, 1988b.

77. Cooper and Hsieh, 1988.

78. Koch, 1990.

79. Ishiguro and Ramey, 1976, 1978; Ramey and Ishiguro, 1978.

80. Norris, Seror, Casaregola, and Holland, 1988; Holland, Casaregola, and Norris, 1990.

81. Cook, Kalb, Peace, and Bernlohr, 1980.

82. Schwarz and Glauner, 1988.

83. Driehuis and Wouters, 1987; Tuomanen and Cozens, 1987.

84. Goodell, Schwarz, and Teather, 1974.

85. De Jonge, Wientjes, Jurida, Driehuis, Wouters, and Nanninga, 1989.

86. Cooper and Ruettinger, 1975.

87. Tsuchido, Van Bogelen, and Neidhardt, 1986.

88. Lutkenhaus, 1983; Jones and Holland, 1985.

89. Robin, Joseleau-Petit, and D'Ari, 1990.

90. Beck and Park, 1976.

91. Norris, 1989.

92. Goodell and Schwarz, 1983; Goodell, 1983.

93. Burman and Park, 1983, 1984.

94. Burman and Park, 1984.

95. Cooper, Hsieh, and Guenther, 1988; Glauner, 1986.

96. Burman and Park, 1983.

97. Helmstetter, Pierucci, Weinberger, Holmes, and Tang, 1979.

98. Woldringh, Mulder, Valkenburg, Wientjes, Zaritsky, and Nanninga, 1990.

99. Woldringh and Nanninga, 1985.

100. Mulder and Woldringh, 1989.

101. Woldringh, Mulder, Valkenburg, Wientjes, Zaritsky, and Nanninga, 1990.

102. Rothfield, DeBoer, and Cook, 1990; Cook, Kepes, Joseleau-Petit, MacAlister, and Rothfield, 1987; MacAlister, Macdonald, and Rothfield, 1983; Rothfield and Cook, 1988.

103. MacAlister, Cook, Weigand, and Rothfield, 1987.

104. Koch and Burdett, 1984; Koch, 1983b, 1985.

105. Woldringh, 1976.

106. Woldringh, Mulder, Valkenburg, Wientjes, Zaritsky, and Nanninga, 1990.

107. Woldringh, 1976.

108. Koppes, Meyer, Oonk, De Jong, and Nanninga, 1980. The cells studied were very slow growing cells (165-minute doubling time), and so it would be expected that if DNA replication initiated invagination, no newborn cells would show invagination.

109. Ohki, 1972.

110. Carty and Ingram, 1981.

111. Daniels, 1969a,b.

112. James and Gudas, 1976.

113. Bauza, DeLoach, Aguanno, and Larrabee, 1976; Churchward and Holland, 1976.

114. Joseleau-Petit, Kepes, and Kepes, 1984.

115. Hakenbeck and Messer, 1977a,b; Pierucci, 1979; Pierucci, Melzer, Querini, Rickert, and Krajewski, 1981.

116. Joseleau-Petit, Kepes, and Kepes, 1984; Joseleau-Petit, Kepes, Peutat, D'Ari, and Kepes, 1987.

117. Cooper 1988b.

118. Pierucci, 1979.

119. Cooper, unpublished experiments.

120. Aldea, Herrero, Esteve, and Guerrero, 1980.

121. Boyd and Holland, 1977.

122. Schaechter, Maaløe, and Kjeldgaard, 1958.

123. Kubitschek, 1987.

124. Meacock, Pritchard, and Roberts, 1978.

125. Cooper, 1988b.

126. Schaechter, Maaløe, and Kjeldgaard, 1958.

127. Zaritsky, 1975a.

128. Schaechter, Maaløe, and Kjeldgaard, 1958.

129. Cooper, 1989. The original data for this table come from Trueba and Woldringh (1980) and Trueba, Van Spronsen, Traas, and Woldringh (1982).

130. Woldringh, De Jong, van den Berg, and Koppes, 1977.

131. Woldringh, Mulder, Valkenburg, Wientjes, Zaritsky, and Nanninga, 1990. It should be noted that Woldringh (personal communication) disputes the parallelism between these lines. Interested readers should undertake their own analyses.

132. Donachie, Begg, and Vicente, 1976; Donachie 1981.

133. Donachie and Robinson, 1987.

134. Helmstetter and Leonard, 1987b; see detailed discussion in Chapter 9.

135. Cooper and Weinberger, 1977; Cooper, Schwimmer, and Scanlon, 1978.

136. Pierucci and Helmstetter, 1976.

137. Helmstetter and Leonard, 1990.

138. Grover, Zaritsky, Woldringh, and Rosenberger, 1980; Woldringh, Grover, Rosenberger, and Zaritsky, 1980.

139. Cooper, 1989.

140. Aldea, Herrero, Esteve, and Guerrero, 1980.

141. Zaritsky and Pritchard, 1973.

142. Zaritsky and Woldringh, 1978.

143. Begg and Donachie, 1978.

144. Trueba, Van Spronsen, Traas, and Woldringh, 1982.

145. Trueba and Woldringh, 1980.

146. Cooper, 1988b.

147. Cooper, 1989.

148. Donachie, 1968; Pritchard, Barth, and Collins, 1969.

149. Woldringh, 1976.

150. Grover, Woldringh, and Koppes, 1987.

151. Trueba and Woldringh, 1980.

152. Trueba and Woldringh, 1980.

153. Aldea, Herrero, and Trueba, 1982.

154. Trueba and Woldringh, 1980.

155. Wientjes and Nanninga, 1989.

156. Woldringh, Huls, Pas, Brakenhoff, and Nanninga, 1987.

157. Cooper, 1988b; Cooper and Hsieh, 1988.

158. Wientjes and Nanninga, 1989.

159. Cooper, 1988a.

160. Cooper, 1988b; Cooper and Hsieh, 1988.

161. Grossman, Ron, and Woldringh, 1982.

162. Nanninga, den Blaauen, Voskuil, and Wientjes, 1985; Nanninga, den Blaauwen, Nederlof, and de Boer, 1983.

163. Grossman, Ron, and Woldringh, 1982.

164. Ryter, Hirota, and Schwarz, 1973.

165. Woldringh, Huls, Pas, Brakenhoff, and Nanninga, 1987.

166. Ryter, Hirota, and Schwarz, 1973.

167. See, for example, Dwek, Kobrin, Grossman, and Ron, 1980.

168. Donachie, Begg and Vicente, 1976.

169. Woldringh, Mulder, Valkenburg, Wientjes, Zaritsky, and Nanninga, 1990.

170. Pritchard, Meacock, and Orr, 1978.

171. Henning, 1975; Henning and Schwarz, 1973.

172. Koch, 1983a.

173. Donachie, 1968.

174. Donachie, Begg, and Vicente, 1976.

175. Cooper, 1988b; Cooper and Hsieh, 1988. Also see Chapter 1.

176. Cooper, 1988b.

177. Kjeldgaard, Maaløe, and Schaechter, 1958; Cooper, 1969.

178. Woldringh, Grover, Rosenberger, and Zaritsky, 1980.

179. Burman and Park, 1984; Park and Burman, 1985.

180. Koch, 1988a.

181. Glauner, Höltje, and Schwarz, 1988.

182. Kubitschek, 1987.

183. Koch and Pinette, 1987.

184. Woldringh, Grover, Rosenberger, and Zaritsky, 1980.

185. Sud and Schaechter, 1964.

186. Donachie, Begg, and Vicente, 1976.

187. Donachie and Robinson, 1987.

188. A variation of the temperature-sensitivity approach, but which uses a temperature-sensitive nonsense suppressor to isolate nonsense mutations in particular functions has been used (Beckman and Cooper, 1973). This approach has not lived up to its initial promise.

189. Donachie, Begg, and Sullivan, 1984; see also Donachie and Begg, 1990.

190. Van de Putte, van Dillewijn, and Rorsch, 1964.

191. Crick, *What Mad Pursuit*, Basic Books, 1988.

7

Density and Turgor during the Division Cycle

I wish that this chapter could be as short as the famous chapter on reptiles in *The Fauna of Ireland* which was, in its entirety, "There are no snakes in Ireland." There is not much to say about cell density and turgor pressure during the division cycle of bacteria, except, in contrast to snakes in Ireland, they do exist, and they must be taken into consideration as a fact of cellular life. But density and turgor have no particular relationship to the division cycle.

Much effort has been expended on the determination of the density of a cell during the division cycle.[1] The cell density did not appear to vary during the division cycle, leading to the suggestion that there is some sort of regulatory mechanism—an isodensity mechanism—that keeps the density constant. Conversely, interest in cell-cycle density increased when theoretical considerations indicated that the density of the cell might vary in a regular manner,[2] and that density variation could play a regulatory role in the division cycle. Similar considerations also led to conjectures that variations in turgor pressure might regulate events during the division cycle.

I. DENSITY DURING THE DIVISION CYCLE

A. *A Priori* Considerations of the Meaning of Cell Density

There are two views of the density of the cell. One view defines density as the weight or mass per unit volume of cell. If we could weigh the cell and independently determine the external dimensions, we could calculate the density of the cell. In the other view, the density of the cell is determined by the average density of the components of the cell. The various cell components have different densities: DNA has a relatively high density of approximately 1.7; lipid has a density less than 1.0; protein has a density of approximately 1.2; and RNA is slightly more dense than DNA. The weighted average sum of the different cell components would give a particular cell density. The difference between these two ways of looking at the cell density depends on how impermeable the cell surface is to the medium in which density is measured. If the cell surface were a porous sieve, then the second concept would be more applicable, while if the surface were completely impermeable, then the first concept would apply.

In order to have an observed density change due to a change in cell composition, there would have to be a large change in the composition of the cell. For example, if the DNA content, relative to the other cell components, doubled during the division cycle, only a slight change in density would be observed. This is because DNA is only 3% of the total cell mass.

The alternative way to look at the density of a cell is to consider the total volume of the cell as determined by its physical boundaries and at the same time consider its total weight or mass. If a given cell volume grew larger while the mass did not change, the density of the cell would decrease. The problem with this *Gedanken* experiment is that it is hard to imagine what to put in the cell while it expands.

In Chapter 6, it was argued that the cell volume expands to just accommodate, in a continuous manner, the expanding cell mass. If the cell grew this way, then cell density would be invariant during the division cycle. The increasing mass of the cell causes the expansion of the surface so that cell density stays constant. If there is too much mass per volume at any particular instant, this information is fed immediately to the cell surface (by pressure and stress considerations; see Chapter 6). The excess *stress* due to the excess mass is relieved by the growth of the cell surface and cell wall. The surface-stress model (Chapter 6) proposes no other way of cell-wall growth except to accommodate the cell mass. This analysis implies that we should not expect to see density changes during the division cycle. The observation of a constant density is not a major observation but merely the concomitant result of the mechanisms of cell growth and division. There is no regulatory mechanism that feels density and then increases or decreases it to some particular defined *isodensity*.

Even if there were density fluctuations during the division cycle, it would be difficult to attribute these to any particular mechanism. As an analogy, consider a car with a full tank of gasoline leaving on a trip. If we measure the density of the car during the trip the density is high at the start (highest weight per constant volume) and decreases continuously with distance. It may be that density is related to some other parameter, in this case, the need to fill the car with gasoline, but as a regulatory element related to running the car, the density has no meaning.

B. Experimental Determination of Cell Density

In concept, the density of the cell is easy to measure. The method of choice would be isopycnic centrifugation in which cells are allowed to float in a density gradient. The density distribution is determined by noting at what density the cells come to rest. A number of different materials have been used to form these density gradients. For example, Ficoll (a synthetic polymer of sucrose) and Ludox (colloidal silica) were used in early studies on the buoyant densities of bacteria. Later experiments were improved with the introduction of colloidal silica coated with polyvinylpyrrolidone (PVP). The PVP coating reduces adherence of the cells to the particles. The main problem with using such methods for density determination is that long periods of centrifugation are needed for the cells to come to an equilibrium density, and the cells may be altered in some way by the centrifugation.

C. Cell Density during the Division Cycle

Three different aspects must be considered when looking at cell density: the absolute density, the variation in density during the division cycle, and variation during different growth conditions. With regard to the absolute determinations of density in different laboratories, there are relatively wide variations; densities of 1.06,[3] 1.10,[4] and 1.13[5] have been reported. These discrepancies are probably attributable more to differences in experimental technique among laboratories than to differences among bacteria. Within a culture, slight variations in density have also been observed. Some of the earliest measurements reported variations on the order of less than 1%.[6] Other experiments have reported fluctuations as small as 0.7%[7] or as large as 5%.[8] Such wide fluctuations obtained in these early experiments have been attributed to problems with overloading, with equilibrium, and with removing cells from a gradient. I know of no attempt to take cells with different densities and place them in the same gradient to see whether there are density differences under identical conditions. Further experimental work reduced the variation to less than 0.15%,[9] and the mean cell densities did not vary by more than 0.1%. This is close to the limit of determining density, and for all practical purposes we can now say that the density of the cell during the division cycle is constant. Even with cells growing at different growth rates, density appears to be invariant. Between growth rates of 0.3 to 2.5 divisions per hour, the variations were no more than 0.03 g/ml.[10] This independence of density from growth conditions depends on a constant osmolarity of the growth media.[11] The basic conclusion drawn from these observations is that cell density is invariant. It appears to be invariant with growth rate, and it is for all practical purposes invariant during the division cycle.

D. The Meaning of Isodensity

There are two ways to look at the invariant density of the cell. Constant density may be due to some regulatory mechanism; this requires a powerful density-regulating mechanism. The alternative is that the isodensity is merely the result of the essentially invariant cell composition. It is important to observe that the cell measures its density only by comparison with the outside world. In the same way that we know some object is hot or cold by the difference between its temperature and our skin temperature, so the idea of an absolute density has no meaning for the cell. A cell in distilled water would feel more dense than a cell in a dense solution. If density were a regulatory mechanism, then we would expect to find that changes in the medium density would cause large changes in the growth and division of cells. The small magnitude in the variation in cell density during the division cycle makes it difficult to imagine what kind of receptor could be used by the cell to determine such small changes. Even if such changes could be detected, it is difficult to know how the cell would use these changes as precise regulators of events during the division cycle. Cells appear to grow in medium of widely varying densities, so the small variations in observed density cannot be used by the cell to determine any particular events during the division cycle.

II. TURGOR DURING THE DIVISION CYCLE

In living cells there is a pressure exerted outward from the contents of the cell against the wall or membrane of that cell; this is called the turgor pressure of the cell. From what does this turgor pressure arise? If we consider a cell in isolation, a column of water in that cell gives rise to a simple hydrostatic pressure. This effect is negligible and may be discounted because the bacterial cells we are concerned with are usually floating in water. Therefore, the hydrostatic pressures are equal inside and outside the cell and cancel out. Active mechanisms that lead to the influx of ions or metabolites into the cell also increase the turgor pressure. If we consider a cell with a particular content, and a moment later the cell has drawn into the interior a molecule by some active mechanism with no corresponding expulsion of any other molecule, the pressure in that cell will have increased by a small amount. An additional source of internal pressure in a cell, and probably the dominant one, is the osmotic pressure. The osmotic pressure is most important because of its magnitude, and the fact that it is a source of differential pressure—i. e., the pressure inside the cell must be greater than the pressure outside the cell.

When a cell is at equilibrium—when the turgor pressure pressing onto the containing walls of the cell just equals the restraining forces of the outer walls of the cell—then we say that the wall pressure equals the turgor pressure. Consider a vessel made of a relatively nonelastic material with a volume of 1.0 milliliter. The vessel contains precisely 1.0 milliliter of water. Although there is a small hydrostatic pressure due to the height of the water, we may define this as a situation where there is no pressure inside the vessel. Now consider that one microliter of water is injected into this vessel. There will be a small increase in the pressure of the water against the vessel; we can now measure this pressure against that in the cell at equilibrium. There will be some compression of the water and a corresponding expansion of the vessel to accommodate the added volume of water. If this expansion did not happen, the pressure would be infinite and we could not have added the small volume of water. At this time we may say that the turgor pressure of the contents of the vessel is balanced by the wall pressure compressing the contents. In many cells, particularly plant cells, the cell is quite rigid. When the turgor pressure increases, there is a slight change in the size of the cell due to the osmotic pressure. In bacteria, this is not necessarily the case. The bacterial wall is expandable. The increased water is accommodated not by an increase in pressure, but primarily by a stretching of the cell wall to accommodate this increased mass of water (see Chapter 6). Thus, an influx of water does not lead to a proportional increase in turgor or wall pressure because of the stretching of the cell wall.

The turgor pressure of a cell is extremely difficult to measure, and it is even more difficult to determine this value during the division cycle. To date there there has been only one reliable set of measurements on bacteria;[12] these were on a gram-negative organism, *Ancylobacter aquaticus*, which has visible gas vacuoles within the cell. As these vacuoles have a gas pressure in equilibrium with the pressure inside the cell, we can apply an external pressure and observe the moment when these gas vacuoles collapse. By using a sophisticated system in which cells were observed as the pressure increased, the turgor pressure was determined in cells of different sizes. It was found that there was no observable change in the measured turgor pressure as a function of cell size. This indicates that there is no measurable variation in turgor pressure during the division cycle.

III. ARE DENSITY AND TURGOR REGULATED DURING THE DIVISION CYCLE?

An interesting philosophical question arises: is the existence of a constancy of a particular phenomenon an invitation to look for a mechanism of its constancy? The notion of regulation embedded in this discussion comes from the regulation of enzyme synthesis. When cells are grown in different conditions, there are changes in the amount of enzyme in a cell. In some well-known cases, such as β-galactosidase, these changes can range over the order of 1000-fold. When a minor perturbation, such as adding a small, nonmetabolizable compound, produces such a large change, the notion of regulation is clear. But if we could not change the expression of β-galactosidase, and no matter what we did or added to the medium, the enzyme concentration did not change, would we conclude that there was a very precise regulatory system, which made sure that the amount of enzyme per mass or per cell remained constant? We cannot have it both ways—that variation proves regulation and nonvariation proves regulation. I conclude, from the constancy of density and turgor during the division cycle, that the cell does not really regulate them in any particular way that affects the division cycle. Density and turgor are properties of cells, inasmuch as the cell is opaque or translucent or has some other irrelevant property. The density and turgor may be objects of study, but I suggest that they are not objects of study with regard to the division cycle.

NOTES

1. Kubitschek, 1987.
2. Pritchard, 1974; Rosenberger, Grover, Zaritsky, and Woldringh, 1978a.
3. Martinez-Salas, Martin, and Vicente, 1981.
4. Kubitschek, 1974.
5. Koch and Blumberg, 1976.
6. Koch and Blumberg, 1976; Woldringh, Binnerts, and Mans, 1981.
7. Koch and Blumberg, 1976.
8. Poole, 1977a.
9. Kubitschek, Baldwin, and Graetzer, 1983.
10. Kubitschek, Baldwin, Schroeter, and Graetzer, 1984.
11. Baldwin and Kubitschek, 1984.
12. Pinette and Koch, 1987.

8

Variability of the
Division Cycle

In the preceding chapters we have treated the cells as though the timing of the various parts of the division cycle were constant and did not vary from cell to cell. There is, however, a significant amount of variability between cells. Cells of the same age will be of different sizes, and cells of the same size will be of different ages. Even newborn cells are not all the same size. Cell-cycle interdivision times are also very variable. The sections of this chapter explore the implications of size and temporal variation during the division cycle.

This variability during the division cycle can be considered either a problem to be minimized, or a source of understanding of the division cycle. In previous chapters, when the observed results did not fit an idealized pattern, this was attributed to the dispersion of interdivision times. If there were no dispersion, the observed results would be much more precise. We will now see how the observed temporal and dimensional variability can be understood, explained, and rationalized in terms of the model of the division cycle described in Chapters 2, 4, 5, and 6.

I. OBSERVED VARIATION DURING CELL GROWTH AND DIVISION

A. Comparing Variations

In order to compare the variation of parameters with different absolute mean values, we shall use the coefficient of variation as a measure of the variability. If two measured parameters of the cell have different mean values, the standard deviation of the values might be expected to differ in magnitude. A mean value of 100 might have a standard deviation of 10 (i.e., 100 ± 10), while a parameter with a mean value of 10 might be ± 1. The coefficient of variation (CV) is the standard deviation expressed as a percentage of the mean value. In these two examples, the coefficients of variation would be the same, 10%, and we would say that the degree of variability in both cases is similar. By using the coefficient of variation, rather than absolute standard deviations, we may compare different types of variables, such as size and time variations, as well as differences in the same type of variable.

B. Size Variations

There are two aspects we may consider variables: the interdivision times (or the times between cell-cycle events), and the sizes of the cells at various times during the division cycle. There is a variation in the size of newborn as well as dividing cells; this variation occurs throughout the division cycle.

An example of the observed variation in the size of cells at various times during the division cycle is presented in Table 8–1.[1] The average size of the newborn cells is half that of the dividing cells, but the coefficient of variation is larger in the newborn cells. This can be explained by a variation, about the value of 0.5, at which cells divide evenly in half. If all cells divided perfectly in half, then the coefficient of variation would be identical in dividing and newborn cells. It could also be that the experimental error in size measurement is a constant, and therefore contributes more to the variation in cell sizes in the smaller cells. There is a degree of variability, however, even about the size of the variation. For example, in slow-growing cells, the CV of the size at initiation varied between 9 and 15%.[2] A reanalysis of the sizes at nuclear and cell division gave values of 11 and 14% respectively.[3] At this time all we need note is that there is a degree of variability.

C. Temporal Variations

Although the variation in interdivision times is the most well-studied and -measured variable of cell division, a number of other time intervals have been studied and may have meaning, as for example, the interiniti-

Table 8.1 Cell Sizes and Their Variation at Particular Events during the Division Cycle

	Cell birth		Initiation of DNA replication		Initiation of cell constriction		Cell separation	
T_D (min)	L_o (μm)	CV_o (%)	L_i^α (μm)	CV_i^α (%)	L_c (μm)	CV_c (%)	L_s (μm)	CV_s (%)
90	1.93	12.8	2.39	22.3	3.63	10.0	3.87	9.9
180	1.78	14.3	2.42	14.8	3.30	10.2	3.55	11.4

T is the interdivision time in minutes, L_o, L_i, L_c, and L_s are the lengths of cells at birth, initiation of DNA synthesis, constriction, and cell separation, respectively. The corresponding coefficients of variation are indicated by the CV and the subscript.

ation time and its variation, or the time for the C or D period. One of the more important variables is the time between initiation and cell division, or the slightly shorter interval between initiation and the start of invagination. Each of these measurements may have a different variance, and the relationship of these variances to size variation and other variables merits some attention.

The coefficient of variation in the age at division is approximately 20%.[4] The analysis of the variation of interdivision times has a relatively long history. Early proposals attributed this variability to the accumulation of small variations in a number of sequential or parallel events composing the division cycle.[5] These models did not account for the relationship of cell size during the division cycle. A more deterministic model was then proposed,[6] which in its broad outlines is similar to the general model described in earlier chapters.

The experiments of Powell[7] should be noted. He observed and followed cells for a number of generations, determining the distribution of interdivision times, and the relationship of interdivision times among mother, daughter, and cousin cells. A number of different statistical distributions fit the observed interdivision-time distribution, but the biological relationship or meaning of these distributions was not apparent.[8] The statistical parameters regarding the variability of the division cycle have been derived from an analysis of the growth of synchronized cultures.[9] Because the pattern of DNA synthesis during the division cycle was not known when these determinations were made, the relationship of size and division was often couched in terms of sizes at division. We will therefore turn to a description of the introduction of variability into the division cycle in terms of the general model proposed in Chapter 2 and detailed in Chapters 4, 5, and 6.

II. ELEMENTS OF VARIATION DURING THE DIVISION CYCLE

A. Model for Variability during the Division Cycle

A general model, incorporating elements of variation, is presented in Fig. 8–1.[10] The model considers the variation of each of the elements in the division cycle, what they mean to the eventual variability in the timing and dimensions of the cells during the division cycle. Each of the major portions of the division cycle is allowed to vary, and the variations are summed to produce the observed variation. Three points should be noted. First, the introduction of the reciprocal–normal distribution of initiation times gives an interdivision time distribution that is skewed to the right, toward higher interdivision times. The reciprocal–normal distribution is a result of a proposed normal distribution of rates of initiator

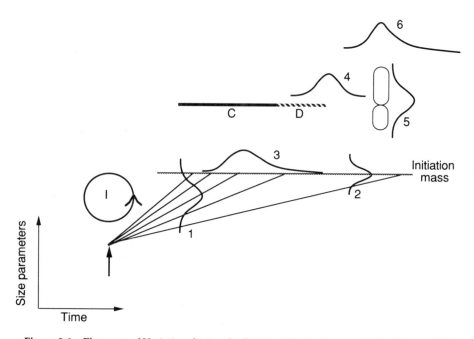

Figure 8.1   Elements of Variation during the Division Cycle. Assume that each part of the division sequence has some variability. If we start with a population of cells of identical mass (at the arrow), there will be a distribution of rates of mass synthesis (1). This distribution is assumed to be normally distributed. Initiation of DNA synthesis occurs when the required initiation mass is reached, but this mass (2) is also variable. Combining the variation in mass synthesis with the variation in initiation mass produces a distribution (3) for the variation in time at initiation. This distribution is approximately a reciprocal–normal distribution. After initiation of DNA synthesis, the cells proceed through the replication–segregation sequence (C and D periods), with variation in the C and D periods. For the total interdivision time of a cell, we may consider the C and D periods are summed to produce a single value for the variability. If we assume that this variation is normally distributed (4), then we have the time to division produced by the addition of a normal and a reciprocal normal distribution. Division is not perfectly symmetrical (5), and in the next division cycle the sister cells do not start off with exactly the same mass. The additional variation (5) due to the failure to divide perfectly in half means that a correct calculation of the size and age distributions must take into account the continuing history of the culture. Variables 1, 2, 4, and 5 are fundamental variables, in that they are the variation of the elements of the division cycle. Variable 3 is a derived variable, as is variable 6, which is the total interdivision time distribution. Other variations produced from this model are variations in the sizes of the cells when particular events such as initiation or termination of DNA synthesis or cell division occur.

synthesis. A normal distribution of rates of synthesis has been proposed previously.[11] This is a general finding for interdivision times. Second, the size at initiation is directly and simply related to the single variable, the cell size required to initiate a round of DNA replication. Third, the asymmetry at division leads to the larger coefficient of variation in newborn cells compared to dividing cells.

B. *A Priori* Qualitative Analysis of Variability

The easiest way to understand the contribution of each of the variable elements to the total cycle and size variability is to consider each of the elements alone.

1. No Variation

If we assume that all of the variables in the general model had no variation, that is, all the CVs were 0.0, then we would expect perfectly regular growth. All cells would have the same interdivision time, cells would be perfectly synchronized, and this synchrony would persist forever. Sizes would be identical in all cells at initiation, at division, and at birth. If we consider a synchronized culture made under these conditions, we would expect that the culture would grow with perfect synchrony and with no decay in the synchronization.

2. Variation in Replication–Segregation Sequence, the C and D Periods

Now consider that the only variation occurring in the culture is found in the replication–segregation sequence. Since all the other elements are invariant, we would find a perfectly regular and synchronous order of interinitiation times. Initiation of DNA synthesis would occur with perfect synchrony. If we started with a population of cells just at the start of DNA synthesis, each with the identical size (variable 2 in Fig. 8–1 would be invariant), then the time until the next initiation would be constant in all cells. There would, however, be some variability in the occurrence of cell division. This is because of variability in the time between initiations of DNA synthesis and cell division. Not all cell divisions would occur at the same time. Whatever variability was found, however, would stay the same in each succeeding generation. For example, if the variation in cell division was such that cells divided over a period of 10 minutes, then they would keep this same variation in succeeding generations. This is because the initiation of DNA synthesis would be perfectly regular and there would be no additional variation other than the regular variation in the replication–segregation sequence. The variation would be noncumulative and would not increase with time. The size at division would also vary, as the rate of mass synthesis would be constant. Cells that divided after a short C+D period would be smaller than cells that divided after a long C+D period. This size variation would not accumulate.

3. Variation in the Rate of Mass Synthesis

If we allow variation only in the rate of initiator synthesis, then we would expect pure cumulative variation. Again consider that a population of cells was selected exactly at the time of initiation of DNA synthesis. There would be a precise and invariant division occurring following the invariant replication–segregation (C+D) sequence. The next division, however, would be determined by the next round of synthesis of initiator, and this would vary. Some cells would initiate DNA synthesis early, and others would wait a longer time. There would be a variation in times over which the next division occurred. In the third division, we can expect cells with a high rate of mass synthesis to occasionally have a rapid mass synthesis in the next division cycle, and some cells would have two successive slow syntheses of mass. This would lead to a broadening in the interdivision time distribution. The independence of initiator synthesis rate in different generations produces a larger and larger spread in the interdivision times. If we knew the precise change in variation from the first precise and synchronized division to the second division, we could predict the third cycle variation. In a sense, the variability of the third cell division is the sum of two separate sequences of variation in the rate of initiator synthesis. The variation in the rate of mass synthesis leads to a cumulative variation in the synchronized culture. Synchrony gets worse and worse with time.

4. Variation in Initiation Mass

Now consider the case where the variation occurs only in the mass at initiation. We would again see a case of cumulative variation. This is because after one variable time for initiation, each cell starts again toward the next round of initiation. Because the masses at which initiation takes place are presumably independent in successive cycles, the next initiation of DNA synthesis again adds to the variability. For example, if one cell initiates DNA synthesis at a relatively small mass, and thus has a shorter interdivision time (because C+D is constant), then if at the next initiation, the daughter cells also initiated at a smaller than average mass, there would be two successive short interdivision times. In other cells there may be two successive long interdivision times. It is clear from this analysis that there will be a steady broadening in the interdivision time distribution. The original population of cells will exhibit less and less synchronized division as the variability accumulates at each interdivision time. Two differences from the pattern of variation caused by variation in the rate of initiator synthesis should be noted. First, in this case the size at initiation would vary, while in the case of variation only in mass synthesis, the size at initiation would be invariant. A second difference is that the variation in mass synthesis would produce a primarily reciprocal–normal distribution of interdivision times, while the variation in initiation mass would produce a symmetrical distribution in interdivision times that depended on how we defined the Gaussian distribution (i.e., normally distributed on a semilogarithmic plot or on a rectangular plot). In any case, it is likely that no major experimental distinctions could be made between initiation mass variation and variation in mass synthesis.

5. Variation in Symmetry at Division

Variation of the position of the septum introduces a cumulative variation attributable to differences between newborn sister cells. Two sister cells may be different owing to the imprecision with which cells divide. As a first approximation, the larger of the two sister cells initiates DNA synthesis earlier, on the average, as it has more initiation mass than its sister. The larger cell reaches the required level of initiation mass sooner. As different division cycles are independent, we will see a cumulative variation in the growth of a synchronized culture.

6. Variation in Total Interdivision Time

All of the variations described above do not occur in isolation. In any culture, all of these different variations contribute to the total variation in the cell interdivision time. By choosing different variations for each of the elements we could, in theory, fit any particular observation of variation in cell size or cell interdivision times. This is not recommended as a fruitful way to study or understand the growth of a cell, particularly as the data are very sparse and have a great deal of variability. This is a way of thinking about the variation so that it becomes understandable within the terms of the general model of the division cycle.

C. Size Homeostasis

During balanced growth, the size distribution of the cells does not broaden. There is a homeostasis of cell size, as illustrated in Fig. 8–2. No matter how variable the synthetic processes or divisions may be, at initiation of DNA synthesis, they all return to the same approximate size.

D. Cumulative and Noncumulative Variation

The distinction between cumulative and noncumulative variation was introduced by Marr, Nilson, and Painter.[12] Their proposal has not been

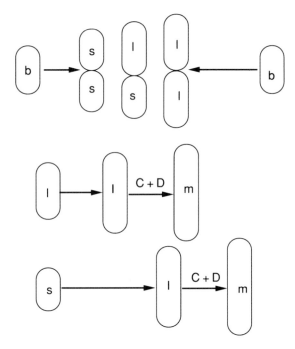

Figure 8.2 Size Homeostasis. Size variability can arise when newborn cells (b) divide after different lengths of time. If the rate of mass synthesis is the same (per amount of mass) in cells with both long and short interdivision times, we would expect the progeny produced at the next division to be either smaller (s) or larger (l) than the original newborn cell. Size variability can also arise from cells dividing asymmetrically and producing a large (l) and a small (s) cell from an intermediate-sized mother cell. A combination of variable sizes at division and asymmetrical division yields the observed size distribution of newborn cells. In the next generation, the larger cells initiate DNA synthesis earlier than average. Thus, the cell sizes at initiation of DNA synthesis are more identical than the sizes at birth. After a relatively constant C + D period, the mother cells (m) that are about to divide are more alike than either of the original cells (l,s) at the start of the division cycle.

widely considered in any general discussion of the variability of the cell cycle. The concept that Marr, Painter, and Nilson introduced was the idea of partitioning any observed variability into either cumulative or noncumulative variation. The difference between these sources of variability is illustrated in Fig. 8–3. In a synchronized culture with no variability, each step takes place over an infinitesimal time interval, and subsequent steps are as steep as the first step. With noncumulative variation, there is variability in each of the steps, but the steps do not change in subsequent division periods; there is a perfect timing mechanism that determines the precise average interdivision time of a cell and

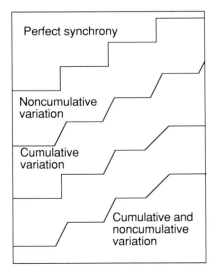

Figure 8.3 Cumulative and Noncumulative Variation. When interdivision times are invariant, perfect synchrony is achieved. With noncumulative variation, whatever variation is observed in one synchronous rise is found in all succeeding synchronous rises. This type of pattern means that there is some underlying precise clock that regulates the division pattern. With cumulative variation, each variation increases over the previous rise. There is a simple summing of the variations. Cumulative and noncumulative variation can combine to give the result that is found in growing cultures.

its descendants. The act of division is distributed, although this distribution does not increase. With cumulative variation, the first step is variable and with each subsequent step the variability increases, so that the slope of the increase at the time of division decreases at each step. It is as though each newborn cell starts the next division cycle completely independent of the previous division cycle. The slope of the succeeding division periods is approximately the product or sum of independent Gaussian distributions. The fourth curve illustrates a culture with a mixture of both cumulative and noncumulative variation. There is a continuous increase in the width of the increase in cell number at division. But the increase in width, or the decrease in the slope, is not so great as would be expected from the initial variability observed. Whatever noncumulative variability occurred at the first generation is subtracted from the variability at the subsequent generations, and the remaining variability—the cumulative variability—gives the subsequent decay curve for synchrony.

The possibility has been suggested that all of the variability in the *Escherichia coli* division cycle is attributable to variation in the D period.[13]

The D period is characterized by a minimal period of 17 minutes followed by an exponential decay period with a half-life of 4 to 6 minutes. This gives an asymmetrical distribution of the frequency of D periods. This model proposes that the initiation of DNA replication is precisely controlled with very little variation in the interinitiation periods.

Studies of synchrony curves obtained from newborn cells eluted from a membrane indicated that the variability could be partitioned so that for *Escherichia coli* B/r two-thirds of the variability is cumulative and one-third noncumulative.[14] These synchrony curves perfectly fit the general model. While this is not proof that we understand the sources of variability, it may serve a heuristic function. Division-cycle variability is explained by the model described in Fig. 8–1. If there are other sources of variability, we would have to demonstrate that any experimental results are not explained by the simple model proposed here.

The difference between cumulative and noncumulative variation may be considered in an analogy with work and vacation. Although the work day may vary, the time that the next work day starts is not affected by how early or late we left the job. If we leave work late at night, we will probably still start work at nine the next morning. The underlying clock that regulates the work day is tightly regulated, and does not accumulate. In contrast, when we go on vacation, and consider the distances traveled, the distances are usually cumulative. A long distance traveled one day does not usually affect the distance traveled the next day. If such distances are randomly chosen, then for a number of different travelers we would see larger and larger variations in distance traveled. The distance variation is cumulative. Of course, if we had a particular destination, then a long day of travel might be inversely correlated with the next day's travel. After a long distance, we may travel only a short distance the next day as we aim for the final destination. This is what is meant by a negative correlation between successive interdivision times.

III. VARIATION IN EQUALITY OF DIVISION

It is possible to determine the equality of division by looking at the placement of the septum in dividing cells. Early studies indicated that division was not symmetrical, in that one cell could be the progeny of the other cell. Not only was this suggested by statistical analyses of sister-cell interdivision times, but it fit into the prevailing idea at that time that there was a unit cell growth of rod-shaped cells.[15] If one cell had all of the older wall, that cell could be considered different from its sister cell. Statistical analysis of constricting cells indicated that division occurred about a mean size of 0.5, with only a random error accounting for the division of cytoplasm between the two daughter cells.[16] A more extended analysis of cells growing at different rates indicated that not only did cells divide evenly, but also that the partition was unbiased, and the precision was related to the cell shape; slender cells divided less precisely than short, squat cells.[17] Other determinations of the symmetry of division are consistent with no asymmetry of division and equivalence of the sister cells produced at division.[18]

A very precise test of the unit-cell model of growth was performed by Verwer and Nanninga.[19] After labeling the murein with diaminopimelic acid, and subjecting the growing cells to autoradiography, they categorized all dividing cells as having a right and a left half; the right half was defined as that cell with the larger number of grains. The mean number of grains in the left cell was then compared to the expected number of grains over the two cell halves, taking into account variability due to the Poisson distribution[20] of grain production in the emulsion. The results were completely in accord with symmetric growth. The lower number of grains in the left cell were shown to be attributable solely to statistical variation. This is a strong demonstration that, in the rod-shaped gram-negative cells, growth does not occur by a unit-cell pattern. The placement of the observed invagination site is presumably determined by events that have not yet been seen, and it is these events that determine the precision of the location of the septum.[21]

IV. CORRELATIONS AMONG DIFFERENT VARIABLES

A. Correlations between Different Interdivision Times

We can watch cells grow and divide under a microscope, and determine the time between divisions. The familial relationships can also be recorded. We can then ask what is the correlation between the the interdivision times of related cells. One of the earliest measurements of mother–daughter correlations[22] revealed a negative correlation. This was supported by other determinations.[23] More recently an analysis of the decay of synchronized cultures led to the conclusion that the interdivision times of mother and daughter cells are negatively correlated.[24]

If all cell interdivision times were uncorrelated, we would still expect to see some correlation. Consider that a cell will divide at a particular time. We record that time (this gives the interdivision time of the mother cell) and follow the two daughter cells to determine their interdivision times. Even if all cycles were biologically uncorrelated, we would expect there to be a negative correlation between mother and daughter interdivision times and a positive correlation between sisters. This is because any experimental error in determining the time of division would produce a correlation. A short mother interdivision time would be correlated with a longer daughter interdivision time. A constant amount cut from the interdivision times of two sister cells would lead to a positive correlation for sister-cell interdivision times. No correlation between interdivision times (a zero correlation coefficient) could be the result of the compensation of a positive mother–daughter correlation or a negative sister–sister correlation by the correlation due to experimental error.

B. Relationship between Size at Initiation and Size at Division

Koch[25] has proposed that if C+D were constant, then the size at initiation should be closely correlated with the cell size at division. The coefficient of variation should also be the same for the two size distributions. On the basis of an analysis of the radioautographic data,[26] Koch concluded that ". . . initiation is much less well controlled than cell division, and therefore initiation cannot control or time cell division." This proposal is limited by the experimental problems in determining size at initiation and size at division. Size at division is a directly observable parameter; all that needs to be seen is the dividing cell. A degree of inherent error is introduced by the fact that no matter how large a cell is, we never know how far away from division that cell may be. The size at initiation depends on two independent determinations. Not only must size be determined by the same experimental processes as the size at division, but an independent determination, with its own error, must be made of the time of initiation of DNA replication. It is therefore expected that the coefficient of variation in size at initiation will be larger than the CV of size at division.

Electron-microscopic autoradiography was used to measure the CV for size at initiation of DNA replication and the smaller CV for size at constriction.[27] The difference was determined by the relationship between the interval between initiation of DNA replication and initiation of constriction. A detailed and extremely complicated analysis of this interval indicated that there was a positive correlation between the length at which a cell initiated DNA replication and the onset of cell constriction.[28] The very high correlation coefficient (up to a value of 1.0), favored a model in which an event related to the initiation of chromosome replication triggered cell constriction.

C. Relationship of DNA Synthesis and Cell Division

Koch[29] has argued that autoradiographic analysis of DNA replication in cells of different sizes suggests that there is no strong correlation between DNA replication and cell division. Although a large array of data was considered by Koch,[30] the type of argument used, and perhaps the strongest set of data analyzed, indicated that for cells with a gap in the synthesis of DNA during the division cycle, " . . . never more than 90% of the cells in any size class were synthesizing DNA, and never more than 60% were not synthesizing DNA in any cell size class." This was interpreted as meaning that even though DNA synthesis may be occurring during only two-thirds of the division cycle in all cells, there is a great deal of variability in the time of initiation and termination within individual cells. It is important to realize that throughout the analysis of cell size and DNA replication during the division cycle, cell size was equated with cell length, and cell length was assumed to be a measure of cell age.

In Chapter 6 it was proposed that there was a variation in the cell width in exponential growth, leading to a variation in cell length and making length an unsuitable measure of cell age during the division cycle. Using this idea, we can reanalyze the predictions of Koch (Fig. 8–4) and conclude that it would be expected that no clear gap set of cells would be seen. Each cell length includes a variety of cell ages, so unlabeled cells can be found within each length class. At midlength of the population, there would be the older cells from the wider population and the younger cells from the narrower population. Extending this reasoning to the entire population, there would be a smoothing of the results. The data arguing against a strong coupling of DNA replication and cell division have an alternative explanation in the variable width proposal.

V. THE INVERSE AGE DISTRIBUTION

As human beings we tend to look at life as a progression from youth to old age. This is reflected in our study of the bacterial division cycle where newborn cells have an age of zero, and then progress through the cycle to dividing cells of age 1.0. While this has a logical consistency, thinking backwards in analysis of the division cycle has its rewards. There are two instances in which an inverted approach to the division cycle yields a clear understanding of the biological phenomenon and leads to improved experimental accuracy.

Upon occasion, we can understand an object or an idea better if we turn it around and look at it from another perspective. If we turn around, literally, the current or classical age distribution, we can gain intuitive insights that are otherwise hidden. Koch[31] has described the history and meaning of the age distribution, but the inverse age distribution is to be preferred over the classical distribution.

The thick line in Fig. 8–5a is the canonical age distribution for an exponentially growing culture with no variation in interdivision times. The thick line in Fig. 8–5b is the canonical inverse age distribution. Both of these curves can, by inspection, explain the exponential growth of a culture (thick line, Fig. 8–5c). As discussed in Chapter 1, the rate of cell increase in a culture is first proportional to the frequency of the older cells in the culture; with time, the rate of cell increase goes up as the younger cells at the start of the growth curve contribute to exponential cell increase. In Figs. 8–5a and b, there are twice as many young cells as old cells.

A difference between the two age distributions appears, however, when we consider cultures with variation in cell interdivision times. The thin lines in Figs. 8–5a and b illustrate what is expected when variation in cell interdivision times are considered. Because negative ages are not plotted in Fig. 8–5a, the variation is all confined to the right side of the distribution, at the older cells. Similarly, because in Fig. 8–5b we do not plot a negative age until division, all of the variation is at the right side of the distribution, at the younger cells. Now consider an experiment in which we compare an ideal culture (with an exponential age distribution with the cells all having precisely the same interdivision time) and a real

---

Figure 8.4 DNA Synthesis as a Function of Cell Length. For cells of constant size with variable widths, it is assumed that DNA synthesis occurs during the first two-thirds of the division cycle with a gap in the last third of the division cycle. This is illustrated in the topmost box where it is shown that 100% of the cells during the first third of the division cycle are synthesizing DNA; 0% of the cells are synthesizing DNA during the last third of the division cycle. The DNA synthetic pattern is presumably not perfectly synchronized with the division cycle, and so there is some deviation from an ideal square wave change in the rate of DNA synthesis during the division cycle. Because of this, the statement that cells initiate DNA synthesis at the start of the division cycle really means that cells initiate about the start of the division cycle. Half of the cells of the culture have started DNA synthesis before cell division, and the remaining half will start after cell division. The shaded boxes below the DNA synthesis pattern are representative of the relative number of cells synthesizing DNA in any size class. The black boxes indicate all of the cells are synthesizing DNA, and the shaded boxes indicate smaller fractions of cells involved in DNA synthesis. The white boxes indicate no cells synthesizing DNA. The shaded boxes are then placed atop each of the cell-size (length) distributions proposed to exist in a culture (see Chapter 6). Below this cell-size pattern is a summed distribution in which the presumed fraction of labeled cells of each length is shown by the shading in each of the boxes. By looking vertically, we can see that it is unexpected that cells of any length class are either 100% labeled or 100% unlabeled. This is summarized in the lowest box, where the DNA synthetic pattern according to length is indicated. This figure is not meant to be a precise representation of the actual numerical features of length, width, and DNA synthetic pattern. Rather, it is a diagram to illustrate how the analysis of the number of labeled and unlabeled cells may be affected by assuming that cell length is an accurate measure of cell age.

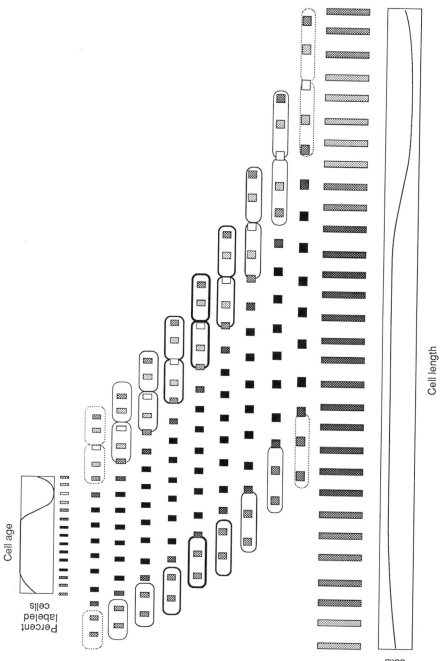

Percent labeled cells

Cell age

Cell length

Percent labeled cells

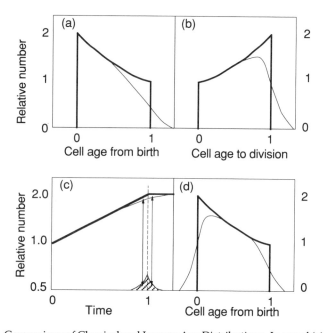

Figure 8.5 Comparison of Classical and Inverse Age Distributions. In panel (a) the canonical age distribution (thick line) and the variable age distribution (thin line) calculated with approximately 20% coefficient of variation are presented. Panel (b) presents the inverse age distribution. As in (a), the thicker line is the canonical distribution for an ideal culture with no variation in cell interdivision times, and the thinner line is the inverse age distribution with variation. The main difference between (a) and (b) is the abscissa, which defines the ages as being either from birth or until division. Panel (c) illustrates the curve expected when an ideal culture (no cell variation) is allowed to grow so that all cells divide just once (thick line). The thin line in panel (c) is the same curve with variation added. The small peak at the bottom of panel (c) is the difference between the two lines. The residual division curve is a straight line because of the ideal distribution of the ages from, or to, division [panels (a) and (b)]. The shaded areas are equal because the number of cells that did not divide by one ideal generation time must be equal to the number of cells that will divide after this ideal generation time. The reason for exponential growth of cell number even when there is a variability in cell interdivision times is that some cells that divided at the start of panel (c) will have a short interdivision time and divide twice before the ideal time is reached producing a smooth growth of cell number. In panel (d), the canonical age distribution is drawn with variation about both birth and division. This curve should be imagined as being derived by knowing the average instant of birth and the average instant of division. The main problem is that in an experiment there is no means to experimentally define the average age at birth.

culture (the cells do not all have the same interdivision times). What would be expected if we ask all of the cells extant at time zero to divide only once? The expected results are shown in Fig. 8–5c. The variable culture (thin line) extends its doubling over a time longer than a mean interdivision time, with a corresponding lowering in the rate of cell increase before the mean interdivision time is reached. The variability appears to be confined to the younger cells; that is, the difference between the ideal and the real occurs at the time when the younger cells (at the time the experiment starts) are producing the increase in cell numbers. The growth curve is ideal at the start of the experiment. This is not seen in Fig. 8–5a, where the variability is associated with the older cells at time zero. It is seen, however, in the dispersed version of the inverse age distribution (Fig. 8–5b). If we consider a real culture, with variable interdivision times, the age distribution as plotted in the inverse manner now differs from the canonical form in the same region as that in which the ideal growing culture differs from the real growing culture.

How can we explain an exponentially growing culture with both an ideal and a nonideal inverse age distribution? How can we expect to get exponential growth curves with two such different age distributions? We might think that the ideal curve would give an exponential curve, and the dispersed curve would have some perturbations. The difference between the two curves (Fig. 8–5c) can be explained as follows. When cells from a culture with variable cell ages at division are allowed to divide only once, some cells must divide at times greater than the mean interdivision time. This means that not all cells will have divided by one interdivision time. What is the meaning of the area between the two curves (Fig. 8–5c) to the left of the mean interdivision time? How can the variable age distribution give us a perfectly exponentially increasing culture during balanced growth? When there is cell-cycle variability, some cells that divide early will divide a second time before the end of the mean interdivision time; these are cells with a short interdivision time. The shaded area at the left (in the lower peak) is a measure of those cells, at time zero, that will divide twice before the mean interdivision time. The shaded area to the right represents those cells that have long interdivision times. The number of cells to the left and the right must be equal. In an exponentially growing culture, the interdivision time distribution produces an age distribution that just allows the culture to grow exponentially. The more variable a culture is with regard to cell interdivision times, the more to the left we will see a deviation between the real and the ideal growth curves of Fig. 8–5c. The earlier a cell was born in this experiment, the more likely it will divide again before one interdivision time occurs. Conversely, for every cell that divides twice within an interdivision time, we might expect that an equal number will divide after the mean interdivision time. This is the explanation of the shaded area in Fig. 8–5c and allows a graphic explanation of why both the ideal and nonideal inverse age distributions give an exponential growth of a culture. No such intuitive explanation is available for the normal, forward, variable age distribution.

The curves in Fig. 8–5c are the simple integral of the curves in Fig. 8–5b. The way Fig. 8–5c is drawn it can be intuitively seen that the slope of the thin line at the time of one doubling is the same as the initial slope. This can be explained by showing that the thin line in Fig. 8–5b crosses the midline at the same cell frequency as the initial cell number.[32] Further, the inverse age distribution is a graph that gives a clear explanation of the elution curves from membrane-elution experiments and explains why the midpoint of the downward sloping line should be used for choosing the time of one generation of elution.

The inverse age distribution was proposed in a different context by Painter and Marr[33] [their distribution r(l)]. They remark that this function of the ". . . remaining life length (life expectancy) . . ." is of interest in some experiments.

VI. THEORY OF BACKWARDS ANALYSIS OF THE DIVISION CYCLE

Historically, synchronized cells have been most commonly used for analyzing the biochemical and biological events during the division cycle. A synchronized population is one in which the cells divide during a relatively short time. By taking samples from newborn cells to dividing cells, it is generally believed that the rates of synthesis of various molecules during the division cycle may be obtained. As reviewed in the previous chapters, few results generally accepted about the division cycle were discovered using synchronized cells. As indicated by the analysis in Chapter 5, the only generally accepted aspect of the division cycle, the pattern of DNA replication during the division cycle, has come from a nonsynchrony method, the membrane-elution method; the membrane-elution method is not a synchrony method (see Chapter 3 for a complete analysis). True, the membrane-elution method makes a synchronized population, but it yields the best results when prelabeled cells are placed on the membrane and newborn cells are eluted from the attached cells. The pattern of incorporation in this case is derived from the radioactivity eluted with the newborn cells. The newborn cells come off the membrane in a particular order, from the oldest cells at the time of labeling to the youngest cells at the time of labeling. Since this older-to-younger order is not the usual way the division cycle is considered, the method is referred to as a backwards method. No further growth of the newborn cells is required, and thus there is no synchronized culture to be analyzed.

There is a reason for the difference in results between the two methods, and I will now present a simple explanation why backwards methods are inherently better, at least for the analysis of events that are correlated with division rather than with the birth of a cell. In this discussion, we will assume that there is no perturbation of the cells by the technique used to produce the synchronously dividing population. We will also assume that the C and D periods are relatively invariant in the population, at least compared to the variation in cell interdivision times. This is what is meant when an event such as initiation or termination of DNA replication is more closely correlated with cell division than with cell birth.

As can be seen in Fig. 8–6, when a measurement of the rate of DNA synthesis during the division cycle is performed on a synchronized population with a variable time between birth and the start and finish of DNA synthesis, we can derive the mean time for the synthesis of DNA during the division cycle (the C period), and the mean time between the end of DNA synthesis and cell division (the D period). In this case, however, we must estimate the midpoint of the rise of three sloping lines; that is, the increase in the rate of DNA synthesis, the decrease in the rate of DNA synthesis, and the increase in the cell number.

The analysis of the division cycle, using the prelabeling method and the membrane-elution technique, gives a different result. Given the conditions of relatively invariant C and D periods within the cells of the population, we can see (Fig. 8–6) that the increase and decrease in the rate of DNA synthesis—now measured by pulse-labeling the culture before placing the cells on the membrane—does not require an estimate of the midpoint of the rises and falls in rate curves.[34] As drawn in this ideal case (with no variation in the C and D periods), we can more easily measure the C and D periods in the fourth panel than in the third panel. In the subsequent generations, either of the synchronized culture experiment or the cells growing in the membrane-elution experiment, there would be a broadening of the curves. We would expect, however, that whatever cumulative variation was added with additional growth, the membrane-elution experiment would always retain a sharper synthetic curve.[35]

It could be said that if the time between birth and initiation of DNA synthesis were relatively invariant, then synchronization, rather than the membrane-elution technique, would be the method of choice. The point

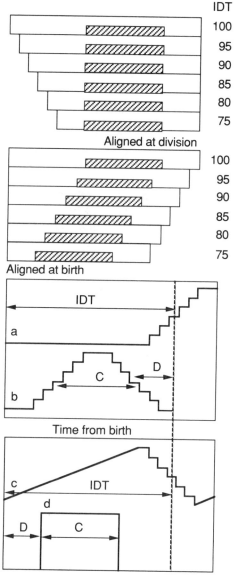

Figure 8.6 Comparison of Division-Cycle Analysis by Forward and Backwards Methods. The uppermost panel is an illustration of a culture of six representative cells that have various interdivision times (IDT) and invariant C and D periods (taken as 40 and 20 minutes, respectively). The cells are aligned at the right, at division. We can see that the cell that has the shortest interdivision time also has the shortest time between birth and the initiation of DNA synthesis. DNA synthesis is indicated by the hatched area. The second panel shows the same cells aligned at birth. A synchronized culture (composed of the six cells) is pulse-labeled at various times during growth, and the rate of DNA synthesis is

to be emphasized here is that the general model, as well as the history of the field, suggest that the C and D periods are relatively invariant, and that the time between birth and initiation of DNA synthesis is the source of most of the cell-cycle variability.

In support of the proposal that the membrane-elution approach is inherently better than the synchrony method, in many laboratories, mine and others that I have discussed this problem with, it is possible to get a measurement of the C and D periods using the membrane-elution experiment, while in a comparable situation, no simple measurement of the C and D periods could be obtained with the synchrony approach. One published example of this contrast is the work of Helmstetter and Pierucci.[36] The classic shift-up experiment[37] indicated a sharp break in the rate of cell increase following a shift-up. This sharp increase, subsequently confirmed using electronic cell counting and a number of shift-up experiments,[38] implies that C and D times are relatively invariant in a population.

It must be strongly reiterated that this discussion has to do with an inherent difference between the two techniques. Extrinsic considerations such as the possible perturbation of the cells in a synchrony experiment and that multiple labelings must be used in a synchrony experiment—in contrast to having all cells labeled for precisely the same time in the membrane-elution approach—suggest that backwards methods such as the membrane-elution technique may be intrinsically better for measurements of biosynthesis during the division cycle.

VII. AGE–SIZE STRUCTURE OF A BACTERIAL CULTURE

Every cell in a bacterial culture has an age and a size. What is the relationship of the age and size for each cell? The answer is not known, and it may be very complicated. But a graphic way of thinking about this problem, within the context of the division cycle of bacteria, is presented

---

measured. The results are seen in the third panel. There is a spread in the time at which cells divide; this gives the observed cell number curve (a). The rates of DNA synthesis observed are shown in curve (b). This curve is obtained by looking up at the second panel and noting the number of cells in the original population that are synthesizing DNA at any particular time. The C and the D periods can now be directly obtained by taking the times from the midpoints of the DNA synthetic curves to the midpoint of the cell-number curve. In the fourth panel we see the result of a membrane-elution experiment where the cell-number curve (c) is shown with the spread indicated by the downward slope. In this experiment, the cells are pulse labeled before being put on the membrane and the amount of label per cell is determined and plotted (d). Because there is no spread in the C and D periods, there is no slope in the radioactivity curve; therefore we can see the rises and falls in the rate of DNA synthesis more clearly than in the synchrony experiment.

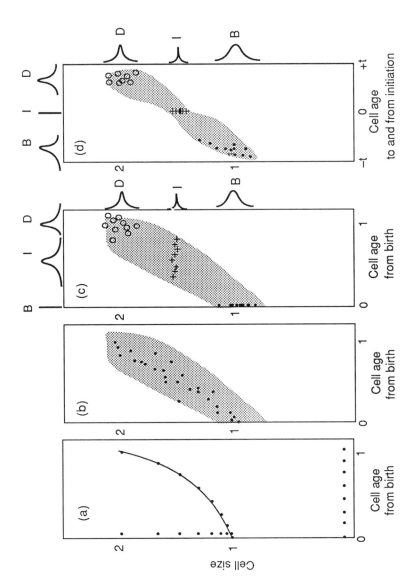

in Fig. 8–7. The age–size structure of a population is a representation of each of the cells in the culture and its age and size. In Fig. 8–7a, a purely deterministic age–size structure is illustrated. The projections of the dots on the panel to the bottom (age) axis and the left (size) axis indicate that when cells are growing exponentially, there are a greater number of smaller cells relative to the population than there are younger cells. Combining this result with the age distribution gives us the size distribution described in Chapter 4.

When variation in interdivision times and sizes at birth and division are introduced, the age–size distribution is as illustrated in Fig. 8–7b. The shaded area indicates a cloud of points preferentially collected around the middle of the shaded area, with fewer cells at the outer edges. It is possible to indicate the cells at particular times during the division cycle, such as birth, division, initiation of DNA synthesis; this is indicated in Fig. 8–7c. There is no distribution in the age at birth, since by definition this is age 0.0. The graphs at the upper and right sides of this panel are representations of the spread of the various distributions. There is some variability in the age at initiation of DNA synthesis and the age at division. The size at birth is not constant. It is expected, however, that the size at division will be slightly less variable than the size at birth. This is attributable to the possibility that unequal division of cells leads to a

---

Figure 8.7 Age–Size Structure of a Bacterial Culture. Panel (a) is the age-size structure for a perfectly deterministic population growing exponentially during the division cycle. The dots on the exponentially increasing line are placed at equal age intervals (shown by their representation at the bottom of the panel). The representation of the dots at the left of this panel indicates that there is a greater concentration of smaller cells than younger cells. This is the graphic explanation for the size distribution of cells described in Chapter 4. In panel (b), the age–size structure for a population with variation in size and interdivision times is illustrated. The cloud of points (indicated by a few points as representative) is one possible age–size structure. In panel (c) the newborn cells are indicated by filled circles, the dividing cells by open circles, and the cells in the act of initiation of DNA synthesis by + signs. It can be seen that the larger cells at birth will, on average, reach the size required for initiation more quickly than smaller cells. This is because any line drawn from a filled circle to a + is shorter (on average) when the filled circle is at a larger size. Panel (d) is a replotting of the pattern in panel (c) with a time scale defined by the time of initiation of DNA synthesis. Cells before initiation have a negative age value and cells after initiation have a positive age value. Initiation takes place, by definition, at age 0.0. There is some variation in the size of cells at initiation. The narrowing of the age–size structure at the time of initiation is a graphic representation of the size-homeostasis mechanism. No matter what size cells are at birth or division, they are returned to their proper age–size relationship at the instant of initiation of DNA synthesis. At the top and right panels of (c) and (d) are representations of the presumed variation of the sizes and ages of cells at particular events. The size at birth is always a little more widely distributed than the size at division due to a slight inequality of division. The size at initiation of DNA synthesis is drawn with a relatively small variability.

broadening of the size distribution. It is clear from panel (c) that larger newborn cells will reach the size at initiation earlier than smaller newborn cells.

An alternate way to look at the distribution is to replot the ages using the age at initiation of DNA synthesis as a starting point (Fig. 8–7d). By defining the age at initiation as 0.0, we get negative ages for cells before initiation and positive ages for cells after initiation. We now see that there is no distribution in the age at initiation, but a variation in the age of newborn cells. The bottleneck at initiation of DNA synthesis enables cells born of different sizes to retain size homeostasis. Since all cells to the left must pass through the bottleneck on the way to division, all cells, of any newborn size, are realigned and assigned a new age and a new size as they pass through the act of initiation.

VIII. PROBABILITY AND DETERMINISM

Is cell regulation deterministic or probabilisitic? This question, which has been asked many times before, is clarified by the discussion presented here. The general model, the framework of our discussion, is a deterministic model. But even a deterministic model may have probabilistic components.[39] In biological systems that are the result of complex biochemical interactions, at some level the deterministic features of the system are obscured by probabilistic features. At the molecular level, the reactions are clearly probabilistic. The sum of the reactions does combine, however, to produce a deterministic result. An analogy might be the process of going to work. This is clearly a deterministic process. The ringing of the alarm clock sets in motion a series of events (washing, dressing, breakfast, commuting) that leads to the eventual arrival at the workplace. But arrival may not be precisely at nine. Some arrive at quarter to nine, and some even arrive at quarter past nine. This variability should not obscure the underlying deterministic processes that lead to work. Similar ideas explain the progress of the cell through the division cycle. There is an underlying deterministic process that necessarily leads to DNA synthesis, and then cell division, but interspersed between these processes are variations that are the results of variations in the components of the individual processes.

NOTES

1. The data are from Nanninga, Woldringh, and Koppes, 1982.
2. Koch and Schaechter, 1962.
3. Koch, 1977.

4. Koch and Schaechter, 1962; Schaechter, Williamson, Hood, and Koch, 1962; Tyson, 1985; Koch, 1987.

5. Rahn, 1931; Kendall, 1948, 1952.

6. Koch and Schaechter, 1962.

7. Powell, 1955, 1956, 1958; Powell and Errington, 1963.

8. Powell (1958) also noted how difficult it is to determine size at division. A number of trivial factors such as the cell orientation in the microscope could give large increases in variation that had no important relationship to the division cycle.

9. Harvey, 1972a,b, 1983.

10. Cooper, 1982b. This paper analyzes the variation in interdivision time of animal cells. The same analysis is applicable to bacteria, but there are more data available on the interdivision time distributions of animal cells. This same model is applied directly to animal cells in Chapter 15 in the discussion of the transition–probability model.

11. Kubitschek, 1962b, 1966.

12. Marr, Painter, and Nilson, 1969.

13. Bremer and Chuang, 1981.

14. Marr, Painter and Nilson, 1969.

15. Powell and Errington, 1963. It should be noted that any differences were found in only a few experiments and were very weak in any case.

16. Marr, Harvey, and Trentini, 1966.

17. Trueba, 1982.

18. Koppes, Woldringh, and Nanninga, 1978; Cullum and Vicente, 1978; Nanninga, Koppes, and de Vries-Thijssen, 1979.

19. Verwer and Nanninga, 1980.

20. For a discussion of the Poisson distribution, see Chapter 9.

21. Staugaard, van den Berg, Woldringh, and Nanninga, 1976.

22. Powell, 1955, 1956, 1958; Powell and Errington, 1963.

23. Kubitschek, 1964b, 1962b, 1966; Schaechter, Williamson, Hood, and Koch, 1962.

24. Plank and Harvey, 1979.

25. Koch, 1977.

26. Chai and Lark, 1970.

27. Koppes, Woldringh, and Nanninga, 1978.

28. Koppes and Nanninga, 1980.

29. Koch, 1977.

30. The published and unpublished data from Forro (1965) and Forro and Wertheimer (1960).

31. Koch, 1987.

32. This is not necessarily true, that the slope at the particular time is exactly the same as the initial slope, but as drawn, it allows a very simple explanation of the curves. A precise mathematical analysis of this type of figure has yet to be derived.

33. Painter and Marr, 1968.

34. In practice there would still be estimates to be made, because the C and D

periods are not perfectly invariant; the point is that the slopes would be much sharper and thus less conducive to error in the estimation of midpoints.

35. Cooper, 1990a.
36. Helmstetter and Pierucci, 1976.
37. Schaechter, Maaløe, and Kjeldgaard, 1958.
38. Cooper, 1969.
39. Koch, 1966b.

9

The Segregation of DNA and the Cell Surface

Distribution of the material of a parental cell to the daughter cells at division is called *segregation*. If the cell were simply an amorphous cytoplasm that divided randomly at division, then a simple equipartition of all material at division would be expected. But the cell is not that simple, and each of the major cell components we have dealt with—the DNA or genetic material, the cell surface, and the cell cytoplasm—has a different segregation pattern. This chapter deals primarily with the segregation of DNA, and in less detail, with the segregation of the surface and the cytoplasm.

I. EARLY STUDIES ON MACROMOLECULE SEGREGATION IN GRAM-NEGATIVE BACTERIA

The first experiments on DNA segregation defined the number of units to be considered during a segregation experiment. When the early experiments were performed, it was just becoming known that there were only a few units of segregating genetic material in the cell; i.e., the cell did not contain a large number of randomly assorting chromosomes.

A. The Results of Van Tubergen and Setlow

Van Tubergen and Setlow[1] studied the segregation of a number of different macromolecules. Cells were labeled with different radioactive precursors of different cell components and the labeled cells were then grown for a number of generations in unlabeled medium. At selected times, the cells were fixed, autoradiographed, and the number of grains per cell was determined. If there were a random distribution of material into the fixed cells, then cells would be expected to be labeled equally and a Poisson distribution of grains would be expected.[2] If a Poisson distribution were found, that would mean that the cells were uniformly labeled. If a distribution different from a Poisson distribution were found, some segregation process other than equipartition of material at each generation was occurring. Material was conserved as the cell grew and divided; no turnover and reutilization of label was observed. Determining the number of generations until a non-Poisson distribution was obtained could measure the number of segregating subunits in the original cell. With regard to DNA, it was found that shortly after labeling a non-Poisson distribution appeared, with a striking excess of unlabeled cells. Van Tubergen and Setlow interpreted this as meaning that there were only a limited number of chromosomes or DNA strands present in the bacterial cell. These segregate as units and do not divide with further growth.

B. Segregation in Microcolonies

Forro and Wertheimer[3] followed these results by studying microcolonies formed by pulse-labeled or long-term–labeled cells. After long-term labeling, they found microcolonies with two heavily labeled cells and two lightly labeled cells. In Chapter 5, these results were interpreted as indicating there were cells with one or two replicating chromosomes— equivalent, in our current understanding, to the younger cells growing with 40- or 60-minute interdivision times. Because the cells were not arranged in any order when the microcolonies were formed, the pattern of segregation of DNA in the microcolonies could not be observed. This is the problem to which we now turn: is there a relationship between the segregation of the DNA and a particular pole of the cell?

This question is subtly related to the proposal by Jacob, Brenner, and Cuzin[4] of the replicon model. From considerations that there was an equipartition of DNA to daughter cells, they postulated an attachment of DNA to the cell surface. The cell surface was proposed to be analogous to the mitotic apparatus of higher cells. One of the predictions of the replicon model is that there might be a nonrandom relationship of the segregation of DNA with regard to one of the poles of the bacterial cell. If the attachment of DNA to the surface were permanent, and if the attachment site were a fixed position in the cell relative to the poles of the cell, then a particular strand would always segregate toward one pole at division.

A large number of papers have presented evidence for the attachment or binding of DNA to the membrane (or more generally, the cell surface) in both gram-negative and gram-positive bacteria.[5] There have been many different conclusions regarding the mode of segregation of DNA in growing cells. Eberle and Lark[6] reported an oriented, nonrandom segregation of radioactively labeled DNA in *Bacillus subtilis*. Chai and Lark[7] observed a co-segregation of DNA and some portion of the cell envelope in *Lactobacillus acidophilus*. Random segregation has been reported by others.[8]

II. *A PRIORI* CONSIDERATIONS OF CHROMOSOME SEGREGATION

Consider that a single cell containing one replicating chromosome is labeled for a short time with tritiated thymidine. As depicted in Fig. 9-1, there will be two labeled strands. Now consider that we allow the cells to grow as illustrated in Fig. 9-1, so that the order of the cells and their arrangement as they grow and divide can be maintained. After two divisions, the expected result is a chain of four cells with the two outermost poles the same as in the original parental cell (generation 2). The problem is to understand the arrangement of the two labeled strands in this chain. To simplify our analysis, we will first consider that the original labeled cells are grown for a number of generations such that the chromosomes are present in cells as single labeled strands. At generation N in Fig. 9-1, we will allow the cell with a single labeled strand to grow into a chain, so that the order of the cells is preserved. After a chain of four (generation N+2) or eight cells (generations N+3) is formed, we fix the chains onto a slide, process the cells for autoradiography, and determine which cell in the chain has the labeled strand. (See Fig. 9-1 for details.)

If the labeled strand went to the left or right at each generation with no preference, i.e., the segregation was random, we would find, among the collection of four cell chains, that half of the cells are labeled in position 1 and the other half labeled in position 2. Eight-cell chains would be equally labeled in positions 1, 2, 3, and 4. The reason that only the lower half of the numbered cells is labeled results from the numbering system of chains. If we consider chains in which only one cell is labeled, then the only cells that are labeled are those cells with numbers less than half of the chain length. In a four-cell chain, only cells 1 and 2 can be labeled. This is because a cell labeled in position 3 is equivalent to labeling in position 2, and a cell labeled in position 4 is the same as position 1. In the microscope there is no differentiation between left and right with regard to chains. In an eight-cell chain, only cells 1 to 4 can be labeled (in chains with only one cell labeled). A corollary of this numbering system is that any observation of an asymmetric distribution of cell label (and there have been many[9]), is the result of not collapsing the right and left halves of a chain so that the label was only in one half of the chain. If 10% of the chains were labeled in position 1, and 5% were labeled in position 8 (in an eight-cell chain), this would not be evidence for asymmetry, since the total labeling in position 1 would be 15%.

If segregation were nonrandom rather than random, in the four cell chains there would be a preponderance of chains labeled in either position 1 or position 2. Although it is not directly relevant to our theoretical analysis, it has generally been thought that if there were some special

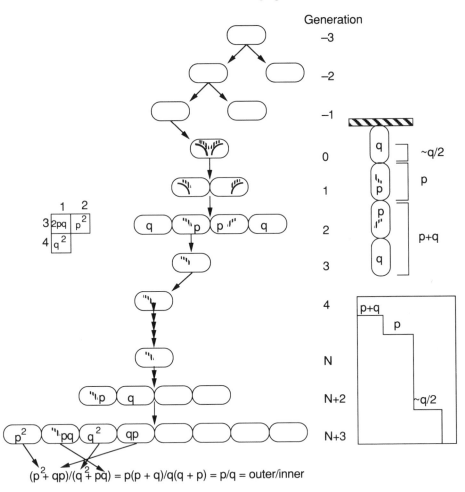

$$(p^2 + qp)/(q^2 + pq) = p(p + q)/q(q + p) = p/q = \text{outer/inner}$$

Figure 9.1 Formal Analysis of DNA Segregation. Consider a culture growing exponentially before and after labeling. At the instant of labeling (generation 0), we may consider that a cell has two labeled strands. This assumes, for simplicity, that the cell has one replicating chromosome, i.e., we are looking at a cell with a 60-minute doubling time (see Fig. 5–1). We may grow these labeled cells for a number of generations (N), and then place the cells into methocel to compel the cells to form chains as they divide (see Fig. 9–2). At N+2 generations (a four-cell chain), we may determine the number of outer and inner cells labeled. Only one cell from each four-cell chain is labeled, because segregation has occurred earlier to separate the two radioactive chains. Assume the degree of segregation to the outer and inner cells is given by the values p and q (the fractions of chains labeled in the outer or the inner cell respectively); and $p + q = 1$. At the next generation, if the cells form an eight-cell chain, any one of four cells is labeled. But these four cells may be considered to be descendants of two of the cells of the four-cell chain. Thus, we may take these four cells as another four-cell chain. Positions 1 and 4 are equivalent to an outer cell, and positions 2 and 3 are equivalent to an inner cell of a four-cell chain. We sum these cells, factor, and derive

mechanism that determined that labeled DNA strands would go to one or another cell, there would be a preference for having the chain labeled in position 1. This would be interpreted as if the DNA strand were attached, even temporarily, to a particular pole, and that this attachment led to the nonrandom segregation pattern. When the original cell with a single strand produces two cells, all that can be seen is that one cell (generation $N+1$) had the DNA strand; call this the leftmost cell. If there is a continuing preference of the DNA strand for the leftmost cell at the next generation (the 4-cell chain), we would expect to see most or all chains labeled in position 1.

Segregation is nonrandom in just the manner we have described. The probability of having a four-cell chain labeled in cell 1 is p, and the probability of having the chain labeled in cell 2 is q. What would we expect for a collection of eight-cell chains? The end-labeled cell (position 1) of the four-cell chain will produce two cells according to the same rules, so the cells will be labeled in positions 1 and 2 of the eight-cell chain in the ratio of p/q. Cells descended from position 2 of the four-cell chain will be labeled in positions 3 and 4 in the ratio of q/p. For the cells descended from the inner cell of the four cell chain, there is a greater probability that an innermost cell, i.e., cell number 4, will be labeled since $p>q$. Once a strand has gone to a particular pole, it follows the general rule that the strand goes toward the same pole with a probability of p. At generation $N+3$, the expected probabilities of finding cells labeled in eight-cell chains are indicated. The eight-cell chain can be considered to be the result of placing the cells into a chain-forming situation

---

the fact that the same ratio of outer to inner values is obtained. Thus, in a steady state of segregation, assuming that the segregation of the cells follows a constant rule, all chain lengths give the same result. If the cells with two labeled chains (generation 0) are allowed to form a four-cell chain immediately after labeling, we see the results in the left-most array of three boxes. Chains can be labeled in either positions 1 and 3, 2 and 3, or 1 and 4. Chains labeled in positions 2 and 4 would be equivalent to, and indistinguishable from, positions 1 and 3. If, in the results at $N + 2$ and $N + 3$, $p > q$, we find that in the four-cell chain produced immediately after labeling (generation 2), the label is preferentially found in the inner cells rather than the outer cells. This is owing to the nonrandom segregation of the unlabeled, and unseen, parental chains to the outer cells. Because bacterial chromosomes have an equipartition mechanism of chromosome segregation (i. e., at division each cell gets a single chromosome), the labeled chains are seen preferentially in the inner cells in the four-cell chain. When the labeled chains are from older cells, the labeled cells will be found in the outer cells of the chains. At the right of this figure is the expected pattern of elution of radioactive DNA from cells bound to a membrane. Note that if $p > q$, the steps between plateaus are different from the factor of 2 expected for random segregation. There is a step that is less than a factor of 2 (because $p/(p + q)$ is greater than 0.5), and then (approximately) less than a factor of 2 (because $(q/2)/p$ is less than 0.5).

one generation later, for example at generation $N+1$, after the cell at generation N had divided one more time. At generation $N+3$, a four-cell chain will have been formed. If we now consider this four-cell chain (composed of cells 1 to 4 of the eight-cell chain), we can sum the two outermost cells (which are indistinguishable in the microscope), and the two innermost cells. If we add the probabilities, factor, and look at the ratios, we find that the outer/inner ratio is p/q (Fig. 9-1). This demonstrates that the segregation pattern is independent of the time after labeling.

In this discussion it is assumed that the segregation pattern of a chromosome is independent of its history—that labeling a strand has no effect on its future segregation pattern, and that whatever rules govern segregation, they govern the strand equally whether it is young or old.

It is interesting to consider the alternative assumption, that the segregation pattern has a historical component and that the degree of nonrandom segregation may be different depending on the age of the DNA strand. For example, a newly synthesized strand may segregate nonrandomly, and with time, segregates randomly. Or a new strand may start out with a random segregation pattern, and with age begin to segregate nonrandomly. The problem with this view is that it leads to instability and the appearance of DNA-less cells. Consider a cell that has a newly synthesized DNA strand made alongside an old DNA strand. If the segregation patterns were different, with one nonrandom and one random, we would not find an equipartition mode of segregation of the chromosomal DNA. Some cells would not have DNA. Since this does not usually occur, segregation must be independent of the age of the DNA strand.

Return to the original labeled cell at generation 0 in Fig. 9-1. This cell has descended from unlabeled cells. The unlabeled strands, which we do not observe with autoradiography, are acting in the same way as the labeled strands we have considered in generations N, $N+1$, $N+2$, $N+3$, etc. At generation 1, these unlabeled strands segregate to the left and right according to the pattern we have described. At generation 2, we find a differential segregation such that the unlabeled strands preferentially segregate to the outermost cells of the four-cell chains. This means that the opposite strand must segregate preferentially into the center two cells of the four-cell chains, cells 2 and 3. We see, paradoxically, that at the second generation after labeling, there is preferential segregation of the labeled strand *away* from the older poles present at the time of labeling. Only at generation 3 or later is a preferential segregation of labeled strands toward the older pole present in generation 2 or any succeeding generation.

If we looked at a number of four-cell chains made immediately after labeling, we would find cells labeled in four different positions. We could calculate the frequency with which different types of labeling patterns are found. This assumes that we know the values of p and q from previous measurements on single cells as described above. Each chain is labeled in either cell 1 or 2 and either cell 3 or 4. At the left in Fig. 9-1 is a matrix with the predicted probability for finding the different pairs (1,3; 1,4; and 2,3; no 2,4 set is found because, according to our numbering scheme, this would be called a 1,3 labeling pattern). The probability of two independent events happening together in sequence is the product of their individual probabilities. The probability of finding the two inner-most cells labeled is p^2, the probability of finding the two outermost labeled is q^2, and the finding of an outer and an inner cell labeled is $2pq$.

Given the particular segregation pattern or ratio of a single labeled strand, if this segregation pattern is independent of the age of the strand beyond generation 2 after labeling, we *must* find the segregation pattern described for generations 1, 2, and 3, with the associated matrices. If we did not have the relationship illustrated here, cells with two DNA strands segregating to one cell and the corresponding DNA-less cells would be produced. If we went to eight-cell chains or sixteen-cell chains (generations 3 and 4 after labeling), similar matrices would be found, but these would be no more enlightening—just more difficult to score and more complicated to analyze.

Now consider what would happen if there were a nonrandom segregation pattern ($p>q$) and the cells of generation 0 were placed on a membrane-elution apparatus (see Chapters 3 and 5). The predicted results are given at the right of Fig. 9-1, where the four-cell chain (two generations after labeling) is turned 90° and illustrated as bound to a membrane. We can imagine that the original cell is composed of four parts, such that each part can be labeled with the probability that a DNA strand will be present in that part of the cell. These probabilities are q, p, p, and q as expected after two generations. At the first division, the DNA is released with a probability of $(p+q)/(q+p+p+q)$; this means that half of the radioactivity is released. At the next division the radioactivity leaves the membrane with a probability of $p/(p+q)$. Since $p>q$, the drop in radioactivity is less than a factor of two. At the next division we find that the probability of a labeled strand's being released is approximately $q/2$.[10] This second drop is greater than a factor of 2.

Consider a numerical example where p is 0.6 and q is 0.4. From a cell with 200 cpm, in successive generations there would be found, in the eluate, 100 cpm per cell, then 60 cpm per cell, and then 20 cpm per cell, to be followed by (approximately) 10 cpm per cell. If there were a completely random segregation pattern, we would expect the radioactivity per cell in successive generations to be 100, 50, 25, etc. A measure of the nonrandomness of segregation can be obtained from the labeled DNA per cell eluting from labeled cells placed upon a membrane. The degree of nonrandom segregation of radioactive DNA strands, with respect to a particular pole, can be determined by either of two methods, autoradiography of chains of cells formed in methocel, or analysis of the radioactivity per cell in eluates of cells placed on a membrane. As we shall see, when cells growing at different rates are measured by the two different approaches, the membrane-elution and the chain-forming approaches, the degree of segregation is constant with membrane-elution and varies with growth rate when chains are analyzed. The surprising conclusion is that these apparently different and contradictory results can both be explained by a single model. This model takes both the constancy of segregation nonrandomness with growth rate and the variation of nonrandomness with growth rate and explains both results. Before describing this unifying model, a brief description of the two methods for analyzing DNA segregation will be presented.

The general analysis is presented before the experimental data because this analysis is *a priori* correct for the conditions described. It could have been drawn out before a single experiment was performed. The major requirement is that there is no change with time in the segregation pattern of DNA strands. Given this, the general analysis must be correct.

III. METHODS OF ANALYZING SEGREGATION OF DNA

A. The Methocel Method

The production of chains of cells preserving the order of cells as they descend from an original parental cell is extremely simple. The method for obtaining such chains is called the methocel method because it involves making the medium viscous with methocel (a methylated cellulose) and spreading out a thin layer of the medium containing labeled cells on a slide. The cells grow in the viscous medium and maintain their arrangement. The cells descending from a single cell are aligned in a chain by the methocel. After fixation of the material to the slide, the cells can be autoradiographed (Fig. 9-2). The methocel method was first developed by Lin, Hirota, and Jacob.[11] They spread the methocel medium containing bacteria along a slide using a slide to which two thin aluminum foils had been taped to form a small trough. The layer of methocel that remained on the slide was just adequate to allow cells to grow in

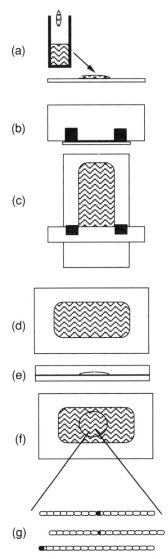

Figure 9.2 The Methocel Method. The methocel method for analyzing the segregation of various cell components begins when (a) a drop of labeled cells is mixed in a vial with growth medium made viscous by the addition of methocel (a soluble polymer similar to cellulose). A drop of the mixture is placed on a slide and then a slide with two rails of aluminum foil taped to one edge is placed perpendicular to the slide (b). The taped slide is run along the methocel mixture to leave a thin layer (equivalent to the thickness of the aluminum foil) on the surface of the slide (c). The resulting layer of bacteria in methocel (d) is then covered (e) with a concave slide to prevent the material from drying. The concavity is wet with sterile methocel medium in order to increase the humidity in the microchamber. After incubation for two to five doubling times, the slide is opened, the cells fixed, and the chains (f) observed after autoradiography. Labeled chains (g) may be observed such as those in which one or more cells have grains associated with the cell.

chains. To prevent the thin layer of medium from drying out, they inverted a small plastic chamber over the cells, or they covered the methocel film with mineral oil. Later the method was simplified by using two slides, one a depression slide moistened with methocel; after forming a layer of methocel, the depression slide was placed over the first slide.[12] This formed a moist chamber that allowed the chains of cells to form. At intervals, the methocel slides were opened, fixed, and autoradiographed.

B. The Membrane-Elution Method

As we have noted in Fig. 9-1, and will analyze in detail below, it is possible to determine the degree of nonrandom segregation of DNA using the membrane-elution method. Because one pole is immobilized as it is attached to the nitrocellulose filter, this serves as a marker for the directionality of segregation. When a large number of elution experiments have been analyzed, it is seen that segregation is nonrandom.[13]

IV. THE OBSERVATION AND EXPLANATION OF NONRANDOM SEGREGATION PATTERNS

A. Random Segregation of DNA

Lin, Hirota, and Jacob[14] were the first to use the methocel method to study segregation. They labeled cells with thymidine and looked at the grains on chains formed in methocel. They concluded that segregation appeared completely random. They observed that no particular cell of a chain was preferentially labeled. One problem with their experiments is that they did not look at individual chains, but they pooled their data on all of the chains. In addition, they did not have knowledge of the chromosome pattern at the time of labeling. As they mainly had chains with up to four labeled cells (two replicating chromosomes), these could not be analyzed by simple two-dimensional matrices as described in Fig. 9-1.

B. Nonrandom Segregation of DNA

The first clear indication of nonrandom segregation came from observations on the elution pattern of thymidine-labeled cells from a membrane. Rather than finding that the plateaus of label were decreasing by halves at each generation of elution, it was found that there were decreases of less than half, followed by a large decrease of more than half at the next generation.[15]

1. Model of Pierucci and Zuchowski

The first detailed analysis of the microscopic pattern of DNA segregation was obtained by Pierucci and Zuchowski.[16] Using the methocel method, they studied the segregation of DNA from cells with known chromosome patterns. They looked only at cells with one replicating chromosome (the younger cells of cultures growing with a 40- to 60-minute doubling time). In four-, eight-, or sixteen-cell chains they found only two labeled cells. They counted many of these chains and tabulated the results in matrices similar to that in Fig. 9-1.

These experiments were followed by an extensive analysis of a number of different models of DNA segregation.[17] In the model that best fit their data (their model IB), one of the two original strands was permanently bonded to the pole present at the time of labeling, and the other strand segregated randomly. A typical result for their experiments is summarized in Table 9-1; this result indicates that the segregation is nonrandom. When their data are compared to the predictions of model IB, the results are very similar to the predictions. Note that the predictions of two other models, the strand inertia and the Helmstetter-Leonard model, are also quite close to the data.

Experiments with *Bacillus subtilis* a number of years earlier led to the proposal that DNA strands were permanently bound to the cell surface shortly after synthesis.[18] These experiments were performed on filamenting cells, and it is difficult to relate these results to exponentially growing cultures.

2. Strand-Inertia Model

An alternative explanation of nonrandom segregation that fits the results of Pierucci and Zuchowski as well as other segregation experiments, is the strand-inertia model.[19] The main experimental innovation at this point was to study chains that had single labeled DNA strands in a cell (see Figs. 9-1 and 9-3). This greatly simplified the analysis of the experiment (Fig. 9-3). To obtain such cells, the culture was grown for a number of generations after labeling, and these cells were then placed in methocel for subsequent segregation analysis. All strands acted similarly, with nonrandom segregation toward the same pole to which they had segregated in the previous generation. One way of looking at this result is to say that when a strand goes in one direction, it continues to go in that direction with a greater probability than going in the other direction; the strands have a measure of *inertia*. The main difference between the strand-inertia model and the model IB of Pierucci and Zuchowski is that at the next division, after the first set of four cells is produced, we would

Table 9.1 The Segregation Results of Pierucci–Zuchowski and Its Comparison with Different Models

	Observed: Four-cell chains			
Position	3	4		
1	43.4	19.1		
2	37.4			

	Observed: Eight-cell chains			
Position	5	6	7	8
1	13.1	12.2	4.3	3.6
2	12.1	12.2	4.3	
3	12.5	12.6		
4	12.8			

	Expected: Random segregation, four-cell chains			
Position	3	4		
1	50	25		
2	25			

	Expected: Random segregation, eight-cell chains			
Position	5	6	7	8
1	12.5	12.5	12.5	6.25
2	12.5	12.5	6.25	
3	12.5	6.25		
4	6.25			

	Expected: Pierucci–Zuchowski model, four-cell chains			
Position	3	4		
1	50	12.5		
2	37.5			

	Expected: Pierucci–Zuchowski model, eight-cell chains			
Position	5	6	7	8
1	12.5	12.5	6.25	3.12
2	12.5	12.5	3.12	
3	18.8	9.4		
4	9.4			

	Expected: Strand-inertia or Helmstetter–Leonard models, four-cell chains (for p=0.65)			
Position	3	4		
1	50	12.5		
2	37.5			

	Expected: Strand-interia or Helmstetter–Leonard models, eight-cell chains (for p=0.65)			
Position	5	6	7	8
1	12.5	12.5	6.25	3.12
2	12.5	12.5	3.12	
3	18.8	9.4		
4	9.4			

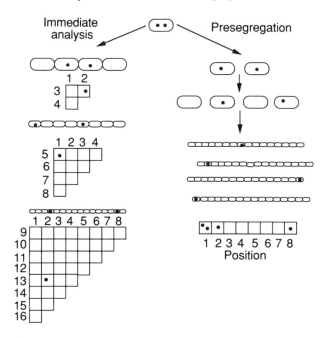

Figure 9.3 Advantages of Presegregation with the Methocel Method. When cells are placed immediately into methocel for segregation analysis (the Pierucci-Zuchowski Method) there are at least two labeled cells in the chains. The accumulated results must be plotted in arrays as indicated below the four-, eight-, and sixteen-cell chains. The dot in each array indicates the pattern observed. In the sixteen-cell chain, the chain-numbering system described in Section II (p. 281) makes the pattern a 2–13 labeled chain, rather than a 4–15 labeled chain; these two labeling patterns are indistinguishable. When a period of presegregation is allowed before chain formation in methocel, only one cell in a chain is labeled. The array below the four chains is a summary of the results obtained for four sixteen-cell chains.

expect to see a bipartite population of eight-cell chains, one from the randomly segregating DNA and the other from the nonrandomly segregating DNA. Segregation values were the same in all generations.[20] Thus, the strand-inertia model fits the theoretical predictions presented in Fig. 9-1.

In addition to simple methocel segregation experiments with presegregation, another experiment demonstrated that there was no permanent attachment of DNA to any particular pole. Cells were labeled, placed on a membrane, and eluted from the membrane for many hours.[21] At each generation of elution, cells were taken for segregation analysis. It was found that for up to six or seven generations there was no change in the segregation pattern of the eluted cells, and when the cells remaining bound to the membrane were analyzed, they had the same segregation pattern. This experiment conclusively demonstrated that both strands were equivalent. If one strand were bound permanently to one pole, then after a number of generations of elution, the bound cells would have given an extremely nonrandom segregation result when the cells were analyzed by the methocel method. Also, the cells eluted from the membrane would have given the converse result, a random segregation pattern.

More important, when the presegregation method was used, it was found that the segregation pattern was not a fixed value, but was a variable, dependent upon the growth rate. The slower the growth at a given temperature, the more nonrandom the segregation (Fig. 9-4).[22] Thus there was some inertial value that varied with growth rate and that could not be accounted for by differences in segregation of different strands.

The most interesting aspect of the entire story is that the segregation as measured by the membrane-elution method gave a constant numerical value to segregation at all growth rates.[23] In a sense, there is a basic contradiction, that one method gave a medium- or growth-rate-dependent variation in the degree of nonrandom segregation, and another method gave a medium- or growth-rate-independent degree of nonrandom segregation. As we shall see, these two results are not contradictory, but can be resolved in an elegant manner by considering a geometrical or topographical analysis of segregation.

V. THE SEGREGATION MODEL OF HELMSTETTER
   AND LEONARD

An interesting mechanistic explanation for the strand-inertia model has been proposed by Helmstetter and Leonard.[24] Although this model is partly a restatement of the strand-inertia model in geometric and physical terms, it is the *type* of model that can explain the segregation data. Helmstetter and Leonard noted that the segregation of new strands preferentially to the center cells in the four-cell chains mimicked the segregation of cell-wall material. If we distinguish between old and new cell-wall material, a greater fraction of cell wall is new in the center of the cell, and a greater proportion of wall material is old in the polar regions of the cell. At the time of division, the cell wall at the side has been intercalated with new cell-wall material (see Chapter 6), and the amount of old and new material in the cylindrical side wall in the two cells that will be formed at division is the same in the two daughter cells. In the center

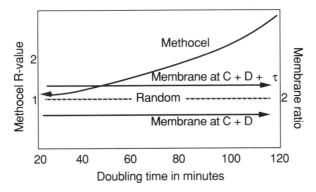

Figure 9.4 Comparison of Segregation Determined by the Membrane-Elution Method and the Methocel Method. The experimental determinations of segregation using the methocel method or the membrane-elution method are summarized by plotting the degree of observed deviation from randomness as a function of growth rate. With the methocel method, the segregation pattern is more nearly random in rapidly growing cells, and becomes nonrandom in slow-growing cells. Random in this case is a ratio of 1 for p/q. In contrast, the nonrandomness of segregation as determined by the ratio of counts in different generations of elution from the membrane-elution apparatus is constant over the same range of growth rates. At C+D minutes following binding of cells to the membrane, the drop is less than 2 and the value is constant at all growth rates; at $C + D + \tau$ minutes (τ is one interdivision time), the drop is greater than 2 at all growth rates. This ratio is also constant over a wide range of growth rates. These results are not contradictory but are in fact complementary; both can be explained by a general model of segregation of DNA at all growth rates.

of the cell that is going to divide the entire pair of new poles has been synthesized as new material. At the opposite end of the cell, the older poles are made up solely of old material. In every *Escherichia coli* cell, one pole is brand new, having been formed at the previous division; the other pole is older. This asymmetry is the basis for the asymmetric segregation of the DNA. Thus, if a new strand binds preferentially to new cell wall, the DNA would segregate randomly when only the side wall was considered, but would segregate preferentially to the newer of the two poles present at the time synthesis was initiated. This model is illustrated in Fig. 9-5.

An alternative formulation proposes that the older strands are bound preferentially to older cell-wall material. The ultimate result is the same, but perhaps the biochemical tests might be different. The binding of the new DNA strand to the new cell-wall material occurs only once, and for one generation. After that, the DNA strand acts in reaction to the segregation pattern of the strand that it made, its complementary daughter

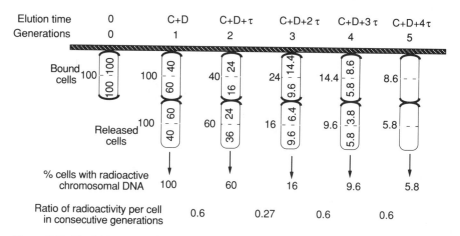

Figure 9.5 The Helmstetter–Leonard Surface-Area Model for Nonrandom DNA Segregation. The predicted pattern of segregation for labeled cells bound to a membrane according to the Helmstetter–Leonard model is illustrated. The numbers in each cell are the radioactivity per cell expected at each generation of elution. The first cells bound have a single replicating chromosome so that all released cells are labeled. At the next division there is a nonrandom elution with the majority of the radioactivity found in the released cell. This is expected by the analysis presented in Fig. 9.1. At the next division, less than half of the remaining radioactivity is released. The ratios of the radioactivity between different generations of elution are presented in the numbers at the bottom. Below the membrane is a diagram of the relative areas of the cell surface. It is assumed that each of the poles is one-sixth of the total cell area, and that the sides form the other four-sixths of the surface area. One of the poles is always old, and one is new, and the area within each of the portions of the side wall is at least half new (because of the intercalation of new material between the preexisting strands), so when a cell divides, all that needs to be considered is the segregation to the cell containing two-sixths of the new cell wall and the cell containing three-sixths of the new cell wall. For a cell with constant shape, the degree of nonrandom segregation is in a ratio of 2:3, or a segregation ratio such that 40% of the time the new strand goes to one side (adjacent to the old pole), and 60% of the time the new strand goes to the other half of the cell, adjacent to the new pole.

strand. This complementary daughter strand is always a new strand, and by following the new strand we see that the old strand segregates in just the manner we have described in Fig. 9-1. Recently, the same model was formulated in even more general terms by discussing the segregation as due to the exclusion of strands of DNA from various cell volumes.[25] All of the alternative formulations are formally equivalent to the description presented here.

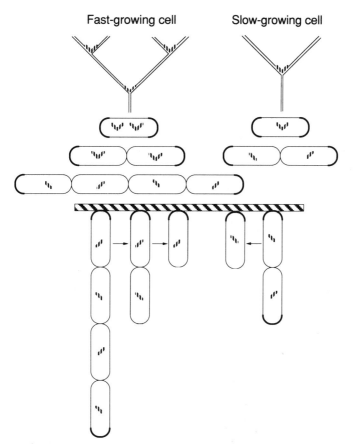

Figure 9.6 Why Segregation Randomness Varies with Growth Rate Using Methocel and Does Not Vary Using the Membrane-Elution Technique. Consider a fast-growing cell with multiple forks and a slow-growing cell with one replicating chromosome. Below the chromosome configuration, with the labeled strands indicated by the dotted band, is an illustration of cells as they form chains. Segregation of strands in the fast-growing cells takes place after the four-cell stage. In the fast-growing cells, at the four-cell stage, the two center cells are composed of new poles at both ends (i.e., the poles were made during or after the existence of the labeled strands). Both ends of the two inner cells of the four-cell chain from the fast-growing cell have new poles. Thus, the two inner cells in that chain have a symmetrical wall-age arrangement, and the two outer cells have an asymmetrical arrangement. The thicker poles indicate that the pole is older than the thinner poles. In the slow-growing cell, the same situation, with one labeled strand in each cell, is reached when all of the cells have an asymmetrical wall-age arrangement. When the fast-growing cells are placed on the membrane, after two divisions a single asymmetrical cell is present; this asymmetry is identical to the asymmetrical cell found in the slow-growing cell after one generation of growth. This can be understood by observing that C + D minutes occurs after three generations of growth of cells with a 20-minute doubling time (assuming C + D = 60), but C + D minutes occurs after only one generation of growth of 60-minute cells. On the membrane, at some time during elution, the strand segregation with regard to the asymmetric cell is the same for both slow- and fast-growing cells. This is not the case for the methocel method. With methocel, we do not lose the symmetrical cells, so there is a decrease in the observed degree of nonrandom segregation at faster growth rates.

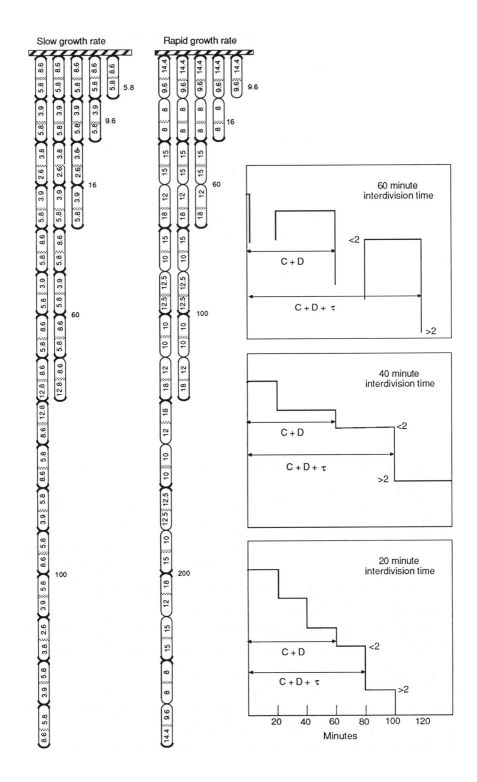

This model is consistent with the finding that there is a decrease in the degree of nonrandom segregation with increasing growth rate when measured using the methocel method with presegregation.[26] Yet the degree of nonrandom segregation is constant when the membrane-elution method is considered.[27] This paradox is explained in Figs. 9-6, 9-7, and 9-8.

An illustration of the expected distribution of labeled strands from fast- and slow-growing cells is presented in Fig. 9-6. With fast growth, the ultimate separation of labeled strands occurs when there is a mixture of cells with asymmetrical pole ages. Some cells are asymmetrical (both poles synthesized during the period of synthesis of the new strand), and some are asymmetrical (with an older pole and a newer pole). In the slower-growing cell, both cells are symmetrical at the time of strands separation. This is so because the initiation of DNA replication occurs a constant time before a cell division that separates DNA strands. At faster growth rates (i.e., with short interdivision times), a number of cell generations intervene between initiation and cell division. This means that cells involved in segregating the DNA will have made both their poles within the time between initiation and cell division. These cells will make the segregation appear more random at rapid growth rates.

If we calculate the predicted results for fast- and slow-growing cells using the membrane-elution technique, we get the results of Fig. 9-7. At both fast and slow growth, there is a decrease in radioactivity per cell less than a factor of 2, and then one greater than a factor of 2. The decreases occur at C+D minutes and C+D+τ minutes after placing the cells on the membrane. In the faster-growing cells, there is an extra division cycle, as

---

Figure 9.7 Analysis of Segregation by the Membrane-Elution Method. The segregation pattern from slow-growing (left membrane) and fast-growing (right membrane) cells are compared. The numbers in the cells indicate the expected amount of radioactivity present at each time. In both cases, there is one division at which the decrease in radioactivty is less than a factor of 2 (100 goes to 60), and at the next division the decrease in radioactivity is greater than a factor of 2 (60 goes to 16). The difference in the number of generations before these two divisions occur is explained in the analysis in Fig. 9–6. When the calculations are made for all growth rates, at C + D minutes of elution there is a drop less than a factor of 2; at C + D + τ minutes of elution there is a drop greater than a factor of 2. This can be seen in the three insets to the right of Fig. 9–1, which depict the observed radioactivity values for elution at three different growth rates. The time for C + D or C + D + τ minutes is indicated by the double-headed arrows in each elution pattern. For example, C + D + τ occurs at 120 minutes in the cells with a 60-minute interdivision time, at 100 minutes for the cells with a 40-minute interdivision time, and at 80 minutes for the cells with a 20-minute interdivision time. It is observed, and expected [because cell shape is constant at different growth rates (see Chapter 6)], that the numerical values of the ratios between plateaus are constant and independent of growth rate.

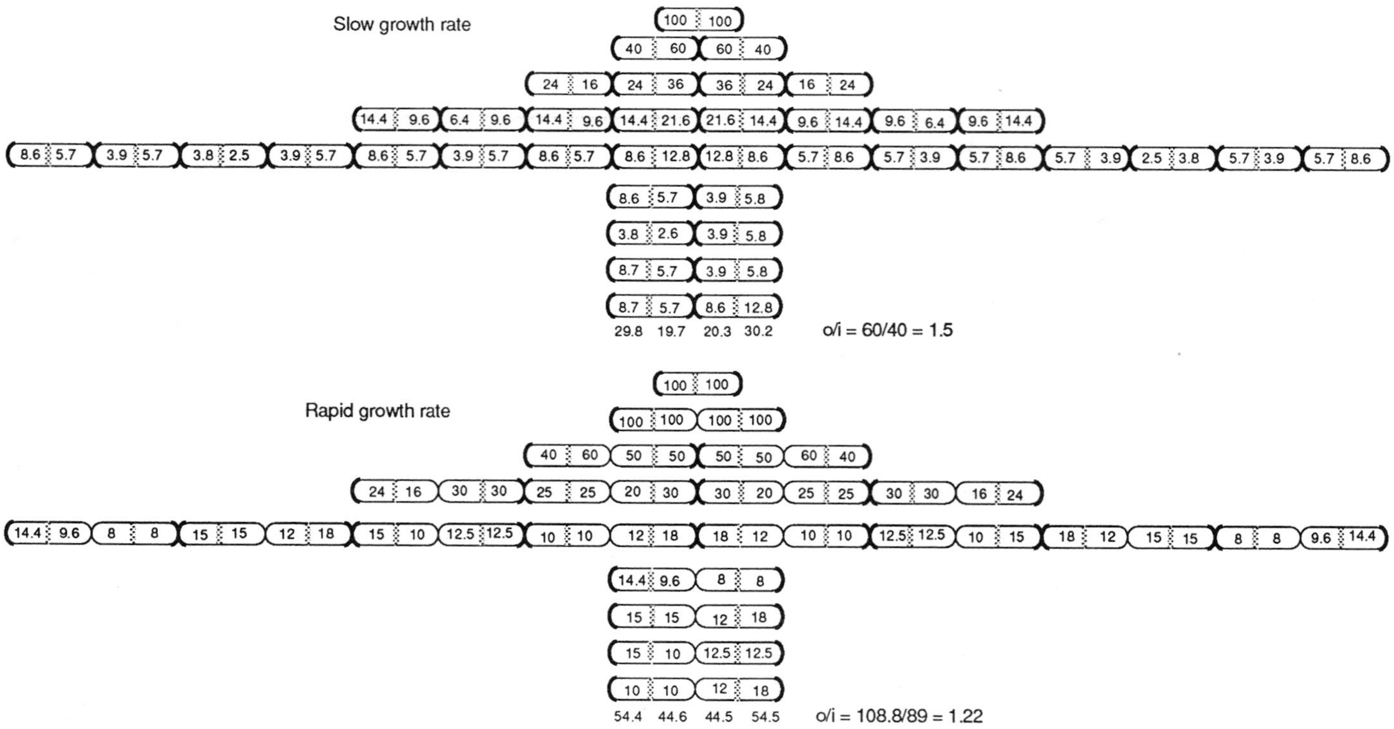
Slow growth rate
29.8 19.7 20.3 30.2
o/i = 60/40 = 1.5
Rapid growth rate
54.4 44.6 44.5 54.5
o/i = 108.8/89 = 1.22

described above, intervening before these decreases in radioactivity per cell. The numerical values of the large and small decreases are independent of the growth rate.

The original results of Pierucci and Helmstetter[28] were not so clear cut as this description implies. In an exemplary analysis, they corrected for the degree of improper binding of the cells to the membrane. They noted that some experiments had a larger number of cells eluted in the second generation than in the first generation. One possible explanation for this is that the cells bind by their sides with both poles attached. The cell grows and divides, but does not release any progeny in the first generation. Then two progeny are released from this single initial cell at the next generation. If this occurred, it would obscure the numerical value of the nonrandom segregation. When experiments with this increase in cell elution were eliminated from consideration, the constant values for the degree of nonrandom segregation were clear.

The most important aspect of the Helmstetter–Leonard model of segregation is that it simply and elegantly explains the variation in the degree of nonrandom segregation with growth rate when the methocel method is used. Consider how the DNA is partitioned between cells growing at different growth rates. When cells are growing slowly, and C+D is less than the interdivision time, a single chromosome replicates in a cell in which both poles are old. In a cell growing more rapidly, for example with a 40-minute interdivision time, replication initiates on two

---

Figure 9.8 Analysis of Segregation by the Methocel Method. Cells growing rapidly or slowly are labeled and allowed to grow into long chains. The numbers indicate the relative fraction of the cells in each position that have label. After a 32-cell chain is produced, the accumulated four-cell chains are grouped, and the probability of finding radioactivity in any particular position is determined. The degree of nonrandom segregation is greater in the slow-growing cells. An outer to inner ratio of 1.5 is found in the slow-growing cells, compared to a value of 1.22 for more rapidly growing cells. The explanation for this discrepancy between the constant degree of nonrandom segregation using the membrane-elution method and the methocel method is that at more rapid growth rates, there is an averaging of the segregation products with methocel analysis. As noted in Fig. 9–6, some cells at the more rapid growth rate appear symmetrical with regard to having an old or a new pole. As the growth rate increases, the proportion of symmetrical cells increases. For example, the 10/10, 15/15 and 8/8 segregation patterns are observed at rapid growth rates but not at low growth rates. We would expect to have the values asymptotically approach 1.0 at infinitely fast growth. Conversely, at infinitely slow growth, with C + D a negligible portion of the division cycle, we would approach an essentially deterministic segregation pattern. This is because all of the side wall would be made before the synthesis of the DNA (would thus be old), and the only new wall that would determine binding would be dominated by the newly synthesized pole made in the last (extremely small) fraction of the division cycle.

chromosomes at the same time as septum formation begins. The replicating chromosomes see an old pole on one side and a new pole being formed on the other side. The difference between these two situations comes when we investigate the progeny cells. In the slow-growing cell, there is no doubt that the daughter cell DNA can choose between an old pole and a new pole; the old pole was present during DNA synthesis, and the new pole was made after DNA synthesis ended. In the more rapidly growing culture, the daughter cells have an old and a new pole, but one of the subsequent progeny cells (the granddaughter cell) has what might be described as two new poles. One is new, as it was made in the first division, and the second is still new, as it is the new septum forming in the daughter cell. Thus we could imagine that some of the DNA would segregate asymmetrically, and some would segregate symmetrically (50/50). In the population as a whole, segregation would appear more random with increasing growth rate (Fig. 9-8). In the methocel method, there is an increase in the *proportion* of cells with random segregation at higher growth rates. This is because the poles present in these cells were both made during the period before strand attachment, and thus, from the strand's point of view, are equivalent. With the membrane-elution method, cells are continuously lost from the membrane, and at one particular division, there is a clear indication of nonrandom segregation without the interference of these symmetrical cells.

The elegant explanatory power of the Helmstetter–Leonard model should not be overlooked. It takes two contradictory results, accepts both, and explains both with a unified model that fits the data. In addition, as the Helmstetter–Leonard model is based on the relative shapes of cells, the constancy of segregation values using the membrane-elution method means that cell shape at different growth rates must be constant. The constancy of cell shape at different growth rates was discussed in Chapter 6. It should merely be noted that the segregation data are compatible with the idea that cell shape is a constant. If the segregation model of Helmstetter and Leonard is proven to be correct, then the segregation data may be taken as a proof that cell shape is constant.

When the degree of segregation at different temperatures was studied using the methocel method, it was observed that segregation was more random as the temperature was lowered.[29] One prediction of this result is that at lower temperatures, the cells should be squatter than at higher temperatures. This change in shape would increase the area of the cell devoted to the pole, and thus make DNA segregation more nonrandom. It is interesting that when the length-to-width ratio was determined in cells growing at different temperatures, there was just this expected change in shape. Cells decreased their length more than their width.[30]

This result supports the shape determination of nonrandom segregation as proposed by Helmstetter and Leonard.

VI. THE ALTERNATE-SEGREGATION MODEL

A model to explain the nonrandom segregation of DNA has been put forward by Canovas, Tresguerres, Yousif, Lopez-Saez, and Navarette.[31] They considered that the nonrandom segregation of the DNA, while not perfectly deterministic, could be described as " . . . nonrandom, with a certain degree of randomness." This was explained by proposing that DNA strands could be either randomly segregated, or nonrandomly segregated; the exact segregation values would depend upon the fraction of cells acting nonrandomly or randomly. By nonrandomly, it is meant that there is a perfect segregation of DNA toward the same pole as it segregated toward in the previous generation. Mathematically, the predictions were derived as follows. Consider that from a cell that contains a single, unreplicated chromosome (that is marked in some way so that we can see it), chains are allowed to form, and we see where these original strands are. Three types of patterns are possible (in the following, the nomenclature of Canovas *et al.* is used): chromosomes in positions 1,4 (outer-outer), called type A segregation pattern; chromosomes in positions 2,3 (inner-inner) called type C segregation pattern; and chromosomes in positions 1,3 (outer-inner), called type B segregation pattern.[32] Consider a population of four-cell chains descended from this original cell. The probability of a DNA chain acting nonrandomly is P, and the probability of the chain acting randomly is 1-P. What is the fraction of four-cell chains with the different segregation patterns? Considering type A segregation, all of the nonrandomly segregating chromosomes would give an A pattern (that is, there would be at least P cells that would give an outer-outer segregation pattern), and one quarter of the remaining cells (i. e., 1-P cells) would also give an outer-outer pattern. This occurs because, of the randomly segregating DNA, one-quarter would be giving type A segregation; one-quarter, type C segregation; and one-half, type B segregation. Thus, type A segregation would be $\{P + [(1-P)/4]\}$ or $(1+3P)/4$ of the cells would be type A. Similarly, type B would be present in a fraction indicated by $(1-P)/2$, and type C segregation would be present in $(1-P)/4$.

The main problem with this model is that it confuses a particular result and a particular mechanism. Consider that P in the alternate segregation model was 0.5. Half of the chromosomes are segregating nonrandomly and half are segregating randomly. This is equivalent to having a bag with two coins. One coin says *left* on one side and *right* on the other, while the second coin says *same* on both sides. After a two-cell chain is formed, and one labeled strand is in each cell, we reach into the bag, pick a coin at random, and then toss it to determine the segregation pattern of each strand. Three-quarters of the time the strands would segregate in the same direction they did before, and one-quarter of the strands would segregate in the opposite direction. This model is exactly the same model as the strand-inertia model, but it confuses the picture, because it ascribes a mechanism that cannot be deduced from the result. Consider a unitary mechanism with a tetrahedral die labeled *change, same, same,* and *same* on the four faces of the die in the bag. In this case, three-quarters of the time, the strands would segregate in the same direction, and one-quarter of the time, they would change direction. By statistical analysis alone, we cannot distinguish the alternate-segregation model from the strand-inertia model.

More important, the original results of the alternate-segregation model do not fit a p^2, $2pq$, q^2 pattern as predicted by the strand-inertia model (see Fig. 9-1). For example, their results for types A, B, and C are 0.64, 0.24, and 0.11. This could be accounted for by p (in the strand-inertia nomenclature) being 0.8, with q being 0.2. This would predict that type C would be 0.04 (rather than 0.11) and type B, 0.32 (rather than 0.24). This suggests that there is some artifact in the observations, possibly from the bromodeoxyuridine labeling.

VII. ALTERNATION OF GENERATIONS FORBIDDEN

There is an interesting side lesson to the alternate-segregation model. The point of this model is that cells, or DNA, can do different things in different generations. In one generation, the DNA may act deterministically, and in the next, randomly. I wish to propose a principle that will eliminate such models from consideration. This principle, the *forbidding of alternation of generations,* means that whatever cells do, they do the same thing in all generations. Whatever the cell could do in either one or the other generation, if one process had even the slightest advantage for the cell, the cell would evolve to do that process all the time.

VIII. SEGREGATION OF PLASMIDS

A. Minichromosomes—Origin, Replication, and Significance

Minichromosomes are small circles of DNA that contain the normal bacterial origin of replication. When placed in a cell they replicate, utilizing the normal machinery involved in the replication of the bacterial chromosome. A number of such minichromosomes have been isolated.[33] The base sequence has been determined as well as the minimal sequence required for proper functioning.[34] These plasmids have been used as a substrate to study the initiation process with purified enzymes.[35] The minichromosomes are present at a relatively high concentration per cell; up to 30 minichromosomes per cell have been measured.[36]

Two important *in vivo* findings indicate that the minichromosome is a good model system for replication. First, the minichromosome does not replicate at random, as each plasmid replicates only once during the division cycle even though there are a large number of minichromosomes.[37] Second, the plasmid replicates during a small portion of the division cycle; this period is the same as the time at which the normal chromosome initiates replication.[38] These two findings mean that *each minichromosome must replicate once per cycle.*

B. The Paradox of High Minichromosome Copy Number

If the minichromosome replicates only once per cell cycle, then how does the high copy number per cell occur? Consider a single cell that is transformed with a minichromosome. The cell has one plasmid. As this cell grows and divides (2, 4, 8, 16, etc.) the plasmid replicates once each cycle and increases in parallel (2, 4, 8, 16, etc.). After the cell grows into a clone or colony we would expect that for every 10^8 cells there would be between 1 and 2 times 10^8 plasmids, depending on when, during the division cycle, replication of the plasmid occurs. There would always be a low copy number per cell.

One way around this paradox is to propose that at the initial introduction of the plasmid, there is an escape from the normal control, and there is a rapid replication of the plasmid up to some high copy number. A transformed cell could have a number of successive replications of the minichromosome occurring in a short time, which would enable the cell to get a high copy number. Later, the observed cell-cycle regulation of minichromosome replication would be introduced, and the minichromosomes would replicate once per cycle in parallel with the chromosome.

C. The Passive-Segregation, Nonequipartition Model of Plasmid Maintenance

An alternative model[39] proposes that there is no equipartition mechanism for partitioning the replicated daughter plasmids at cell division, and this nonequipartition mechanism, if it can be called a mechanism, leads to the production of cells with a high copy number. This passive model and its consequences for plasmid copy number are illustrated in Fig. 9–9. Consider a cell transformed with a single plasmid. Assume that

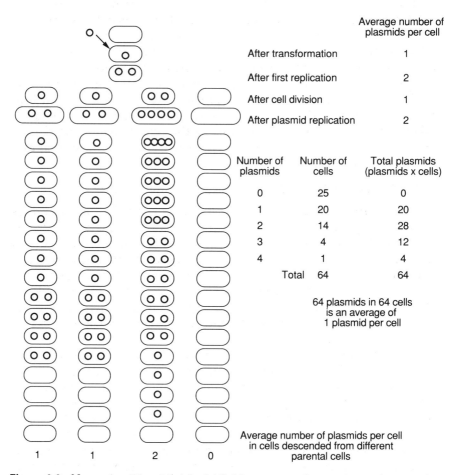

Figure 9.9 Nonequipartition Model of Minichromosome Segregation. After transformation of a recipient cell, there is a replication to produce two plasmids per cell. At division, these segregate randomly with a 1:2:1 pattern; one cell has no plasmids. At the next generation, the columns indicate the expected distribution of plasmids in the cells. The table gives the total plasmids in each cell, classified according to the number of plasmids per cell. After two generations, there is one cell with four plasmids per cell. When selection is applied to these cells, and cells with no or few plasmids are eliminated, the observed plasmid copy number increases.

the plasmid comes immediately under control of the normal replication system, and thus there is only one time during the division cycle that plasmid replication occurs. The plasmid replicates in the middle of the division cycle to produce two plasmids. When the cell divides, the plasmids can segregate at random such that there is either one plasmid in each daughter cell, or there are two plasmids in one daughter and no plasmids in the other. The resulting probability distribution predicts that half of the daughter cells produced would have one plasmid per cell, one-quarter would have two plasmids, and one-quarter would have no plasmids. This last cell would be a plasmidless segregant.

Let us follow these daughter cells for one more round of replication and cell division. Each plasmidless cell produces plasmidless cells. Sixteen such cells are drawn in a vertical column (column 4) in Fig. 9–9. Each of the cells with one plasmid replicates that plasmid to produce two plasmids per cell, and at division, we get a 1:2:1 distribution of cells with zero, one, or two plasmids. This is drawn for each of the two daughter cells with 16 representative cells (columns 1 and 2). Now consider the cell that has two plasmids. Replication occurs to produce four plasmids in the dividing mother cell. At division, the plasmids would segregate at random to produce, with a probability of $\frac{1}{16}$, a nonplasmid-containing segregant. This would occur if, by chance, all four of the plasmids went into one of the daughter cells, leaving the other daughter cell with no plasmids. This complementary sister cell, also produced at a frequency of $\frac{1}{16}$, is a cell with four plasmids per cell. Simple probability analysis leads to the prediction that one-quarter of the cells would have 1 plasmid per cell, one-quarter would have 3 plasmids per cell, and three-eighths of the cells would have two plasmids per cell (column 3). When the total distribution of the 64 cells is considered (columns 1-4), there is an average of 1 plasmid per cell, but some cells have more than one plasmid per cell. We have thus produced some cells with a higher copy number than would be expected on the basis of a strict rule of one replication per cell and a strict rule of equipartition of the plasmid at cell division. After an initial distribution of one per cell in the initial transformant, there is a successive broadening of the distribution to accommodate, after two divisions, cells with up to four plasmids per cell.

Consider the normal laboratory process by which cells are transformed and plasmid-containing cells are produced. After transformation, and a period of expression, cells are plated onto a selective medium. This eliminates the untransformed cells and allows the transformants to appear. Consider that the minichromosome contains a resistance marker to a particular antibiotic, and further, that different concentrations of the antibiotic can be used to select different degrees of

Table 9.2 Selection for High Plasmid Copy Number

Plasmid Number Selected Against	Plasmid Number in Surviving Cells	Number of Cells Killed	Number of Cells Remaining	Plasmids Remaining	Average Plasmids Per Cell After Selection
0	1,2,3,4	25	39	64	64/39=1.64
0,1	2,3,4	45	19	44	44/19=2.31
0,1,2	3,4	59	5	12	12/5=2.4
0,1,2,3	4	63	1	4	4/1=4.0

resistance. Assume further, that the degree of resistance is proportional to the number of plasmids or resistance markers per cell. Table 9–2 lists the results for different levels of selection. At the lowest antibiotic concentration we could imagine that only plasmidless cells are eliminated. Any cell, even with only one plasmid, is allowed to grow. The 25 plasmidless cells would be eliminated and the number of plasmids per *surviving* cell would be 1.64. There is an increase in the average plasmid concentration per cell caused by eliminating plasmidless cells. With further increases in antibiotic concentration there is an increase in plasmids per surviving cell. At the extreme, when all cells with 0, 1, 2, or 3 plasmids are prevented from growing, and only that cell with 4 plasmids is allowed to grow, the culture produced has four plasmids per cell.

The paradox of high minichromosome copy number can be explained by a passive, nonequipartition pattern of plasmid segregation with a strong selection for those cells with a high copy number. Rather than a mechanism for segregation, there would be no mechanism, and segregation would occur by probabilistic means.

D. Copy Number and Stability of Minichromosomes

Circumstantial evidence for a random segregation model may be gleaned from the literature. For example, cells with a minichromosome copy number between 12 and 38 (depending on the growth rate) had up to 64% of the cells segregating as plasmidless cells.[40] This result is hard to evaluate as there may have been differences in the growth rate of the segregants and the plasmid-containing cells, but at a minimum, this result demonstrates that there may be a wide distribution of copy number content in a population of minichromosome-containing cells. Other examples of minichromosome instability have been reported.[41]

Another aspect of minichromosome segregation is the possibility that the plasmids are arranged in some sort of segregated grouping that makes the effective number of segregating units less than the measured copy number. If minichromosomes segregated in larger packets of size 10, then if there were 60 copies per cell, there would be effectively only 6 segregating units. The frequency of appearance of minichromosomeless segregants would be higher than expected under the notion of randomly assorting, individual plasmids.

E. Directed Nonrandom Segregation of Minichromosomes

The segregation of minichromosomes was described above as if they were randomly assorted at each division. It should be noted, however, that the segregation of minichromosomes, like their replication (see Chapter 5), is similar to that of the chromosome. When a membrane-elution experiment is carried out and the minichromosomes are observed, there is a nonrandom pattern of minichromosome radioactivity eluted from the membrane.[42] The pattern fits the chromosome pattern with a ratio between drops of less than a factor of 2 followed by a drop of more than a factor of 2.

IX. MECHANICAL SEGREGATION MODELS

The initial proposal of the replicon model required that growth of the cell surface occur between the attachment points of DNA. We have seen (Chapter 4) that there is no zonal growth of the cell surface. This led to modifications of the replicon model in an attempt to segregate the DNA strands by binding them to the cell surface in consonance with the pattern of wall growth. One of the simplest modifications was the revised replicon model, which had as its central component the proposal that the DNA binds to the poles of the cell.[43] This modification could fit any mode of cylinder growth in a rod-shaped bacterium.

Other models with more or less complexity have been proposed. For example, a model has been proposed that requires the attachment of the daughter chromosome termini to the cell wall in a position adjacent to the new cell poles, the displacement of the parental chromosome terminus by an origin, the movement of the chromosome terminus to a new location by the tension imparted by supercoiling, and the determination of the future septum site by the chromosome terminus.[44] Needless to say, a model of this type is very difficult to prove by direct experimentation, and it is primarily a speculative model. Of course, as it is speculative, it is just like most of the models that try to explain segregation. A model has even been proposed that suggests that the segregation of DNA is due to the physical separation of the nucleoids by the associated cytoplasmic components (ribosomes, RNA, proteins), thus making the septum of the cell fall between the separated nucleoids.[45]

The general characteristic of all of these proposals is that they seek to explain the regular partition of DNA to the two daughter cells, but none of them deals with the central numerical data of segregation as detailed above. The main problem with all of these proposals is that any attempt to analyze the segregation process must take into account the constant nonrandom segregation of strands. None of the proposals for segregation has done this.

Koch[46] has proposed that larger cells allow nuclei to rotate about their axes more than smaller cells, and this explains the change in segregation pattern with growth rate. What was not taken into account here is that the degree of nonrandom segregation is really constant, with the methocel result demonstrating only an apparent change in the segregation pattern with growth rate.

X. SEGREGATION OF CYTOPLASM

We would expect the cytoplasm of the prokaryotic cell to segregate at random. No visible compartmentation of the cell would allow any departure from a random distribution of material to the daughter cells. There is evidence to support this *a priori* statement. The classic work of Van Tubergen and Setlow[47] demonstrated the random distribution of protein and RNA into daughter cells, and the exponential decrease in label from membrane-elution experiments (see Chapter 4) confirms the proposal. It is interesting, however, to note two possible points of exception. Proteins that bind to the DNA could segregate along with the DNA to give a nonrandom distribution into daughter cells as described in this chapter. A similar result could also arise by the association of proteins with portions of the cell surface. There is evidence that various heat-shock proteins segregate in a nonrandom manner, with all of the protein going into one of the two daughter cells in *Caulobacter* (see Chapter 11).[48] The analysis in Chapter 4 indicates that there is no evidence that proteins segregate nonrandomly when analyzed in bulk. It may be that some particular protein does segregate nonrandomly.

XI. SEGREGATION OF THE BACTERIAL SURFACE

The segregation of the peptidoglycan was considered in Chapter 6. The evidence supported a completely subdividable and randomly segregating cylindrical side wall, with a conserved pole segregating nonrandomly. The morphological constraints of surface synthesis are consistent with the evidence, and from *a priori* as well as biochemical considerations, we can rationalize the pattern.

A. Unit-Cell Models of Cell Growth and Division

The antithesis of a nonconserved cell surface is the proposal of a unit cell. This proposal suggests that cells have a conserved cell surface, and grow such that half of the cell surface is made of completely new material. A cell is conserved, and a new cell grows from only one pole. This model was proposed for *Escherichia coli* based on microscope observations of growing bacteria. It was observed that cells placed on agar appeared to grow from only one end. The other end was stationary.[49] When autoradiographic analyses were made of the distribution of grains in cells with labeled cell walls, it was shown that the label in both halves was equivalent.[50] This entailed correcting for the Poisson distribution of grains that makes one half of almost every cell always have more grains than the other half. The correction entailed calculating the number of grains expected in the cell half with fewer grains, and demonstrating that the lower number of grains were present by chance and not owing to some asymmetrical distribution of cell-wall growth. This experiment eliminates any unit-cell models. Also, the analysis of peptidoglycan synthesis using the membrane-elution method indicated that there was no apparent conservation of this part of the cell wall (Chapter 6).

In retrospect, we can see that the problem with the original unit-cell hypothesis was that a cell can be stuck in the agar such that one end is adhering to the agar surface more strongly than the other. Even though the cell would be growing equally from both ends, we would see one end as stationary and all of the movement of a pole would be confined to the other nonadherent end. A unit-cell hypothesis for *Schizosaccharomyces pombe* has been proposed, based on microscopic evidence of one stationary and one moveable end.[51] Such evidence must be treated with skepticism because of the problems now evident with the unit-cell model of *Escherichia coli*.

B. Segregation of Cell Membranes

The earliest work on the segregation of membranes followed the segregation of different permeases using biochemical tests for permeases as a measure of the heterogeneity in a culture.[52] A nonrandom, zonal, unit-cell type of model emerged from these studies. Direct visualization of the markers in the cytoplasmic membrane[53] or in the outer membrane[54] have also been studied, although here the results indicate a nonrandom pattern of membrane synthesis.

The results have moved away from the zonal, conserved type of growth for the membrane. A very good example of the random insertion is provided by studies of the cell surface using electron-microscopic analysis of antibodies binding to newly synthesized cell wall.[55] The results are clear in demonstrating, at a relatively gross level, that there are no conserved units. A similar result was obtained when the protein that is covalently attached to the peptidoglycan is studied. There was random insertion over the entire surface of the outer membrane, and there was no conservation of these molecules in any fixed spatial location.[56]

The number of segregating membrane subunits have been determined by autoradiography. When the cell lipids are labeled, and allowed to grow for many generations in unlabeled medium, all cells are labeled for approximately 7-8 generations, after which unlabeled cells appear.[57] This is consistent with the presence of approximately 256 conserved membrane subunits in the cell surface. It also explains why we see no conserved units using techniques such as immunolabeling to see growth zones.

NOTES

1. Van Tubergen and Setlow, 1961.
2. To intuitively understand the Poisson distribution, assume we have a collection of 100 uniformly labeled cells. After autoradiographic analysis consider that a total of 100 grains are observed on the 100 cells. These 100 grains are not distributed so that there is one and only one grain per cell. Statistical variation produces some cells with more than one grain, and so some cells are unlabeled. But because of the initial assumption that the cells were uniformly labeled, these ungrained cells do not mean that some cells are unlabeled. The cells with no grains just happened to be cells that did not get a grain above the cell. How do we account for the statistical variation in grains over the cell population? The Poisson distribution, $p_r = \dfrac{n^r}{r!} e^{-n}$ allows us to calculate the fraction of cells, p_r that received r grains when there are an average of n grains per cell. The fraction of cells that have zero grains is given by $p_0 = \dfrac{n^0}{0!} e^{-n}$ which reduces to e^{-n}. For $n = 1$ (an average of 1 grain per cell), the value of e^{-1} is 0.37. With 100 grains distributed over 100 cells, 37% of the cells would be unlabeled. If more than 37% of the cells were unlabeled, we could conclude that there was not a uniform distribution of label among the cells and that there were some unlabeled cells in the culture. The number of cells with 2, 3, or 4 grains per cell could also be calculated and compared with the assumption of uniform labeling.

3. Forro, 1965; Forro and Wertheimer, 1960.

4. Jacob, Brenner, and Cuzin, 1963.

5. Fielding and Fox, 1970; Ganesan and Lederberg, 1965; Ivarie and Pene, 1973; O'Sullivan and Sueoka, 1972; Ryter, 1968; Smith and Hanawalt, 1967; Sueoka and Quinn, 1968; Ogden and Schaechter, 1985; Hendrickson, Kusano, Yamaki, Balakrishnan, King, Murchie, and Schaechter, 1982; Kusano, Steinmetz, Hendrickson, Murchie, King, Benson, and Schaechter, 1984; Jacq and Kohiyama, 1980; Jacq, Kohiyama, Lother, and Messer, 1983.

6. Eberle and Lark, 1966.

7. Chai and Lark, 1967.

8. Ryter, 1968; Ryter, Hirota, and Jacob, 1968; Lin, Hirota and Jacob, 1971; Chai and Lark, 1967.

9. For example, Lin, Hirota, and Jacob, 1971.

10. Actually, it is not precisely an equipartition of label. The details are presented below in the analysis of the Helmstetter–Leonard Model.

11. Lin, Hirota, and Jacob, 1971.

12. Cooper and Weinberger, 1977.

13. Pierucci and Helmstetter, 1976.

14. Lin, Hirota, and Jacob, 1971.

15. Pierucci and Helmstetter, 1976.

16. Pierucci and Zuchowski, 1973.

17. It should be noted that these experiments were prompted by earlier results, using the membrane-elution method, that indicated that segregation was nonrandom. What they wished to know was not whether segregation was nonrandom, but the mechanism of this nonrandom segregation.)

18. Eberle and Lark, 1966.

19. Cooper and Weinberger, 1977.

20. Cooper and Weinberger, 1977.

21. Cooper, Schwimmer, and Scanlon, 1978.

22. Cooper and Weinberger, 1977.

23. Pierucci and Helmstetter, 1976.

24. Helmstetter and Leonard, 1987b.

25. Helmstetter and Leonard, 1990.

26. Cooper and Weinberger, 1977.

27. Pierucci and Helmstetter, 1976.

28. Pierucci and Helmstetter, 1976.

29. Cooper and Weinberger, 1977.

30. Trueba, van Spronson, Traas, and Woldringh, 1982.

31. Canovas, Tresguerres, Yousif, Lopez-Saez, and Navarette, 1984.

32. There is no 2,4 segregation pattern as this is classified as type 1,3 using the standard nomenclature.

33. Meijer, Beck, Hansen, Bergmans, Messer, von Meyenburg, and Schaller, 1979; Messer, Meijer, Bergmans, Hansen, von Meyenburg, Beck, and Schaller, 1979; Yasuda and Hirota, 1977.

34. Zyskind, Cleary, Brusilow, Harding, and Smith, 1983.

35. Fuller and Kornberg, 1983.

36. Løbner-Olesen, Atlung, and Rasmussen, 1987.

37. Koppes and von Meyenburg, 1987.

38. Leonard and Helmstetter, 1986; Helmstetter and Leonard, 1987a.

39. Helmstetter and Leonard, 1989.

40. Løbner-Olesen, Atlung, and Hansen, 1987.

41. Yasuda and Hirota, 1977; Messer, Bergmans, Meijer, Womack, Hansen, and von Meyenburg, 1978; Leonard, Weinberger, Munson, and Helmstetter, 1980.

42. Helmstetter and Leonard, 1987b.

43. Koch, Mobley, Doyle, and Streips, 1981.

44. Cavalier-Smith, 1987.

45. Woldringh, Mulder, Valkenburg, Wientjes, Zaritsky, and Nanninga, 1990.

46. Discussion following Koch, 1990.

47. Van Tubergen and Setlow, 1961.

48. Reuter and Shapiro, 1987.

49. Donachie and Begg, 1970.

50. Verwer and Nanninga, 1980.

51. Streiblova and Wolf, 1972.

52. Autissier and Kepes, 1971, 1972; Kepes and Autissier, 1972.

53. Donachie and Begg, 1970.

54. Ryter, Schuman, and Schwarz, 1975; Begg and Donachie, 1977; Vos-Scheperkeuter, Hofnung, and Witholt, 1984.

55. Jaffe and D'Ari, 1985.

56. Begg, 1978.

57. Green and Schaechter, 1972. One extremely important technical innovation should be noted in this paper. When we look at the grain distribution from cells produced by growth in unlabeled medium, we must also take into account not only the Poisson distribution from the decay of grains, but we must also consider that the random partitioning of material may not divide the material in equal portions and further that the initial cells are part of a population that has a diverse size distribution. When all of these factors are taken into account, the resulting equation from Koch, as described in this paper, is one that should be considered in future studies on segregation.

10

Transitions and the Bacterial Life Cycle

I. A SHORT HISTORY OF THE BACTERIAL LIFE CYCLE

The first reported determination of the rate of bacterial growth was that of Buchner, Longard, and Riedlin[1] who studied *Vibrio cholerae asiaticae* growing in broth. Very large differences in growth rate were reported (19–40 minutes per doubling). These differences were then shown by Muller[2] to be due to variations in the *lag* for a newly inoculated culture. Coplans[3] analyzed the lag further and expressed it as *restraint* of growth. Penfold[4] appears to have been the first to examine the various factors affecting the lag phase as measured by viable cell counts. As with his predecessors, he visualized the lag as a required part of the life history of bacteria. The most important observation of Penfold was that when cultures were subcultured from maximal growth (presumably what is now called exponential growth), there was no lag. This result was also observed by others,[5] although not all workers agreed with this result.[6] Penfold attributed these discrepancies to the probability that the cultures in the previous experiments had actually passed the point of maximal growth.

All of this early work relied solely on viable counts on solid medium. No spectrophotometric analysis was available at this time,[7] and an electronic cell counter such as the Coulter Counter was beyond imagination. But the work was far from unsophisticated. Penfold and Ledingham[8] were quite advanced in their mathematical analysis of the lag phase. The main conclusion from this early work was that the lag phase was a definite period of the life cycle of bacteria, and different laws governed the different phases. The codification of this view, which in some form has lasted to this day, was performed by Buchanan, to whose work we now turn.

If there is an award to go to the most well known but mostly unread (and uncited) paper, it must surely go to the paper by Buchanan on the *life phases* in a bacterial culture.[9] Buchanan distinguished seven different phases of growth (a record, for Lane-Claypon[10] had distinguished only four), and his description has been passed down, for many generations, in numerous textbooks. His seven phases were the initial stationary phase, the lag phase, the logarithmic-growth phase, the negative-growth-acceleration phase, the maximum-stationary phase, the phase of accelerated death, and the logarithmic-death phase. Some of these phases are illustrated in Fig. 1-4. Although Buchanan is not cited, it appears that his diagram has been passed down from generation to generation, and the ideas of lag phase, early-, middle-, and late-exponential phase, stationary phase, and death phase are attributable to his work.

Soon it was revealed that there are precise size changes in cells when an overgrown culture is inoculated into fresh medium. Clark and Ruehl[11] studied a large number of bacteria and showed that the largest cells were present in the culture after about 4–9 hours of growth. In a series of papers, Henrici[12] extended this view and proposed that there was a regular variation in sizes during the growth cycle of a bacterial culture.

The basic theme of studies on the lag phase was that they represented a regular succession of developmental phases in much the same way that a multicellular organism develops. This is seen in a summary paragraph from Henrici.

> The cells of bacteria undergo a regular metamorphosis during the growth of a culture similar to the metamorphosis exhibited by the cells of a multicellu-lar organism during its development, each species presenting three types of cells, a young form, an adult form and a senescent form; that these variations are dependent on the metabolic rate, as Child has found them to be in multicellular organisms, the change from one type to another occurring at the points of inflection in the growth curve. The young or embryonic type is maintained during the period of accelerating growth, the adult form appears with the phase of negative acceleration, and the senescent cells develop at the beginning of the death phase.

The life cycle of bacteria was believed to be similar to the life cycle of higher organisms,[13] the only difference being that there is no physical association between the bacterial cells as in the tissues of a higher organism. As Winslow and Walker have stated it:[14] ". . . the study of the bacterial life cycle is, then, the bacteriological equivalent of the study of embryology, adolescence, maturity, and senescence in the higher forms."

Yet for the last fifty years, since the incisive series of papers by Hershey,[15] we have known that the lag phase is not a time during which nothing happens; rather it is a time of rapid cell growth. This growth is best measured spectrophotometrically or chemically. The introduction of these newer technologies[16] opened up a new era in the study of the lag phase. Hershey concluded that the lag phase is merely due ". . . to initially unfavorable conditions of growth" which we would now recognize as the inheritance from the period of growth cessation in the stationary phase.

II. THE BACTERIAL LIFE CYCLE AS SHIFT-UPS AND SHIFT-DOWNS

It was pointed out in Chapter 1 that the end of this idea of a required life cycle came with the analysis of Schaechter, Maaløe, and Kjeldgaard of steady-state growth. We now realize that the life cycle is no more than an anthropomorphic construct. The bacterial life cycle is merely the result of the particular limited growth conditions of a particular culture. There is no obligatory sequence of phases of a bacterial life cycle. This analysis will now be extended with a description of the life cycle of bacteria in the terms that have been developed in preceding chapters.

The lag phase may be thought of as a shift-up from a culture growing with an infinite doubling time to a culture with a shorter, finite doubling time (Fig. 10-1). The lag phase is the rate maintenance of the original stationary phase culture at zero growth rate. The cell number does not immediately increase because there must be a passage of the cells through a C and a D period. In the ideal case, the transitions from minimal size to maximal size or back would be C + D minutes. Any extensions in this time may be considered to be owing to variations in the rates of C or D or both and variations in the rate of mass increase. The rate of mass increase may not be immediately maximal. Only when the mass achieves a particular amount per origin of DNA or per cell will DNA synthesis start. Further, DNA synthesis will start in only a fraction of the cells, and with time all cells will eventually initiate DNA synthesis. The variation is attributable to the cell-size variation produced when the cells stop growing.

The stationary phase must therefore been seen as something far from specific and universal, but merely a historical artifact, dependent upon specific local conditions, and one that must be placed, if used at all, in its proper context. The stationary phase is the result of cells growing successively slower—as though they were passing through a succession of media of slower and slower growth rate—and when the cells stop, they are at some minimal size. Cell size decreases because cells continue to divide after cell mass ceases to increase. Because there is a slow removal of growth potential from the culture, the cells are able to finish C and D

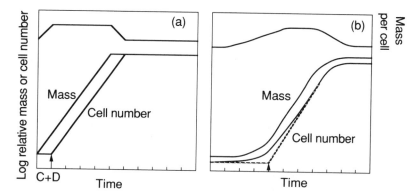

Figure 10.1 The Life Cycle of a Culture as Shift-Ups and Shift-Downs. At the left we see the ideal expectations for an overgrown culture placed into fresh medium. The previous growth rate may be assumed to be 0, and so we have a period of rate maintenance (zero rate of cell division) for C+D minutes as was described in Chapter 5. When mass synthesis stops after a period of growth, this is equivalent to a shift-down to zero growth rate; there is a period of rate maintenance of cell division (see Chapter 5) for C+D minutes. The result is that cells grow smaller as indicated by the upper graph (panel (a)). This is because division continues while mass synthesis has stopped. In practice the results are different and illustrated on the right. If we imagine that the overgrown culture has starved for many days in its nongrowing state, the cells may have been reduced in size so that they are below the minimal size that would occur from an ideal situation. Further, whatever has happened to the cells, there is no immediate attainment of the maximal rate of increase in cell mass (panel (b)). Rather, there is a relatively slow accommodation of the growth rate to the final maximal growth rate. This slow increase of mass, extending over a period longer than C+D minutes leads to the start of the increase in cell number—the end of the lag period—appearing much later. The increase in cell number is spread out over time because the increase in mass was spread out over time. When we look at such a curve, we draw the best lines through the data (dotted line) to get the lag time (indicated by arrow).

periods and to divide. Therefore the cells in the stationary phase are small.

This analysis also leads to an understanding of the terms early-log, mid-log, and late-log to distinguish cells in different parts of the life cycle. Early-log phase is the phase when the cells just emerge from the lag phase and grow with what appears to be an exponential kinetics. The late-log phase is the end of exponential growth just before the cells re-enter stationary phase. Mid-log is the phase between these two phases. If we wished to purify a bacterial enzyme from cells, for all practical purposes and assuming that the growth rate did not change the properties of the enzyme itself, it would make no difference when we harvested the cells. Presumably waiting until the cell concentration was highest, as in late-log phase, would be best. But when we study the division cycle, or wish to study the physiology of cell growth, it is improper to make the distinction between the different phases of exponential growth. The only reproducible phase of growth is long-term, steady-state, exponential growth. It may be possible, by using extreme precautions and reproducible protocols, to produce stationary-phase cells, inoculate these cells into fresh medium, and after a precise time obtain a reproducible set of cells in early-log phase. But these conditions would be extremely difficult to reproduce among different laboratories and to use as a system to investigate bacterial growth. As noted in Chapter 1, the terms early-, mid-, and late-log phase should not be used to describe cells used for precise physiological experiments. Rather than use these terms, we should state that an experiment was done with cells "four hours after inoculating an overnight culture into fresh medium." Logarithmic phase is not something subject to qualification. A cell is either in logarithmic phase or not. Either its properties are constant or its properties are changing. Only logarithmic phase, exponentially growing cells are useful in the study of the bacterial cell cycle.

NOTES

1. Buchner, Longard, and Riedlin, 1887.
2. Muller, 1895.
3. Coplans, 1909.
4. Penfold, 1914.
5. Barber, 1908; Chesney, 1916.
6. Rahn, 1906; Coplans, 1909.
7. Longsworth, 1936.
8. Ledingham and Penfold, 1914.
9. Buchanan, 1918.
10. Lane-Claypon, 1909.
11. Clark and Ruehl, 1919.
12. Henrici, 1923–24a,b, 1925, 1928.
13. Hegarty, 1939; Sherman and Albus, 1923.
14. Winslow and Walker, 1939.
15. Hershey, 1938, 1939, 1940; Hershey and Bronfenbrenner, 1937, 1938.
16. Longsworth, 1936.

11

The Division Cycle of
Caulobacter crescentus

The previous chapters have dealt primarily with symmetrically dividing, rod-shaped bacteria. In this chapter and the following two, the model that has been developed for the symmetrically dividing, rod-shaped, gram-negative bacteria will be applied to an analysis of the division cycle of bacteria with different shapes, surface structures, and division cycles. As will be seen, these differences are actually superficial, for there is a deep and fundamental similarity between the division cycles. This similarity allows the different division cycles to be analyzed in terms of a common model.

In this chapter we turn to a bacterium with an unusual division cycle, *Caulobacter crescentus*.[1] We will emphasize how the division cycle of *Caulobacter crescentus* is different from, and similar to, the division cycle of the rod-shaped, gram-negative bacteria. By looking at *Caulobacter* through the lens of the ideas on *Escherichia coli* and *Salmonella typhimurium*, we may discern and understand the differences and show that there is a greater similarity to the bacterial division cycles than is at first apparent. Some have interpreted this proposal as meaning that it is not worth studying *Caulobacter*, or that this is an attempt to force *Caulobacter* into the mold of *Escherichia coli*, with a fit that is not comfortable. This chapter does not conclude or imply that either organism is better or more valuable for the study of growth and cell division. By seeing the different organisms as part of one underlying and general plan of the division cycle, a simpler explanation of the complex pattern of *Caulobacter* may be seen. Ideas from *Escherichia coli* may be useful to the study of *Caulobacter*, and conversely, ideas from *Caulobacter* may be useful for the study of *Escherichia coli*; the relationship between the two organisms is reciprocal.

I. THE GROWTH AND DIVISION PATTERN OF
CAULOBACTER CRESCENTUS

Caulobacter crescentus is a gram-negative bacterium that has a relatively unusual division cycle. It has attracted many workers because it is held up as a model system for studying cell division, cell differentiation, and the cell cycle.[2] At division, two different daughter cells are formed; each goes through a slightly different developmental program with distinct morphological appendages appearing at the poles at specific times during the different division cycles. One of the purposes of this chapter is to show that *Caulobacter* may be more a model for the study of bacterial surface synthesis than a model for developmental events or cell differentiation during the division cycle.

A. The Division Cycle of *Caulobacter*

Cells of the *Caulobacter* species have a stalked appendage (the prefix *caulo* means stalked) that allows them to attach to surfaces. At division, a motile daughter cell is produced, different from the stalked sister cell (Fig. 11–1). The stalked cell continues to grow whether free or surface-bound. The motile cell eventually loses the flagellum that made the cell motile, and a stalk appears at the pole that had the flagellum. This pattern is then repeated, with the stalked daughter cells releasing motile swarmer cells which then develop into stalked cells. These stalked cells grow and release motile cells, repeating the growth and division pattern.

The idea that *Caulobacter* may be a model for cell differentiation derives from the observation that at division two different cells are produced (Fig. 11–1). In higher organisms, where the final products are cells of different tissues, there is a time, or a particular division, when two daughter cells go off in different biochemical and structural directions; it is this process that appears similar in *Caulobacter*. The observed changes in the individual *Caulobacter* cells are also proposed as models for cell change and pattern formation during the division cycle. Flagella in the newborn swarmer cells are lost and replaced by stalks. *Caulobacter* may also be a model for positional and temporal information. How does the cell replace the flagellum at one pole rather than at the other pole? How does the cell make proteins at some particular time during the cycle and not at other times during the cycle? These questions will be explored in the framework of an alternative view of the division cycle of *Caulobacter*.

B. DNA Replication during the Division Cycle of *Caulobacter*

1. DNA Synthesis during the Division Cycle of Stalked and Swarmer Cells

With regard to our thematic interest in growth and division during the division cycle, *Caulobacter crescentus* has two interesting properties. First, two different patterns of DNA replication are observed during the division cycles of the stalked and swarmer cells. Second, there is a periodic expression of specific genes during a portion of the division cycle. The pattern of DNA replication in *Caulobacter* is shown in Fig. 11–2. The

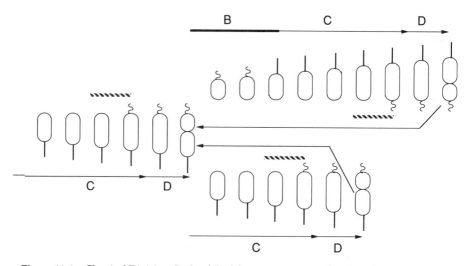

Figure 11.1 Classical Division Cycle of *Caulobacter crescentus*. A flagellated motile cell and a stalked (adherent) cell are produced at division. Flagella are indicated by the curved line and the stalk is indicated by the straight line. These two cells have different patterns of DNA replication and different interdivision times. This is the classic, or standard, description of the life cycle or division cycle of *Caulobacter*. In the stalked cell (lower cycle), the synthesis of DNA (C period) begins immediately upon division, while in the swarmer cell, DNA synthesis is delayed; a B period exists, referred to by most *Caulobacter* workers as a G1 period. The arrows indicate two different cycles, one for the stalk cycle and one for the swarmer cell cycle. In one cycle, the stalked cell cycles throw off swarmer cells, and in the other cycle, the swarmer cells throw off stalked cells. The thick bars indicate the approximate time during the division cycle when there is synthesis of flagellin and hook proteins.

stalked cells, upon releasing daughter swarmer cells, begin DNA replication immediately.[3] DNA replication starts at cell division in these stalked cells. In the terminology of the other gram-negative bacteria, there is no B-period in stalked cells (see Chapter 5). After 90 minutes of DNA synthesis (the C period), there is a 30-minute period between termination of DNA replication and cell division (the D period). The total interdivision time in the stalked cells is 120 minutes.

The released swarmer cells exhibit a slight modification of this pattern. The C and D periods are the same, 90 and 30 minutes respectively, but there is a 60-minute B period before the start of DNA replication (or as many workers studying *Caulobacter* describe it, a G1-period[4]). The observation that the C and D periods are the same in the two cells fits the idea that these periods are constant, even though cells may exhibit two different interdivision times. In *Caulobacter*, the interdivision time differences

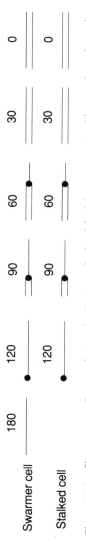

	180	120	90	60	30	0

Swarmer cell

Stalked cell

Figure 11.2 Chromosome Pattern during the Division Cycle of *Caulobacter crescentus*. The numbers above the chromosome diagrams indicate the minutes before division. The difference between the swarmer and the stalked cell is the period devoid of DNA replication, the B period, in the beginning of the cell cycle. Although the chromosomes are drawn as linear, unidirectionally replicating structures, replication is actually bidirectional from a unique origin, as in *Escherichia coli*.

are inherent in the cell-division pattern. The analysis of DNA replication in *Escherichia coli* growing with different interdivision times was based on varying the interdivision time by changes in growth medium (see Chapter 5). Rather than having the interdivision time altered by growth conditions, *Caulobacter* presents a natural condition in which the two products of division, the stalked cell and the motile cell, have two different interdivision times (Fig. 11–2). This difference in chromosome pattern would be puzzling if division produced equal-sized daughter cells. But in *Caulobacter*, cell division is unequal.[5] At division, the swarmer cell is smaller than the stalked cell owing to an asymmetric placement of the new poles.[6] This size difference may explain the different DNA synthesis patterns. The stalked cell presumably has enough mass per origin to initiate DNA replication immediately, while the smaller swarmer cell must grow to a size—presumably the same size as the newborn stalked cell—before attaining the critical mass required for initiation of DNA replication.

The relative sizes of the swarmer and stalked cells at initiation of DNA replication are not known. If size were the sole determinant of initiation, we would expect that the measured size, determined by the external dimensions of the cell, would be the same in the two cells at initiation of DNA replication. This leads to a qualification about what is meant by size or mass as they relate to the regulation of DNA replication. The external size of a cell, determined either by microscopy or by electronic size measurements using a Coulter Counter, is not necessarily related to the material that attains the requisite level to initiate DNA replication. Therefore, it is not a critical test to measure the size of swarmer cells at initiation and find that they are precisely the same size as newborn stalked cells. The initiator may be something other than external size. For example, the initiator may be only material within the inner membrane, the cytoplasm of the cell, or it may be the inner membrane itself, or it may be some particular subfraction of the cytoplasm. In any case, the requirement for additional growth until the smaller newborn swarmer cell starts DNA replication fits our ideas on the regulation of DNA replication (see Chapter 5). From this perspective, the different patterns in the two cells arising from the division of a stalked cell is understandable in the same way that the regulation of DNA replication is understood in the symmetrically dividing, gram-negative rods. Besides the similarities in DNA synthetic pattern and its possible regulation—based on the observation of constant C and D periods and the observation that cell-size differences can account for the observed patterns—the chromosome structure and its pattern of replication is also similar. There is a single circular chromosome that replicates in a bidirectional manner.[7] This suggests that the regulation of chromosome replication in *Caulobacter* is similar to that in *Escherichia coli*.

There is no evidence for multiple forks in *Caulobacter* at more rapid rates of growth.[8] At faster growth rates, the cycle changes, with a shortening of the B and C periods. This should not be taken as evidence that the pattern of regulation of DNA synthesis in *Caulobacter* is different from that in *Escherichia coli*. The lack of multiple-fork replication may be attributable to the inability of the *Caulobacter* to synthesize mass at a rapid rate, even in rich medium. *Caulobacter* is found primarily in fresh water and other nutrient-poor situations, and it may have evolved in such a way as to not grow rapidly even when presented with a complete array of nutrients. Because a cell is given a nutritious diet does not mean that the cell will take advantage of it—particularly if it is not used to such a surfeit of nutrients. If, however, the growth rate were to be increased by some procedure so that cells had a doubling time of 45 minutes, we would expect multiple forks to be found if the C period did not change in length. The shortening of the C and D periods as growth rate increases can be viewed as analogous to the change in the C and D periods as the growth rate is shortened from very slow growth to a 60-minute doubling time in *Escherichia coli* (see Figs. 5–1 and 5–5). We may envision the standard growth pattern of *Caulobacter* as similar to that of the slow-growing *Escherichia coli*, where a relatively long C + D period changes in direct proportion to the change in growth rate. As we saw in Chapter 5, when a cell with a 90-minute interdivision time speeds up to a 60-minute interdivision time, the pattern of DNA replication during the division cycle does not change. Similarly, the range of growth rates studied in *Caulobacter* may be such that the C and D periods change with growth rate, keeping the pattern of DNA replication similar at different growth rates.

2. Is the DNA Regulation in *Caulobacter* Circular?

A circular organization—the end of a sequence of events is required for, or is involved in, the regulation of the start of that sequence of events—has been proposed for the regulation of DNA synthesis in *Caulobacter*. In the discussion of DNA synthesis in *Escherichia coli*, it was noted that the regulation of DNA synthesis is linear, not circular. Termination of DNA replication is not a prerequisite for initiation of a new round of DNA replication. In *Caulobacter*, evidence has been presented that there is a requirement for the completion of some part of the previous round of replication before a new round of DNA synthesis can be initiated in the next cycle.

The experiment that led to the proposal of circular regulation involved adding hydroxyurea, an inhibitor of DNA replication, to a synchronized culture at various times during the DNA replication period.[9] This was done in a mutant that was temperature-sensitive for the initiation of DNA replication. The cells were incubated in the presence of hydroxy-urea for 90 minutes. After the hydroxyurea was removed, DNA synthe-sis was measured under conditions permissive or nonpermissive for initiation of DNA replication. When DNA synthesis was blocked by adding hydroxyurea at any time up to and including 0.83 of a round of DNA replication (i.e., when replication had proceeded less than 83% of the way along the genome), replication could not be initiated in the next cycle at the elevated temperature. If replication was beyond 0.83 of a C period, or even into the D period, then the next round of rep-lication could be initiated at either the permissive or the nonpermissive temperature. The conclusion was that there is a point during the DNA replication cycle that must be passed in order for the cell to initiate the next round of replication.[10] If there were truly such a point, then one would not only expect this to preclude multiple-fork replication, but it would also be a very different regulatory situation from that in *Escherichia coli.*

While the data in support of circular regulation are exemplary, another interpretation is possible. In one experiment, the rate of DNA synthesis was elevated immediately after removal of the hydroxyurea. This might indicate rate stimulation as described by Zaritsky[11] (see detailed analysis of rate stimulation in Chapter 5). If this increase in DNA synthesis were owing to rate stimulation, this would be evidence that initiation could occur even if the previous round of replication had not passed a particu-lar point. The exact quantification of this result must be left to experi-ments in which the mass increase is measured during hydroxyurea inhi-bition. Although it was suggested that RNA and protein synthesis continue during hydroxyurea treatment, in fact there is a significant inhibition of RNA synthesis with hydroxyurea.[12] It is also possible that turnover might increase the incorporation of label even when there is a large inhibition of net synthesis. A complete set of experiments around the *execution point* of the C period is not available for analysis. Only two experiments are presented, one with hydroxyurea added at 0.52 of the C period and another at 0.92 of the C period. In a large number of experi-ments about these points, we might expect to observe synthesis occur-ring at a continuously increasing fraction of origins of replication. We would see different levels of residual synthesis with the increasing frac-tion of genomes that had passed the particular restriction point. It is not clear, therefore, that the results observed are not owing to leakage of initiation during the period of inhibition.

The experiment just described is similar to that of Lark and Renger[13] (described in detail in Chapter 5). The Lark and Renger experiment demonstrated two different events before initiation that were sensitive to different concentrations of chloramphenicol. It was subsequently shown that leakage of protein synthesis at different intermediate concentrations of chloramphenicol could explain the appearance of two regulation points.[14] It was concluded that any experiment of this type was acceptable only if a plateau of escape synthesis was found (in this case, the initiation of DNA synthesis) as the concentration of inhibitor was varied. If such a plateau was not found, the results could not be distinguished from leakage.[15] Until such a plateau is reported for the genomic restriction point in *Caulobacter crescentus*, the conclusion that there is a circular regulatory mechanism must be accepted with caution.

C. Specific Protein Synthesis during the Division Cycle of *Caulobacter*

The *Caulobacter* division cycle is of particular interest because of the observation that some proteins are made at specific times during the division cycle. This is in strong contrast to the *Escherichia coli* results that indicate there is no cell-cycle–specific protein synthesis (see Chapter 4). The proteins made periodically during the *Caulobacter* cycle are primarily those associated with the production of the flagella, although other proteins also appear at various times during the division cycle.

1. Synthesis of Flagella-Related Proteins during the Division Cycle

Flagella appear at a particular time during the division cycle of the stalked cells (Fig. 11–1). If the interdivision time of the motile cell has a length 1.0, then flagellum formation occurs during the ages 0.7–0.8. Expression of the hook protein and flagellin genes is initiated somewhat earlier in the division cycle, with expression being observed in the middle of the C period.[16] Measurements of mRNA from the hook protein gene[17] or the flagellin genes[18] indicate that these genes are also transcribed at this time in the cell cycle.[19] This transcriptional regulation occurs at the same time as the genes are replicated.[20]

2. Synthesis of Other Proteins during the Division Cycle

Other proteins are made at specific times during the division cycle. Two methyl-accepting proteins involved in chemotaxis, a methyltransferase and a methylesterase, are synthesized coincident with the synthesis of flagellum components. They are localized in the membrane of the incipient swarmer cell before division, in the same manner as the flagellum is sequestered in the incipient swarmer cell.[21] Just before cell division, surface receptors for a *Caulobacter*-specific phage appear.[22] Shortly after cell division, pili are formed at the base of the flagellum.[23]

The biosynthesis of a number of different proteins in *Caulobacter crescentus* is therefore periodic in the cell cycle, and appears to be controlled at the level of transcription.[24] In Chapter 4 it was concluded that there are no observable cell-cycle–specific syntheses of any proteins in the simpler rod-shaped, gram-negative bacteria. How do we explain this conflicting result in light of the general model for other gram-negative bacteria such as *Escherichia coli* and *Salmonella typhimurium*? The answer to this question lies in looking at the overall development of *Caulobacter crescentus* from a slightly different perspective.

D. The Pattern of Pole Development—The Predestined Path to the Stalk

The usual manner of considering the cycle-specific synthesis of flagella protein is to think of the synthesis occurring at a particular time during the division cycle. In order for this to happen, there must be a particular trigger that starts a cascade of regulatory events at a particular time during the division cycle. An example of this approach is seen in the following passage:[25]

> The "clock" that initiates the regulatory cascade and ultimately determines the timing of *fla* gene expression in the cell cycle has not been identified. In contrast to *E. coli*, where there is no temporal control of flagellum biosynthesis and the master control genes for *fla* expression are under positive regulation by the cyclic AMP/CAP complex, it seems reasonable to speculate that in *C. crescentus* a gene(s) at the top of the hierarchy [i.e., those which respond to the initial 'trigger'] is expressed in response to a cell-cycle signal and thereby initiates the regulatory cascade. One candidate for the cellular clock activating this crucial *fla* gene(s) is chromosome replication. Evidence for this proposal is provided by the observation that DNA synthesis is necessary for *fla*K (hook protein gene) expression, but that duplication of the gene is not sufficient for its expression. Thus, the DNA synthetic requirement may be for expression of a gene above the hook protein in the regulatory cascade.

The viewpoint expressed in this passage is that periodic synthesis of particular proteins is due to an event initiating a sequence of protein syntheses or regulatory events at a particular time in the division cycle. At division, or sometime after division, the clock starts; whether it is the replication of the chromosome or some other event that starts the clock is not known.[26] The starting of the clock initiates a sequence of syntheses with the final product a visible flagellum.

1. Linear Model of the *Caulobacter* Division Cycle

It is interesting to look at the developmental process of the *Caulobacter* division cycle in terms of the history of pole development. The life histories of two newly made poles, one in the stalk cell and one in the swarmer cell, are illustrated in Fig. 11–3. The two older poles (thin poles), one stalked and one with a flagellum, descend from poles made in some prior generation. The flagellated pole was made in the previous cycle, while the stalked pole may have been made two, three, or more cycles previously. Let us look at the development of these two new smooth and undifferentiated poles from their formation, at the left (division 0), to their ultimate disposition in the form of stalked poles, at the right.

First, consider the pole attached to the stalked cell (lower cell, Fig. 11–3). The stalked cell grows and a flagellum appears just before division (division 1b). The swarmer cell we are following (because it contains the new pole) leaves the stalked cell. This cell, driven by the motility forces of the flagellum at the new pole, moves elsewhere. The flagellum at the new pole is soon shed and a stalk develops in its place. This stalked cell continues its development whether attached to a surface or free floating in the medium; we can visually distinguish a stalked cell, and the stalked end, from other types of poles. After the stalk develops, there is no further change in the stalked pole. There has been a conjecture[27] that there are rings appearing at each division marking the ages of the stalk as rings mark the ages of a tree. This conjecture was not supported by an examination of cells of known age.[28] The pole we are following does not change any further, and new cells are now released from this permanently stalked cell at subsequent divisions (division 2b, etc.).

Now consider the motile swarmer cell (upper cell) that has the other newly formed pole. The flagellated end of this motile cell forms a stalk. After the formation of the stalk at the older end, the new pole on the swarmer cell forms a flagellum at the pole opposite the newly formed stalk. After a division (1a), we follow the motile swarmer cell and not the stalked cell. The flagellum of the motile cell is shed, and a stalk is produced on the pole. This stalk is now the end of pole development. For both poles formed at division (division 0), the end of development is a stalked pole.

The ultimate end of pole development is a stalked pole. Each pole must develop a flagellum, even temporarily, between the formation of a smooth pole at division and the final stalked structure; this flagellum gives way to a final and permanent stalk.[29] The way to understand pole formation is to observe that pole formation is not complete until a stalk is formed. The pole is predestined to pass through the stages of development from a smooth pole, to a flagellated pole, to a stalked pole. If we imagine this sequence as the inevitable life history of a pole, and the occurrence of any biosyntheses associated with the pole as part of this developmental sequence, then we see that the periodic protein syntheses

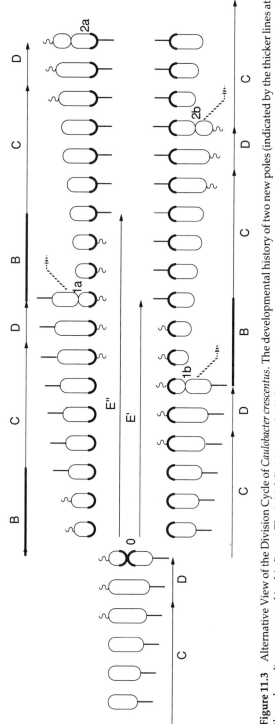

Figure 11.3   Alternative View of the Division Cycle of *Caulobacter crescentus*. The developmental history of two new poles (indicated by the thicker lines at the pole end) are presented in this figure. The initial division at which the new poles are formed is division 0. These two smooth, bald poles can be followed as they develop in subsequent cycles. In the stalked cell (lower cell), there is flagella development on the new pole. After another division, producing a swarmer cell, the flagella on the new swarmer cell are lost and replaced by another stalk. From this point, no further development of the *Caulobacter* pole occurs. In the pole associated with the motile cell (upper cell), there is a slightly longer period before flagella appear. As in the stalked cell, after a division there is a loss of the flagella and the appearance of a stalk. The period between cell division and the final formation of a stalk is termed the E period (this is done for mnemonic purposes with E following after the succession of B, C and D periods). Note that two different E periods are designated E' and E''; they relate to the shorter and longer interdivision times of the daughter cells produced at division 0, or more properly, the shorter and longer times before stalk formation is complete.

are pole specific rather than cell-cycle specific. They are related to the sequential changes in the pole. These developmental processes (i.e., the periodic protein synthesis during a division cycle and the production of different cell types because of asymmetric division) may be seen as initiated by the start of the invagination process, continuing with pole development, and finally ceasing with the final formation of a stalk.

Although we could speculate on how the developing pole causes, induces, initiates, or regulates the synthesis of the various components required for its sequential development, an understanding of the mechanism of the regulation of these biosyntheses is not required for the arguments presented here. But for illustrative purposes, let us consider a simple way of regulating this developmental sequence. Rather than a clock proceeding from division and initiating flagellin synthesis at the right time, it is the continuous development of the pole that initiates flagellar protein synthesis at the particular time in the division cycle. There is a cell-cycle clock that starts with cell division. Consider the new pole that is attached to the stalk (lower cell, Fig. 11–3). At some time during the development of a pole, there is a requirement for flagellin; perhaps a signal from the new pole starts flagellar protein synthesis. When enough flagellin is made, a feedback mechanism shuts off specific protein synthesis. It is not suggested that there are no sequential regulatory paths following this request by the pole for flagellar materials. It may be that there is the induction of a single particular message by the pole, and that this induction leads to a cascade of syntheses leading to the complete complement of polar proteins. This would account for many of the genetic observations of sequential requirements for proteins in a particular order. The model proposed here implies that the observed periodic synthesis is no different from the cell-cycle–specific pole development from a smooth pole, to a flagellated pole, to a stalked pole. The periodic synthesis is the biochemical analogue of the morphogenetic sequence from newborn pole to stalked pole. The periodic sequence is thus part of the replication–segregation sequence, and leads to the end of the division process; it is not part of a recurring cyclic process. Division leads to the stalked cell, which is the ultimate morphological stage of the polar developmental process. The formation of the stalk does not initiate any particular processes in the growing cell.

The development of *Caulobacter* may be considered similar to the simpler *Escherichia coli*. To understand this, consider that every *Escherichia coli* cell has two poles. Though they appear similar,[30] the two poles of a cell are always of different ages. One pole of every cell has been formed at the last cell division; the other pole is older. It may be only one generation older, but with some probability decreasing with the number of generations, the pole may be two, three, or more generations older. The main difference between *Caulobacter* and the symmetrical rod-shaped bacterium such as *Escherichia coli*, is that we cannot tell which pole is older in *Escherichia coli*. In a dividing *Caulobacter*, the oldest poles are the stalked poles, the younger poles are the flagellated poles, and the youngest poles at the septum are those just being formed by division. It may be that the poles of *Escherichia coli* keep developing and are completed a generation or so after visible separation into two daughter cells.[31] With *Caulobacter*, we can easily see this continuing development after division.

That *Caulobacter* appear to make two different cells is the consequence of the different lengths of time it takes to complete the poles that appeared at some previous division. The older pole of the two outer poles of a dividing cell is more developed than the younger pole. It is this difference in developmental age, as well as the size difference, that distinguishes the two products of division. The difference in pole ages leads to the differences in the daughter cells as they pass through a flagellated stage and a stalked stage. The periodic syntheses that are observed as cell-cycle specific are now seen as a result of the completion of pole synthesis after the cells have separated. In terms used to describe *Escherichia coli*, division follows after a C and D period.[32] In *Escherichia coli*, the division process appears to end at division. In *Caulobacter*, we see that there must be an additional period, the E period, that is required for the completion of the processes that began with the initiation of DNA synthesis in the previous cycle (Fig. 11–4).

The linear model for *Caulobacter* development accomplishes two purposes. It explains the periodic synthesis of proteins in terms that fit the general model of mass, DNA, and surface synthesis (presented in Chapters 4, 5, and 6). It also replaces the idea of a differentiation pattern in which cells develop into two different cell types. The proposal that *Caulobacter* has two different developmental or differentiation pathways, which is one of the main justifications for studying this organism, can be explained as the result of pole differentiation. The poles follow slightly different timings due to the initial unequal cell division occurring at the start of the developmental program. Rather than thinking of *Caulobacter* developing into two cell types from a single cell, it is easier to see that at all times we are looking at different phases in the development of a pole. There is only one cell type and this cell type is characterized at different times by having a stalk and a smooth pole, a flagellum and a smooth pole, or a stalked end and a flagellated end. But in all cases, the poles are following the same program. *Caulobacter* growth and division can be discussed in the same terms as we have discussed the interrelationships of mass, DNA, and surface in other gram-negative bacteria.

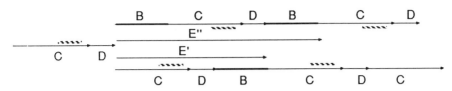

Figure 11.4 Schematic Analysis of *Caulobacter* Growth and Division. The linear *Caulobacter* division cycle is illustrated. Because of the intervening B period in the swarmer cells, there is a rapid loss of synchrony in the synthesis of DNA. DNA synthesis starts immediately in the stalked cells (lower lines) and is delayed in the swarmer cells (upper lines). The important point to notice is that the E periods overlaps the B, C, and D periods, but the E periods are merely the ends of the prior division cycles. Two different E periods are drawn, a longer one in the swarmer cell (E″) and a shorter one in the stalked cell (E′). The hatched lines indicate the synthesis of pole-specific proteins at particular times during the E period. These times, however, coincide with particular times during the division cycle that is then in progress.

The division cycle of *Caulobacter* can be presented in the form of an I + C + D[33] pattern (see Chapter 5) as illustrated in Fig. 11–4. Here we see the complex growth pattern broken down into a series of overlapping linear developmental sequences with a series of I + C + D + E patterns accounting for each of the division cycles. The complexity of the *Caulobacter* pattern is owing to both the completion of pole formation after division and the unequal division cycles.

2. Arguments against the Linear Model of Regulation of Periodic Protein Synthesis

Some experiments do not fit the model presented here. When DNA synthesis was interrupted by raising the temperature in synchronous cultures of a cell with a temperature-sensitive mutation in DNA replication, the synthesis of various flagellum proteins (hook, flagellins A and B) was inhibited. A simple explanation of this result is that the inhibition of DNA synthesis, or the termination of DNA replication, prevented cell division,[34] and the lack of new poles prevented pole-specific protein synthesis. This explanation is in accord with the model presented above. Further experiments were done to distinguish between DNA synthesis *per se* as a requirement for flagellar protein synthesis and DNA synthesis as the actual timer of expression. For example, DNA synthesis was inhibited by raising the temperature for 60 or 120 minutes in the middle of the C period in a mutant that was temperature-sensitive for DNA synthesis. It was observed that hook protein synthesis, and presumably flagellar protein synthesis, was delayed for a time equal to the time that DNA replication was inhibited.[35] In this experiment, however, we cannot distinguish between the prolongation of DNA synthesis and the delay of division as the proximal cause of the inhibition of the periodic protein synthesis. (See the discussion above on circular regulation of DNA synthesis.)

A more crucial experiment is the effect of inhibition of cell division on the periodic synthesis of flagellin.[36] When penicillin is added to *Caulobacter* at a concentration that inhibits cell separation, the periodic synthesis of protein still takes place. The DNA pattern in the cells was not disturbed, so this supports the notion that DNA synthesis, rather than pole formation, regulates the cell-cycle–specific syntheses. A reanalysis of that experiment (noting that the ordinate scales are different in the control and in the penicillin-treated cells) reveals that the synthesis of the flagellin is reduced over 80%. Further, the remaining synthesis, which is periodic, can be accounted for by a small amount (perhaps 10-15%) of continuing residual division, unaffected by the penicillin. The amount of residual division was not reported. Either (a) any small level of periodic synthesis is indicative of a functioning regulatory mechanism, or (b) the larger amount of inhibition indicates that the control mechanism leading to synthesis was affected by the inhibition of cell division, with the residual synthesis due to leakage past the inhibition of the antibiotic.[37]

Another problematic result is the demonstration that various heat-shock proteins that appear to be unrelated to pole formation are made in a cell-cycle–specific manner and segregate at division in a cell-type–specific manner.[38] For example, the homolog to the *lon* gene (as determined by antigenic relatedness) and part of the RNA polymerase were preferentially synthesized in the stalked cell, while the *gro*EL homolog was enhanced in the progeny swarmer cells. Four of the heat-shock proteins, synthesized before division, were segregated in a specific manner at cell division. The stalked cell received the greater share of the *lon* homolog, the *dna*K homolog, and the RNA polymerase subunit, while the *gro*EL homolog was segregated equally to stalked and swarmer cells. Although some of the experiments were complicated by an associated heat shock, particularly in the experiments demonstrating cell-type–specific synthesis after division, it does appear that proteins that at first sight do not relate to pole development or function are segregated preferentially into different cells.[39] This cell-cycle–specific synthesis, as well as cell-specific segregation, is not predicted by the linear model presented above. Nevertheless, the linear model offers an alternative view of these proteins. If the asymmetric segregation results are correct, then when the function of the different proteins is understood, we will be

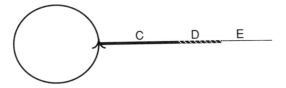

Figure 11.5 Icon of the *Caulobacter* Division Cycle. The summary of *Caulobacter* cell cycle regulation is presented in an icon similar to that for the symmetrical rod-shaped bacteria (see Chapter 5). Continuous initiator synthesis regulates DNA synthesis and wall synthesis. After division there is a continuation of the linear sequence with the E period's being devoted to completion of the poles formed at division.

able to apportion them to the different cell types according to their function. If the *lon* homolog binds to replicating DNA, we would predict that segregation would be preferentially to the stalked cell. This, of course, might be expected for the *dna*K protein product.

E. Icon for the *Caulobacter* Division Cycle

The growth and division of *Caulobacter* can be described in terms of continuous, presumably exponential, mass synthesis driving the synthesis of DNA and the growth of the cell surface as in *Escherichia coli.* An icon that describes, in a shorthand way, the logic of the *Caulobacter* division cycle is presented in Fig. 11–5. This should be compared to the icon for the division cycle of the gram-negative, rod-shaped bacteria developed in Chapter 5. The main addition to the *Caulobacter* icon is that the developmental process does not end at division as we have seen or assumed to be the case with simple rod-shaped, gram-negative bacteria such as *Escherichia coli* and *Salmonella typhimurium.* In those cases, it was assumed that at division the process of pole formation was essentially finished, although there may have been some final but limited extensions of the new pole. In contrast, the process of surface synthesis does not end with the division of *Caulobacter* into two cells, but continues from a smooth pole, to a flagellated pole, to a stalked pole. These processes may now overlap, in the way that multiforked chromosomes have overlapping cycles as expressed in the I + C + D description of the pattern of DNA replication. The formation of flagella may occur during the growth or invagination process of other cycles, and stalk formation may occur at times when other processes are also occurring and overlapping the developmental sequence.

II. APPLICATIONS OF THE *CAULOBACTER* DIVISION CYCLE

A. Methodological Implications of *Caulobacter* Growth

The production of free-swimming swarmer cells from attached stalked cells produces a synchronized culture of swarmer cells. The biosynthetic processes could be studied in synchronized cells, including the disappearance of the flagellum, the formation of a stalk, and the reappearance of a flagellum at the other pole. This approach, in fact, is one of the main methods for studying the division cycle of *Caulobacter*. In addition to the forward synchronization approach, *Caulobacter* can be studied by the same method that was used to study *Escherichia coli* and *Salmonella typhimurium*, the backwards method. The difference is that we need not use a membrane-elution method, for the natural adherence of stalks to a substrate produces a bound population of cells releasing newborn cells into the medium. We could take a growing culture attached to a substrate and pulse-label it with a protein precursor. The cells would then be allowed to continue growing, and the sequential selection of newborn cells would take place. This would be the equivalent of the membrane-elution method. The first cells would be cells with labeled proteins synthesized at the end of the division cycle, and with time, the swarmer cells collected would be those with proteins synthesized earlier and earlier in the division cycle. This technique has been used by some workers, and their determinations of the pattern of DNA synthesis during the division cycle are extremely clear.[40] This result is support of the proposal that backwards methods may be superior to forward methods for cell-cycle analysis (Chapter 8).

B. The Age and Size Distributions of *Caulobacter crescentus*

Unlike the gram-negative, rod-shaped bacteria described in previous chapters, *Caulobacter* have an asymmetrical division that produces two daughter cells with different birth sizes and different interdivision times. What are the size and age distributions of these cells? The size distribution will probably be broader than the size distribution for symmetrically dividing *Escherichia coli* cells. At a minimum, without any variation in size at particular events such as birth and division, the ratio between the smallest and the largest cells in the culture is greater than two. This is because the swarmer cell produced at division is less than half the size of the mother cell. The age distribution is also unusual in that the cells have ages that vary over two different mean interdivision times. The mathematics of the age distribution has been worked out by Tax.[41] A more important application of this derivation may be for asymmetrically dividing (budding) yeast, which should have a pattern similar to that described by the age distribution of *Caulobacter* species.

The age distribution for asymmetrically dividing cells has been worked out[42] and a simple representation of the results is presented in Fig. 11–6. It can be seen that the classical age distribution shown in Fig. 1–2 is a special case where the two daughter cells are equal and the interdivision times of the daughter cells are equal.

C. DNA Segregation in *Caulobacter crescentus*

The segregation pattern for DNA has been determined in *Caulobacter crescentus* using the biological equivalent of the membrane-elution method. This organism binds to surfaces by a stalk and releases newborn cells into the medium, so we can determine the pattern of segregation by measuring the radioactivity in the released cells in the same manner as we determine the radioactivity in cells released from a membrane. The analysis reveals that the segregation is primarily random,[43] but the actual results indicate a nonrandom component. For example, the radioactivity

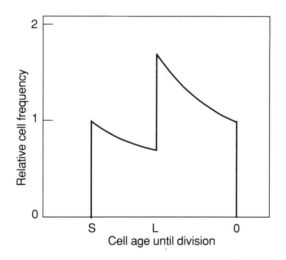

Figure 11.6 Age Distribution for Cells with Unequal Interdivision Times in Unequal Progeny. This age distribution is derived simply by considering that the number of progeny of each type must be the same as the number of dividing cells. Thus the two vertical lines at the ages of the smaller (S) and the larger (L) cells are equal to the number of cells at division (O). The numbers decrease with time (as for the age distribution in Fig. 1–2), giving this age distribution. The case for equal division and equal interdivision times in the progeny is a special case of this age distribution, with the two vertical lines moving toward each other and in the limit they are at the same place. At this time there are twice the number of cells at birth as the number of dividing cells.

released during the first generation, defined as 1.0, was in the subsequent generations 0.43 and 0.26, rather than the expected values of 0.5 and 0.25 for random segregation. This minor difference was not important with regard to the major question that was asked, whether or not progeny cells were determined in their differentiation pattern by the perfect association of a strand of DNA with a particular progeny cell (see Chapter 9 for details). This model was clearly eliminated by the segregation data. The evidence still leaves open the possibility that there is a small nonrandom component to the segregation of DNA in *Caulobacter*. This nonrandom segregation could be similar to that observed in *Escherichia coli* (Chapter 9).

III. *CAULOBACTER* AS A GRAM-NEGATIVE ROD

A number of prepublication readers of this chapter have argued that the model proposed here is a procrustean attempt to force the unique division cycle of *Caulobacter* into the simpler model offered by *Escherichia coli*. The question is sometimes posed asking whether *Caulobacter* is more like *Escherichia coli* or whether *Escherichia coli* is more like *Caulobacter*. Which organism should be taken as the model system that the other is compared to? In an even stronger attempt to dissociate the two organisms, it is argued that they may in fact be quite different in their patterns of growth and division.

What we see here, on a small scale, is the general argument regarding the direction of biological research. Are we to look for the broadest generalizations and minimize the differences as historical accidents? Or are we to emphasize the myriad approaches that organisms have taken to solve biological problems, and not subsume these differences in a general model? It is my personal approach to opt for the broad generalization and the search for general laws to explain a large variety of phenomena. I feel the same way about models of the cell cycle. The unity that is observed in diverse cell cycles (see Chapters 12 and 13) is to be welcomed. While there may be differences, it is best to express these differences as covering, and sometimes obscuring, a deeper and fundamental unity of logic and organization. Others may see some value in the linear model of *Caulobacter*. It may be that both the linear and circular characterizations of the division cycle have utility, and in time there may be some experiments that can distinguish between the two views of *Caulobacter*. Looking back on the history of microbiological research, it is astonishing to see that various models are proven to be true when considered alone, yet yield different conclusions when alternative models are available for comparison. Perhaps the explicit statement of the linear model will lead to experiments that test whether this explanation has a greater validity than the classical picture of *Caulobacter* growth and division.

In this chapter I have proposed that the growth and division cycle of *Caulobacter* can be drawn in a manner similar to the pattern of growth and division of the simpler gram-negative rods. To a large extent, it may be argued that this recasting of the division cycle does not lead to any experimental differences in what questions are asked of *Caulobacter*. We can still look at the molecular biology of biosynthesis of the DNA, RNA, protein, and cell wall; the way in which the division cycle is drawn does not affect that approach. The way in which the *Caulobacter* division cycle is drawn will be useful in the analysis and interpretation of many other experiments. It is not intended that the classical view be replaced by this linear view; rather it is proposed that they should stand equal, side by side, and be used for analysis as appropriate. As the field develops, it may turn out that one or the other view is superior for analysis; then perhaps we may choose one and discard the other.

NOTES

1. Poindexter, 1964.

2. Newton, 1984; Shapiro, 1976, 1985; Kurn and Shapiro, 1975; Ely and Shapiro, 1984; Shapiro, Agabian-Keshishian, and Bendis, 1971; Wood and Shapiro, 1975; Ely and Shapiro, 1989.

3. Degnen and Newton, 1972; Iba, Fukuda, and Okada, 1977.

4. I propose that the letters C and D be reserved for the prokaryotic period of DNA replication and the period between termination of replication and cell division. In eukaryotic cells, the alternative names S and G2 have been used. The distinction between the two types of organisms should be maintained. Also the B period should be reserved for the period devoid of DNA synthesis in newborn prokaryotic cells, and G1 should be used only for eukaryotic systems (see Chapter 15). It appears that the postsynthetic gap in DNA synthesis, the D period, is obligatory in *Caulobacter crescentus* and not obligatory in *Escherichia coli*. DNA synthesis in *Escherichia coli* may be continuous at fast growth rates, while it has thus far been impossible to eliminate the D period in *Caulobacter*. But the D period in the *Enterobacteriaceae* is as obligatory as the postsynthetic gap is in *Caulobacter*, although sometimes, at faster growth rates, it is obscured in *Escherichia coli*. Just because the D period cannot be seen, because of overlapping rounds of DNA replication, does not mean that the D period is not present.

5. Terrana and Newton, 1975.

6. Newton, 1984.

7. Lott, Ohta, and Newton, 1987; Dingwall and Shapiro, 1989.

8. Iba, Fukuda, and Okada, 1977.

9. Nathan, Osley, and Newton, 1982.

10. The possibility that the regulatory point might even be completion of replication, or some point later than 0.83 in the cycle, is pointed out by Nathan, Osley, and Newton (1982), but in the analysis presented here, the exact time of the restriction or regulatory point is not important. The important question is whether or not there is something about the previous cycle that exerts a regulatory or veto effect on the subsequent initiation of DNA replication.

11. Zaritsky, 1975b.

12. Degnen and Newton, 1972. See their Fig. 2 for the data on inhibition of RNA synthesis.

13. Lark and Renger, 1969.

14. Cooper and Wuesthoff, 1971.

15. Cooper, 1974.

16. Sheffery and Newton, 1981.

17. Ohta, Chen, Swanson, and Newton, 1985.

18. Minnich and Newton, 1987.

19. Newton, 1989. The situation is slightly more complicated, for there are actually three different flagellins of different molecular weights that make up the final visible flagellum. Although the precise relationship of these three proteins to flagellum synthesis, or for that matter any of the other proteins or mRNAs of flagellum biosynthesis, it appears that all these proteins are made in a cycle-specific manner. In our discussion we shall assume that the findings for any one protein will hold for all related proteins.

20. Lott, Ohta, and Newton, 1987.

21. Gomes and Shapiro, 1984.

22. Huguenel and Newton, 1982.

23. Sommer and Newton, 1988.

24. Newton, 1989.

25. Newton, 1989.

26. The replication of the chromosome has been proposed as a temporal regulator of this differential protein synthesis (Osley, Sheffery, and Newton, 1977; Sheffery and Newton, 1981) although it has been proposed that phospholipid synthesis underlies the observed regulation of these synthetic processes (Shapiro, Mansour, Shaw, and Henry, 1982).

27. Staley and Jordan, 1973.

28. Swoboda and Dow, 1979.

29. Whether the pole continues to be modified after it is visibly complete is not relevant to the central point. All this would mean is that the final point in the linear developmental process is later than we would normally believe it to be, based on microscopic analysis.

30. Woldringh and Nanninga (1985) have noted that it may be possible to distinguish the older and new poles of *Escherichia coli*. The older poles appear more rounded and the new poles have a decided point.

31. Koch, Verwer, and Nanninga (1982) have noted that diaminopimelic acid incorporation into extant poles indicates there may be some postdivision completion steps in pole formation.

32. Cooper and Helmstetter, 1968.

33. As described in Chapter 5, it should be observed that I is not the B period or the G1 period, but is the period between initiations.

34. Degnen and Newton, 1972.

35. Sheffery and Newton, 1981.

36. Sheffery and Newton, 1981.

37. Another complication that should be noted is the fact that there is a slowing down of incorporation of labeled precursors into protein and DNA after one generation of growth in the presence of penicillin. It may be that we should normalize the flagellin label to the total incorporation. This could increase the amount of flagellin synthesized, but it would not eliminate the arguments made here.

38. Reuter and Shapiro, 1987.

39. The asymmetric segregation of proteins related to the cell surface, which is expected from the morphological differentiation of cells at division, has been described by Agabian, Evinger, and Parker, 1979.

40. Iba, Fukada, and Okada, 1977. This method was also used by Degnen and Newton (1972). The results are more clear in the experiment of Iba *et al.*, because they used a differential method, rather than an integral method (see Chapter 3 for a detailed analysis) to assay DNA synthesis; they removed all of the released cells each time rather than allowing the cells to continuously accumulate in the medium. Mansour, Henry, and Shapiro (1980) also used this backwards approach to study phospholipid synthesis during the division cycle.

41. Tax, 1978.

42. Hartwell and Unger, 1977; Tyson, Lord, and Wheals, 1979.

43. Osley and Newton, 1974.

12

Growth and Division
of *Streptococcus*

Streptococcus faecium is an organism that looks very different from *Escherichia coli*. Not only is it structurally different, but the biochemical and morphological events during its division cycle also appear to be different. *Streptococcus faecium*[1] is a gram-positive organism with an almost spherical shape; the cells grow in chains, characteristic of *Streptococci*. The most important difference between *Streptococcus* and *Escherichia* is that the gram-positive cell wall does not grow with a diffuse intercalation pattern observed in gram-negative cells. There are also some minor differences reported in the pattern of cytoplasm and DNA synthesis during the division cycle. It is proposed here that, despite these differences, the fundamental pattern of *Streptococcus* growth during the division cycle is similar to that reported for rod-shaped, gram-negative cells.

I. THE DIVISION PATTERN OF *STREPTOCOCCUS*

A. Surface Growth in *Streptococcus*

A growing *Streptococcus* cell may be envisioned as the growth of a new pole on an old pole. The old pole is conserved and the new pole, made up of all new cell-wall material, appears between the two older poles (Fig. 12–1). The basic pattern of surface growth in *Streptococcus faecium* was originally obtained from studies of the distribution of fluorescent antibody on growing cells.[2] The new wall grows out from the central edge of the spherical cell. A pole, once synthesized, is conserved. Since there is no analogue of the cylindrical side wall in *Streptococci*, the growth of the cell may be considered one of new pole synthesis between two conserved poles.

The pattern of cell growth is, however, more complex than this simple description. During normal growth, new pole synthesis starts before the older pole is complete. Long chains result from to a large number of poles' being synthesized on one physical unit. A single round of wall growth will be defined as the synthesis of new pole from the central ridge

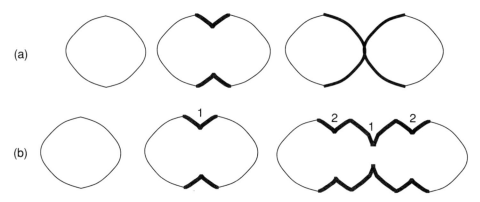

Figure 12.1 Growth of *Streptococcus* Cell Surface with Overlapping Rounds of Wall Growth. Cells in (a) show the growth of the *Streptococcus* cell surface from a single *unit* cell to two daughter unit cells. New wall growth occurs from the break between the two conserved poles. Surface growth occurs by the separation of the annular ridges present on the parental cell by new wall growth between the ridges. Growth appears to be at the leading edge of the growing pole. Analysis of sections of growing bacteria have suggested that the growth of the surface is actually due to the separation of a cross-wall that is growing between the two putative cells. As the cross-wall grows inward, it is separated at the outer edges of the cross-wall to form what is seen as new cell wall. The simplest growth pattern has cross-wall twice the thickness of the final wall. The new poles are produced by the tearing apart of the cross-wall peptidoglycan to form two independent layers. There may be some diffuse insertion of peptidoglycan into the newly growing pole, but this is probably a minor contribution to the final pole material. At the end of a single round of replication, the cells separate and two progeny, equivalent to the parental cell, are produced. The older poles are stable, and no new material intercalates diffusely into the peptidoglycan. When cells start new rounds of wall growth before the old round is finished (b), giving chain formation typical of streptococci. Additional initiations of rounds of replication can now occur at the borders between the old and new pole material. It is possible that growth at new regions of the wall relieves the cell of the need to expand at the old growth region.

to the final separation of the two new poles. Chains of cells are formed when new rounds of wall growth are started before old rounds of wall growth are completed. A single round of wall growth is analogous to a single round of DNA replication, with multiple rounds of wall growth analogous to the appearance of multiforked chromosomes where initiation takes place before the completion of an old round of replication. An example of this overlapping pattern of wall growth is illustrated in Fig. 12–1b. The long chains are due to the initiation of new rounds of pole formation before the completion of pole synthesis that had started earlier.

The experimental analysis of wall growth has been pursued primarily by electron-microscopic analysis[3] and chemical measurements of peptidoglycan synthesis during the division cycle. Direct measurements of the rate of peptidoglycan synthesis during the division cycle of an individual unit cell are difficult to analyze. The complex pattern of growth makes it difficult to distinguish to which pole we should attribute a particular incorporation measurement. When synchronized cells were studied,[4] it was observed that the rate of wall synthesis declined during most of the period of chromosome replication; the rate increased during the latter part of chromosome replication and between termination of DNA replication and cell division. The conclusion was that an increase in cell septation—i.e., the completion of the invaginating new pole—was correlated with an increased rate of peptidoglycan synthesis.[5] It is difficult to relate this observation to the rate of growth of a pole during a single unit division cycle.

Cell-wall growth in *Streptococcus* appears to be at the leading edge of the invaginating cell wall. A detailed analysis of thin sections of growing cells[6] led to the proposal that wall growth occurred primarily by inward growth of a cross-wall. As cross-wall formed, the separation of the cross wall at the outer edge of the septum led to the appearance of the curved coccal cell wall. More detailed analysis of three-dimensional reconstructions of *Streptococcus* wall sections[7] led to the modification of this proposal to include the insertion of new peptidoglycan into the pole directly from the cytoplasm without having to pass through the growing cross-wall.[8]

How are cell-wall growth zones initiated? When it was believed that there was zonal growth of the gram-negative cell surface (see Chapter 6), and that the regulation of cell-surface increase was determined by the initiation of new zones or changes in the rate of growth of these zones, there might have been some relationship between the regulatory mechanisms of the gram-negative rods and *Streptococci*. For example, if new zones in *Escherichia coli* were inserted at the completion of a particular event, such as initiation or termination of DNA replication, then we might have searched for a similar mechanism in *Streptococcus*. Many papers on *Streptococci* have attempted to make this analysis. Since zonal cell-surface growth does not occur in gram-negative cells, the question of the regulation of initiation of new growth zones is a completely open question in *Streptococcus*. At the present time, it does not appear that there is any relationship between the regulation of cell-surface synthesis in gram-negative bacteria and *Streptococcus*. We will return to a consideration of surface growth after describing DNA and cytoplasm synthesis during the division cycle.

B. An Icon for Cell-Wall Growth of *Streptococcus faecium*

An icon representing cell wall growth of *Streptococcus faecium* is presented in Fig. 12–2. The central idea is that there are overlapping sequences of wall synthesis that produce the complex patterns observed during single-cell growth.

C. DNA Synthesis during the Division Cycle of *Streptococcus faecium*

The first studies on the pattern of DNA synthesis used residual synthesis times and residual division times (see Chapter 5) to estimate the C and D values.[9] For slow-growing populations, with doubling times of 70 to 80 minutes, C times of 50 to 52 minutes were reported. The D period was estimated to be 25–28 minutes by measuring the residual cell division in the presence of mitomycin, an inhibitor of DNA synthesis. The D period was subdivided into two parts (D1 and D2). Cells within 5 minutes of division could divide in the absence of protein synthesis, while cells further than 5 minutes from division could not. D1 is the period from termination of DNA synthesis until the event that made cell division independent of protein synthesis. Cells before this event were prevented from dividing when protein synthesis ceased. When the cell cycle parameters were determined, using synchronized cells,[10] it was reported that the C period was approximately 51 minutes and the D period was approximately 24 minutes. A gap in DNA synthesis of 19 minutes was proposed for these cells growing with a 70-minute interdivision time.[11]

All of these values must be taken cautiously. When chloramphenicol was added to cells, there was no clear cessation of DNA synthesis after 50 minutes; it appeared that there was a substantial amount of residual synthesis. This suggests that there may be some leakage of protein synthesis even though the measured value is quite low. In Chapter 5, in

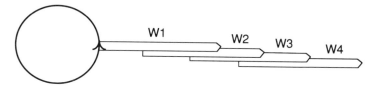

Figure 12.2   Icon for Streptococcal Cell-Wall Growth. Cell-wall synthesis is driven by a continuous increase in mass. Overlapping rounds of cell-wall synthesis occur at rapid growth rates. Just as with DNA synthesis, this occurs because new initiations of wall synthesis occur before rounds of wall synthesis that are in progress have terminated. If there were a constant time for a round of wall synthesis, then the more frequently wall synthesis was initiated, the more complex the cell chain would appear.

the discussion of the bacterial restriction point as defined by chloramphenicol, it was pointed out that a slight amount of leakage could make a large difference in determining the execution point of a particular event. This criticism applies to the results reported on *Streptococcus*. More problematic are the determinations of the D value, for in the presence of mitomycin, there is still a large amount of residual DNA synthesis. This could give more residual cell division than might otherwise be expected.

A published example of the problems of using residual division to determine various cell-cycle parameters comes from the use of mitomycin to measure the D period; in cells with an interdivision time of 64 minutes, the residual division time (i.e., the time until cells stopped dividing) was 48 minutes.[12] But a reanalysis of the results indicated that the cell number doubled after addition of the mitomycin. This result indicated a D value of 64 minutes or a D time equal to the interdivision time.[13]

Slightly different values for slow-growing cells have been reported.[14] The C value was constant and therefore the same in slow-growing cells as in more rapidly growing cells; the D period increased with slow growth. In slow-growing cells with a doubling time of 83 minutes, the C period was 52 minutes and the D period was 47 minutes. In the slow-growing cells, a 31-minute gap in DNA synthesis would be observed, and DNA synthesis would be initiated 16 minutes before cell division and be completed after 36 minutes into the subsequent cycle (Fig. 12–3). (Qualita-

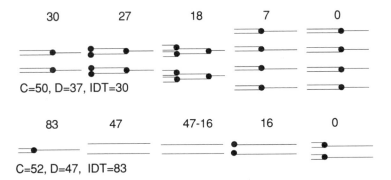

Figure 12.3 Chromosome Replication during the Division Cycle of *Streptococcus*. The upper figure is the pattern expected for a rapidly growing cell (interdivision time of 30 minutes, C of 50 minutes and D of 37 minutes); the lower figure is the pattern in slow-growing cells (interdivision time of 83 minutes) and C and D periods of 52 and 47 minutes. The times above the chromosome figures are minutes before division. When the D period is greater than the interdivision time (upper pattern), there are four terminalized chromosomes present at division. This is one of the main distinctions between the *Escherichia coli* pattern and that of *Streptococcus*.

tively, although not quantitatively, the DNA synthesis pattern in slow-growing *Streptococcus* would resemble the 50-minute cells described for *Escherichia coli* in Fig 5–1).

A different approach to determining the pattern in slow-growing cells is to analyze autoradiographic grains on cells labeled with tritiated thymidine.[15] Exponentially growing cells with a 90-minute doubling time were labeled for a short time and then fixed and autoradiographed. By assuming that there were populations with zero, one, and two replicating chromosomes (as expected from the results with synchronized cells), the data were analyzed. It was determined that the C period was 55 minutes and the D period was 43 minutes, in good agreement with the previous results. The main experimental point in this autoradiographic approach is that the analyzed cells were unperturbed cells; no synchronization was needed. But the cell ages were determined from the size classes of cells using microscopy. If there were a precise relationship between cell age (size) and chromosome replication, then we would have observed a unique replicating chromosome content in each class of cells. This was not observed. Rather, a large degree of variability complicated the results. For example, at cell birth, 60% of the cells were replicating one chromosome, while the remainder were still in the gap. If cells had initiated, on the average, 8 minutes before division, we might have expected that a larger fraction of the cells at birth would be replicating DNA and be finished with the gap phase of the cycle. More disturbing was the finding that a number of cells in the gap increased in the newborn-size class. This is clearly not predicted by the C and D values proposed, as the newborn cells should not have a gap. It was also noted that there was a great deal of variability. Thus, there might not be a very precise relationship between the replication of the chromosome and the division of the cells.[16]

For rapidly growing *Streptococcus* with a doubling time of 31 to 34 minutes, the C period was estimated to be 52 minutes and the D period was 37 minutes.[17] The methods used were determinations of residual division after inhibition of DNA synthesis and residual DNA synthesis when mass synthesis was inhibited.[18] In these fast-growing cells, we would expect multiforked chromosomes, as the interdivision time is less than the time to replicate the genome. The expected pattern of DNA replication is illustrated in Fig. 12–3. The expected rate of DNA synthesis during the division cycle in rapidly growing cells, would change in proportion to 2 to 6 to 4. There is nothing that prevents this pattern from existing, but it should be noted that if the D period is as reported, this is the first case in which the D period is greater than the interdivision time. This means that throughout the division cycle, cells have at least two complete genomes, with four genomes at the instant of division.

D. The Synthesis of Mass

Studies with synchronized cells indicated that RNA and protein synthesis was exponential.[19] When exponentially growing cells were studied by autoradiography, exponential mass synthesis was found, but there appeared to be some deviations from exponentiality at the beginning and end of the division cycle.[20] Cells were pulse-labeled with either leucine or uracil, autoradiographed, and the grains per cell were counted. The cell size was determined from electron-micrographs. The rate of incorporation was proportional to cell size over most of the cycle, but there was less-than-expected incorporation in cells of the largest and smallest size classes.

As argued before (Chapter 4), it is expected that cytoplasm would increase exponentially with no deviation, even at the instant of cell division. We might expect such an exception with regard to *Streptococcal* growth for the following reasons. As cells grow, the leading edge allows less and less volume growth. The cell may be hindered from synthesizing cytoplasm because the cell density[21] increases at the larger cell sizes. At some point, new cell-wall growing points are inserted into the cell wall as a result of the start of growth at the annuli present on the progeny cells. This relieves the stress that had been built up by cytoplasm increase. This, in turn, relieves the hindrance to growth caused by the cell wall.

This deviation from exponentiality can be dealt with in two ways. In the first approach, it is proposed that the data are consistent with exponential growth and that any deviations are minimal. When this type of autoradiographic approach was used to determine RNA synthesis in *Escherichia coli*,[22] deviations were also observed at the extreme size classes. There are a number of possible explanations for any difficulties with the autoradiographic method. For example, the method for determining cell sizes may not be accurate enough, or the relationship between cell size and cell age not precise enough, or there may be an exaggeration of the contribution of large cells to the growth pattern.[23] Whatever the ultimate explanation, it appears that autoradiography is not able to accurately measure the cell-cycle synthetic pattern, and thus it is still most likely that cytoplasm synthesis is exponential.

An alternative approach to the observations of deviations from exponential cytoplasm synthesis in *Streptococcus* is to note that even if there were actually deviations from exponentiality, perhaps due to the confining nature of cell wall expansion, these deviations have no relationship to the regulation of cell growth or cell division. They are a consequence of the pattern of volume expansion in these cells. The important point is that there is no apparent relationship of cytoplasmic events to the division cycle of *Streptococcus*. No specific syntheses are stimulated or caused by the division cycle of the cells, nor are any cytoplasmic syntheses regulating the division cycle.

E. Analysis of a Shift-Up Experiment

The growth rate in *Streptococcus* is regulated by using a defined medium with different concentrations of glutamate. There have been no published reports showing a well-behaved pattern of DNA, RNA, protein, and cell surface per cell as a function of the growth rate analogous to the Schaechter, Maaløe, and Kjeldgaard result with *Salmonella*. In the absence of this information, the results observed at different growth rates may be due to the specific biochemistry of the limiting agent. For example, if glutamate at low concentrations led to a preferential inhibition of cell wall growth, we might expect thinner cell walls at slower growth rates. Perhaps we might even expect cells to be larger (i.e., the chains will be longer) than they would have been if the growth rate were varied by adding different nutrients to a poor starting medium. As we have seen in Chapter 5, it is not the actual growth rate that is important, but the ratio of DNA to cytoplasm synthesis that is important. If the chain growth rate of all cell components slows down equally, no change in the cell composition is obtained. It is important to have increasing levels of repression so that the rate of cytoplasm synthesis increases under conditions of excess precursors for DNA synthesis and protein synthesis.

Despite these problems, it is interesting to look at a shift-up experiment with *Streptococcus*. Cells growing in limiting glutamate (22 μg/ml) at a 76-minute doubling time were shifted to a higher concentration of glutamate (300 μg/ml) to give a 33-minute doubling time.[24] The absorbance increased to the new rate after 10 minutes (this is a major difference, as in *Salmonella* the absorbance begins to increase almost immediately). There is a similar delay until uracil incorporation increases, but the incorporation of a precursor of protein had a complex kinetics with a period of rapid uptake followed by a slowing to the steady-state rate. The pattern of DNA synthesis was difficult to discern owing to variability in the measurements, but there was a delay in the synthesis of peptidoglycan as determined by lysine incorporation. In addition to the biosynthetic pattern, using electron-microscopic analysis, the size and pattern of growth zones in cells at different growth rates and following a shift-up were determined.

II. THE PROPOSAL OF THE FUNDAMENTAL CELL

The growth of *Streptococcus faecium* appears to be quite different from that of the gram-negative rods, *Escherichia coli* and *Salmonella typhimurium*. Not only are there major differences in cell-wall growth, but the pattern of DNA replication has a major difference in that the D period appears longer than the interdivision time. It has been pointed out in previous chapters that these differences are merely variations on a larger theme. Here this concept is formalized with the proposal of the *fundamental cell*. This view of the division cycle suggests that the complex patterns of *Streptococcus* are attributable to a minor change in the growth or biosynthetic pattern from the basic, underlying, fundamental cell.

The idea behind the fundamental cell is illustrated in Fig. 12–4. Consider the *Escherichia coli* cell with a 40-minute doubling time, with a C period of 40 minutes, and a D period of 20 minutes. The expected chromosome pattern (Fig. 12–4a) is a half-replicated chromosome in the newborn cell and two half-replicated chromosomes in the dividing cell. Now consider the same cells with D periods of 60 or 100 minutes. The expected chromosome replication patterns are multiples of the initial pattern with a 20-minute D period. There are either two or four half-replicated chromosomes going to four or eight half-replicated chromosomes in the dividing cell. If the cells appeared as drawn in Fig. 12–4a,b,c, as a simple, expanding rod-shaped cell, we would have a choice between accepting this pattern as a new type of replication pattern or as merely a cell with a longer D period. As the figure is drawn, when there are more than two complete chromosomes in the dividing cells (b,c), it is not stated which chromosomes segregate with which chromosomes at division. In cells with D periods less than the interdivision time, this problem never arises, as there are only two complete genomes at division, and one appears in each of the two daughter cells.

Contrast the patterns a, b, and c with the similar patterns d, e, and f. The same chromosome configurations as in patterns a, b, and c are shown as filamentous cells in patterns d, e, and f. In the case of filamentous growth, we would conclude that some mutation or environmental condition led to the change in D period. This change could be the result simply of the inability of the cells to separate because of minor alterations in septum completion. This complex pattern, with four chromosomes going to eight chromosomes, can be attributed to a simple change in the operational definition of the basic cell. Although the operational definition would change, our understanding of the underlying or fundamental cell would not change. Even in the cell with the 100-minute D period, the cell would be made up of connected cells, each with a 20-minute D period.

At this point it must be stated that this proposal of the fundamental cell, and the analysis of the patterns in Fig. 12–4, have elements of aesthetic choice as well as scientific roots. Cells with a long D period (patterns b and e, or c and f) are operationally indistinguishable on the

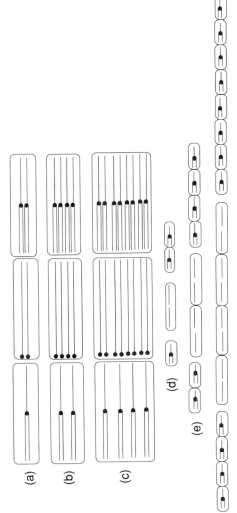

Figure 12.4 Concept of the Fundamental Cell. Consider the growth of an *Escherichia coli* cell with a 40-minute doubling time; C is 40 minutes with D periods of 20, 60 and 100 minutes respectively. The expected DNA replication patterns during the division cycle are shown in the upper half of the figure (a-c), with the cells shown as unit cells, and in the lower half of the figure (d-f) with the *Streptococcal* division pattern included. The chromosome patterns in the upper and lower figures are the same; the only difference is the visualization of the chromosomes in a dividing cell. If patterns b or c were now found in cells due to mutation or environmental conditions, we say that the pattern of cell growth was quite different from that in our standard C and D model (Chapter 5). But if these same patterns were found in cells as drawn in patterns e or f, we would say that they were due to changes in the ability of cells to separate. We could retain the concept of a fundamental cell, with the simple modification that the cells are unable to separate cleanly at the end of the D period.

basis of DNA contents and cell sizes. Trying to explain the differences as due to a minor modification of a basic pattern has no scientific foundation. This is therefore a rather personal suggestion. It is more satisfying to minimize the complexity of division-cycle analysis and combine the patterns together in a common mold. Rather than describe the variety of patterns that are found, it is better to ascribe them to modifications that do not change the underlying motif of the division cycle. It is possible to reject the proposal of a fundamental and basic cell, and to classify the different patterns that are found without fitting them into a unified and basic model. It is suggested that the concept of a fundamental cell is a simplification and aid to understanding the bacterial division cycle. Let us now apply this concept to the division cycle of *Streptococcus*.

In one of the earliest papers on the growth of *Streptococcus*, the conditions affecting the mass of an individual cell were determined.[25] It was found, for example, that the concentration of serine, proline, and particularly tryptophan in the medium affected chain length. In subsequent work[26] on *Streptococcus*, conditions were chosen to minimize the formation of chains and to study the smallest cell available. There may be a condition that is not known, and may never be known, that would allow cells to grow as a single fundamental or unit[27] cell. There would be a single round of wall synthesis and a single round of DNA replication with a D period less than the interdivision time.

From this basic cell, we would then be able to derive the more complex patterns simply by extending the D period. This proposed fundamental cell accounts for both the observed pattern of wall growth and the observed pattern of DNA replication during the division cycle. Both aspects of cell-cycle biosynthesis are explained by the concept of a fundamental cell from which other more complex patterns are produced by extending the D period.

III. EVENTS IN SURFACE GROWTH

The D period of *Streptococcus* has been subdivided into two periods. One period lies between termination of DNA replication and an event that allows cells to divide even when protein synthesis is inhibited. The other is the time between that event and cell division. This partitioning was based on the observation of a slight residual division when protein synthesis was inhibited.[28] While the observation can be accepted, it is suggested that there is actually no partitioning of the D period. The residual division is probably due to leakage of division under conditions of mass inhibition. This measurement of residual cell division is an execution-point determination. As such, it is subject to our general criticism of execution-point experiments, and in this case there is no evidence—electron-microscopic or biochemical—to support the occurrence of any specific event during pole formation.

There have been other proposals of events during the division cycle defined by various antibiotics. For example, cerulenin, an inhibitor of fatty acid synthesis in bacteria,[29] allowed 43 minutes of cell division over a wide range of growth rates. The D period was 33 minutes determined in the same cultures with mitomycin inhibition of DNA synthesis and measurements of residual cell division. This was interpreted as the existence of a cerulenin-sensitive event occurring approximately 10 minutes before the termination of chromosome replication.

Two different times at which particular execution points occur have been defined by β-lactam antibiotics.[30] One point during the division cycle, sensitive to N-formimidoyl thienamycin and methicillin, occurred before the completion of chromosome replication. A different point in the division cycle was sensitive to cefoxitin and cephalothin, and occurred later in the division cycle. These experiments reveal another problem with the interpretation of events in the division cycle using residual cell division after addition of an antibiotic. If there were an event occurring at a particular point in the division cycle, x minutes before cell division, then at a minimum we would expect there to be a continuation of division at the control rate for approximately x minutes before a sudden change to no cell division. Actually observed was an immediate slowing of the rate of cell division, and a rise in cell number to some plateau. Extrapolation of the plateau value back to the control curve gave the time of residual division. An alternative interpretation proposes that the addition of the antibiotic inhibits cell division, and cells can leak through with various probabilities and divide. The longer the wait, the more leakage is observed, and there is really a wide range of operational points at which cells are affected. There is no unique event, but the entire process of cell-wall synthesis is affected.

Temperature-sensitive mutants have also been isolated in *Streptococcus* and their residual division at the nonpermissive temperature analyzed in order to see when in the division cycle the mutation affects cell division.[31] More than 200 temperature-sensitive mutants of *Streptococcus* were isolated, and the block in the division cycle was determined. Approximately half were described as blocked in the terminal stages of division, with another large group blocked at different stages of septation. When the entire group of mutants were plotted on a histogram with their relative locations with respect to division, chloramphenicol-sensitive point, initiation of septation, termination of DNA synthesis, and the penicillin-sensitive point during the division cycle, there was a continuous range of mutants. Residual division could be observed for every time before cell division. This observation suggests that the results should be interpreted as indicating no specific events during the division cycle. As we have noted in our analysis of the results of Lark and Renger (Chapter 5), on the proposal of two events sensitive to chloramphenicol before initiation of DNA replication, if there is a continuous series of points, we either have to postulate a large or infinite number of events during the division cycle, or accept the results as due to leakage. In the case of the temperature-sensitive mutants, it is most likely that the differences are allele specific, with different mutants, for various reasons not known at this time, allowing different amounts of residual division. The finding of the large number of different sensitive points in the division cycle that argues that the results are owing to leakage rather than to the existence of any particular cell-cycle events specifically affected by the growth at the nonpermissive temperature.

IV. THE SURFACE-STRESS MODEL AND *STREPTOCOCCAL* GROWTH

One of the earliest applications of the the surface-stress model of Koch was to successfully explain the growth and shape of the *Streptococcal* cell wall.[32] The basic observation upon which the surface-stress model was applied to *Streptococcus* was the observation that the shape of two soap bubbles, when fused, formed a shape strikingly similar to that of the streptococcal pole. Further mathematical analysis and computer-simulation studies indicated that the thickness of the septal wall relative to the ultimate hemispherical pole could account for the final pole shape. The surface-stress model gave a slightly different explanation for the observed overlapping of rounds of cell-wall synthesis. When the cross-wall cutting through the cell was twice as thick as the final wall, then pole completion could occur. When the ratio of the thicknesses was less than two, the mathematical analysis yielded an incomplete pole and the formation of chains. In the ideal state, this model could account for either growth of a fundamental or unit cell, or infinitely long chains. By introducing some variability in the thickness of the invaginating septum, we could account for the observed distribution of chains in a *Streptococcal* culture. The final simulation result suggested that the septum grew in thickness as it formed, reaching a thickness twice that of the pole in the immediate region of pole splitting.

V. SEGREGATION OF DNA IN *STREPTOCOCCUS*

There has been only one experiment on the segregation of DNA in *Streptococcus*. The labeling–presegregation method in methocel was used (see Chapter 9) in order to simplify the analysis. The results showed a very slight nonrandom segregation, but segregation did not vary with growth rate. As described in Chapter 9, when the growth rate was varied in *Escherichia coli* a change in the degree of nonrandom segregation was observed. At the higher growth rates, segregation appeared more random. As we have already noted, the variation in growth rate by change in the concentration of a single substrate may not be the same as changing the growth rate by the addition of a number of new metabolites. It not yet possible to conclude that the pattern of DNA segregation in *Streptococcus* is different from that in *Escherichia coli*.

VI. DENSITY DURING THE DIVISION CYCLE
OF *STREPTOCOCCUS*

Cell density could change as the cell surface failed to keep up with the increasing cell cytoplasm. If the initiation of new rounds of wall growth were to be slightly delayed, we would expect a variation in cell density during the division cycle. Investigations of the density of *Streptococcus* during the division cycle have led to the conclusion that there are cell-cycle–specific variations in cell density.[33]

It was argued in Chapter 7 that cell density does not vary during the division cycle, or more precisely, if it did vary, it was not a regulatory factor regarding the division cycle. The variation of density (or something related to cell density such as an internal cell pressure) has been proposed as a regulatory factor in *Streptococcus*.[34] It should be noted, however, that when cells of widely different densities were classified according to their invagination pattern, the differences were minimal.[35] If density were a regulatory factor, then placing cells in media of different densities would lead to large variations in the pattern of cell growth. Until such a study is undertaken, the proposal that density affects the cell cycle must be accepted with caution.

VII. COMMENTS ON THE GROWTH PATTERN
OF *STREPTOCOCCUS*

Having described an alternative view of synthesis and growth during the division cycle of *Streptococcus*, a series of caveats must be added to this analysis. These warnings are added in spite of the fact that the basic results agree with the general model of growth and division of the most well known organism, *Escherichia coli*. When the data on *Streptococcus* are analyzed, the results are unable, by themselves, to support the conclusions described. The synchronized cultures were poorly synchronized, with almost half of the division cycle allocated to the rise in cell number.[36] More important, internal evidence from the synthetic patterns indicates perturbations from normal growth. For example, when cells from a synchronized culture are allowed to grow, there is a variation in the pattern of synthesis in consecutive cycles.[37] Further, growth did not appear to be balanced as the cell composition changed during the course of synchronous growth. With regard to the autoradiographic studies on unsychronized, pulse-labeled cells, the data alone cannot distinguish among different models. This is most evident in the studies of DNA synthesis in which an independent determination of the pattern of synthesis could not be obtained from the data. If the pattern was not known from previous experiments, the data would not be able to give, independently, the pattern of DNA synthesis.

It should also be noted that the determinations of synthesis in synchronized cells were not done by a differential determination, but by an integral or cumulative determination. Radioactive precursors were present throughout the period of growth before and following synchronization; thus the variations are minimal and subject to errors of interpretation.

What are we left with? It is comforting that the existing data, no matter how tenuous, agree with the basic model proposed in the previous chapters. Less comforting is the knowledge that these data are only the start of the need for better and more accurate determinations of synthesis during the division cycle. Let me therefore propose a future program for the analysis of *Streptococcus*. In Table 12–1, there is a reanalysis of the three basic parameters that are expected for the chromosome configurations (Fig. 12–2). These particular predictions have not yet been com-

Table 12.1 Cell Composition Expected for *Streptococcus faecium* at Different Growth Rates

	Fast Growth	Slow Growth	Ratio (F/S)
Doubling time	30	83	
DNA/cell	4.55	1.85	2.46
Mass/cell	7.82	2.29	3.42
Mass/genome	1.72	1.24	1.39

For the relevant formulae see Chapter 5. These are the predicted values for the parameters (C and D values) given in Figure 12–3.

pared with the proposed chromosomal patterns. The ratios are such that if the predicted patterns are correct, then the predicted results will be found.

Regarding the mass per cell, it should be noted that if the initiation of DNA occurs, as in the *Enterobacteriaceae*, at a size that is constant, then we would expect a particular ratio of the mass per cell at the different growth rates. The predicted ratio of 3.4 does not appear to be found for size measurements of cells at different growth rates.[38] A number of possibilities exist. First, the volume measurements determined by electron-microscopic analysis of cells may not be the same as the mass of the cells determined by dry weight or other direct chemical measurements. Another possibility is that the initiation size may not be constant at different growth rates. Whatever the explanation, further analysis of these basic parameters, which are equivalent to the Schaechter, Maaløe, and Kjeldgaard measurements, should go a long way toward clarifying the situation.

NOTES

1. This organism was originally called *Streptococcus faecalis*. Recent changes in nomenclature have changed the name of this organism to *Enterococcus hirae*, but as this name is not well known this chapter will continue to use the older name.
2. Cole and Hahn, 1962.
3. Higgins, 1976; Higgins and Shockman, 1976.
4. Hinks, Daneo-Moore, and Shockman, 1978a.
5. My analysis of the data suggests that we cannot conclude anything from these results, as the synchrony was not adequate and the results, although a differential measurement, were not able to distinguish different patterns of synthesis.
6. Higgins and Shockman, 1970.
7. Higgins, 1976.
8. Higgins and Shockman, 1976.
9. Higgins, Daneo-Moore, Boothby, and Shockman, 1974.
10. Hinks, Daneo-Moore, and Shockman, 1978b.
11. It should be noted that the rise in cell number occurred over a time equal to about half of the interdivision time. Although this is a very low degree of synchrony, total measurements of DNA in this culture indicated that there were periods of increased and linear rates of DNA synthesis interspersed with lower rates of DNA synthesis. It was reported that the DNA nearly doubled during this period, and thus it was assumed determination of this period of rapid synthesis could give the C time. It should be noted that the synchronized cultures were not in balanced growth. Over three generations there is a divergence in the cell composition.

12. Gibson, Daneo-Moore, and Higgins, 1983b.

13. This criticism can be also applied to the recent results of Bourbeau, Dicker, Higgins, and Daneo-Moore, 1989.

14. Gibson, Daneo-Moore, and Higgins, 1983a.

15. Higgins, Koch, Dicker, and Daneo-Moore, 1986.

16. The basic problem with this autoradiographic approach is that the results can only be observed to be consistent with a particular description of chromosome replication, and not prove that that is the pattern. The data may be consistent with other patterns not tested, and therefore there is a degree of ambiguity about this result.

17. Reported by Gibson, Daneo-Moore, and Higgins, 1983a, although the data are from an abstract (Daneo-Moore, Bourbeau, and Carson, *Abstr. Annu. Meet. Am. Soc. Microbiol. 1980*, **166**, p. 95) so it is not possible to judge the original data.

18. These results are subject to the reservations expressed above.

19. Hinks, Daneo-Moore, and Shockman, 1978b.

20. Higgins, Koch, Dicker, and Daneo-Moore, 1986.

21. I use the word density very loosely here. A crowding effect is visualized that might hinder cytoplasm synthesis by restricting the free diffusion of material through the cytoplasm. It is not clear that this crowding would lead to any increase in the physical density of the cells.

22. Ecker and Kokaisl, 1969.

23. When we study the rate of synthesis in cells of different sizes, we measure the cell size and determines the autoradiographic activity of the cell. As we know from our analysis of cell cycle variability (Chapter 8), there are very few cells in the smallest and largest cell-size classes. Let us assume, for the moment, that we analyzed cells from a population and obtained, by selective counting, an equal number of cells in each cell-size class. This would give equal weight to the contribution of cells that might be only a very minor component of the entire life cycle of the cells. If there are variations in the cytoplasmic growth of a cell, then we might be determining the pattern in those cells that just happen to be the slowest-growing cells in a normal distribution of cell synthetic rates. Or these slow cells may be slow for unusual reasons not related to the basic growth law of the cytoplasm. For this reason, it is correct to conclude that mass synthesis in *Streptococcus* is exponential. There may be some odd cells at the extremes, but these may be accidentals that have no relevance to the basic growth law of the cells. In a sense the argument here is for a majority view of the cell cycle, with only a small minority of cells contributing to any deviations from exponentiality.

24. Gibson, Daneo-Moore, and Higgins, 1984.

25. Toennies, Iszard, Rogers, and Shockman, 1961.

26. The work on *Streptococcus* that is analyzed here has essentially all come from a single group (Shockman, Higgins, Daneo-Moore and their associates) located at Temple University in Philadelphia.

27. A unit cell was hypothesized for *Streptococcus* by Edelstein, Rosenzweig, Daneo-Moore, and Higgins (1980); however, this dwelt mainly on the constant size of the poles at different growth rates. When the doubling times of the cells were varied from 30 to 110 minutes by changing the concentration of glutamate

over a range of 14 to 300 μg/ml, the size of the older pole of growing cells was seen to be a constant. This was proposed as the unit cell on which other cell patterns were built. The relationship of this unit cell to the formation of chains of cells was not stressed.

28. Higgins, Daneo-Moore, Boothby, and Shockman, 1974.

29. Vance, Goldberg, Mitsuhashi, Bloch, Omura, and Nomura, 1972.

30. Pucci, Hinks, Dicker, Higgins, and Daneo-Moore, 1986.

31. Canepari, Del Mar Lléo, Fontana, Satta, Shockman, and Daneo-Moore, 1983.

32. Koch, Higgins, and Doyle, 1981, 1982.

33. Glaser and Higgins, 1989; Dicker and Higgins, 1987; Bourbeau, Dicker, Higgins, and Daneo-Moore, 1989.

34. Glaser and Higgins, 1989.

35. Dicker and Higgins, 1987.

36. If the pattern of DNA replication in *Escherichia coli* were to be analyzed with cells with this variability, the pattern of DNA replication could not be determined.

37. Hinks, Daneo-Moore, and Shockman, 1978b.

38. Gibson, Daneo-Moore, and Higgins, 1984.

13

Growth and Division
of *Bacillus*

We now turn our attention to the division cycle of gram-positive, rod-shaped organisms, the *Bacilli*. Although there are differences between the division-cycle pattern of these organisms and that of the gram-negative rod-shaped organisms, we shall see that the discussion is simplified if we analyze these organisms in terms of the general model made for division-cycle regulation described in the previous chapters.

I. SURFACE GROWTH DURING THE DIVISION CYCLE OF *BACILLUS SUBTILIS*

A. Structure of the *Bacillus subtilis* Cell Wall

Bacillus subtilis is a rod-shaped, gram-positive bacterium. It thus has the overall shape of the *Enterobacteriaceae*, but with a thicker cell wall more like the *Streptococci*. The peptidoglycan of the *Bacilli* is much thicker than that of *Escherichia coli*. This thick wall produces the defining gram-positive phenotype, and also leads to a different mode of wall growth during the division cycle. The distance from the cytoplasmic membrane that contains the enzymes for cell-wall growth to the outer surface of the cell wall is greater than the reach of the enzymes responsible for peptidoglycan synthesis. Thus it is not possible to have the thick wall synthesized by direct addition of material throughout the wall thickness. In particular, the outer extremities of the peptidoglycan are inaccessible to the action of the enzymes that add subunits to the resident peptidoglycan. Because of this thickness, the peptidoglycan is synthesized close to the cytoplasmic membrane, and then migrates out from the membrane by having new peptidoglycan inserted underneath the pre-existing peptidoglycan (Fig. 13–1).

B. Turnover of Peptidoglycan

The peptidoglycan of *Bacillus subtilis* is not stable but turns over. When cells are labeled with a short pulse of a peptidoglycan specific label, the

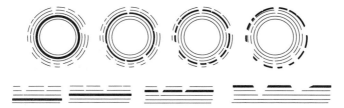

Figure 13.1 Inside-to-Outside Growth of *Bacillus subtilis* Cell Wall. A cross-sectional view of the growth of the gram-positive cell wall in *Bacillus* reveals that the newly synthesized wall (indicated by the thick line) is laid down relatively loosely next to the cytoplasmic membrane. With further growth, new wall is inserted between the membrane and the wall synthesized earlier. This forces the wall toward the outside of the cell. Because the circumference increases with distance from the cell, the wall stretches and eventually ruptures from the stretching force, as the ring of peptidoglycan moves toward the outside of the cell. In addition to the circumferential growth, the extension of the side wall leads to rupturing of the wall in the axial direction. As the bonds are continuously broken, both by cylindrical extension and inside-to-outside movement, the wall being followed (thick line) is eventually lost to the medium.

incorporated label is stable for a time, and then there is an approximately exponential decay process, during which the label leaves the cell and is excreted into the medium.[1] To summarize these results in terms of the currently accepted model of wall growth, the label is incorporated close to the cytoplasmic membrane. With further growth, new layers of peptidoglycan are synthesized between the labeled material and the cytoplasmic membrane. The label is thus pushed farther and farther away from the membrane. There is a lag period owing to the time required for the labeled peptidoglycan to move from the cytoplasmic membrane out to the extremities of the peptidoglycan. Bonds are then broken by enzymatic and nonenzymatic mechanisms, and eventually the label sloughs off the cell at the outer extremities. Rather than a stretching of a rubberized wall, as described in Chapter 6, a better metaphor for *Bacillus* wall growth is that of a molting of the outer wall. Old wall leaves the cell as new wall moves from the inside to the outside of the cell.

C. Inside-to-Outside Growth of Peptidoglycan

The new aspect of wall growth that is exhibited in the *Bacilli* is inside-to-outside growth. A new layer of peptidoglycan is constructed in an unstressed conformation between the cytoplasmic membrane and the adjacent peptidoglycan. As this layer is completed, new layers are then constructed below this layer. This forces the previously synthesized peptidoglycan sheet to move out and to enclose a larger volume. In moving out from the cytoplasmic membrane, the peptidoglycan sheet becomes stressed, and becomes the load-bearing layer protecting the cell from internal pressures. With further synthesis of more peptidoglycan the stress-bearing layer is pushed out and stressed further until bonds are enzymatically broken by hydrolases sequestered within the peptidoglycan or present in the medium.[2] With continued growth and the formation of new lower layers of peptidoglycan, the older peptidoglycan moves away from the center of the cell, bonds are continuously broken, and eventually enough bonds are broken so that material is sloughed off.[3]

The kinetics of release of wall material is marked by an initial lag phase, a period of exponential release of material, and a final plateau value where release stops. The lag period is accounted for by the time required for material to move out from the membrane. The exponential decrease is due to material sloughing off from the cell surface, and the resistant fraction is presumably due to the pole material that does not turn over rapidly. The poles are more stable than the cylindrical side wall (Fig. 13–2).

As the peptidoglycan moves to the outside, two types of stress develop. One is stress in the circumferential direction as the diameter of the hoops of peptidoglycan grows larger. In addition to this there, are the axial stresses that occur as the cell elongates. These two stresses can combine to give a helical torsion to the growing cell wall, and it is this helical stress that gives the growing cell a twisting morphology when cells are grown as filaments.[4] Koch[5] has reinterpreted these proposals for helical wall growth in terms of the stresses that occur as the wall rips or is cut as inside-to-outside growth occurs.

D. Zonal or Diffuse Surface Growth

Just as with the *Enterobacteriaceae* (see Chapter 5), there has been a continuing argument about zonal and diffuse surface growth in *Bacilli*. Early studies of the distribution of immunofluorescent antibody on the cell surface led to the conclusion that there were discrete zones of cell-surface synthesis.[6] Analysis of nucleoid segregation led Sargent[7] to propose zonal surface growth. Radiolabeling experiments on synchronized cells also were interpreted on the basis of zonal surface growth that increases at a constant rate.[8] The doubling in the number of growth zones was proposed to be dependent upon DNA replication. It was also reported that there was a sudden change in the rate of wall synthesis at the end of the division cycle.[9]

Biochemical experiments have supported the notion of diffuse cell-surface growth. Evidence was presented that old and new material were intermingled, which is consistent with an intercalation mode of cell-wall

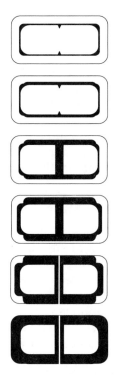

Figure 13.2 Side Wall and Pole Growth in *Bacillus subtilis*. The displacement of old wall by new wall growth appears to be more rapid in the side of the cell than at the poles. The black areas, indicating newly synthesized wall, expand more rapidly in the cylindrical portion of the cell, with slower wall replacement in the pole area. The septum that forms may not be the final pole thickness, and after cell separation there is continued incorporation in the new septum. The septum reaches its final hemispherical form after being formed, and there is a tearing of the two halves of the new pole to produce two cells.

synthesis.[10] Detailed autoradiographic analysis of cell-wall growth supported a diffuse mode of surface growth.[11] De Chastellier, Frehel, and Ryter have observed that there is diffuse side-wall growth but a defined zone of incorporation at the site of invagination.[12] The possibility that turnover could have obscured the presence of specific zones of growth has been tested. The turnover does not seem to account for the absence of zones,[13] thus supporting a diffuse growth pattern. Electron-microscopic analysis was extremely supportive of the proposal that cell-wall growth occurred over the entire surface of the rods, with wall material gradually moving from a position next to the cell membrane to a position at the outer surface of the cell.[14]

E. Topography of Surface Growth

In *Escherichia coli* there is a decrease in the label in the cylindrical wall when cells start to invaginate. As described in Chapter 6, this is a strong support for the stress model of surface growth. The same phenomenon of a decrease in side wall labeling when cells invaginate has been found with *Bacillus subtilis*.[15] This suggests that the stress model applies also to gram-positive rods, and that the topographical arrangement and partitioning of synthesis is identical to that found in the gram-negative cells (see Chapter 6). The kinetics of appearance of bacteriophage binding sites also supports the proposal that there are no growth zones along the cylindrical wall.[16]

F. Variability of Growth and Filamentation in *Bacillus subtilis*

One of the most interesting observations regarding growth of *Bacillus subtilis* is the dependence of cell size or filamentation on cell density. When bacteria are grown in a very dilute culture, septa develop, but cell separation does not occur, and cells grow as long filaments. When the same cells are placed in the same medium and allowed to grow to a higher density, the filamentation is suppressed.

The explanation for this phenomenon appears to be the differences in the distribution of autolysins in cells growing at different cell densities. An autolysin is released from the cell surface of *Bacillus*, and this enzyme is responsible for the splitting of bonds at the surface of the cell. In dilute solution more of the enzyme is in the medium and not bound to the extremities of the cell surface. This means that there is a deficiency in the cutting of peptidoglycan bonds. As the cell elongates, and septa are formed, these septa do not go through the splitting process at low cell densities, as there is not enough of the released enzyme in the vicinity of the cell to effect the septation process. Higher densities of bacteria lead to a higher amount of enzyme in the neighborhood of each cell, and this enzyme more readily binds to the cell surface. This leads to cutting of peptidoglycan bonds and the separation of cells.

G. Multiseptate Growth of *Bacillus subtilis*

Whereas the growth of the gram-negative cells has been described in terms of single cells at all times, with filamentation or chains of cells appearing as an aberration, in *Bacillus subtilis* the formation of chains or multiseptate bacteria appears to be a normal aspect of their growth. But this chain formation is related to the growth rate and cell density of a culture. As cells increase their growth rate, the average number of cells per chain increases from approximately 2 per chain to up to 30 subunits per chain. As we shall see, this explanation of this multiseptate growth is that the sequence of events in the cell cycle is not delimited by the act of cell division. With an overlap of septum formation, because the period between termination and the final separation of cells is much greater than the interdivision time, multiseptation is a necessary result.[17] Microscopic examination indicated that 138 minutes elapse between the formation of a septum and the final separation of cells.[18] There appear to be an additional 12 minutes between termination of DNA replication and septum formation for a value of 150 minutes for the D period.

Of course, this entire discussion is rendered problematic by the variability of filament formation. It comes down to a definition of what is the basic process. If the nonfilamenting cell is the basic cell, with filamentation occurring when there is some decrease in the availability of autolysins, then we see the difference between the two as a trivial difference.

II. CYTOPLASM SYNTHESIS DURING THE DIVISION CYCLE OF *BACILLUS SUBTILIS*

A. The Synthesis of Enzymes during the Division Cycle of *Bacilli*

Bacillus subtilis was one of the first organisms for which cell-cycle–specific enzyme synthesis was reported. Using filtration to obtain a synchronized culture, investigators observed steps in the synthesis of sucrase, histidase, and aspartate transcarbamylase.[19] One possible explanation of this proposed step-wise synthesis was that enzyme steps were correlated with the sequential replication of the genome. When four enzymes (histidase, aspartate transcarbamylase, ornithine transcarbamylase, and dehydroquinase) were studied, the steps observed in their synthesis corresponded well, but not perfectly, with the order of the genetic markers on the *Bacillus subtilis* genome.[20]

While the bursts of synthesis were seen when synchronized cells were not induced, it was possible to induce the synthesis of these enzymes at any time during the division cycle. This meant that the genetic material was available at all times to direct the synthesis of the enzymes. Various models have been proposed to explain these bursts of synthesis in the uninduced case. For example, it was proposed that a doubling in potential to synthesize an enzyme, by the duplication of a gene, would lead to bursts of synthesis.[21] The periodic gene replications would lead to periodic fluctuations of repression and enzyme synthesis. Another explanation of the bursts of enzyme activity depended on the proposed instability of the enzymes.[22] Evidence against the sequential replication of DNA accounting for steps in DNA synthesis came from studies where DNA synthesis was inhibited, yet the steps in enzyme synthesis were unaffected.[23] This led to the further elaboration of models in which the oscillations of negative feedback loops could lead to the appearance of bursts of enzymes during the division cycle.[24]

The observations of steps in enzyme synthesis were supported by the parallel observation of steps in enzyme synthesis during outgrowth of spores of *Bacillus cereus* or *Bacillus subtilis*. In the case of *Bacillus cereus*, experiments indicated that although the major period of DNA replication occurred just before cell division, there was a small but significant amount of DNA replication within 5 to 10 minutes after germination.[25] The synthesis of enzymes during spore outgrowth in *Bacillus subtilis* paralleled the genetic order of the genes.[26]

How shall we accommodate this evidence for cell-cycle–specific enzyme synthesis into our model of cell-cytoplasm synthesis which argues that cytoplasm synthesis is independent of the division cycle? There are two different responses, both extensions of the arguments made in Chapter 4. First, we can accept that the measurements are accurate, and that there are variations that occur with some periodicity in the division cycle. If this be the case, then I would merely reiterate the argument that there is no relationship between these oscillations and the induction of DNA synthesis or cell division. This argument proposes that the oscillations are secondary phenomena in which various regulatory circuits respond to the cell-cycle events. For example, the draining of various precursors of DNA synthesis from the pool at the start of new rounds of DNA replication could set up some periodic circuit that leads to the observed bursts of synthesis.

The second response, to which I am partial, is to say that there is actually no evidence for such oscillations and that the evidence may be disregarded. When we apply the criteria for judging synchrony experiments (Chapter 3), we can disallow essentially all of the evidence for cell-cycle–specific enzyme synthesis. It is possible to take many, if not all of the graphs, and reinterpret the data to support the absence of any cell-cycle–specific synthesis. The evolutionary arguments made in Chapter 3 also apply here. There is no logical support for the maintenance of any of the cell-cycle–specific synthetic patterns. None of the enzymes studied can be arguably shown to have a logical connection with the division cycle such that a cell-cycle–specific synthetic pattern would be advantageous to the cell. As argued in Chapter 3, such periodic syntheses are most likely inefficient and detrimental to the growth of the cell. It is therefore suggested that there is no evidence for cell-cycle–

specific synthesis in the *Bacilli*, and we can assume that synthesis is continuous and exponential. It is of interest that for the last two decades there has been essentially no confirmation or follow-up on the original reports. If such periodicities were reproducible then such efforts would have been expected.

B. Macromolecular Synthesis and Cell Growth during the Division Cycle of *Bacillus subtilis*

Regarding the synthesis of categories of macromolecules such as protein or RNA, there are no experiments that have investigated these syntheses during the division cycle of *Bacillus*. There have been efforts, however, to determine the general pattern of growth of a cell. The very first analysis using the Collins–Richmond methodology (see Chapter 3) was performed on *Bacillus cereus*,[27] and the conclusion was that cell growth is exponential during the middle part of the division cycle. Technical limitations made it more difficult to determine the growth pattern at the start and finish of the division cycle. As with *Escherichia coli*, there have been proposals of linear growth during the division cycle.[28] Sargent has also suggested that synthesis of the cell is linear during the division cycle.[29] Other, more complex, patterns of growth have been proposed for *Bacillus subtilis*.[30] One reading of the sparse data on the *Bacilli* suggests that the cell grows continuously and exponentially in the same way as *Escherichia coli*. There is no evidence so strong as to disallow this conclusion.

III. DNA SYNTHESIS DURING THE DIVISION CYCLE OF *BACILLUS SUBTILIS*

A. Dichotomous Replication in *Bacillus subtilis*

Dichotomous DNA replication, the appearance of multiply forked chromosomes, was first discovered in *Bacillus subtilis*. This organism is transformable, and therefore it is possible to determine the relative frequency of different genes by a simple assay. When the relative frequency of different genes was determined from cells with one genome (either spores, or resting cells), and compared to germinating or growing cells, it was found that some genes increased by a factor of four.[31] This was explained by presuming that the DNA in spores is present as unreplicated genomes. The ratio of any gene to any other gene is taken as unity.

Although our discussion has been directed almost entirely toward the pattern of DNA replication in *Escherichia coli* and related bacteria, in fact the first indication that there were multiple forks—referred to as dichotomous replication—was obtained in *Bacillus subtilis*. When the relative frequency of transforming activity for particular genes was compared in spores and cells that had germinated from spores, it was found that for some genes there was a regular progression in the ratio of transforming activity for two genes from one, to two, to four.[32] This could be explained as follows. In spores, all DNA is present as unreplicated genomes. The ratio of any gene to any other gene is defined as unity. When spores germinate, the genes at the origin immediately double, and their ratio, compared to the genes at the end of the replicating chromosome, increases. If, before the first replication points reach the end of the genome, there is another initiation, then the ratio of transforming activity of genes at the origin to genes at the terminus will increase to four. When spores of *Bacillus subtilis* are germinated in the presence of rich medium containing 5-bromouracil in place of thymine[33] chromosomes may have up to 15 replication points. This is as though the cells had up to four levels of initiation.

B. Bidirectionality of *Bacillus subtilis* DNA Replication

Evidence for bidirectionality of DNA replication in *Bacillus subtilis* has been presented.[34] The finding of bidirectional replication is apparently ubiquitous for all bacterial genomes.

C. Rate of DNA Synthesis during the Division Cycle of *Bacilli* in Balanced Growth

1. Determinations of the C Period

There have been a number of determinations of the C period in *Bacillus subtilis*. The general result, while not so precisely defined as with *Escherichia coli*, suggests that the pattern of DNA replication is quite similar to that of the gram-negative bacteria (Fig. 13–3). Donachie[35] reported a C value of approximately 49 minutes based upon DNA determinations in synchronized bacteria. Marker frequency values have indicated that the C period is 53 minutes between growth rates of 20- to 60-minute doubling times, and with decreasing growth rate, the C period increased.[36] This result, of course, is extremely reminiscent of the *Escherichia coli* results. When the thymine concentration was varied, the C value increased with decreasing thymine concentrations, in just the same manner as in *Escherichia coli*. Morphological analysis of outgrowing spores indicated that the C period was 50 minutes at 30° and 40 minutes at 37°C.[37] Measurements of DNA synthesis following spore germination indicated that 50 minutes was required for a round of DNA replication.[38] A value of 45 minutes was obtained for C from the study of DNA synthesis in a temperature-

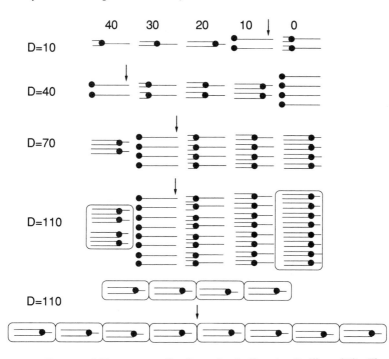

Figure 13.3 Presumed Chromosome Configuration in Growing *Bacillus subtilis*. The chromosome configurations for cells with a 40-minute doubling time are illustrated. For the first four sets C is 40 and D varies over 10, 40, 70, and 110 minutes. For D=110, the concept of the *fundamental cell* (Chapter 12) is used to distinguish between cells with increasing chromosome contents, or separate physical entities made up of many cells that have not yet separated. It may be that a culture is composed of a mixture of each of the particular cells if there is a great deal of variability in the size of the D period. We may consider a culture with a variable D period to be equivalent to a mixture of cells, each of which can be characterized by the particular fundamental cell with a particular D period. The arrow indicates the time of termination.

sensitive mutant.[39] Occasionally longer C values have been reported in *Bacillus subtilis*. At elevated temperatures, in a temperature-sensitive initiation mutant, a C value of 80 to 90 minutes was reported.[40] Studies of residual DNA synthesis in an initiation mutant gave a C value of 60 minutes.[41]

A completely different approach to the *Bacillus subtilis* replication pattern was reported by Holmes, Rickert, and Pierucci,[42] who used the membrane-elution technique in the same way that it has been used for *Escherichia coli*. This appears to be the only time that the membrane-elution technique has been used with any gram-positive organism. They obtained very clear elution patterns, and were able to measure the C and D values in the standard way. They concluded that C was approximately 40 minutes, and was relatively constant over a wide range of growth rates. As we shall see below, they concluded that the D period was variable. It was not so variable, however, as to preclude the use of the membrane-elution technique from giving satisfactory results.

2. The D Period in *Bacillus*

In contrast to the constancy of C, the D period in *Bacillus subtilis* appears to be variable. Two different laboratories have reported the D period to be constant, but the values they obtained were very different. The multiseptate chain formation in *Bacillus* has been interpreted as owing to a constant D period of approximately 138 minutes.[43] Sargent reported D values of 20 minutes.[44] A variability of D within a single culture has also been reported. When DNA synthesis was inhibited, and the residual division observed, division continued for approximately 250 minutes, but the rate of cell division began to slow almost immediately after the start of the experiment.[45] The amount of residual division was consistent with an average value for the D period of 55 minutes. These results were interpreted as indicating that D was very variable, and could vary from a few minutes to almost 250 minutes (Fig. 13–4).

If D were extremely variable within a culture, the membrane-elution method could not work. A variable D period within a culture would lead to the obscuring of the discrete generations of elution so as to make the elution pattern uninterpretable. Let us now pursue this variable D period further to present one type of analysis that may explain this relatively confused situation.

When *Bacilli* are grown at different growth rates, as in the experiment of Schaechter, Maaløe, and Kjeldgaard, various results are obtained that require a slightly different interpretation. Early measurements of the change in size with growth rate revealed that the size of the individual unit increased with growth rate (similar to that with *Salmonella* or *Escherichia coli*). At a given growth rate, when cell density was varied, chains were formed at low cell density. While cell size did not change, the number of cells in a chain varied from 2 to 30 cells per chain.[46]

The simplest explanation, as pointed out earlier, is that with increasing growth there is a change in the D period. There is an essential cell (see Chapter 12) that composes longer and longer chains as the D period becomes a larger fraction of the interdivision time. In its most basic terms, this increase in D can be stated to be a failure of cells to separate after septum formation.

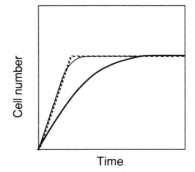

Figure 13.4 Residual Division in a Culture with a Variable D Period. In a culture with an invariant D period, the dotted line is the expected result for residual division after inhibition of DNA synthesis. The actual result with *Bacillus subtilis* is that the final residual division is consistent with a 55-minute D period, but the residual division is extremely variable. Residual division diverges almost immediately from the ideal curve, and continues to rise for a period longer than 55 minutes.

3. Determination of Cell Mass at Different Growth Rates

When the mass per cell is measured at different growth rates, equivalent to the fundamental experiment of Schaechter, Maaløe, and Kjeldgaard, an unusual result is obtained. Rather than a straight line on a semilogarithmic plot of the mass per cell against the reciprocal of the doubling time, a bimodal pattern is found (Fig. 13–5). The data cluster around two different slopes, one with a C+D period of 142 minutes and the other slope with a C+D period of 62 minutes.[47] This result means that a culture can take two different and discrete values at a given growth rate. For example, if we repeated a growth rate of 30-minute doubling time for many successive days, we might observe that different cell sizes are determined based on the concentration of the culture. As we have noted above, this may be due to the differences in availability to the cell of an autolysin that is released to the medium. At high cell densities, the immediate concentration of the enzyme is high enough to allow cutting and ripping of the septa. This allows for a low C+D value. At low cell densities the filaments form, and this leads to the higher C+D value. It is not clear whether we should expect this bimodal pattern, or whether we should expect a continuous range of values in size per cell at all different growth rates.

When the mass per cell was determined over a large number of growth rates, there was not a monotonic function as with the Schaechter,

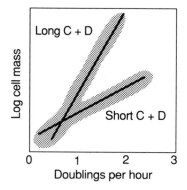

Figure 13.5   Cell Mass as a Function of Growth Rate in *Bacillus subtilis*. When the mass per cell is measured over a large number of growth rates, the data collect along two slopes. This means that at some particular growth rate, on different days, or in different cultures, the mass per cell will be found to be either of two values. If C is a constant, then the difference between the two slopes is due to a variation in the value of D.

Maaløe, and Kjeldgaard experiment. At growth rates faster than 60-minute doubling time, individual determinations gave more than one value for a particular doubling time. For example, cells with a 120-minute doubling time could have values that varied over a factor of 5. On separate days, or in separate experiments, determinations of cell size could therefore be variable. Closer examination of the results indicated that there were two types of results. One cluster of size determinations as consistent with a C+D value of about 142 minutes, while the other cluster was consistent with a C+D of approximately 30 minutes. The D period is not variable in a particular culture, but the growth of *Bacillus subtilis* is so sensitive to external variables such as cell concentration that a culture has *either* a long D period *or* a short D period. When further analysis was performed based on the DNA contents of the cellls, it was estimated that some D periods were larger than 80 minutes and some were smaller than 20 minutes. This analysis was consistent with the proposal that the initiation mass as well as the C period were constant over a wide range of growth rates.

There may be a simple biochemical explanation for this variability. Koch[48] has proposed that an exoenzyme is present on the surface of *Bacillus subtilis* that is associated with the peptidoglycan. It is this enzyme that facilitates the release of peptidoglycan with cell growth.

IV. THE SURFACE-STRESS MODEL AND GROWTH OF *BACILLUS*

Koch[49] has applied the surface-stress model to the growth of *Bacilli* with great success. Burdett and Koch have shown that the measured shapes of the *Bacillus subtilis* pole are consistent with growth according to the surface-stress model.[50] One of the major differences between gram-positive wall growth and gram-negative wall growth is that in the gram-positive cells, during septum formation, there is the formation of a double-thick wall which is then split into two by the forces acting on the cell surface.

V. THE SEGREGATION OF CELL WALL AND DNA IN *BACILLUS*

The observed pattern of segregation of DNA in gram-negative cells has been interpreted in terms of the pattern of wall growth and the relationship of DNA binding to the cell wall (Chapter 9). It was proposed that newly made DNA is bound to newly made wall, and by having this repeat in successive generations, we could obtain the observed pattern of nonrandom segregation. This model is supported by the circumstantial evidence that there is a binding, apparently specific, of particular regions of DNA to the cell wall. In *Escherichia coli*, however, there is no evidence that there is a specific association of the DNA and a particular piece of cell wall during growth. Such an assocation has been observed in *Bacillus subtilis*. When both the cell wall and the DNA were labeled, and the cells examined by either light microscopy[51] or electron microscopy,[52] it was concluded that there was an association between conserved cell-wall material and the DNA. Although the the statistical analysis of this information is not as refined as we would like, it may be that *Bacillus* is a superb organism for the analysis of the segregation of DNA.

Another interesting aspect of the segregation problem is the observed nonrandom positioning of incipient spore formation in chains of cells.[53] It would be interesting to see what happens to the DNA strands in this situation.

VI. GROWTH AND REGULATION DURING THE DIVISION CYCLE OF *BACILLUS*

The results and conclusions described in this chapter are consistent with the general model for the gram-negative division cycle. In the gram-positive organism, we see an exponential synthesis of cell cytoplasm or mass driving the growth of the cell surface and leading to the initiation of DNA synthesis in the same manner as in *Escherichia coli*. Although there is not a one-to-one correspondence of the experiments in the two types of organisms, there is enough evidence to warrant such an overall conclusion. The differences between the organisms, primarily the differences in cell-wall synthesis and turnover, are merely superimpositions of the requirements for synthesizing a thick wall upon the general model of bacterial growth during the division cycle.

VII. A UNIFIED VIEW OF BACTERIAL GROWTH DURING THE DIVISION CYCLE

This and the preceding two chapters have described apparently diverse division-cycle patterns in terms of a general model derived primarily from the study of *Escherichia coli*. The general model could be applied to a large number of other bacteria as well—e.g., *Proteus*,[54] *Synecococcus, Klebsiella, Myxococcus*—although generally in a more limited way, as there is less information available. I am not aware of any data in disagreement with this general model. In the future, as more work is done on a diverse array of organisms, I expect that the findings will be incorporated into the general framework presented here.

NOTES

1. Anderson, Green, Sturman, and Archibald, 1978; Frehel and Ryter, 1979; Glaser and Lindsay, 1977; Mauck, Chan, and Glaser, 1971; Pooley, 1976a,b; De Boer, Kruyssen, and Wouters, 1981; Jolliffe, Doyle, and Streips, 1980; Cheung, Vitkovic, and Freese, 1983; Mauck and Glaser, 1970.
2. Koch, 1988b. The enzymes responsible have been identified as an amidase and a gluosaminidase.
3. This discussion is primarily attributable to Koch, 1990; Koch and Doyle, 1986; Mobley, Koch, Doyle, and Streips, 1984.
4. The original observation and analysis of helical growth of *Bacilli* comes from Mendelson, 1976, 1978, 1982; Mendelson, Favre, and Thwaits, 1984.
5. Koch, 1989.
6. Hughes and Stokes, 1971; similar conclusions were reported by Chung, Hawirko, and Isaac, 1964; Cole, 1965.
7. Sargent, 1974.
8. Sargent, 1975a,b.
9. Dadd and Paulton, 1968.
10. Mauck and Glaser, 1972.

11. Mauck, Chan, Glaser, and Williamson, 1972.

12. De Chastellier, Frehel, and Ryter, 1975; De Chastellier, Hellio, and Ryter, 1975.

13. Frehel and Ryter, 1979.

14. Fan, Beckman, and Gardner-Eckstrom, 1975.

15. De Chastellier, Hellio, and Ryter, 1975.

16. Archibald and Coapes, 1976.

17. Paulton, 1970a.

18. Paulton, 1970b; 1971.

19. Masters, Kuempel, and Pardee, 1964.

20. Masters and Pardee, 1965.

21. Kuempel, Masters, and Pardee, 1965.

22. Donachie, 1965.

23. Donachie and Masters, 1966.

24. Goodwin, 1966; Theoretical objections to this type of model have been raised by Tyson, (1979).

25. Steinberg and Halvorson, 1968.

26. Kennett and Sueoka, 1971. See detailed discussion in Chapter 4.

27. Collins and Richmond, 1962.

28. Siccardi, Galizzi, Mazza, Clivio, and Albertini, 1975.

29. Sargent, 1975b.

30. Burdett, Kirkwood, and Whalley, 1986.

31. Yoshikawa and Sueoka, 1963a,b; Yoshikawa, O'Sullivan, and Sueoka, 1964.

32. Yoshikawa, O'Sullivan, and Sueoka, 1964; Yoshikawa and Sueoka, 1963a,b.

33. Yoshikawa and Haas, 1968.

34. Matsushita, O'Sullivan, White, and Sueoka, 1974; Wake, 1972, 1973, 1975; O'Sullivan, Howard, and Sueoka, 1975.

35. Donachie, 1965.

36. Ephrati-Elizur and Borenstein, 1971.

37. Siccardi, Galizzi, Mazza, Clivio, and Albertini, 1975.

38. Keynan, Berns, Dunn, Young, and Mandelstam, 1976.

39. Mandelstam and Higgs, 1974.

40. Sargent, 1975b.

41. Upcroft, Dyson, and Wake, 1975.

42. Holmes, Rickert, and Pierucci, 1980.

43. Paulton, 1970a,b, 1971; Sedgewick and Paulton, 1974.

44. Sargent, 1975a,b.

45. Holmes, Rickert, and Pierucci, 1980.

46. Sedgewick and Paulton, 1974; Paulton, 1970a,b.

47. Holmes, Rickert, and Pierucci, 1980.

48. Koch, 1988.

49. Koch, 1983a; Koch, Higgins, and Doyle, 1981.

50. Burdett and Koch, 1984.

51. Schlaeppi and Karamata, 1982.

52. Schlaeppi, Schaefer, and Karamata, 1985.

53. Hitchins, 1975, 1976, 1978, 1980.

54. Gmeiner, Sarnow, and Milde, 1985. This paper gives a general overview of the cycle of *Proteus mirabilis*.

14

The Growth Law and Other Topics

I. THE CELLULAR GROWTH LAW

For almost 100 years, since Ward[1] studied the growth of bacteria between divisions, there have been discussion, analysis, controversy, and confusion about the *growth law*—the way individual bacteria grow during the division cycle. How do bacteria grow between birth and division? The usual approach to understanding bacterial growth during the division cycle (i.e., the growth of the entire cell rather than of any one of its components) has been to find a simple mathematical formula that best describes the growth pattern of cells. One of the earliest modern studies was that of Bayne-Jones and Adolph[2] who concluded, from the analysis of time-lapse films, that *Bacillus megaterium* and *Escherichia coli* exhibited exponential growth rates when the length of the cells was used as the criterion of growth. Similarly, a steady, accelerating growth rate for length was observed by Knaysi[3] for *Bacillus cereus*.

The introduction of the static analysis of cell growth by Collins and Richmond[4] marked a new approach to the study of bacterial growth. In concept, the Collins–Richmond method is simple (see Chapter 3). If we can determine the size distribution of the existing population of bacteria, the size distribution of the cells at division, and the size distribution of the cells at birth, then we can calculate the rate of cell-size increase during the division cycle. During steady-state growth the number of cells in every size class increases exponentially. This exponential increase in cell number is the result of (1) cells leaving the size class by growth to a larger size class; (2) cells entering the size class by growth from a smaller size class; (3) cells leaving the size class by division (primarily from the larger size classes); and (4) cells entering the size class by birth from the division of larger cells (primarily into the smaller size classes). If the distributions of newborn and dividing cells are relatively narrow, then these cells make a negligible contribution to the middle range of the total size distribution. In this case, the distribution of extant cells accurately reflects the growth rate of cells in the middle of the division cycle. When Collins and Richmond applied their method to *Bacillus cereus*, they concluded that the growth rate in the middle range of cell sizes continuously increased. The growth rates at the start and finish of the division cycle, however, were so dependent upon assumptions about the birth and division distributions that no statements could be made about the growth rates at the ends of the size distribution.

Kirkwood and Burdett[5] prepared electron-microscope pictures of steady-state cells of *Bacillus subtilis* and determined the size and morphological state (mononucleoid, binucleoid, or septate) of individual cells. They proposed that by applying the Collins–Richmond equation to cells partitioned according to these morphological states, we could obtain a more accurate picture of the growth rate, particularly if the global growth rate of the cell was influenced by the particular morphological changes observed. If at the time that cells changed from mono- to binucleate, the growth rate changed, we could derive the growth pattern more accurately. After classifying and measuring over 2000 cells, Kirkwood and Burdett proposed that (1) the growth rate of mononucleate cells was greater than the growth rate of binucleate cells of the same size; (2) the growth rate of binucleate cells was greater than the growth rate of septate cells of the same size; (3) there were major transitions in growth coincident with the major transitions in cell morphology (chromosome duplication and the initiation of cross-wall formation); and (4) there was a steadily increasing growth rate within each class that approximated a linear increase. Regarding the total population, Kirkwood and Burdett concluded that growth corresponded to an exponential rate with a constant additive term.

Koppes, Woldringh, and Grover[6] have applied the Collins–Richmond formula to study the growth rate of *Escherichia coli*. They inverted the method and calculated the cell-size distributions expected for various growth laws. They hoped that statistical comparisons of the theoretical predictions and the experimental data would allow them to choose among sixteen different models of cell growth. It is quite likely that their observed distributions may be the best available data.[7] The list of models considered by Grover, Woldringh, and Koppes[8] is an indication of the major underlying assumptions regarding the growth of cells during the division cycle. Two exponential models and six linear models were analyzed. One was a simple exponential model and the another model was exponential with a minimal-size parameter below which cells failed to grow. The linear models were derived by assuming that the rate of cell growth was constant until a given cell age or cell size was reached, at which time the growth rate doubled. In one cell-age–control model, the growth rate doubled at a constant cell age during the division cycle. In another, the growth rate doubled at a constant time before division. A special case considered growth linear throughout the division cycle, with the rate of mass increase doubling at the instant of division. The linear size models were similarly partitioned. One model assumed that cells grow to a minimal size, after which there is a constant probability of growth-rate doubling. In another model, the cells above a minimal size doubled their growth rate; the probability that the doubling occurred increased linearly with size. The final model assumed the growth rate doubled at some time during the division cycle with a constant probability.

The application of these eight models to the experimental data[9] led to the analysis of sixteen different comparisons, as both the area and the length of the cells were considered. Statistical analysis allowed the rejection of all the length models (i.e., models concerned with a change in cell length, irrespective of the total volume or area of the cells). Cell area was the parameter that could best be associated with active control of cell growth. Of the eight models in which surface area alone was considered, three linear models were rejected: the model with simple linear growth throughout the cycle, the model in which doubling in growth rate occurred with a constant probability from cell birth, and the model in which the growth rate doubled at a particular cell age. Even after these rejections, there were five contending models that could not be distinguished or eliminated by the data on the cell-size distributions of a bacterial culture. This disappointing[10] result suggests that the Collins–Richmond method, despite its being described as the most powerful[11] method for understanding the growth laws, is not able to solve the growth problem. This was suggested earlier by Koch in a prophetic and insightful paper.[12]

The discussion above illustrates the main thematic aspects of the growth-law problem. One theme is that the growth law can be obtained by measurements or observations made on the whole cell; another is that interest has centered on distinguishing between linear and exponential growth laws or subvariants; and a third is the search for better methods for measuring cell growth. Harvey, Marr, and Painter[13] studied the growth of *Escherichia coli* and *Azotobacter agilis* by using electronic sizing determinations. In support of an exponential growth law, they found an increase in the absolute and specific growth rate between divisions. Cullum and Vicente[14] performed essentially the same analysis as Grover, Woldringh, and Koppes,[15] except that they measured cell lengths in a light microscope which has a lower resolution than the electron microscope. Cullum and Vicente concluded that *Escherichia coli* grew bilinearly between divisions. Kubitschek and Woldringh[16] used a different mathematical analysis of electron-microscopic measurements and determined that the growth law was exponential. Finally, Trueba, Neijssel, and

Woldringh[17] proposed that whatever the growth law, it was the same for a number of different bacterial populations.

All of the discussion above has a subtext: the belief that there is a global growth law that can be stated and that the cell will follow. This is a top-down approach to understanding the growth of the cell; it assumes that once we understand the growth law of the whole cell, the rules governing synthesis of the different components of the cell sum up to that particular growth law. An alternative approach to understanding bacterial cell growth is to reverse this perspective and derive the growth law from an analysis of the biosynthesis of the constitutive components of the cell; this approach leads to a new growth law.[18]

In Chapter 4 we saw that RNA and protein, the main components of the cytoplasm, are synthesized exponentially during the division cycle. The other components of the cytoplasm also increase exponentially. In Chapter 5, the synthesis of DNA was described as composed of a series of linear rates of synthesis during the division cycle. These linear rates are proportional to the integral number of replication points in a cell. In Chapter 6, it was concluded that the rate of surface synthesis is a complex function that is close to, though not precisely, exponential.

The total mass of the cell can be calculated by adding the weighted synthesis rates of the different cell components. The growth of the cell is described as

$$M_{tot_\alpha} = w \cdot M_{exp_\alpha} + x \cdot M_{dna_\alpha} + y \cdot M_{sur_\alpha} + z \cdot M_{oth_\alpha}$$

M is the mass (either total [tot], mass of material increasing exponentially [exp], material increasing in a manner similar to DNA [dna], material increasing as the surface increases [sur], or material increasing in some other manner [oth]). The coefficients w, x, y, and z are weighting factors proportional to the average weight fraction of each component, and α is the cell age during the division cycle.

The calculated global growth pattern and the separate biosynthesis patterns are illustrated in Fig. 14–1.[19] The various components are plotted in proportion to their presence in the cell. The two major components of the cell, the protein (55% of total dry weight) and the RNA (20.5% of total dry weight), increase exponentially during the division cycle. The glycogen, the polyamines, and the soluble salts of the cytoplasm are assumed to increase exponentially. These minor components sum to 6.4% of the total dry weight, giving approximately 81.9% of the mass of the cell increasing exponentially. (Even if the synthesis pattern of these minor components were different, for example, linear, there would be no major change in the conclusions drawn below.) The cell-surface components aggregate to approximately 15% of the total dry weight; the DNA accounts for the remaining 3.1% of the cell mass.

Total cell synthesis is not exponential (Fig. 14–1), but it is so close to exponential that it is experimentally indistinguishable from an exponential-growth law. This is because the RNA and the protein, two major components of the cell, dominate the calculation. The other com-

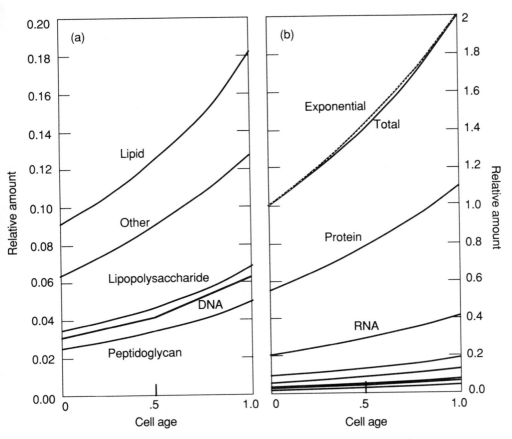

Figure 14.1 Synthesis of Cell Components during the Division Cycle. The curves are drawn proportional to their relative weight for *Escherichia coli* grown in minimal glucose medium. The percentage dry weights are peptidoglycan, 2.5%; DNA, 3.1%; lipopolysaccharide, 3.4%; other (including polyamines, salts, glycogen, etc.), 6.4%; lipid, 9.1%; RNA, 20.5%; and protein, 55.0%. The RNA, protein, and other materials were assumed to follow an exponential increase. The synthesis rates of lipid, lipopolysaccharide, and peptidoglycan are shown as proportional to the peptidoglycan synthesis rate. The DNA synthesis rate is bilinear with a doubling in rate in the middle of the division cycle. The dotted line is an exponential increase; it indicates the difference between the calculated total mass increase (solid line) and exponential increase. Panel (a) is an enlarged portion of panel (b) to illustrate the biosynthetic rates of the less prominent material.

ponents have a minor effect on the total cell mass increase during the division cycle. We may therefore state the growth law as

$$M_{tot_\alpha} \simeq M_{exp_\alpha}$$

The growth of the bacterial cell is approximately exponential during the division cycle. It is unlikely that any foreseeable experiment will be able to distinguish the actual global rate law for cell mass from exponential. In contrast to approaches looking for a global growth law, the method used here is to sum the growth laws of the components of the cell. The components have been aggregated, producing three major groups to be considered: material synthesized exponentially during the division cycle, material synthesized proportional to DNA, and material synthesized as cell surface (see Chapter 2). The growth law of the cell is approximately exponential.

The individual growth laws used in this calculation share two properties. First, all have been obtained by the use of the pulse-labeling, membrane-elution method,[20] a method that has been independently supported as valid in the analysis of the bacterial division cycle.[21] Second, the growth laws are logical outgrowths of our understanding of how macromolecules are synthesized. The exponential-growth law for cytoplasm stems from the assumption that newly made cytoplasmic components are immediately involved in the synthesis of additional cytoplasm (Chapter 4); the linear-DNA–rate law comes from our understanding that DNA is synthesized during the division cycle in proportion to a small number of replication points moving processively along the DNA template (Chapter 5); and the surface-synthesis–rate law is derived from *a priori* considerations of the rate of surface synthesis required by the constant density of the cell during the division cycle (Chapter 6). The combination of these logical ideas, and the experimental supports of these ideas, lead to confidence in the statement that the growth law of bacteria is essentially exponential. Cell growth between divisions is simple and continuous and total cell growth can be understood as derived from the synthesis of the individual cell components.

II. THE LENGTH GROWTH LAW

We have just seen that there is no simple mathematical growth law for bacteria; the growth of a bacterium is the sum of the individual growth patterns of the cell components. It is of interest to analyze the length distribution of cells. In Chapter 6 the shape of bacteria was analyzed. It was concluded that there was an inverse length-to-width relationship;

thin cells were longer and thick cells were shorter. The length distribution of a population can be derived by considering (1) the ideal length distribution of cell populations of given widths growing during the division cycle; (2) the variation from this growth law due to statistical variation including the deviation from a constant size at birth and at division; and (3) the presence of subpopulations with different widths in a given cell population. (See Chapter 6 for a detailed discussion of the inverse length-width distribution). Although the length distribution will not be presented in analytical detail, Fig. 14–2 describes and presents an example of how one would go about calculating a length distribution.

The canonical length distributions for each of the separate width populations within a growing population is presented in Fig. 14–2a. There are four times as many short cells as long cells within a population (Chapter 4).[22] If each subpopulation were ideal, and if the relative contribution of each width were also known, then we could calculate the size distribution exactly. First, however, the distribution for each width must be modified to include statistical variation of the sizes at birth and division and for experimental error. This is shown in Fig. 14–2b. In addition, we must make an assumption regarding the frequency of each of the widths within the population. The cells of different widths would not be present equally, but would be approximated by a probability distribution with the mean width cell most common, and cells of other widths less common as they deviate from the mean. The analysis and the final result for a sample calculation is presented in Fig. 14–2c; the result is similar to the patterns in the literature.[23]

Given the number of variables that could be used to fit the experimental data, what is to be said about the length growth law? The conclusion proposed here is that any observed length distribution can be simulated as accurately as necessary by the proper choice of weightings, variations in the growth of individual cells, and variations in cell width. This means that the length distribution may be accounted for, if not actually solved, by the proposal that there is an inverse length–width relationship.

III. REGULATION OF SYNTHESIS AT INITIATION

In the discussions of the three major categories of material synthesized during the division cycle, DNA synthesis was characterized by sharp changes in the rate of synthesis at different times during the division cycle. In contrast, the other patterns of synthesis, for cytoplasm and cell surface, were much smoother and had fewer discrete changes during the division cycle. Cytoplasm synthesis was exponential, and the synthesis

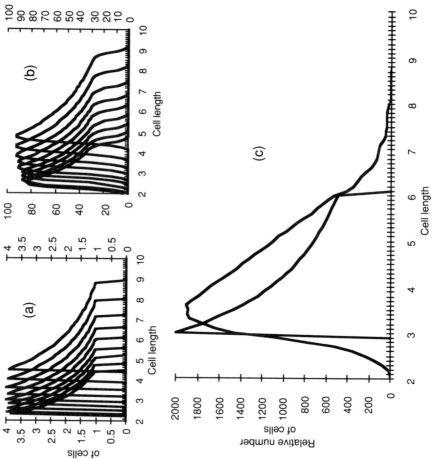

of the surface was approximately exponential. In both cases there was no sharp break in synthesis at any point in the division cycle. If we look again at these syntheses, we can see a greater generalization of the pattern of macromolecular synthesis. The generalization is that once initiated, the macromolecular syntheses of the chains of material that make up the cytoplasm and the cell surface go at constant rates. The elongation of chains, whether the chains are protein, DNA, RNA, or peptidoglycan strands, takes place at a constant rate of chain extension. It is difficult to conceive that every chain in the cell is subject to regulation of the rate of chain elongation. All chains in the cell are regulated at the level of initiation. The rates of chain elongation of the major cell components are summarized in Table 14–1. All of these macromolecular syntheses are similar in that they are made up of a series of linear rates of synthesis. As shown in Fig. 14–3, even protein synthesis may be considered at some level to be a linear synthetic process. The reason that we see protein synthesis as exponential is that so many proteins are being synthesized—equal to the number of ribosomes in the cell—that at any practical level it is impossible to observe or measure the individual initiation processes. The reason that DNA stands out as having a linear synthetic pattern is that a small number of individual chains are being synthesized at any one time. It is this quantitative difference that leads to the qualitative difference between protein synthesis and DNA synthesis during the division cycle.

IV. THE FUNDAMENTAL EXPERIMENT OF BACTERIAL PHYSIOLOGY REANALYZED

Let us now return to the fundamental experiment of Schaechter, Maaløe, and Kjeldgaard with which we began this book. Schaechter, Maaløe, and Kjeldgaard plotted straight lines on semilogarithmic paper for various

---

Figure 14.2 Calculation of Cell-Length Distribution with Variable Cell Widths. For each cell cycle, the length distribution was computed for cells with widths as illustrated in Fig. 6–12. It is assumed that the length of the cell increased exponentially during the division cycle. [The distribution function is thus determined as c/L^2. The actual length distribution is slightly different, as length does not increase precisely exponentially (Chapter 6)]. The result for each cell width is presented in (a). Each of these distributions is smoothed out in (b) by taking a running average about each size interval. The nine widths of Fig. 6–12 are assumed to be present in the relative frequencies of 0.5, 1.0, 2.0, 5.0, 10.0, 5.0, 2.0, 1.0, and 0.5. The final calculated length distribution is shown in (c). The narrower distribution in (c) is the expected distribution with no variation in cell width.

Table 14.1 Rates of Chain Extension for
Macromolecules

Macromolecule	Rate of chain growth
DNA	10,000/second
Protein	15–20/second
RNA	45/second
Peptidoglycan	160/second

properties of the cells such as the RNA, DNA, nucleoids, and mass per cell. The determination of the DNA and mass per cell were described in Chapter 5. The observed nucleoids are determined by the termini of chromosomes; their number per cell at different growth rates is consistent with a constant time for termination before cell division. We will now

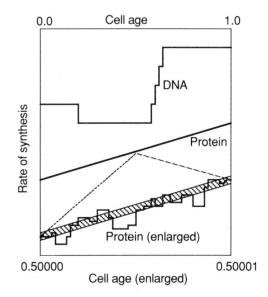

Figure 14.3 Microscale Pattern of Protein Synthesis during the Division Cycle. If we were to enlarge the scale of protein synthesis during the division cycle, we would see a pattern of horizontal lines in the same way that DNA synthesis is composed of horizontal lines. Each time the synthesis of a new protein is initiated, there is an increase in the rate of synthesis. Occasionally a protein synthetic round is terminated, and this is indicated by a decrease in the rate of protein synthesis. Although the increase on a larger scale is exponential, this exponential is made up of many increases and decreases that occur at a microscale.

look at all of the cell components on the Schaechter, Maaløe, and Kjeld-gaard graph, and see how they are interrelated.

First, consider that the number of poles per cell is always 2. Imagine that this result is due to the formula

$$2^{P/\tau}$$

where P is the time between cell division and the formation of a pole and τ is the interdivision time of the cells. The time between the end of pole formation and cell division is, by definition, zero, so the relative value for the number of poles per cell is constant.[24]

Consider now that we move back through the division cycle from division and see what happens at earlier times. The next event is termination, which occurs D minutes before division. In this case the number of termini per cell is given by the formula

$$2^{D/\tau}$$

As we move further back, we encounter initiation, which occurs C+D minutes before a division. The amount of mass per cell is proportional to

$$2^{(C + D)/\tau}$$

These results are plotted in Fig. 14–4. The slopes of the lines increase as we move further back from division to the time that the event occurs. If we include the plot for DNA replication in Fig. 14–4 (calculated from the formula in Chapter 5), the slope for the DNA per cell can be approximated by

$$2^{44/\tau}$$

This means that the DNA is synthesized *as though* it were all made 44 minutes before the end of the division cycle. Because DNA is made between 60 and 20 minutes before division, this is a reasonable conclusion.

Although area was not measured in the Schaechter, Maaløe, and Kjeldgaard experiment, it is possible to consider what would be expected if it were measured. The area of the cell varies as the two-thirds root of the total volume of the cell. This follows from the constant density (Chapter 7) and constant shape (Chapter 6) of cells at different growth rates. The formula for the area is

$$2^{(2/3)(C + D)/\tau}$$

The exponent in this case is $40/\tau$. The slope for area appears between the slope for DNA and the slope for termini per cell.

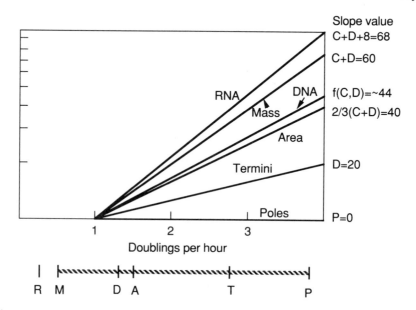

Figure 14.4 Idealization of Schaechter, Maaløe, and Kjeldgaard Experiment. The expected values of various components per cell at different growth rates are graphed. Poles do not vary; there are always two poles per cell at all growth rates. The most rapid change in cell composition occurs in mass and RNA per cell. The poles per cell are arrived at by considering that all cells have two poles per cell. The termini are derived from considerations of constant values for the D period. The mass per cell is a function of the fact that cells initiate DNA synthesis at a fixed mass per origin; the time before a division when that initiation occurs is fixed. The different exponents at the right are arranged in linear form at the bottom, to illustrate the sequence of events that occurs during the division cycle. The steeper the slope, the earlier during the division cycle an event occurs.

The slope for RNA is given as

$$2^{(C + D + 8)/\tau}$$

The slope for RNA is higher than the mass curve by the value of 8 minutes. Where does this 8 minutes come from? As noted in Chapter 4, it takes approximately 8 minutes for a ribosome to make all the protein required for ribosomes. Consider a shift-up. The rate of protein synthesis does not reach its maximal rate until at least 8 minutes following the shift-up because it takes that long for ribosomes to be made. Although the process is complicated, we can consider an analogy to rate maintenance, in which there is a delay until the rate of protein synthesis is maximized. During a shift-up, the ribosomal RNA must be made before a newly functioning ribosome can be produced. It may be 8 minutes until the new steady state of mass and RNA synthesis is reached. These considerations give a delay of 8 minutes before the rate of protein synthesis is maximized. Only those ribosomes that are in process (taken as a group) will become ribosomes in the first 8 minutes after a shift-up, and only after this period will the rate of protein synthesis be maximal. If the initiation of DNA synthesis were determined by the increase in protein, then we would expect to have the slope for the protein line equal to $(C+D)/\tau$ and the slope of the RNA curve to be steeper by the time required to make completed ribosomes.

Each of the slopes in Fig. 14–4 can be approximated by a formula of the form

$$2^{X/\tau}$$

The X value is either a value that is taken from other experimental or theoretical considerations, or that fits the data. This type of numerology has an interesting interpretation. Figure 14–4 presents the various slopes for the RNA, mass, DNA, area or cell surface, termini, and poles per cell. At the bottom of Fig. 14–4, the values for the different slopes are summarized in a linear pattern indicating the order in which various biosynthetic processes occur before cell division. There is a logical order to the slopes. First, RNA synthesis increases, leading to an increase in cell mass as protein synthesis increases. This leads to an increase in the amount of DNA, which then leads to an increase in the nucleoid content of the cell. Finally, the number of poles increases when cells divide. The slopes of the Schaechter, Maaløe, and Kjeldgaard experiment are logical outcomes of our understanding of the biosynthetic processes of the cell.

The analysis of the Schaechter, Maaløe, and Kjeldgaard experiment deals primarily with results obtained on *Salmonella typhimurium* and *Escherichia coli*. But the implications of this experiment extend beyond these organisms. If these measurements were applied to a large number of organisms, we could see the range of possible physiological states that bacteria could achieve. By simple measurements of the cell content of DNA, RNA, protein, cell wall, nucleoids, and any other components that may be bacterium specific, we could determine the C, D, and initiation mass values. In this way, a simple overview of the physiology of an organism can be obtained. Such comparative studies may be as important as in-depth analyses of a single organism.

NOTES

1. Ward, 1895.
2. Adolph and Bayne-Jones, 1932; Bayne-Jones and Adolph, 1933.
3. Knaysi, 1941.

4. Collins and Richmond, 1962.

5. Kirkwood and Burdett, 1988.

6. Koppes, Woldringh, and Grover, 1987.

7. They were scrupulous in eliminating bias from the results. For example, they corrected for the probability that the larger cells would be more likely to touch an edge of the photograph and thus not be counted; uncorrected, this would have given an increased frequency of smaller cells.

8. Grover, Woldringh, and Koppes, 1987.

9. Grover, Woldringh, and Koppes, 1987.

10. Grover, Woldringh, and Koppes, 1987.

11. Burdett and Kirkwood, 1983; Grover, Woldringh, and Koppes, 1987; Koppes, Woldringh, and Grover, 1987.

12. Koch, 1966a.

13. Harvey, Marr, and Painter, 1967.

14. Cullum and Vicente, 1978.

15. Grover, Woldringh, and Koppes, 1987.

16. Kubitschek and Woldringh, 1983.

17. Trueba, Neijssel, and Woldringh, 1982.

18. Cooper, 1988d.

19. Neidhardt, 1987.

20. Helmstetter, 1967, 1969.

21. Skarstad, Steen, and Boye, 1983, 1985.

22. Koch, 1987.

23. Grover, Woldringh, and Koppes, 1987.

24. In this discussion, only the relative values are considered. It is true that the formula appears to give a value of 1 pole per cell, but the important point is to show that the values change in a particular way. No normalization factor has been added to make the values absolutely correct.

15

The Eukaryotic
Division Cycle

---

*The two amoebae may be called young animals in the sense that they have
just come into existence as new individuals, but nothing in their tissues or
characters distinguishes them from their parent. So far as the period of
youth has any interest or significance, these animals escape it.*

P. C. Mitchell, in *The Childhood of Animals*, 1912

In 1968 it was suggested that ". . . there is considerable evidence which
suggests a similarity between the regulation of growth in animal cells and
bacteria, for it appears [in a pattern similar to bacterial DNA synthesis]
that the time for the synthesis of DNA (the S period), and the time
between the end of DNA synthesis and mitosis (the G2 period) are quite
constant in animal cells with different generation times."[1] That attempt
to bridge the gap between the prokaryotic and eukaryotic division cycles
with a unified model did not have an immediate impact on studies of the
eukaryotic division cycle. The two fields diverged, and two different
views of the division cycle emerged.[2] In recent years, however, there has
been a gradual melding of the study of bacterial and animal cells, yielding
a more unified view of the division cycle. This chapter will describe how
two different descriptions emerged, and more importantly, how a unified model is now developing to account for the regulation of division in
both prokaryotes and eukaryotes.[3]

I. THE EUKARYOTIC DIVISION CYCLE

In contrast to the bacterial division cycle described in the preceding
chapters, the eukaryotic cycle is usually described as composed of three
phases: the G1 period, between cell birth and DNA synthesis; the S
phase, the period of DNA synthesis, and the G2 period, the period
between the end of DNA synthesis and cell division.[4] After the initial
description of these phases, a great deal of work appeared describing the
time spans of the phases in different cells, in different tissues, and under
different conditions of growth in cell culture. It was observed that the S

and G2 periods were less variable than the G1 period. When a cell was grown at different growth rates by varying the growth conditions (medium, pH, growth factors), the S and G2 periods were relatively constant and the G1 period varied.

How was this generalization of a variable G1 interpreted? Because S and G2 were invariant, if cell growth slows, producing a longer interdivision time, the G1 period increases. If the growth rate increases, the interdivision time is shorter, and the G1 period decreases. These observations were interpreted as indicating that the G1 period regulated the interdivision time. A long G1 period was correlated with a long interdivision time, and a short G1 period was correlated with a short interdivision time. This viewpoint of the regulatory abilities of passage through G1 is illustrated by the following passage:

> The time taken for such transit, i.e., the length of G1, presumably depends on the efficiency with which these [G1] requirements are met.[5]

The initial idea that in animal cells the G1 period was the important variable was soon supported by other lines of research: the isolation of mutants that were affected in "passage through the G1 period," the ubiquitous finding of G1 arrest, the identification of "G1 events," and recently, the cloning of genes that are "required for passage through G1." In the field of cell differentiation, it is proposed that cells differentiate from a point in the G1 phase of the cell cycle. Collectively, these results implied that there are points in the G1 period at which cells could be arrested. These G1 points, sometimes called restriction points, were consistent with the original conclusion that there are G1-regulatory events.

A succinct summary of the regulation of the eukaryotic division cycle is found in Becker's textbook on cell biology:[6]

> Much evidence points to G1 as the most critical phase in the cell cycle for regulatory purposes. We have already seen that this is the phase that varies most among cell types. Moreover, cells that have stopped dividing are almost always arrested in the G1 phase. For example, we can stop or slow down the process of cell division in cultured cells by allowing the cells to run out of either nutrients or space or by adding inhibitors of vital processes such as protein synthesis. In all such cases, the cells are arrested in G1.

> These findings suggest that when a cell leaves G1 and enters the S phase, it is committed to completing the cycle. Therefore, the release of cells from G1 appears to be a critical control mechanism. More specifically, researchers have identified a "point of no return" in late G1 called the restriction point. Cells that have passed this point are committed to division, whereas those that have not passed this point can remain in G1 indefinitely.

This description of the current view of the eukaryotic division cycle is accepted by a large number of practicing cell biologists.[7] Despite this consensus, there exists an alternative view of the division cycle of higher cells.

II. AN ALTERNATE ANALYSIS OF THE EUKARYOTIC DIVISION CYCLE

First, a description of the eukaryotic division cycle will be presented from the perspective of the division cycle of bacteria. This viewpoint will then be applied to a number of experimental situations, allowing the reader to apply the model to other systems not covered here.

A. The Definition of a G1 Cell

What is a G1 cell? The problem is vocabular as well as experimental. For exponentially growing cells in an unlimited medium, the answer is simple: G1 cells are cells with a G1, or 2n, DNA content. Temporally, they are cells between mitosis and the start of the S phase. When S phase starts, the amount of DNA immediately increases, and the cells do not have a 2n DNA content; the cells are now in S phase. When the DNA content has doubled, S phase ends, and G2 cells (with a 4n DNA content) are produced. Temporally, G2 cells have completed S phase but have not yet entered mitosis. Mitosis (M phase) is the phase in which the configuration of the 4n DNA changes, the nuclei dissolve, and the DNA condenses to form chromosomes that are segregated into daughter cells. When the M phase is finished, the cycle is completed, and two daughter cells are formed that are in the G1 phase of the cell cycle.

But there is more to a G1 cell than its DNA content or DNA configuration. G1 cells have an additional property that distinguishes them from cells in other parts of the cycle. Between cell divisions, there is a continuous variation in cell size; G1 cells are smaller, on the average, than cells in S, and even smaller than the cells in G2.

Consider the following *Gedanken* experiment: cells are treated with an inhibitor or environmental change that inhibits mass increase and prevents the initiation of S phase. That treatment allows cells in S, G2, or M to finish the cycle, producing two daughter cells.[8] How do the cells in different parts of the division cycle react to this condition? The G1 cells neither increase cell mass nor initiate S phase; they remain cells with a G1-DNA content. S-, G2-, and M-phase cells are able to complete their division cycles and therefore produce two daughter cells, each with a G1-DNA content. S-phase cells would, in the absence of the inhibitor, increase their size between S phase and cell division. Mass synthesis is inhibited under the proposed conditions, so the mother cell and the two new daughter cells are smaller than normal.

Consider two cells, one just before, and one just after, the start of S phase (Fig. 15–1). The cells have almost the same size; they differ by an infinitesimal amount because the G1 cell is infinitesimally younger than the cell that just started S phase. The daughter cells descending from the S-phase cell have a G1-DNA content and are approximately half the size of the cell arrested in the G1 phase. This is because the cell in S phase divides once, while the cell in the G1 phase did not divide. Similar considerations apply to the G2 and M cells. Their daughter cells will be larger than the daughter cells descending from the S-phase cells, but they will be smaller than the cells that were in the G1 phase.

We now have a collection of cells, all with a G1-DNA content. Are these cells "in G1?" A distinction should be made between G1 cells found in exponentially growing cultures and cells that have as one of their properties a G1-DNA content. This distinction leads to the paradox that *cells may have a G1-DNA content and not be "in G1."*[9] These cells all have a G1 amount of DNA but many have a non-G1 amount of cell mass. They are not like the normal G1 cells in a growing population. These cells are usually called "cells arrested in G1." These cells should be referred to as "cells with a G1-DNA content."

B. The Continuum Model

The alternative model of the eukaryotic cell cycle is called the continuum model (Fig. 15–2). This name is derived from the idea that the preparations for division occur continuously during the division cycle and are not confined to any particular phase. The name also arises from the continuity in the division-cycle patterns of bacteria and eukaryotes.

To illustrate the continuum model, assume that there are no cell-cycle–specific signals regulating the start of S phase. Between two sequential starts of S phase a signal accumulates continuously; this signal eventually triggers the start of DNA synthesis. The increase of this trigger in the G1 phase is the completion of an accumulation process begun at the start of the previous S phase. Now suppose that an inhibitor of mass synthesis is added to these growing cells. Assume the inhibitory treatment also inhibits the synthesis of the trigger.[10] Either the trigger is some constant fraction of cell mass or is a measure of cell mass. Eventually all the cells will have a G1-DNA content but varying amounts of the trigger. What would be the most likely conclusion drawn from these cells in terms of the G1-event model of the cell cycle? It would be that the treatment that inhibited mass synthesis was cell-cycle specific because

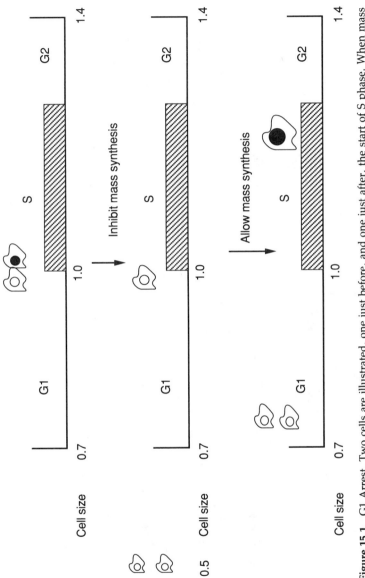

Figure 15.1 G1 Arrest. Two cells are illustrated, one just before, and one just after, the start of S phase. When mass synthesis is inhibited, the cells either arrest (if they are in the G1 phase) or proceed to divide (if they are in S phase). The S phase is indicated by the solid nucleus showing that DNA synthesis is occurring. The cells with a G1-DNA content are not synchronized at a particular point in the G1 phase of the cycle following starvation, but are approximately one doubling time apart. The mass does not increase during the period of inhibition, so when the cells coming from the S-phase cells divide, the resulting daughter cells are half the size of the G1 cells that did not divide. The cells that arose by division must grow for one extra doubling time before they initiate DNA synthesis. Initiation of DNA synthesis is not synchronized.

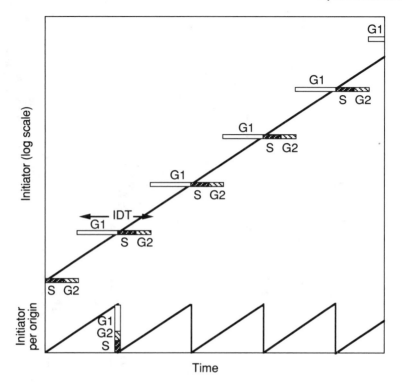

Figure 15.2 The Continuum Model. The lower sawtooth graph shows the amount of initiator per origin, varying from 0.5 to 1.0. A unit (1.0) amount of initiator is required for the initiation of DNA synthesis. The vertical bar on the sawtooth graph indicates the amount of initiator present at each time. At zero time, the cells have just entered an S phase and are progressing through S and G2. The amount of initiator per origin is increasing from 0.5 to 1.0 over one doubling time (τ). When the initiator per origin reaches 1.0, another S phase is initiated. Since there is a sudden doubling in the number of available origins of replication, the amount of initiator per origin decreases to 0.5. (The term origin is defined formally; actually multiple origins appear to initiate rounds of DNA synthesis during S phase.) The G1 phase is the time between the end of G2 and the start of S (omitting the short M period). If the rate of initiator synthesis decreases, and S and G2 remain constant, then G1 increases. Conversely, if the rate of initiator synthesis increases, the G1 phase decreases. There is no G1-specific synthesis of initiator; synthesis of initiator occurs continuously in all phases of the division cycle. As shown in the upper graph, this formal statement of the continuum model does not envision any sudden appearance or disappearance of particular molecules.

the cells were arrested in the G1 phase of the cell cycle. Thus, the inhibition of the production of a noncell-cycle–regulated protein or substance led to the conclusion that it was cell-cycle specific. This is the basic result that supports G1 events and G1-specific syntheses in eukaryotic cells. A continuous synthesis of the regulatory molecule (whatever that molecule may be), leads to the same results—a collection of cells with a G1-amount of DNA—with a different explanation of how this result came about.

C. Formal Statement of the Continuum Model

The continuum model, illustrated in Fig. 15–2, may be formally described:

1. The initiator is synthesized continuously during the division cycle.
2. DNA synthesis starts when a threshold amount of initiator per origin of DNA is achieved. The term origin is used very loosely here. Upon initiation of DNA replication, the start of S phase, the number of origins formally doubles. This is a formal concept, and should not be taken as a statement regarding the number of origins of DNA replication in a eukaryotic cell. At initiation of DNA synthesis, there is a halving in the initiator per origin. This change in the ratio occurs because of the doubling in the denominator, the number of origins. There is no continuous initiation of DNA replication after a cell achieves a unit content of initiator. The cell responds to a ratio, and the ratio is correct only for an instant; after that instant, in the formal model proposed here, the initiator per origin decreases to less than the amount per origin required for initiation of DNA replication.
3. Cells in S, G2, or M complete their phases and divide even under conditions where the initiation of DNA synthesis is inhibited. Although there are cases where cells appear to be arrested in other phases of the cell cycle (G2 arrest has been reported a number of times, and arrest in the middle of the S phase may be viewed simply as a result of the inhibition of DNA replication), the most common observation is that cells proceed through the later stages of the cell cycle.
4. The pattern of DNA synthesis during the division cycle of any cell is a function of the rate of synthesis of initiator relative to the rate of passage of the cells through the S and G2 periods.

D. The Continuum of Division-Cycle Patterns

The patterns of DNA replication described in Chapter 5, and the patterns of DNA replication found in animal cells, form a continuum. As we slow the growth of bacteria, a G1 phase appears. In bacteria this is called the B phase (see Chapter 5). Conversely, there are examples of animal cells that do not have a G1 phase. The continuum model does not imply that animal cells are simply slow-growing bacterial cells, or that there is any direct functional relationship between the two types of cells. What is proposed is that the constancy of the S and G2 periods (or C and D periods) indicates that the same type of regulatory logic applies to both types of cells.

E. Progression through the Eukaryotic Cell Cycle

The eukaryotic division cycle is generally presented as a circle (Fig. 15–3). Around the circle the phases of the cycle are arranged in sequence. The cycle starts with mitosis, followed by a G1 phase, an S phase, a G2 phase, and once again, mitosis. At mitosis, the sequence of G1 events, if they exist, are presumably initiated. A sequence of events occurs in the G1 phase leading to the initiation of the S phase.

It is instructive to look at the pattern of DNA replication in bacteria in terms of the eukaryotic model. Consider bacterial cells growing at different growth rates with C and D constant at 40 and 20 minutes respectively (see Chapter 5, Fig. 5–1). If one studied cells with a 90-minute interdivision time—that is, cells with a 30-minute B or G1 phase—the eukaryotic model proposes that some process started at the cell division before the start of DNA synthesis. As the growth rate is increased, the interdivision time shortens, and the G1 phase decreases in length. At some growth rate, initiation occurs at cell division. With even faster growth rates, initiation occurs *before* cell division. As the growth rate increases, initiation of DNA synthesis moves further back so that eventually initiation occurs before the previous termination of DNA replication. With ever-increasing growth rates, initiation moves back and occurs more than one generation before cell division. Thus it was concluded, early in the analysis of the bacterial division cycle (Chapter 5), that no sequence of events starting at division led to the initiation of DNA replication in bacteria. The bacterial model has the synthesis requirement for the initiation of DNA replication occurring in all phases of the division cycle. The ubiquitous presence of a G1 phase in animal cells, and the initial difficulty

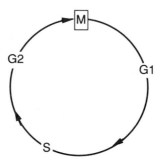

Figure 15.3 Icon for the Classical Eukaryotic Division Cycle. The phases of the cycle are arranged in a circle, with the end of the division cycle, mitosis (M), starting the G1 phase. Following the G1 phase is the S phase, which is followed by the G2 phase. This leads to the resumption of the cycle with another M phase.

of demonstrating such a phase in bacterial cells,[11] led to the divergence in the two views of the division cycle. The purpose of this chapter is to produce a unified way of looking at the division cycle.

F. Analysis of the Variation in G1 Phase

It is a common observation that the G1 period is the most variable phase of the eukaryotic division cycle. This is true when comparing the same cells growing under different conditions, or comparing all cells in a culture. This is a corollary of the fact that the S and G2 phases are constant; a constant S and G2 necessarily imply a variable G1.

The interpretation usually given to the relationship of the G1 phase to the total interdivision time of a cell is that the G1 phase is the cause of the particular interdivision time. If a cell is growing fast, with a short interdivision time, the short G1 phase is the regulatory element that allows the cell to grow faster. Whatever events occur in the G1 phase occur faster or more efficiently at faster growth rates, leading to a shorter G1, and thus faster-growing cells. Slow-growing cells are thus caused by the cells' progressing through the G1 phase more slowly; this produces a longer G1 phase, which in turn leads to a longer interdivision time. The difference between the eukaryotic and bacterial viewpoints is illustrated in Fig. 15–4. We see that the variability in the size of G1 and its relationship to the total interdivision time can be explained by a series of G1 events or by a continuous process between the starts of S phases.

Another analysis of the variation in G1 comes from work on mutants that are altered in the size of the G1 phase. Liskay and Prescott have studied a strain of cells, V79, that does not appear to contain a G1 phase.[12] They mutagenized the cells and selected cells that had a G1 phase. As S and G2 were constant, they observed that the G1 phase in different variants was equal to the increase in the interdivision time in the variants. This is an experimental support of the idea that the G1 period is produced when the doubling time of the culture is greater than the sum of the S and G2 periods.

How were these mutant or variant strains obtained? Mutants were grown with radioactive thymidine for a short time, and then allowed to grow without label; the procedure was repeated many times. The labeled cells were then frozen. It was assumed that the incorporated label would kill the labeled cells, while those cells that did not incorporate label, or incorporated less label, would be relatively resistant to killing by the incorporated radioactive thymidine. The cells that survived this suicide experiment would be enriched in variants that had gaps in DNA synthesis during the division cycle. G1-containing variants were isolated using this selection procedure.

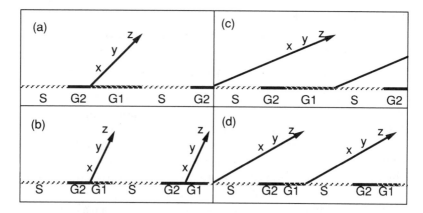

Figure 15.4 Comparison of the Eukaryotic and Prokaryotic Views of G1-Phase Variability. Panels (a) and (c) are two different interpretations of the observed preparations for DNA synthesis as experimentally measured during G1. In (a) these preparations (labeled x, y, and z) are all part of a period of preparation for S which takes place in G1. In (c) they are part of a larger period of preparation for S taking place in the period between initiations, but with x, y, and z only fortuitously occurring during the G1 period. Panels (b) and (d) show how each of the models varies the preparatory events when the growth rate increases and the G1 period decreases. Although this figure is drawn as though there are different events occurring during division cycle, there is no firm evidence that the initiation of DNA synthesis occurs after a sequence of discrete reactions. The alternative view, that initiation occurs when some unitary factor has accumulated to some crucial level, is also compatible with this figure, with the events x, y, and z indicating the achievement of a particular level of the hypothetical initiator substance.

The continuum model reinterprets the selection process as follows. When the cells were selected that had a G1 period, selection was not specifically for cells with a G1 period, but for cells that grew more slowly. If cells retain a constant S phase, then any slowing of growth increases the fraction of cells that do not synthesize DNA during a short labeling period. The procedure thus selects for cells that grow slowly. The appearance of the G1 period is attributable to the selection of mutants that had a slower growth rate without any concomitant change in the time required for S and G2.

In order to see whether the mutants obtained were altered in different functions, different mutant cells were fused to produce hybrid cells. These hybrids were analyzed for the presence or absence of a G1 phase. All combinations of different variants produced hybrids that lacked a G1 phase (Fig. 15–5). The interpretation of the result was that the mutants were affected in different functions controlling the G1 phase. In contrast, the continuum model proposes that the fused cells had minor deficien-

G1(long) x G1(long) ⟶ G1(absent or short)

Slow growing x Slow growing ⟶ Fast growing

Mutant (a,b,c) x Mutant (d,e,f) ⟶ Wild type (no G1 phase)

Figure 15.5 Explanation of Complementation of G1 Mutants. When different mutants exhibiting a G1 phase are tested for complementation, there is an increase in growth rate. The interdivision time decreases while S and G2 remain relatively constant, so the size of the G1 period decreases.

cies in any of a number of cell functions that were not cell-cycle specific. These mutations caused the rate of mass synthesis to decrease, resulting in the appearance of a G1 phase. When two cells with different mutations are fused, the resulting cell has a complete complement of wild-type functions, and the growth rate increases; the interdivision time decreases, and the G1 phase decreases or disappears. To give a concrete but hypothetical example, consider that one of the variants has a slightly impaired lysyl-tRNA-synthetase so that the synthesis of mass or protein is slowed down. This leads to the appearance of a G1 phase, because the S and G2 are unaltered. Consider that another variant has a slightly impaired hexokinase so that it too has a slower rate of mass synthesis and therefore has a G1 phase. When two mutant cells are fused, the two minor deficiencies are complemented by the normal activity of the other cell. The growth rate returns to normal and the G1 phase disappears. Rather than have the appearance of the G1 phase be a recessive trait, it was the result of variants altered in the rate of mass synthesis. None of the impairments, in the continuum-model view, affects the regulation of the division cycle. The change in the length of the G1 period was due to the alteration of the mass synthesis rate by mutations in functions unrelated to division-cycle regulation.

To summarize, the eukaryotic results are also consistent with the time between initiations of S phases regulating the division cycle. From this viewpoint, the variation in the G1 period is not the *cause* but the *result* of the variation in interdivision time. A longer time between initiations would produce a longer G1, and a shorter time between initiations would lead to a shorter G1.

G. Analysis of G1 Arrest

One of the most important supports for the concept of G1 events is the finding that inhibiting cell growth leads to the arrest of cells in the G1 phase of the division cycle. When cell growth is stopped, it is observed that cells produced after a number of hours all have a G1 amount of DNA.

The resting cells appear to be arrested in the G1 phase of the division cycle. When cells are placed in deleterious conditions, they do not initiate any more S phases. Cells that have started S, or are in G2 or M, continue through the S, G2, and M phases to produce two daughter cells with a G1 amount of DNA. These cells are not in G1, although they have a G1 amount of DNA; they are not synchronized at a particular point in the division cycle (Fig. 15–6).

H. Release from G1 Arrest

Consider cells arrested with a G1-DNA content, released from their inhibited condition, and allowed to resume growth. We consider daughter cells descended from cells that were just about to leave G1 and start an S phase or descended from cells that had just started S phase. The G1 cells about to enter S phase were inhibited just short of having the amount of material that would allow them to enter the S phase (Figs.

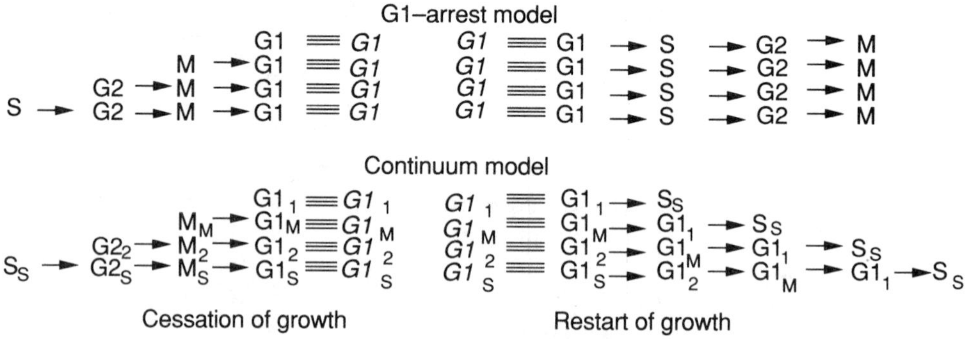

G1–arrest model

$$
\begin{array}{llllll}
 & & G1 \equiv G1 & G1 \equiv G1 \to S \to G2 \to M \\
 & M \to G1 \equiv G1 & G1 \equiv G1 \to S \to G2 \to M \\
 & G2 \to M \to G1 \equiv G1 & G1 \equiv G1 \to S \to G2 \to M \\
S \to G2 \to M \to G1 \equiv G1 & G1 \equiv G1 \to S \to G2 \to M
\end{array}
$$

Continuum model

$$
\begin{array}{llll}
 & G1_1 \equiv G1_1 & G1_1 \equiv G1_1 \to S_S \\
 & M_M \to G1_M \equiv G1_M & G1_M \equiv G1_M \to G1_1 \to S_S \\
 G2_2 \to M_2 \to G1_2 \equiv G1_2 & G1_2 \equiv G1_2 \to G1_M \to G1_1 \to S_S \\
S_S \to G2_S \to M_S \to G1_S \equiv G1_S & G1_S \equiv G1_S \to G1_2^M \to G1_M \to G1_1 \to S_S
\end{array}
$$

Cessation of growth   Restart of growth

Figure 15.6 G1 Arrest According to the Continuum Model. In the upper panel, the classical view of cell-cycle arrest is illustrated. When deleterious conditions are presented to growing cells, the cells in S, G2, and M proceed through the cycle to join the extant G1 cells at a particular point in the G1 phase of the division cycle. After restimulation of growth, the cells leave this arrest point as a uniform cohort of cells proceeding through the division cycle in a synchronous manner. In the lower panel, the same observations are reinterpreted in terms of the continuum model. Although cells are arrested with a G1 amount of DNA, these cells are not in the same cell-cycle position; they have different amounts of initiator depending on where they were when starvation began. The subscripts indicate the amount of initiator or mass per cell at each time during the starvation period. The cells arrested with a G1 amount of DNA have different amounts of mass, indicated by the subscripts 1, m, 2, and s. Those cells closer to initiation when growth was inhibited are the first to resume DNA synthesis, and the remaining cells start DNA synthesis at different times. No synchrony of initiation of DNA synthesis is expected.

15–1 and 15–6). Assume that the amount of trigger does not decrease during the period of inhibition. If the synthesis of the trigger resumed immediately upon resuming growth, the G1 cells that had not entered S phase would enter S phase almost immediately. In contrast, the cells descended from those that just started S phase are almost one full cell-division cycle from having enough trigger to initiate S phase. They are approximately half the size of cells that normally enter S phase, and must double in size before the initiation of S phase. The daughter cells will wait until one generation has passed, enough time for the cell masses to double, before sufficient trigger will be present to allow the initiation of S phase. According to the continuum model, the cells with a G1-DNA content are heterogeneous with regard to the resumption of DNA synthesis; DNA synthesis does not resume synchronously.

There are many examples of such a nonsynchronous recovery from arrest in the G1 phase of the division cycle, but it is instructive to look at the analysis of a temperature-sensitive mutant arrested with a G1 amount of DNA. Liskay[13] isolated a temperature-sensitive cell that was described as defective for some function in G1. At the nonpermissive temperature, the cell divided for a time, then stopped. All of the cells had a G1 amount of DNA. When inhibited cells were returned to the permissive temperature, the cells synthesized DNA before they divided, indicating they were in the G1 phase rather than in the G2 or S phases of the division cycle. A decided lack of synchrony in DNA synthesis was observed when the cells returned to the permissive temperature. For example, nuclei were labeled over a period of 14 hours, and cell division appeared to take place over at least 12 hours. The G1 event explanation assumes that there is a point in G1 at which cells are arrested. Some cells arrive at this point early during growth at the nonpermissive temperature, and some cells do not arrive there until much later, so there is a wide range of times at which cells are required to remain at the arrest point. It was suggested that ". . . the lack of synchrony could be simply due to an increasing 'pathology' incurred by cells at they 'sit' at their cell-cycle block."[14]

According to the continuum model, cells closest to initiation at the time that the temperature was raised will have more mass or initiator per origin than cells that have just started DNA replication.[15] The remaining cells will be spread out between these two extremes. When the permissive temperature returns, the cells that were almost at initiation initiate first; the last cells to initiate are descended from those that just started DNA replication. Approximately one doubling time is required for all cells to initiate DNA replication and divide (Fig. 15–6). This is in accord with the experimental results.

I. Analysis of G(0) Cells

Now consider cells incubating in some inhibited condition for an extended period of time. The cells may not act ideally. The amount of trigger may decrease (by decay, for example), or the general metabolism of the cell may be debilitated and the cell's ability to resume mass synthesis may be diminished. If either or both of these conditions prevail, a delay will be observed before any S phase begins after growth is resumed. If the time before the start of S phase were longer than the normal G1 of growing cells, the cells have been proposed to be in G(0). To understand the hypothesis of G(0) in numerical terms, assume that a growing cell has a G1 period of 10 hours. After the inhibitory treatment and starvation for many days, the S phase starts after 20 hours. The conclusion is that the inhibited cell entered a resting state, the G(0). When stimulated to grow, the cell takes 10 hours to leave the G(0). The cell leaves G(0) and enters the normal cell cycle in the G1 phase. The 10 hours to leave G(0) and the 10 hours to pass through G1 explain the 20-hour delay until DNA synthesis begins. Longer starvation conditions lead to a number of different times for resumption of S phase and thus there are different or deeper G(0) states.[16] The continuum model proposes that these differences are the result of differences in the ability of the starved cells to restart the synthesis of mass or initiator when inhibition ceases. No G(0) need be postulated. The deeper G(0) state reported after progressive increases in starvation time is the result of a progressive slowing in the ability of the cell to recover mass synthesis after starvation.

The support for a G(0) phase is bolstered by the current *Zeitgeist* in cell biology that assumes that G1 is a complex phase of the cell cycle, and that the controls for cell growth reside in G1. The strengthening of this G(0) view has taken place largely by the general use of the term G(0). Many biologists studying resting cells describe the resting phase as G(0) or perhaps more cautiously as G1/G(0). This referencing of G(0) has made G(0) appear ubiquitous. In contrast, the continuum model suggests that G(0) does not exist at all. Future writings on cells arrested with a G1-DNA content might use a more neutral terminology such as "cells arrested with a G1-DNA content in low serum" or "a mutant arrested with a G1-DNA content at high temperature."

One difficulty with the analysis of G(0) is that it is hard to define. It is a shorthand description of cells that are resting, quiescent, not growing, confluent, starved, or in some other nongrowth situation. It is difficult to know whether cells of one type, starved in one way, are in the same G(0) as other cells starved in another way. The term G(0) has made no distinction between different cells and different starvation protocols. The wide use of the term G(0) has led to the general belief that there is a particular position in the division cycle at which cells stop growing. Let us analyze one particular experiment that bears on the existence of G(0). This experiment is particularly useful because it is explicit in its presentation, clear in its experimental base, and as we shall see, leads to a clear application of the continuum model.

J. The G(0) Model of Zetterberg and Larsson

Zetterberg and Larsson presented experiments in support of a precise definition of the G(0) period. Zetterberg and Larsson[17] divided the G1 period into two parts, a G1pm (post-mitotic) phase and a G1ps (pre-S) phase. The G1pm period, which in the 3T3 cells studied by Zetterberg and Larsson lasted for 3.5 hours after mitosis, is defined as the period of the division cycle from which cells can leave the cycle and enter the G(0) period (Fig. 15-7).

Zetterberg and Larsson used a time-lapse video recorder to observe cells growing in monolayers. After determining the ages of the growing cells by noting the time since the last mitosis, all of the cells were subjected to varying periods in serum-free medium. The division of the cells

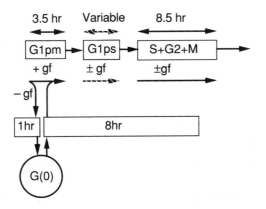

Figure 15.7 Schematic Description of the G(0) Model of Zetterberg and Larsson. The original caption to this figure reads: "During the first 3.5 hr after mitosis (G1pm), the cell makes the decision whether or not to progress through the cell cycle. This decision depends on the presence of growth factors (gf). If the cell senses a lack of growth factors (−gf) in G1pm, it will leave the cell cycle within 15 to 60 minutes and enter a state of quiescence [G(0)] from which it takes 8 hours to reenter the cycle after the growth factor level in the environment again becomes optimal (+gf) for proliferation. Once the cell has entered G1ps, it will eventually initiate DNA synthesis. However, G1ps is highly variable in length and in fact responsible for most of the variability in the duration of the G1 and of the whole cell cycle."

was then followed during the treatment and after the serum was replaced. To explain their results with numbers, consider that the interdivision time was 16 hours, that the G1 period was 7 hours, and that the S, G2, and M phases accounted for the remaining 9 hours. Let us look at the experiment in which the cells were starved for 1 hour. Zetterberg and Larsson observed that only cells that were within 3.5 hours of the last mitosis, the youngest cells, were affected in the next division. The division of these youngest cells was delayed or set back 8 hours. Those cells that were older than 3.5 hours, and thus were within 12.5 hours of the next mitosis, divided on schedule, whether or not the cells were starved for between 1 and 8 hours. A cell that was 2 hours postmitosis would have the next division occur 9 hours later than normal. This means that with a 1-hour starvation, the next division was set back 8 hours. Division occurred not at 15 hours as expected if there was a 1-hour delay, but 23 hours after the previous division.

Zetterberg and Larsson proposed that during the first 3.5 hours after division the cells are in a G1pm (postmitotic) phase. When cells are starved of serum only the cells in the first part of the division cycle can enter the G(0) phase (Fig. 15-7). Return to the G1 period from the G(0) period takes 8 hours; this accounts for the 8-hour setback in cell division. At 3.5 hours after mitosis, the cells leave the G1pm phase and enter the G1ps (pre-S) phase. Cells in this phase go on to divide even if incubated in serum-free medium. A point in G1 is now defined by this experiment. This control point is a G1-regulatory point.

A reanalysis of the model of Zetterberg and Larsson is best put in the form of a question: What does this model imply or predict for subsequent cell cycles? The answer is illustrated in the three panels of Fig. 15-8. Panel (a) illustrates exponential cell growth. Cells of different ages exist in the exponentially growing culture at time zero. Label the different age groups 1, 2, 3, 4, 5, and 6 in order of ascending age. The cells of group 1 are the youngest cells, with ages from 0.0 to 0.16, the cells of group 2 have ages from 0.16 to 0.32, and so on. Although the groups are divided into equal fractions by cell age, there are not equal numbers of cells in each group. There are more cells in the youngest group than in the oldest group because of the exponential age distribution (Chapter 1). Because the oldest cells in the culture at time zero divide to give the cell increment during the first 16% of the division cycle, and the youngest cells divide during the last 16% of the division cycle, the age distribution allows the exponential increase in cell number (Fig. 15-8a). For this *Gedanken* experiment, the variability in cell interdivision times is eliminated. The predicted order of cell division during unperturbed exponential growth is illustrated in Fig. 15-8a.

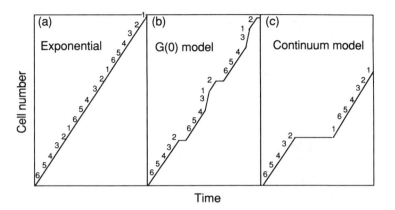

Figure 15.8 Reanalysis of a G(0) Model. The predicted order of division of different cell-age cohorts during exponential growth [panel (a)], after starvation for 1 hour according to the model of Zetterberg and Larsson [panel (b)], and according to the continuum model [panel (c)]. The numbers in each of the panels are the cells of different ages at time zero, with group 1 being the youngest and group 6 being the oldest cells in the culture. In panels (b) and (c), it is assumed that a period of serum starvation took place for approximately 1 hour at time zero. The cell number increases by divisions occurring first in group 6 (the oldest cells at time zero), then groups 5, 4, 3, 2, and 1. In the next division cycles, this order repeats, as the daughter and granddaughter cells now divide. According to the set-back model of G(0) [panel (b)], an age cohort is delayed in division while other cells, unable to enter G(0), are not delayed. As analyzed by the continuum model [panel (c)], all cells are delayed, but some cells do not exhibit this delay until after one division that occurs at the normal time.

What would we expect from the set-back model for G(0) (Fig. 15-7)? The predictions are illustrated in Fig. 15-8b. We can approximate the 3.5 hours in the G1pm phase as the first 16% of the division cycle; the cells in group 1 are in the G1pm phase. Assume that we give a short incubation in serum-free medium so that only cells in group 1 are set back into G(0). By considering only a short treatment, we eliminate problems resulting from cells dividing and newly entering G1pm. As the only cells with an affected division pattern are the cells in group 1, only groups 6, 5, 4, 3, and 2, which compose 83% of the division cycle, proceed to divide normally with a 16-hour interdivision time. When group 1 is expected to divide, at the end of the first division cycle, there is no cell division. Group 1 cells have been set back, as their first division is delayed 8 hours, approximately one-half of an interdivision time. The cells in group 1 divide in the middle of the second division cycle. The cells of the other age groups are not able to enter into G(0) because they are not in the G1pm phase of the division cycle. The daughter cells of groups 6, 5, 4, 3, and 2 continue to divide normally. The expected pattern of cell-number increase is shown in Fig. 15-8b. There is a cessation of cell division at the end of the first division cycle and a relative increase in the rate of cell division as the cells of group 1 divide at the same time as the cells of group 3. If the cells in G1pm are set back, and the other cells are not affected, this is the predicted cell-division pattern in the second division cycle.[18] This pattern has not been observed despite the large amount of work on cells starved in serum.

The continuum model predicts that cells past a point in the division cycle (i.e., the cells in S, G2, or M) will divide on schedule for the remainder of the division cycle. Even with prolonged serum starvation, cells in the S, G2, and M phases will divide normally to produce cells with a G1-DNA content. If there is residual leakage of mass synthesis during serum starvation, there would be some leakage of cells into the S phase; these cells would also subsequently divide. The imperfect inhibition of initiation of S phase accounts for the observation that cells in the G1ps phase are able to enter S phase; cells at the end of the G1 phase would preferentially be able to enter the S phase by leakage. In addition, the continuum model predicts that it is impossible to affect the division of the cells in group 1 without also affecting the subsequent divisions of the cells in the other five age groups. The same starvation that caused the delay in cell division of the youngest cells will also cause a delay in the division of cells in the other groups, but the delay will be exhibited only in the second and succeeding division cycles. Whatever decrease in mass synthesis delayed the division of the youngest cells also affects the other cells, because the synthesis of initiator occurs at all times during the division cycle. The synthesis of initiator is inhibited in cells of all ages when serum is removed from the growing cells. The continuum model predicts that if there were a setback of the cells of group 1, there would also be setbacks of the other age groups. The pattern of cell increase would be as found in Fig. 15-8c. Experimental results support this prediction.[19]

Two conservation laws for the division cycle will be presented in Chapter 16. These conservation laws propose that it is impossible to rearrange the order of cells within the division cycle or produce a synchronized culture by batch treatment of all of the cells. The proof of these conservation laws is this analysis of a proposed G(0) phase illustrated in Fig. 15-8.

K. Analysis of c-*myc* Synthesis during the Division Cycle

The study of cellular oncogenes, genes that are homologous to the transforming genes of certain viruses, is popular for both intellectual and pragmatic reasons. An altered regulation of these genes presumably leads to an alteration in the growth patterns of cells; this alteration is expressed in the malignant growth of cells. Kelly, Cochran, Stiles, and Leder[20] have suggested that "It seems likely that some oncogenes may deregulate normal proliferation by acting at control points in the cell cycle." They report that there is a cell-cycle control of the cellular oncogene, c-myc, and conclude that ". . . c-myc is an inducible gene that is shown to be modulated by a specific type of growth signal and expressed in a cell-cycle–dependent manner." In contrast to this conclusion, Thompson, Challoner, Nieman, and Groudine,[21] and Hann, Thompson, and Eisenman[22] presented evidence that neither the mRNA nor the protein determined by the c-myc gene was synthesized in a cell-cycle–specific manner. Both the mRNA and the protein are synthesized at a constant rate during the cell cycle.

A unified view of the prokaryotic and eukaryotic division cycles allows the analysis of the results of Kelley et al.[23] in terms of the continuum model. In contrast to the conclusion that the c-myc gene product is synthesized in a cell-cycle–specific manner, specifically in "the early G(0)/G1," the continuum model proposes that the observed synthetic pattern is a result of starving the cells. The finding that c-myc mRNA synthesis occurred in cells with a G1-DNA content does not mean that the synthesis occurred in the G1 phase of the cell cycle. The phase specificity is due to the fortuitous production of cells with a 2n or G1-DNA content. These resting cells are not like the G1 cells of exponentially growing culture (see Figs. 15-1 and 15-6).

Kelly et al. conclude that ". . . the asynchrony of mitogen-stimulated lymphocytes in the late G1 phase of the cell cycle . . . precludes any conclusions concerning the subsequent relative levels of c-myc mRNA following early G1 induction." They cite 3T3 fibroblasts as being able to ". . . synchronously enter S phase approximately 12 hours after incubation in platelet-poor plasma." The evidence for this is that ". . . seventy percent of the cell nuclei in the above experiment were labeled after a 24-hour incubation in platelet-poor plasma. . . ." If the cells were synchronized after the resumption of growth, we would expect that all cells would become labeled shortly after the first cell was labeled. If they were not synchronized, we would expect cells to be labeled over a period of approximately one doubling time from the time the first cell started DNA synthesis. That the labeling required 24 hours of incubation is an indication that the cells were not synchronized. This experiment supports the proposal that cells arrested with a G1 amount of DNA are not synchronized with regard to the division cycle. It was also observed that the c-myc mRNA levels decreased after the initial burst of synthesis. After 18 hours the cells were entering S phase (the "G1/S border"), and the c-myc mRNA had been reduced nearly to background levels. Kelly et al.

concluded that ". . . the c-*myc* mRNA expression in fibroblasts is confined to the early G1 phase of the cell cycle."

How do we explain the observed kinetics of c-*myc* mRNA synthesis? Two observations should be explained. The first is the initial higher specific rate of synthesis; the second is the cessation of synthesis that occurs coincidentally with the start of S phase. The continuum model explains these observations as follows. When cells are starved, inhibited, or cease growing for unknown causes (as with lymphocytes), the cell cytoplasm is different from that in growing cells. Some products are reduced and some are elevated. When the cell resumes growth and proliferation, the cell returns to the previous steady-state composition. This may involve the over-synthesis of some products because of the possible prior decay of some regulatory molecules. A particular mRNA may be oversynthesized at the start of the new growth period (e.g., owing to the low concentration of a negative regulatory molecule), but synthesis will cease when the regulatory molecule is resynthesized. This explanation according to the continuum model means that these syntheses bear no relationship to the cell cycle; c-*myc* mRNA synthesis is an event that occurs independent of the DNA cycle (S, G2, M, and G1). The cessation of c-*myc* mRNA synthesis after stimulation occurs only coincidentally with the start of S phase.

L. Criteria to Demonstrate Cell-Cycle Regulation

The results on c-*myc* synthesis emphasize two criteria that must be met in order to prove the cell-cycle specificity of a particular synthesis. First, if we believe that the starved or resting cells are a synchronized population in the G1 phase of the cycle, then the G1-specific synthetic pattern should be found in two successive G1 periods following stimulation of growth. Synthesis must be found not only in the early period after growth resumes, but in the second G1 period following the first S, G2, and M periods as well. This criterion eliminates artifacts due to syntheses caused by the original starvation condition.

The second criterion is that the experiments should be repeated with cells that are not perturbed by the starvation conditions. The work of Thompson et al.[24] and Hann et al.[25] applies the second criterion to c-*myc* mRNA and protein synthesis. Exponentially growing cells were separated into fractions by counterflow centrifugation, producing cells in different parts of the cell cycle. Cells were measured for DNA content and were shown to have DNA contents that varied as expected with cell size. The smallest cells had a 2n DNA content, indicating they were G1 cells; the largest cells had a 4n DNA content indicating a G2 or M phase. Cells with intermediate DNA values indicated cells at different stages of the S phase. The c-*myc* mRNA content of these cells was measured using probes specific for c-*myc* mRNA sequences. Controls that measured β-actin mRNA, thymidine kinase mRNA, and histone H2b mRNA were included. When either MSB-1 cells (a transformed chicken T-cell line) or chicken embryo fibroblasts were studied, the level of c-*myc* mRNA was constant throughout the cycle. Additional experiments showed that the level of c-*myc* mRNA was highest in growing cells and lowest in density-arrested cells. This, of course, confirms and explains the results of Kelly *et al.* The growth-arrested cells of Kelly *et al.* had low levels of c-*myc* mRNA; when the cells were stimulated the level increased. The increase occurred shortly after growth stimulation; thus the synthesis occurred while the cells had a G1-DNA content. Thompson *et al.* conclude that ". . . previously described variations in levels of c-*myc* mRNA could be the result of this activation process, rather than the result of regulation during the proliferative cycle." The level of c-*myc* mRNA remained a constant fraction of the ribosomal RNA. The ribosomal RNA was continuously increasing during the cell cycle, so the c-*myc* RNA must have been synthesized continuously during the cycle. Studies on the c-*myc* protein yielded the same result: there is no cell-cycle regulation of the production of the c-*myc* protein. Studies of arrested cells, after release of inhibition, indicated that there was a transient increase in c-*myc* protein. It was concluded that the ". . . transient changes in c-*myc* protein synthesis observed 1 hr after release would seem to be a serum-dependent effect on synthesis rather than a cell-cycle–regulated event."

A number of control experiments were included in these studies. β-actin was continuously synthesized during the cell cycle. This pattern of synthesis is expected for a protein that "is an essential structural protein required for cell integrity and therefore should be transcribed in all viable cells." What is most interesting is that the other two genes studied, TK and H2b, were transcribed in a cell-cycle–specific manner. There is a large body of evidence demonstrating that histones are transcribed only during S phase.[26] This leads to a fundamental problem. How can the observation of S-phase–specific mRNA (and presumably protein) synthesis be reconciled with the continuum model? The continuum model does not rule out the synthesis of proteins specifically associated with the replication–segregation sequence (see Chapter 4). For example, the slight but definite increase in thymidine kinase mRNA may be due to the release of a feedback control when DNA synthesis starts. The requirement for activated thymidine triphosphate, a precursor of DNA, increases. Similarly, the histone gene product may accumulate along with the DNA as part of the S phase. These increased syntheses, however, have no causal relationship to the start of the next cell cycle.

The interval between starts of S phases is determined solely by the rate of initiator synthesis occurring continuously during the cycle. These cell-cycle–specific syntheses do not regulate the starts of S phases. The replication–segregation sequence is the end of a process and the beginning of none.[27]

The experiments on c-*myc* synthesis are important because they highlight the changes occurring in the analysis of the eukaryotic division cycle. These experiments demonstrate that continuous synthesis is common and that periodicities could be artifacts of the experimental situation.

M. The Cloning and Identification of G1 Genes

In a paper entitled "Molecular Cloning of a Gene That Is Necessary for G1 Progression in Mammalian Cells," Greco, Ittmann, and Basilico[28] describe the isolation, cloning, sequencing, and regulation of the DNA that complements the mutant allele of a temperature-sensitive strain of BHK cells. At the nonpermissive temperature, mutant *ts*11 produced a population of cells all with a G1 amount of DNA. It was concluded that the mutation affected some function involved in the progression of the cells through the G1 period of the cell cycle. Another cloned gene for a different function, *ts*BN51, was also isolated. Both mutants were described as "blocked in G1 progression" or "blocked in G1."[29]

The current approach to analyzing the cell cycle suggests that there are G1-specific events regulating cell growth and division.[30] Because the analyses of such G1 mutants are typical, and are presented in excellent papers, these studies are good models to demonstrate the continuum-model view of the cell cycle. Because so much effort is being devoted to finding the putative G1 events,[31] it is important to recognize that the results can also be explained by a model where no G1-specific events are postulated.

Are *ts*11 and *ts*BN51 G1 mutants? The continuum model proposes that both temperature-sensitive mutations affect some function of cell metabolism involved in the synthesis of cell mass. The function operates throughout the cell cycle, and the protein is synthesized throughout the cell cycle. When the temperature is raised, mass synthesis ceases, and initiation of DNA synthesis stops. Cells that have initiated DNA synthesis or are in the G2 or M phases of the division cycle proceed to division. Thus, the final cell population is composed of cells with a G1 amount of DNA. We should not conclude that these G1-DNA-containing cells are related to the normal G1 cells of the division cycle (Figs. 15-1 and 15-6).

In support of the idea that the gene analyzed is involved in G1 regulation, Greco *et al.*[32] measured the mRNA synthesized when cells were stimulated to grow after a period of starvation in medium with a low concentration of serum. As these cells all had a G1 amount of DNA, it was concluded that the cells were in the G1 period of the cell cycle. Immediately after stimulation with fresh serum, the mRNA was synthesized from the gene that was cloned. Greco *et al.* concluded that the mRNA was made during the G1 phase of the division cycle. According to the continuum model, the cells are not in G1, but have a G1 amount of DNA. The arguments that applied to c-*myc* gene products also apply here. If the cells are not arrested in the G1 phase, we cannot conclude that the mRNA is synthesized specifically in the G1 period. A delay before DNA synthesis resumes leads to the conclusion that the mRNAs are synthesized in the G1 period.

So as not to be misunderstood, I wish to reiterate that these genes may be important for studying cell growth. The work described is important because the genes have been sequenced. It is not argued here that these genes are unimportant, or that there is any problem with the gene isolation or sequencing. The biochemical approach to gene structure may tend to obscure problems with the original definition of gene function. The mutation may define an important gene, but it does not necessarily define a gene for a G1-specific event.

N. The Restriction Point

The restriction point is defined as a particular time during the cell-division cycle when a particular controlling function or event occurs. Inhibiting the performance of this function prevents the cell from passing the restriction point. The restriction point is operationally defined by experiments in which growth is reduced by the use of suboptimal growth conditions or inhibitors. When cells are placed in these growth conditions, those cells that have passed the restriction point will be able to proceed to perform a measurable event such as DNA synthesis or cell division. Occurrence of the event indicates that the cells have already performed the required function at the restriction point. Cells that cannot perform the measured event are defined as before the restriction point. The first postulation of a restriction point for a eukaryotic cell was that of Pardee[33] in 1974. Since that time, a number of additional studies have elaborated the concept,[34] culminating in the proposal that B cells have 3 restriction points during the division cycle.[35]

Before the proposal of restriction points in animal cells, such restriction points were described in bacteria (see Chapter 5). Those regulatory points were not named restriction points, and thus the experiments are largely forgotten. Reanalysis of the bacterial experiments demonstrated that the principles on which restriction points were identified were weak.[36] This analysis is presented in detail in Chapter 5.

Pardee[37] was the first to use the term restriction point as defined above, although there were a few earlier papers that touched on similar ideas.[38] The purpose of the experiments was to analyze the nature of quiescence. Evidence was provided that cells in quiescence, no matter how they achieved quiescence, were at a single, unique position in the division cycle. This model of quiescence suggests there is a ". . . single switching point, the restriction point (or R point) in G1, that regulates the reentry of a cell into a new round of the cell cycle."[39] Cells whose growth is arrested by "nonphysiological means" (such as hydroxyurea or colchicine), do not stop at the R point. Pardee's experiment is simple. Exponentially growing cells are placed in different media suboptimal for growth and allowed to reach quiescence. The cells are then placed in fresh growth medium, and the time until DNA synthesis resumes is measured. When cells were allowed to achieve quiescence by any of three different treatments (low serum, isoleucine deprivation, and glutamine deprivation, all for 64 hours), in all cases DNA synthesis resumed after 8 hours in complete medium. This result was interpreted as indicating that all cells were stopped at a point eight hours before the start of the S phase. The conclusions were summarized: "In each experiment, the quiescent cells required the same length of time to recommence DNA synthesis. These results are consistent with each of the cell populations' being blocked at the same point."[40]

Because of problems in measuring the precise time DNA synthesis resumed, sequential inhibitions of growth were studied. Cells were allowed to achieve quiescence under one condition, then transferred to a different condition that independently produced quiescence. The ability of the cells to synthesize DNA was then determined. The reasoning was: "If the block applied second stops the cell at an earlier time in G1 than does the block applied first, the cells should make DNA. But if the block applied second acts later or at the same point as the one applied earlier, the cells should not progress on to DNA synthesis."[41] A number of such double-block experiments were carried out using five different physiological conditions. There was no significant DNA synthesis in any case. It was concluded that all five of these blocking conditions arrested the cell at the same point, the R point.

Consider these experiments from the perspective of the continuum model. DNA synthesis (S phase) starts when enough cell mass or protein has accumulated in a cell. This accumulation occurs continuously between starts of S phase and does not occur in any particular part of the division cycle. Cells that have just started S phase have the lowest amount of initiator per origin; other cells have different amounts that vary continuously throughout the cell population. Inhibition of protein synthesis prevents or inhibits the initiation of DNA synthesis. Cells that have initiated DNA synthesis, i.e., are in S, G2, or M phases, continue to complete DNA synthesis, pass through the G2 period, and proceed through mitosis, producing two daughter cells each with a G1 amount of DNA. Inhibition of protein synthesis, by whatever means, will produce cells all with a G1 content of DNA.

If protein synthesis is allowed to resume by replacing the deficient medium with a complete medium, the S phase will begin when enough protein, mass, or initiator has accumulated. The time required for initiation will differ in different cells. Now let us reconsider the double-block experiments. If protein synthesis has been inhibited with one suboptimal growth condition, and the cells are then transferred to another, suboptimal growth condition, S phase will not start. Two different ways of inhibiting protein synthesis are equivalent with regard to the start of the S phase. Where no escape from the sequential blocks was observed, both treatments were inhibitors of protein or mass synthesis. A double-block experiment that does not allow resumption of S phase does not indicate that the two conditions block cells at the same point.

How we explain the constant 8-hour delay before the start of S phase under different conditions of inhibition? We could expect that only a few hours of incubation under starvation conditions would allow cells to accumulate at the postulated R point. One possibility is that a long incubation period (64 hours) led to difficulties in the cells resuming protein synthesis. Therefore, for approximately 8 hours, there was little protein synthesis. When protein synthesis finally resumed, DNA synthesis was initiated.

The timing experiments were affected by the fact that cells initiate DNA synthesis at different times. It was noted that ". . . different cells begin thymidine incorporation at different times. Thus, measurement of the time of initiation of DNA synthesis by a cell population depends upon the behavior of an early initiating subclass of the population. We can only conclude, therefore, that this subclass is at the same point in different quiescent cultures."[42] If there were a unique restriction point, cells would be expected to proceed toward S in a synchronous manner. The observed asynchronous initiation of DNA synthesis is predicted by the continuum model. The continuum model proposes that the cells arriving with a G1-DNA content are not all at the same point in the division cycle (Figs. 15-1 and 15-6). Cells descended from G1 cells at the start of quiescence are the cells closest to initiation of DNA synthesis, while cells descended from cells in the S phase at the time of starvation are furthest from initiation. The experiments do not support the existence of a unique restriction point in the division cycle. The experiments do support the existence of a collection of cells that are classified by virtue of one parameter (DNA content), but that are different in other parameters (size, or ability to initiate DNA synthesis).

One of the ironies of the history of the restriction point in animal cells is that the original postulation of the restriction point emphasized the strongly argued conclusion that the restriction point is unique. It was the uniqueness of the restriction point that gave the original postulation its great power. If there were no unique time delay until resumption of DNA synthesis, or switching of inhibition conditions allowed DNA synthesis, a different conclusion might have emerged. It might have been concluded that there was no single restriction point. If each condition of inhibition were able to give a different time before resumption of DNA synthesis, it might have been concluded that there were an unlimited number of physiological restriction points. The finding of a large number of restriction points, while referred to as an extension of the restriction-point hypothesis, was in fact an undermining of the fundamental logic of the restriction point. A specific isoleucine restriction point was found, and other restriction points have been reported.[43]

The importance of finding many restriction points should not be overlooked. Rather than taking a large number of restriction points as support for the restriction-point proposal, multiple restriction points are a problem. A number of restriction points are compatible with the suggestion that leakage of mass synthesis allows cells to enter S phase during different starvation procedures. Further, the analysis of the bacterial restriction point as described in Chapter 5 is very similar to the analysis presented here of eukaryotic restriction points. The initial postulation of two restriction points gave way when it was shown that if two existed, then an infinite number existed. A better conclusion was that the restriction points were due to leakage.

O. The Transition-Probability Model

The transition-probability model proposed, in its original form,[44] that there were two phases that regulated the interdivision time distribution of cells. There was a probabilistic phase and a constant phase. The probabilistic phase was thought to be associated with the variable G1 phase, while the constant phase was associated with the more constant S and G2 phases. When a cell divides, the newborn daughter cells enter the probabilistic part of the division cycle and remain there until they leave in a purely stochastic manner. They leave the probabilistic phase with the same kinetics as the decay of a radioactive atom. The probabilistic phase is followed by a constant phase. The minimal interdivision time is determined by the length of the constant phase. After leaving the probabilistic phase of the division cycle, cells enter a second, constant phase of the division cycle. The total interdivision time would be equal to the sum of the times that cells resided in the variable and constant phases.

The interdivision time distribution predicted by the transition-probability model could be analyzed and compared to experimental determinations of the interdivision time distribution. The results have been plotted in the form of an α-curve, a plot of the fraction of cells that have not yet divided against time (Fig. 15-9). When plotted on semilogarithmic paper, a flat line at 100% for a period of time equal to the constant phase would be expected. This flat line would then break sharply to a line with a slope proportional to the decay constant for the cells in the variable phase. In practice, the curves are not ideal, and the main evidence for the transition-probability model has been a straight-line segment for a portion of the curve. This straight line on a semilogarithmic graph has been taken as evidence for a random transition.[45]

In addition to the α-curve, the β-curve has been determined. If the constant portion of the division cycle were the same in two daughter

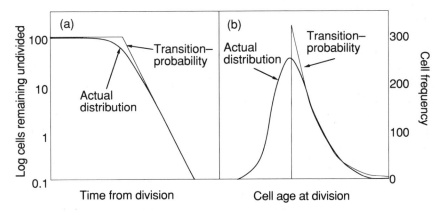

Figure 15.9 The Transition-Probability Model. The α-curve (a) is the fraction of cells that have not yet divided, plotted on semilogarithmic paper. We start out with a set of newborn cells, then record when each divides. A prediction of the transition-probability model is that the decreasing curve is a straight line on a semilogarithmic graph. The experimental data do not show a sharp break. When the same data are plotted as a frequency graph (b), there is a clear discrepancy between the data and the predictions of the transition-probability model. One reason that the discrepancy does not show up on the semilogarithmic paper is that the number of points in between 100 and 10 takes up as much vertical space as the number of points between 10 and 1. This means that the values expand in the lower regions and compress in the upper regions.

cells, and if there were a random component to the division process, we would expect that a plot of the difference in interdivision times between sister cells would give a straight line through the origin with no curvature. When the first cell divides, assuming that the constant periods are the same in both sister cells, the cell with the longer interdivision time has also left the constant phase and is now in the probabilistic phase. One measures, by this difference determination, only the rate of leaving the variable phase. If the transition-probability model is correct, this gives a straight line for the β-curve. The experimental evidence for the transition-probability model has been a large number of measurements on the interdivision times of cells. This entails taking time-lapse movies of cells in culture, noting when cells divide, and accumulating data on the interdivision time distributions. Although there is no sharp break in the curve as predicted, a portion of the curve is a straight line on the semilogarithmic plot. This has been taken as support for the transition-probability model.

The continuum model makes predictions regarding the variability of cell-cycle interdivision times that are strikingly similar to the experimental results. A general model of cell-cycle variability was presented in Chapter 8. This model of cell-cycle variation applies to animal cells: cell mass accumulates at a rate that varies about some mean; initiation takes place when the cell mass reaches some amount that varies about a mean; passage through the S and G2 periods varies about a mean value; and cell division is approximately equal, but there is some statistical variation in the equality of cell division. The predictions of the continuum model have been studied by computer-simulation experiments; the predictions fit the data.[46] The α-curves and β-curves are easily derived by the continuum model. In addition, the quartile test, which has been proposed to be specifically valid only for the transition-probability model,[47] is also fit by the continuum model.[48]

The transition-probability model has been an influence on the field of cell-cycle studies. It is widely believed that the transition-probability model has something to add to our understanding of the eukaryotic division cycle. The transition-probability model has one major problem. In order for the cell to follow a random transition, each cell must be born with one, and only one, molecule or element that decays. When this decay occurs, it must signal the cell to progress through the division cycle. It is difficult to imagine how a cell makes *one* of anything that can regulate the division cycle.[49] To summarize, the data that support the transition-probability model can be explained by the simple rules of the continuum model. There is no need to invoke any probabilistic mechanism for the regulation of the division cycle.

P. Mitogenesis or Cytogenesis

When resting cells are stimulated to grow by the addition of some substance, cells eventually go through mitosis. This process of stimulation is termed mitogenesis. In practice, the measurement of the stimulation of division has given way to the measurement of the stimulation of DNA synthesis. It is generally true that cells must synthesize DNA before division as cells are arrested with a 2n amount of DNA. The measurement of DNA synthesis is easier than the measurement of cell division, so DNA synthesis is generally used as a measure of mitogenesis.

One problem with using DNA synthesis as a measure of cell growth is that according to the continuum model, initiation of DNA synthesis is a symptomatic expression of the increase in cell cytoplasm. More important, DNA synthesis may not be linearly related to the strength of the mitogenic response. [50] For this reason it has been proposed that a better measure of the response of resting cells to a stimulatory compound is the measurement of the increase in mass synthesis (Fig. 15-10). Even when the rate of mass synthesis is proportional to the dose of mitogen, it is not necessarily true that the DNA synthetic response is proportional to the concentration of mitogen. Depending on when DNA synthesis is measured (and it is generally measured approximately 24 or more hours following stimulation), we can get unusual nonlinear determinations of the potency of a mitogen.[51]

The continuum model proposes that mitogens should be considered cytogens; the most direct measurement of mitogenesis is the stimulation of cytoplasm synthesis. It would be preferable to use leucine incorporation rather than thymidine incorporation as a measure of growth stimulation. Thymidine has been popular because the background levels of incorporation are relatively low, and we can also determine the number of cells that have been stimulated by using autoradiography. It must be realized, however, that there are problems with thymidine incorporation, and these must be understood in any assay of mitogenesis.

Q. Competence and Progression

The complexity of the G1 phase is given support by the proposal that mitogenesis can be divided into two processes, competence and progression. Competence factors are compounds that do not act as mitogens, but which when added to cells before a second substance, a progression factor, can allow the second substance to act as a mitogen. The second substance, by itself, would not act as a mitogen. An example of a competence factor is platelet extract; an example of a progression factor is platelet-poor plasma.[52] Neither one is a mitogen alone, but when the

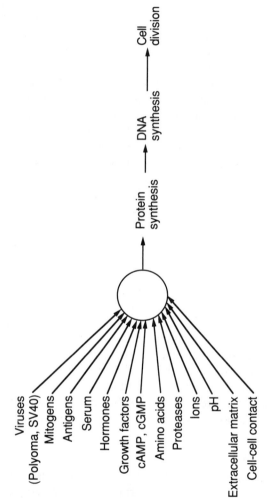

Figure 15.10 External Conditions Induce DNA Synthesis by First Activating Mass Synthesis. Listed at the left are a number of compounds or conditions that affect growth. Those that stimulate growth from resting cells are termed mitogens. Their primary effect is proposed to be on mass or cytoplasm synthesis. As mass or cytoplasm, or some specific molecule that is a part of cell mass accumulates, DNA synthesis is initiated, and cell division soon follows. The original definition of mitogen referred to the final mitoses observed when a mitogen was added. This gave way to the assay of DNA synthesis. The continuum model proposes that the most immediate measure of a mitogen is its effect on cytoplasm synthesis. No matter how complex the additions, all that is required is that the cell integrate the signals and decide whether or not to increase mass synthesis. If there is an increase in mass, then there will be DNA synthesis and mitosis. A competence factor is proposed to be a very weak stimulator of mass synthesis, or one that when present in concert with another weak stimulator, gives enough stimulation of mass synthesis so that there is an initiation of DNA synthesis.

platelet extract is added to cells and then removed, addition of plasma can cause cells to synthesize DNA.

The continuum model interprets this type of experiment by proposing that both substances are weak cytogens. When added together, there is enough stimulation of mass synthesis, producing DNA synthesis and mitosis. The reason each one alone does not give a mitogenic response is that usually we do not wait long enough. If we measured mass synthesis directly, we would see the stimulation of mass synthesis with both the progression and the competence factors. A progression factor needs a competence factor because in some way, the two together stimulate enough mass synthesis to allow DNA synthesis to start. There is no biochemical distinction between competence and progression. This distinction between mitogens should be discarded in favor of a classification of mitogens in terms of the strength with which they stimulate mass synthesis.

R. The Future of the Animal Cell-Division Cycle

Although there remains a separation between the prokaryotic and the eukaryotic views of the division cycle, there are some indications that the arguments against the functionality of G1 and G(0) are beginning to have some effect. Pardee, in a recent review of the status of G1 events, has accepted the possibility that "G1 events necessary for eventual onset of S phase can begin during the previous cycle, at the same time and in the same cell as other events such as DNA synthesis or preparation for mitosis."[53] The separation between the different views of the division cycle does not rely on the timing of events, but now rests primarily on whether there are discrete events.

Another hopeful sign that a reconsideration of the concept of a G(0) phase is underway is the decision by the editors of a book on the cell cycle to omit ". . . the enormous amount of work on growth factors, cell signalling, oncogenes, and the G(0)→G1 transition almost completely."[54] This decision is a recognition that these factors may be only peripherally related to the cell cycle. These aspects of cell growth are related to the production of more cell mass, with the cell growth leading to an initiation of DNA synthesis and cell division. The increase in cell mass, as noted in Chapter 4 as well as in this chapter, is not a cell-cycle–specific event.

A third hopeful sign is the proposal of a unified view of the cell cycle based on the conservation of the mechanisms that induce mitosis and other transitions in the cell cycle.[55] The proposal is consistent with the continuum model. Thus we see another step toward a unified view of the division cycle of prokaryotes and eukaryotes.

III. THE DIVISION CYCLE OF *SCHIZOSACCHAROMYCES POMBE*

The yeast, *Schizosaccharomyces pombe*, is a rod-shaped eukaryotic cell that has received a great deal of attention as a model system for understanding the eukaryotic division cycle.[56] At first glance, this yeast cell grows similarly to the rod-shaped, gram-negative bacteria. Cells elongate, then invaginate or constrict, and two new daughter cells are formed. A fundamental difference between the division cycles of *Schizosaccharomyces pombe* and *Escherichia coli* appears when we consider periodic variations in biosynthesis during the division cycle. It has been proposed that there are periodic variations in the biosynthesis of cytoplasmic components during the division cycle of *Schizosaccharomyces pombe*. This periodicity has been proposed to be an important area of work for understanding the eukaryotic division cycle. As described in detail in Chapter 4, such periodic variations do not exist in *Escherichia coli*.

The evidence and ideas regarding periodicities in the division cycle of this yeast cell will first be reviewed, then an explanation of why I believe that the prokaryotic and yeast division cycles have more in common than is otherwise believed will be presented.

A. Periodicities in the Growth of *Schizosaccharomyces pombe*

Mitchison[57] has recently reviewed the evidence for periodicities in *Schizosaccharomyces pombe* and has concluded, ". . . without exception, cell cycle growth in *Schizosaccharomyces pombe* is not a smooth exponential process but instead shows periodicities when the measurements have been made with sufficiently sensitive techniques." Different periodicities have been reported, but these results should be accepted with caution.

DNA synthesis is periodic, with DNA synthesis occupying approximately 10% of the division cycle and initiation occurring in the early part of the division cycle. This periodicity of DNA synthesis is easy to understand and is similar to the division-cycle pattern of other eukaryotic cells.

Measurements of the length and volume of *Schizosaccharomyces pombe* cells indicates that there is a periodicity in the overall growth of the cell. There is also a variation in the growth pattern with a change occurring in the pole from which growth occurs, reminiscent of the unit-cell model described in Chapter 6. The major conclusion is that there are discrete times during the division cycle at which growth changes from one linear growth pattern to another. Dry mass measurements also indicate a bilinear pattern. The rate of protein synthesis also changes during the division cycle, with an acceleration point and a plateau, before another acceleration point begins. RNA synthesis also shows such steps and periodicities.

When global properties of the cell metabolism such as the rate of carbon dioxide production or the rate of oxygen uptake were analyzed, it was also concluded that there were periodicities. The acid-soluble pool of metabolites also appeared to follow a periodic pattern. In addition, enzyme activity showed steps or changes with at least two enzymes shown to have periodic steps. Magnesium has been shown to vary during the division cycle. There are also steps in the ability to synthesize various enzymes.

B. A Critique of the Proposal of Periodicities in *Schizosaccharomyces pombe*

I am concerned about the cytoplasmic periodicities in *Schizosaccharomyces pombe*. I acknowledge that the literature in this field is enormous, and it would be impossible to refute or analyze all of the experiments. In the following sections, I shall summarize the cause of my difficulty in accepting the cytoplasmic periodicities in *Schizosaccharomyces pombe*.

1. Synchronization and the Experimental Evidence

All of the experiments demonstrating periodicities use synchronization to obtain the results. In the previous chapters we have seen the problems with synchronization (see Chapter 3). Even the use of controls in which the culture is put through the process of synchronization and the total population is analyzed do not make the results more acceptable.

2. The Proposal of Quasi-Linear Synthesis

Mitchison proposes that some of the periodicities are not precisely linear but are quasi-linear. But what is quasi-linearity? Consider a bilinear synthesis pattern (see Chapter 4) with the rate of synthesis doubling in the middle of the cycle. In a synchronized culture we would not see instantaneous doubling, because the cell cycle is variable. There would be a smooth increase from one rate to another. Now suppose the doubling in rate is not instantaneous even within a single cell. This means that the linear pattern becomes even less sharp, until all we are left with is a small blip at the time of cell division. Mitchison proposes the idea of quasi-linearity to explain the almost exponential experimental results. If we take the variable division cycles, make the bilinear pattern variable during a single cell's division cycle (i.e., not a sharp break but a curving change in rate), then we have a proposal very close to what is observed. But what if we turn the explanation around and propose the concept of quasi-exponentiality? The idea of quasi-exponentiality is that biosynthesis is exponential during the division cycle (for all of the arguments and analyses, see Chapter 4), but there may be factors that intervene to give some structure to this pattern. For example, if at mitosis mRNA synthesis were inhibited because the chromosomes were condensed, we might see a flattening in some synthetic patterns at cell division. The underlying biosynthetic pattern, however, would have no relationship to the division cycle.

To summarize the criticism of the proposed quasi-linearity, it is suggested here that as the data fit exponential kinetics better than linear kinetics. If we wish to argue for linear kinetics, then we must propose mechanisms by which the underlying linear kinetics are obscured from the experimenter. I suggest that it is better to accept the experimental results as exponential, and fit them into a simple mechanism for exponential synthesis during the division cycle as described in Chapter 4.

3. Protein Synthesis during the Division Cycle of *Schizosaccharomyces pombe*

Experimental evidence has been published showing that most of proteins in the *Schizosaccharomyces pombe* are made at a constant rate during the division cycle.[58] The conclusion derived from these experiments is tempered by the usual suggestion that there may be minor proteins that are made in a cell-cycle–specific manner that are too few to be detected or that even though synthesis is invariant, there may be cell-cycle–specific modifications of the activities of the proteins. I suggest that these experiments should be accepted as support of an invariant and exponential synthesis of cytoplasm during the division cycle.

4. Evolution and Periodicities

The major evolutionary problem concerning periodicities is their inefficiency (see Chapter 4). If there is a sudden doubling in the amount of an enzyme, or there is a doubling in its rate of synthesis, there will be times when the amount of that protein is relatively abundant. If we take the minimal concentration as that amount that the cell needs to grow efficiently, why then does a cell have an excess at other times?

A second evolutionary problem has to do with the genetic information required to maintain the periodicities. There must be some proposal to give one an understanding of why the cell would devote so much energy to the maintenance of these periodicities.

5. Problems of Mechanism

One of the central themes of this book is that the best work fits a large amount of biochemical and biological data in a meaningful and explanatory way. The choice of particular cytoplasm, DNA, and cell-surface synthetic patterns in bacteria resulted as much from considerations of the mechanism of their synthesis as from experimental data (see Chapters 4, 5, and 6). An understanding of the mechanisms producing a particular pattern gives support to the belief that a particular pattern is the correct pattern. That is not to say that any newly discovered pattern, not yet explained by mechanisms, cannot be correct. But an entire field should not be based on experiments that have shown periodicities for over 20 years with not one inkling of how these periodicities arise.

6. Historical Considerations

When I read the work on *Schizosaccharomyces pombe* I get a feeling of *deja vu* all over again. The entire argument has been played out in bacteria over the past twenty years. As described fully in Chapter 4, the cytoplasm of the bacterial cell increases continuously, exponentially, and invariantly during the division cycle; there are no cytoplasmic events in bacteria. This is despite a number of experiments in bacteria showing the same periodicities as proposed for yeast. For example, instead of magnesium's being the single stimulus for cell-cycle variation and regulation, calcium has been proposed to regulate the bacterial division cycle. RNA synthesis during the bacterial division cycle was proposed to be periodic before being shown to be exponential. Enzymes were thought to vary before it was shown they did not vary. Further, the observation of a doubling in the rate of induced enzyme synthesis is consistent with a doubling in the number of genes coding for that enzyme (see discussion of F*lac* replication in Chapter 6). In yeast this does not appear to be the case. No change in the rate of enzyme synthesis appears to have any relationship to the replication of the genes during the S phase. I hope that when the experiments on *Schizosaccharomyces pombe* are repeated with a nonsynchrony approach (see Chapter 3), it will be seen that the cytoplasmic syntheses are not periodic, and that cytoplasm increases invariantly and exponentially during the division cycle.

7. The Division Cycle of *Saccharomyces cerevisiae*

There have been a number of reports of periodic synthesis during the division cycle of a different yeast, *Saccharomyces cerevisiae*.[59] The evidence is impressive, and a number of different ways of synchronizing the cells have been used. Analysis of these experiments reveals problems that weaken the experiments. For example, in one set of experiments[60] the patterns of synthesis obtained after different synchronization procedures were not the same. It is hoped that these experiments will be repeated with a nonsynchrony approach to confirm the periodic synthetic patterns. More likely, the patterns observed may be owing these particular molecules' being associated with DNA synthesis and the S phase. In this way the molecules are more like the histones of animal cells that are synthesized in a cell-cycle–specific manner, because they are regulated as part of the S phase.

In a different experiment with *Saccharomyces cerevisiae*, the evidence suggests that there are no major changes in the rate of protein synthesis during the division cycle.[61]

One of the few reports of a change in cell size with growth rate has been performed with *Saccharomyces cerevisiae*.[62] The data look quite similar to the classical Schaechter, Maaløe, and Kjeldgaard experiment, and it may be that this is evidence for a similar pattern of regulation.

IV. BACKWARDS ANALYSIS OF THE EUKARYOTIC DIVISION CYCLE

It is historically interesting that the method first used to obtain the eukaryotic pattern was also a backwards method, the method of labeled mitoses.[63] Many sophisticated methods are now available for cell-cycle analysis and the determination of the length of the various phases of the eukaryotic division cycle. But the original determination of the phases of the eukaryotic division cycle was obtained by the method of labeled mitoses, a cousin to the membrane-elution method. The method of frequency of labeled mitoses is a backwards method, just like the membrane-elution method. As described in Chapter 8, the backwards approach has many advantages over synchrony methods. In the method of labeled mitoses (Fig. 15-11), a radioactive label specific to DNA is added to a growing, unperturbed culture. After a short period of labeling, the cells are fixed at intervals and subjected to autoradiography. The fraction of those mitotic cells that are labeled is determined and plotted with time. As in the membrane-elution method, the amount of radioactivity in dividing cells is determined as a function of time after labeling. As in the membrane-elution technique, the first cells to be examined (by being mitotic figures) are those cells that were the oldest cells in the culture at the time of labeling. With time, mitoses are seen are from cells that were younger and younger at the time of labeling. The G2 period is that period when unlabeled mitoses are first seen after the labeling period. The S phase is then indicated by the appearance of labeled mitotic cells; the length of time that labeled mitotic figures are seen is equivalent to the S phase. A period of unlabeled mitoses then occurs that is equal, in time, to the sum of the G1 and G2 periods. By subtracting the previously determined G2 period, we can obtain the G1 period.

It is interesting that two different methods first used to identify the pattern of DNA replication during the division cycle in two very different

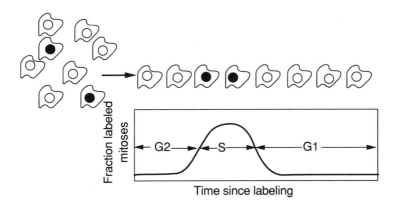

Figure 15.11 The Frequency-of-Labeled-Mitoses Method: A Backwards Method for Eukaryotic Cell-Cycle Analysis. Exponentially growing, unperturbed cells are pulse-labeled with tritiated thymidine and after the label is removed, the cells are allowed to grow. At various times cells are collected, fixed to a slide, and autoradiographed. The fraction of labeled mitotic figures is determined. (Black nuclei are labeled.) The phases of the cycle emerge in reverse order as the G2 phase is revealed first, then the S phase, and then the G1 phase.

types of cells were backwards methods. The arguments in support of the inherent strength of backwards methods (Chapter 8) receive circumstantial support from these results.

V. TERMINOLOGY OF THE DIVISION CYCLE IN PROKARYOTES AND EUKARYOTES

Different terms for different phases of the division cycle have emerged. These are illustrated in Table 15-1. There are no eukaryotic equivalents for the E, U, and T periods. It is proposed here that with prokaryotes the

Table 15.1 Terminology of Cell-Cycle Phases in Prokaryotes and Eukaryotes

Phase	Prokaryotic Term	Eukaryotic Term
First gap phase	B	G1
DNA synthesis	C	S
Postsynthetic gap	D	G2
Postdivision synthesis	E	na
Initiation to constriction	U	na
Constriction to division	T	na
Chromosome separation and division	na	M

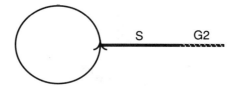

Figure 15.12   Continuum-Model Icon for the Eukaryotic Division Cycle. As with bacteria, a continuous synthesis of mass is the driving force that regulates surface and DNA synthesis. The main point is that cell division is the end of this process and does not feed back to start any processes, or any clock, during the division cycle.

letters B, C, and D (and even E, if that is useful) be retained, while with eukaryotes the names retained are G1, S, G2, and M.

VI. THE EUKARYOTIC DIVISION-CYCLE ICON

An icon summarizing the eukaryotic division cycle is presented in Fig. 15-12. It has the same form as the bacterial icon, with the end of the division cycle not feeding back or affecting prior events. Mass synthesis occurs continuously, starts S phases at particular times, and the cell proceeds through the S, G2, and M phases to produce two new daughter cells.

NOTES

1. Helmstetter, Cooper, Pierucci, and Revelas, 1968.
2. Pardee (1968) has used bacteria as a model for the analysis of the animal cell-division cycle, but no unification of ideas in the two systems was proposed.
3. This chapter is a general summary of ideas that have appeared previously in a number of different articles. For more details on the background, relevant references, and particular applications, see Cooper (1979, 1981, 1982a,b, 1984a,b, 1987, 1988c); Guiget and Cooper, 1981; Okuda and Cooper (1989).
4. Mitosis, when cell division and chromosome segregation occur, is a fourth phase that may be considered for analysis, but it is relatively short and therefore is generally eliminated from discussion or is considered to be part of the G2 phase of the cell cycle.
5. Liskay and Prescott, 1978.
6. Becker, Wayne, *The World of the Cell*, Benjamin-Cummings, Menlo Park, CA. 1986. p. 475.
7. Baserga, 1985; Hochhauser, Stein, and Stein, 1981; Pardee, Dubrow, Hamlin, and Kletzien, 1978.

8. For experiments on this particular point see Wharton, 1983.

9. Cooper, 1979.

10. Cooper, 1982a. In this earlier analysis the trigger was called the *hypothetical initiator*.

11. Helmstetter and Pierucci, 1976.

12. Liskay and Prescott, 1978.

13. Liskay, 1974.

14. Liskay, 1974.

15. Cooper, 1979.

16. Augenlicht and Baserga, 1974.

17. Zetterberg and Larsson, 1985.

18. This is not explicitly stated or experimentally studied by Zetterberg and Larsson. Their experiments and model deal only with the first division cycle after the serum starvation. The predictions of their model for the divisions of the second cycle are deduced from their model.

19. Larsson, Zetterberg, and Engstrom, 1985. When this proposal was first made (Cooper, 1987), I was unaware of this experimental fact. It was only pointed out to me in a rebuttal article by Fantes (1986, 1987).

20. Kelly, Cochran, Stiles, and Leder, 1984.

21. Thompson, Challoner, Nieman, and Groudine, 1985.

22. Hann, Thompson, and Eisenman, 1985.

23. Kelley, Cochran, Stiles, and Leder, 1984.

24. Thompson, Challoner, Nieman, and Groudine, 1985.

25. Hann, Thompson, and Eisenman, 1985.

26. Alterman, Ganguly, Schulze, Marzluff, Schildkraut, and Koultchi, 1984.

27. Cooper, 1979.

28. Greco, Ittmann, and Basilico, 1987.

29. Ittmann, Greco, and Basilico, 1987.

30. Baserga, 1985; Hochhauser, Stein, and Stein, 1981; Prescott, 1976.

31. Cochran, Reffel, and Stiles, 1983; Hirschorn, Aller, Yvan, Gibson, and Baserga, 1984; Lau and Nathans, 1985.

32. Greco, Ittman, and Basilico, 1987.

33. Pardee, 1974.

34. Rubin and Steiner, 1975; Holley and Kiernan, 1974a,b; Holley, Baldwin, and Kiernan, 1974; Bolen and Smith, 1977; Stiles, Capone, Scher, Antoniades, Van Wyk, and Pledger, 1979; Leof, Wharton, Van Wyk, and Pledger, 1982; Prescott, 1976.

35. Melchers and Lehrnhardt, 1985.

36. Cooper and Wuesthoff, 1971, Cooper, 1974.

37. Pardee, 1974.

38. Temin, 1971.

39. Pardee, 1974.

40. Pardee, 1974.

41. Pardee, 1974.

42. Pardee, 1974.

43. Chen and Wang 1984; Wynford-Thomas, LaMontagne, Marin, and Prescott, 1985; Allen and Moskowitz, 1978a,b.

44. Smith and Martin, 1973.

45. One problem with analyzing the transition-probability is that it has changed with time. In order to fit the original data better, the model was modified to have two different random transitions (Brooks, Bennett, and Smith, 1980). This modification makes it more difficult to give a precise critique of the transition-probability model.

46. Cooper, 1982b.

47. Shields, 1977, 1978.

48. All of the simulations and the details of the analysis are presented by Cooper, 1982b.

49. This argument was proposed by Arthur Koch.

50. Guiget and Cooper, 1981.

51. Guiget and Cooper, 1981.

52. Pledger, Stiles, Antoniades, and Scher, 1977; Stiles, Capone, Scher, Antoniades, Van Wyk, and Pledger, 1979.

53. Pardee, 1989.

54. Brooks, Fantes, Hunt, and Wheatley, 1989.

55. Murray and Kirschner, 1989.

56. Fantes, 1984a,b; Fantes and Nurse, 1981; Mitchison, 1964, 1989.

57. Mitchison, 1989. The following data are taken from this review by Mitchison.

58. Wain, 1971; Wain and Staatz, 1973.

59. Johnston, Eberly, Chapman, Araki, and Sugino, 1990; Chapman and Johnston, 1989; White, Green, Barker, Dumas, and Johnston, 1987.

60. Johnston, Eberly, Chapman, Araki, and Sugino, 1990.

61. Elliot and McLaughlin, 1978.

62. Tyson, Lord, and Wheals, 1979.

63. Howard and Pelc, 1951a,b.

16

Conservation Laws of the Division Cycle

I will now propose two conservation laws, analogous to the conservation laws of physics (conservation of energy and conservation of mass). These laws will offer criteria that can be used to test the validity of different proposals relating to the division cycle. Experiments have been published that conflict with these conservation laws. But just as in physics, where some results contradict the conservation laws—witness the fission and fusion weapons that do not follow the original conservation laws—experiments in biology may also violate or appear to violate proposed fundamental laws. The occasional announcement of a perpetual motion machine is no reason to jettison the law of conservation of energy. Unlike physics, however, biology does not welcome the promulgation of laws that supersede experimental results. When the conservation laws of physics were violated, the laws were modified to state that the total of energy and mass—because they were seen to be interconvertible—was constant. Thus, any experimental discrepancy with these proposed division-cycle conservation laws will have to be justified in terms that explain exactly why the laws appear to have been violated. I hope that it is a sign of the maturity of this field that such conservation laws can be proposed, and that differing claims can be tested by their consistency with these fundamental laws.

I. CONSERVATION OF CELL AGE ORDER

The first law is the law of conservation of age order.[1] The law can be stated:

> *For any genetically pure bacterial culture growing in balanced exponential growth, and ignoring statistical variation, there is no batch treatment of the culture that can lead to an alteration of the cell age order. A corollary of this law is that no batch treatment of an exponentially growing culture can produce a synchronized culture.*

An exponentially growing culture contains cells from age 0 to 1. Consider two cells, one at age x and the other at age x+δx. Because of the principle of continuous and exponential increase in cell mass, these two cells have different masses. They cannot be the same cell size (on average). As the culture grows, the older cell divides first, and then the younger cell divides. This occurs again and again in future generations, except for statistical variation. Variation in interdivision times may lead to some changes in the order, such that the younger cell (originally age x) divides twice before the older cell divides once. This occurs if some descendants of the younger cell have two successive short interdivision times, while the descendants of the older cell have a longer-than-average interdivision time.

The law of conservation of age order states that it is impossible to treat these two cells of different ages in the same way, i.e., with batch treatments, and produce a synchronized culture with both cells the same age, much less reverse the age order and make the younger cell older than the older cell. Let us consider the meaning of *batch* treatment. If we froze the older cell, and then waited until the younger cell was exactly the same age as the older cell before placing the two cells together, we would have two cells of the same age. This is because, in this case, there is no batch treatment; the cells were treated differently. A more rigorous example is seen when thymine starvation produces a pseudosynchronous culture (see detailed analysis in Chapter 5). Consider a culture in balanced growth that is starved of thymine so DNA synthesis stops in place. As the culture grows, new initiation points are inserted in each of the cells so that after one doubling time, all cells have the initiation potential to start a new round of replication. When thymine is added back to the culture, DNA synthesis resumes. At some later time, there will be a synchronous division as all the cells in the culture will divide at the same time, approximately $C+D$ minutes after the thymine is restored. But this is not a synchronized culture; the cells at the time of the synchronous division are not all physiologically alike. The cells retain a memory of their order, because the cells that were larger at the time that starvation started will still be larger. This retention of age order occurs because of exponential mass synthesis. If a cell starts with a large mass, at the end of a fixed period of thymine starvation, it will have more mass than in the original smaller cells. After the synchronous division, there will be no further synchronous divisions, as the cells resume an exponential mode of cell increase. This emphasizes the idea that there is a difference between a synchronized population of cells and a culture in which one particular aspect of the division cycle is synchronized. Only the synchronization of the entire cell and all of its components should be termed a synchronized culture.

II. CONSERVATION OF SIZE DISTRIBUTION

The law of conservation of size distribution may be considered a corollary of the law of conservation of cell age order, but it needs to be stated explicitly so that any experimental tests or criticisms of the laws may be accurately directed.

The law of conservation of size distribution states:

For any genetically pure bacterial culture growing in balanced exponential growth, there is no batch treatment that can alter the cell size distribution so that the width of the distribution decreases. A corollary of this law is that no batch treatment of an exponentially growing culture can produce a synchronized culture.

The law derives from two facts: first, the cytoplasm is synthesized in response to extant cytoplasm; and second, it is impossible to treat cells so that unequal-sized cells are affected differently. For example, if we could inhibit mass synthesis only in large, dividing cells, then we could imagine all the smaller cells proceeding to that point and ceasing to grow. We would now have a culture with a narrow range of sizes, all large and about to divide. Similarly, if there existed a unique event in the division cycle with regard to cytoplasm synthesis, a particular inhibitor might be developed to arrest cells at that cytoplasmic event. Adding such an inhibitor to a culture would allow all cells to grow to a particular cytoplasmic size and stop. All cells beyond that point would continue through the cycle and divide. The new daughter cells would grow and be arrested at the inhibition point. If such a specific cytoplasm inhibitor existed, then we could synchronize cells by batch treatment. The law of conservation of size distribution states that such a compound cannot exist because there are no such cytoplasmic events.

It could be argued that these laws derive from an inability to imagine a model that could allow a batch treatment to alter the age or size distribution. If any experimental model for batch synchronization can be proposed, it will lead to advances in our understanding of the division cycle.

NOTES

1. This conservation law was previous called the principle of age order invariance (Cooper, 1987).

Epilogue

Scientists believe there is a hierarchy of facts and that among them may be made a judicious choice. . . . This shows us how we should choose: the most interesting facts are those which may serve many times; those are the facts which have a chance of coming up again. We have been so fortunate as to be born in a world where there are such. Suppose that instead of 60 chemical elements there were 60 milliards of them, that they were not some common, others rare, but that they were uniformly distributed. Then, every time we picked up a new pebble there would be great probability of its being formed of some unknown substance; all that we knew of other pebbles would be worthless for it; before each new object we should be as the new-born babe; like it we could only obey our caprices or our needs. Biologists would be just as much at a loss if there were only individuals and no species and if heredity did not make sons like their fathers.

H. Poincaré, in *The Foundations of Science*, 1913

I. THE UNITY OF CELL BIOLOGY

Biologists, like lawyers, are fond of making distinctions. Different species and even different strains of the same species are reported to do different things. This book describes the fundamental cell cycle, and looks beyond the superficial differences between different strains to present a unified model of the division cycle. If this book has any single theme, it is that there is a unity throughout biology with regard to the division cycle. Although there are many differences between division cycles—witness the differences between *Caulobacter* and *Escherichia coli*, not to mention the larger differences between prokaryotes and eukaryotes—I believe that beneath the differences there is a fundamental unity. For many aspects of biology, cell biology, and biochemistry, there is a strong unity. We all believe, for example, that DNA is the genetic material for growing cells, and the structure of DNA is the same for all organisms. Although there may be differences in the sequence of DNA, and the precise mechanisms of DNA replication, and even in the packing of DNA, despite these differences there is a unity. Complementary strands allowing the genetic material to be replicated, and the sequence of DNA as a code for the synthesis of proteins are also unifying ideas amid surface differences. At some level we can see unity amid the differences. What I have proposed is that the logic of the division cycle is the same throughout biology, and upon this unity there are differences that tend to obscure it.

The unity of biochemistry was succinctly stated by Arthur Kornberg in his recent autobiography:[1]

> The unity of biochemistry in Nature, first seen clearly in these studies, applies not only to glucose metabolism, but also to the other basic features of cellular operations. From the genetic code to the encyclopedic details of the thousands of cellular proteins, these evolutionary triumphs achieved 2 to 3 billion years ago have withstood countless encounters with forces that could mutate or replace them. At these fundamental levels, there are few distinctions between a human, a mouse, and a yeast cell.

And I would add, between different bacteria. If there is any unity of biochemistry and biology, I would offer that there is a unity in the fundamental aspects of cell growth as well. This book has emphasized this unity in these fundamental aspects. These are DNA synthesis, surface synthesis, cytoplasm synthesis, and the general logic of the regulation of the division cycle.

It is not proposed that unity arises by evolution from a common ancestor. Many different modes of regulation of the division cycle may have been tried in the past. But it is proposed that there is something so structurally fundamental to the logic of the division cycle as described here that cells evolve to take advantage of this logic. It could have been that the replication of DNA was the driving force and it triggered, in some way, mass and surface synthesis; we could have a regulatory system based on DNA replication rather than mass expansion. When we understand the logic of the division cycle at its deepest level, we will be able to explain why the cell evolved the pattern it has, and why all cells have this particular pattern.[2]

II. BIOSYNTHESIS DURING THE DIVISION CYCLE

In Chapter 2, it was proposed that the components of the cell could be aggregated into three groups, the cytoplasm, the genome, and the cell surface. In the chapters that followed, the patterns of synthesis during the division cycle were described. Although there are many complicated aspects to the patterns, the basic ideas are simple. Cytoplasm is synthesized exponentially during the division cycle, DNA is synthesized as a series of linear rates of synthesis, and the surface of the cell increases to just enclose the cytoplasm and the genome. The formulas describing these biosynthetic patterns may be more or less complicated, but the ideas are simple, logical, and supported by the existing evidence.

III. THE LOGIC OF THE DIVISION CYCLE

The basic rules governing cell growth and division can be summarized in the logical construction of the division cycle. From the vast number of experiments, a simple picture emerges. The synthesis of the cytoplasm, or mass, or something that increases similar to cell mass, regulates the initiation of DNA synthesis. Once DNA synthesis is initiated, there is a period of DNA synthesis and a subsequent cell division. The driving force of mass increase is implicated in the regulation of cell surface increase in bacterial cells; it may also be involved in surface synthesis in eukaryotes.

IV. THE REMAINING PROBLEMS

This simple picture, which summarizes to a great extent the preceding chapters, does not mean that we have solved the problems of the division cycle. There are some points of uncertainty, and I would like to list these to highlight where I foresee the fruitful areas of investigation in the next decade.

The unusual and perhaps disappointing property of all of the unknowns about the bacterial division cycle is that these questions have been known almost since the beginning of the study of the division cycle. The years of analysis have highlighted the unknowns and have eliminated many possible mechanisms; the four central problems still remain.

A. The Initiation Problem

What regulates the initiation of DNA synthesis? In bacteria a great deal is known of the elements and reactions involved in the initiation of DNA replication. What we do not know is how the cell responds to signals that cause the initiation to occur at a particular time during the division cycle.

B. The Termination Problem

The discovery that there was an origin of replication, and replication proceeded bidirectionally from this origin, posed the termination problem. Aside from the question of the biochemistry of termination itself, the most important question is whether the act of termination of DNA replication triggers some event or sequence of events that leads to or affects cell division.

C. The Invagination Problem

What is the trigger for invagination of the bacterial cell? And once invagination is initiated, what force or mechanism causes the cell to invaginate and form a new set of poles? If we understood that process, then we would understand the continuation of pole formation as well.

D. The Division Problem

How does the division process end? At the last instant, when the new poles are formed, how do cells separate? What is the chemistry of that last attachment between cells, and is there something unique about pole formation as the cell divides?

V. CARTESIAN AND DEDUCTIVE SCIENCE

This volume takes a strongly deductive or Cartesian approach to the analysis of the cell cycle. Although there are many experiments, the basic understanding of the cell cycle has come from a deductive process rather than an inductive or experimental process. Since essentially every problem in this field has experiments proving at least two different models, the inductive approach fails. I do know that we need experiments to keep the thought processes moving ahead, but the thought processes are the final arbiter of the experimental results.

VI. THE BABY MACHINE

There is one additional theme that runs throughout this book. It is the importance, utility, and value of the membrane-elution method. This method, referred to affectionately as "the baby machine," is of central importance to our understanding of the bacterial division cycle. The baby machine has given important results for cytoplasm, DNA, and cell-surface synthesis; it has been used to measure the stability of cell wall and other parts of the cell; it has been used to measure the randomness or nonrandomness of segregation; and it can determine the pattern of biosynthesis during the division cycle of almost any cell component. It is surprising, in light of its importance, that the baby machine is used by so few laboratories in the world. I believe that the baby machine has not begun to live up to its full potential and will reveal much more about cell growth and division in the future. I hope that the use of the baby machine, as described in this book, convinces many researchers to investigate this method, and to use the baby machine to study the division cycle.

NOTES

1. Kornberg, A. 1989. For the Love of Enzymes. Harvard Univ. Press. Cambridge, Mass.

2. Cooper, 1991. This article contains the complete discussion of the principle of the Unity of Cell Biology.

Bibliography

Abbo, F. E., and Pardee, A. B. 1960. Synthesis of macromolecules in synchronously dividing bacteria. *Biochim. Biophys. Acta* **39**:478–485.

Adler, H. I., and Hardigree, A. A. 1964. Cell elongation in strains of *Escherichia coli*. *J. Bacteriol.* **87**:1240–1242.

Adolf, E. F., and Bayne-Jones, S. 1932. Growth in size of micro-organisms measured from motion pictures. II. *Bacillus megatherium*. *J. Cell. Comp. Physiol.* **1**:409–427.

Agabian, N., Evinger, M., and Parker, G. 1979. Generation of asymmetry during development: Segregation of type-specific proteins in *Caulobacter*. *J. Cell Biol.* **81**:123–136.

Aldea, M., Herrero, E., and Trueba, F. J. 1982. Constancy of diameter through the cell cycle of *Salmonella typhimurium* LT2. *Curr. Microbiol.* **7**:165–168.

Aldea, M., Herrero, E., Esteve, M. J., and Guerrero, R. 1980. Surface density of major outer membrane proteins in *Salmonella typhimurium* in different growth conditions. *J. Gen. Microbiol.* **120**:355–367.

Allen, R. W., and Moskowitz, M. 1978a. Arrest of cell growth in the G1 phase of the cell cycle by serine deprivation. *Exp. Cell Res.* **116**:127–137.

Allen, R. W., and Moskowitz, M. 1978b. Regulation of the rate of protein synthesis in BHK21 cells by exogenous serine. *Exp. Cell Res.* **116**:139–152.

Altenbern, R. A. 1968. Chromosome mapping in *Staphylococcus aureus*. *J. Bacteriol.* **95**:1642–1646.

Alterman, R., Ganguly, B. M., Schulze, S., Marzluff, W. F., Schildkraut, C. L., and Koultchi, A. I. 1984. Cell-cycle regulation of mouse H3 histone mRNA metabolism. *Mol. Cell. Biol.* **4**:123–132.

Anagnostopoulos, G. D. 1971. Unbalanced growth in semicontinuous culture system designed for the synchronization of cell division. *J. Gen. Microbiol.* **65**:23–26.

Anderson, A. J., Green, R. S., Sturman, A. J., and Archibald, A. R. 1978. Cell wall assembly in *Bacillus subtilis*: Location of wall material incorporated during pulsed release of phosphate limitation, its accessibility to bacteriophages and concanavalin A, and its susceptibility to turnover. *J. Bacteriol.* **136**:886–899.

Anderson, P. A., and Pettijohn, D. E. 1960. Synchronization of division in *Escherichia coli*. *Science* **131**:1098.

Andresdottir, V., and Masters, M. 1978. Evidence that F*lac* replicates asynchronously during the cell cycle of *Escherichia coli* B/r. *Mol. Gen. Genet.* **163**:205–212.

Archibald, A. R., and Coapes, H. E. 1976. Bacteriophage SP50 as a marker for cell wall growth in *Bacillus subtilis*. *J. Bacteriol.* **125**:1195–1206.

Atlung, T., Clausen, E. S., and Hansen, F. G. 1985. Autoregulation of the *dna*A gene of *Escherichia coli*. *Mol. Gen. Genet.* **200**:442–450.

Atlung, T., Rasmussen, K. V., Clausen, E. S., and Hansen, F. G. 1985. Role of the *dna*A protein in control of replication. *In* Molecular Biology of Bacterial Growth. M. Schaechter, F. C. Neidhardt, J. Ingraham, and N. O. Kjeldgaard, (eds.), pp. 282–297. Jones and Bartlett, Boston, Mass.

Augenlicht, L., and Baserga, R. 1974. Changes in the G0 state of WI-38 fibroblasts at different times after confluence. *Exp. Cell Res.* **89**:255–262.

Autissier, F., and Kepes, A. 1971. Segregation of membrane markers during cell division in *Escherichia coli*. II. Segregation of Lac-permease and Mel-permease studied with a penicillin technique. *Biochim. Biophys. Acta* **249**:611–615.

Autissier, F., and Kepes, A. 1972. Segregation de marqueurs membranaires au cours de la croissance et de la division d'*Escherichia coli*. *Biochimie* **54**:93–101.

Baldwin, W. W., and Kubitschek, H. E. 1984. Evidence of osmoregulation of cell growth and buoyant density in *Escherichia coli*. *J. Bacteriol.* **159**:393–394.

Baldwin, W. W., and Wegener, W. S. 1976. Selection of synchronous bacterial cultures by density sedimentation. *Can. J. Microbiol.* **22**:390–393.

Barber, M. A. 1908. The rate of multiplication of *Bacillus coli* at different temperatures. *J. Infect. Dis.* **5**:379–400.

Barner, H. D., and Cohen, S. S. 1956. Synchronization of division of a thymineless mutant of *Escherichia coli*. *J. Bacteriol.* **72**:115–123.

Baserga, R. 1985. Biology of Cell Reproduction. Harvard University Press, Cambridge, Massachusetts.

Bauza, M. T., DeLoach, J. R., Aguanno, J. J., and Larrabee, A. R. 1976. Acyl carrier protein prosthetic group exchange and phospholipid synthesis in synchronized cultures of a pantothenate auxotroph of *Escherichia coli*. *Arch. Biochem. Biophys.* **174**:344–349.

Bayne-Jones, S., and Adolf, E. F. 1933. Growth in size of microorganisms measured from motion pictures. III. *Bacterium coli*. *J. Cell. Comp. Physiol.* **2**:329–348.

Beacham, I. R., Beacham, K., Zaritsky, A., and Pritchard, R. H. 1971. Intracellular thymidine triphosphate concentrations in wild type and in thymidine requirement mutant of *Escherichia coli* 15 and K12. *J. Mol. Biol.* **62**:75–86.

Beachey, E. H., and Cole, R. M. 1966. Cell wall replication in *Escherichia coli*, studied by immunofluorescence and immunoelectron microscopy. *J. Bacteriol.* **92**:1245–1251.

Beck, B. D., and Park, J. T. 1976. Activity of three murein hydrolases during the cell division cycle of *Escherichia coli* K-12 as measured in toluene-treated cells. *J. Bacteriol.* **126**:1250–1260.

Becker, W. 1986. The World of the Cell. Benjamin-Cummings, Menlo Park, California.

Beckman, D., and Cooper, S. 1973. Temperature-sensitive nonsense mutations in essential genes of *Escherichia coli*. *J. Bacteriol.* **116**:1336–1342.

Beckwith, J. 1987. The operon: An historical account. *In Escherichia coli* and *Salmonella typhimurium:* Cellular and Molecular Biology. F. C. Neidhart, J. L. Ingraham, K. B. Low, B. Magasanik, M. Schaechter, and H. E. Umbarger, (eds.), pp. 1439–1443. American Society for Microbiology, Washington, D. C.

Begg, K. J. 1978. Cell surface growth in *Escherichia coli*: Distribution of matrix protein. *J. Bacteriol.* **135**:307–310.

Begg, K. J., and Donachie, W. D. 1973. Topography of outer membrane growth in *E. coli*. *Nature (London) New Biol.* **245**:38–39.

Begg, K. J., and Donachie, W. D. 1977. Growth of the *Escherichia coli* cell surface. *J. Bacteriol.* **129**:1524–1536.

Begg, K. J., and Donachie, W. D. 1978. Changes in cell size and shape in thymine-requiring *Escherichia coli* associated with growth in low concentrations of thymine. *J. Bacteriol.* **133**:452–458.

Bellino, F. L. 1973. Continuous synthesis of a partially derepressed aspartate transcarbamylase during the division cycle of *Escherichia coli* B/r. *J. Mol. Biol.* **74**:223–238.

Bezanson, G. 1980. Replication of the multicopy mini-R1 plasmid Rsc11 in *Escherichia coli* K12: Possible nonrandom selection of participants. *Plasmid* **3**:319–327.

Bird, R. E., and Lark, K. G. 1968. Initiation and termination of DNA replication after amino acid starvation of E. coli 15T⁻. Cold Spring Harbor Symp. Quant. Biol. 33:799–808.

Bird, R. E., and Lark, K. G. 1970. Chromosome replication in Escherichia coli 15T⁻ at different growth rates: Rate of replication of the chromosome and the rate of formation of small pieces. J. Mol. Biol. 49:343–366.

Bird, R. E., Louarn, J., Martuscelli, J., and Caro, L. 1972. Origin and sequence of chromosome replication in Escherichia coli. J. Mol. Biol. 70:549–566.

Bird, R., Chandler, M., and Caro, L. 1976. Suppression of an Escherichia coli dnaA mutation by the integrated R factor R.100.1: Change of chromosome replication origin in synchronized cultures. J. Bacteriol. 126:1215–1223.

Bleecken, S. 1969a. Das duplificationssystem der bakterienzelle. I. Beziehungen zwischen DNS-replikation und zellteilung bei balanciertem zellwachstum. Zeit. Alg. Mikrobiol. 9:415–435.

Bleecken, S. 1969b. Das duplificationssystem der bakterienzelle. II. Beziehungen zwischen DNS-replikation und zellteilung bei übergängen zwischen stationären wachstumzustanden. Zeit. Alg. Mikrobiol. 9:499–430.

Bleecken, S. 1969c. Das duplificationssystem der bakterienzelle. III. Topologische untersuchungen der DNS-replikation under der kernteilung: Gaesamtmodel der zeitlichen und raumlichen organisation des duplificationssystems. Zeit. Alg. Mikrobiol. 9:587–601.

Bleecken, S. 1969d. Duplication of the bacterial cell and its initiation. J. Theor. Biol. 25:137–158.

Bleecken, S. 1971. Replisome controlled initiation of DNA replication. J. Theor. Biol. 32:81–92.

Blumenthal, L. K., and Zahler, S. A. 1962. Index for measurement of synchronization of cell populations. Science 135:724.

Bolen, J. B., and Smith, G. L. 1977. Effects of withdrawal of a mitogenic stimulus on progression of fibroblasts into S phase: Differences between serum and purified multiplication stimulating activity. J. Cell. Physiol. 91:441–448.

Bonhoeffer, F., and Gierer, A. 1963. On the growth mechanism of the bacterial chromosome. J. Mol. Biol. 7:534–540.

Botsford, J. L. 1981. Cyclic nucleotides in prokaryotes. Microbiol. Rev. 45:620–642.

Bourbeau, P., Dicker, D., Higgins, M. L., and Daneo-Moore, L. 1989. Effect of cell-cycle stages on the central density of Enterococcus faecium ATCC 9790. J. Bacteriol. 171:1982–1986.

Boyd, A., and Holland, I. B. 1977. Protein d, an iron-transport protein induced by filtration of cultures of Escherichia coli. FEBS Lett. 76:20–24.

Boye, E., Steen, H. B., and Skarstad, K. 1988. Flow cytometry of bacteria: A promising tool in experimental and clinical microbiology. J. Gen. Microbiol. 129:973–980.

Boyle, J. V., Goss, W. A., and Cook, T. M. 1967. Induction of excessive deoxyribonucleic acid synthesis in Escherichia coli by nalidixic acid. J. Bacteriol. 94:1664–1671.

Bramhill, D., and Kornberg, A. 1988. A model for initiation at origins of DNA replication. Cell 54:915–918.

Braun, R. E., O'Day, K., and Wright, A. 1985. Autoregulation of the DNA replication gene dnaA in E. coli K-12. Cell 40:159–169.

Braun, V., Gnirke, H., Henning, U., and Rehn, K. 1973. Model for the structure of the shape-maintaining layer of the Escherichia coli cell envelope. J. Bacteriol. 114:1264–1270.

Bremer, H., and Chuang, L. 1981. The cell cycle in Escherichia coli B/r. J. Theor. Biol. 88:47–81.

Bremer, H., and Churchward, G. 1977a. Deoxyribonucleic acid synthesis after inhibition of initiation of rounds of replication in *Escherichia coli* B/r. *J. Bacteriol.* **130**:692–697.

Bremer, H., and Churchward, G. 1977b. An examination of the Cooper-Helmstetter theory of DNA replication and its underlying assumptions. *J. Theor. Biol.* **69**:645–654.

Bremer, H., and Churchward, G. 1978. Age fractionation of bacteria by membrane elution: Relation between age distribution and elution profile. *J. Theor. Biol.* **74**:69–81.

Brock, T. D. 1988. The bacterial nucleus: A history. *Microbiol. Rev.* **52**:397–341.

Brooks, R. F., Bennett, D. C., and Smith, J. A. 1980. Mammalian cell cycles need two random transitions. *Cell* **19**:493–504.

Brooks, R., Fantes, P., Hunt, T., and Wheatley, D. 1989. The cell cycle. *J. Cell Sci. Suppl.* **12:** i–vii.

Brostrom, M. A., and Binkley, S. B. 1969. Synchronous growth of *Escherichia coli* after treatment with fluorophenylalanine. *J. Bacteriol.* **98**:1271–1273.

Buchanan, R. E. 1918. Life phases in a bacterial culture. *J. Infect. Dis.* **23**:109–125.

Buchner, H., Longard, K., and Riedlin, G. 1887. Ueber die vermehrungsgeschwindigkeit der bacterien. *Zentr. Bact. Parasitenk.* **2**:1–7.

Burdett, I. D. J., and Kirkwood, T. B. L. 1983. How does a bacterium grow during its cell cycle? *J. Theor. Biol.* **103**:11–20.

Burdett, I. D. J., and Koch, A. L. 1984. Shape of nascent and complete poles of *Bacillus subtilis*. *J. Gen. Microbiol.* **130**:1711–1722.

Burdett, I. D. J., Kirkwood, T. B. L., and Whalley, J. B. 1986. Growth kinetics of individual *Bacillus subtilis* cells and correlation with nucleoid extension. *J. Bacteriol.* **167**:219–230.

Burman, L. G., and Park, J. T. 1983. Changes in the composition of *Escherichia coli* murein as it ages during exponential growth. *J. Bacteriol.* **155**:447–453.

Burman, L. G., and Park, J. T. 1984. Molecular model for elongation of the murein sacculus of *Escherichia coli*. *Proc. Natl. Acad. Sci. USA* **81**:1844–1848.

Burman, L. G., Raichler, J., and Park, J. T. 1983a. Evidence for diffuse growth of the cylindrical portion of the *Escherichia coli* murein sacculus. *J. Bacteriol.* **155**:983–988.

Burman, L. G., Raichler, J., and Park, J. T. 1983b. Evidence for multisite growth of *Escherichia coli* murein involving concomitant endopeptidase and transpeptidase activities. *J. Bacteriol.* **156**:386–192.

Burns, V. W. 1959. Synchronized cell division and DNA synthesis in a *Lactobacillus acidophilus* mutant. *Science* **129**:566–577.

Cairns, J. 1963a. The bacterial chromosome and its manner of replication as seen by autoradiography. *J. Mol. Biol.* **6**:208–213.

Cairns, J. 1963b. The chromosome of *Escherichia coli*. *Cold Spring Harbor Symp. Quant. Biol.* **28**:43–46.

Campbell, A. 1957. Synchronization of cell division. *Bact. Rev.* **21**:263–272.

Campbell, A. 1964. The theoretical basis of synchronization by shifts in environmental conditions. *In* Synchrony and Cell Division. E. Zeuthen (ed.), pp. 469–484. Wiley/Interscience, New York.

Canepari, P., Del Mar Lléo, M., Fontana, R., Shockman, G. D., and Daneo-Moore, L. 1983. Division blocks in temperature-sensitive mutants of *Streptococcus faecium* (*Streptococcus faecalis*) ATCC 9790. *J. Bacteriol.* **156**:1046–1051.

Canovas, J. L., Tresguerres, E. F., Yousif, A. E., Lopez-Saez, J. F., and Navarette, M. H. 1984. DNA segregation in *Escherichia coli* cells with 5-bromodeoxyuridine-substituted nucleoids. *J. Bacteriol.* **158**:128–133.

Carl, P. L. 1970. *Escherichia coli* mutants with temperature-sensitive synthesis of DNA. *Mol. Gen. Genet.* **109**:107–122.

Carty, C. E., and Ingram, L. O. 1981. Lipid synthesis during the *Escherichia coli* cell cycle. *J. Bacteriol.* **145**:472–478.

Cavalier-Smith, T. 1987. Bacterial DNA segregation: Its motors and positional control. *J. Theor. Biol.* **127**:361–372.

Cerda-Olmeda, E., Hanawalt, P. C., and Guerola, N. 1968. Mutagenesis of the replication point by nitrosoguanidine: Map and pattern of replication of *Escherichia coli* chromosome. *J. Mol. Biol.* **33**:705–719.

Chai, N.-C., and Lark, K. G. 1967. Segregation of deoxyribonucleic acid in bacteria: Association of the segregation unit with the cell envelope. *J. Bacteriol.* **94**:415–421.

Chai, N.-C., and Lark, K. G. 1970. Cytological studies of deoxyribonucleic acid replication in *Escherichia coli* 15T⁻: Replication at slow growth rates and after a shift-up into rich medium. *J. Bacteriol.* **104**:401–409.

Chakraborty, T., Yoshinaga, K., Lother, H., and Messer, W. 1982. Purification of the *E. coli* *dna*A gene product. *EMBO J.* **1**:1545–1549.

Chaloupka, J., and Strnadová, M. 1972. Turnover of murein in a diaminopimelic acid-dependent mutant of *Escherichia coli*. *Folia Microbiol.* **17**:446–455.

Chandler, M., Bird, R. E., and Caro, L. 1975. The replication time of the *Escherichia coli* K12 chromosome as a function of cell doubling time. *J. Mol. Biol.* **94**:127–132.

Chandler, M., Funderburgh, M., and Caro, L. 1975. Replication fork velocity in *E. coli* at various cell growth rates. *In* DNA Synthesis and Its Regulation. M. Goulian, P. Hanawalt, and C. F. Fox (eds.), pp. 695–700. W. A. Benjamin, Inc. Menlo Park, CA.

Chang, C.-F., Shuman, H., and Somlyo, A. P. 1986. Electron probe analysis, x-ray mapping, and electron energy loss spectroscopy of calcium, magnesium, and monovalent ions in log-phase and in dividing *Escherichia coli* B cells. *J. Bacteriol.* **167**:935–939.

Chapman, J. W., and Johnston, L. H. 1989. The yeast gene, DBF4, essential for entry into S phase is cell cycle regulated. *Exp. Cell Res.* **180**:419–428.

Chen, D. J.-C., and Wang, R. J. 1984. Cell division cycle in mammalian cells. VII. Mapping of G1 into six segments using temperature-sensitive cell cycle mutants. *Exp. Cell Res.* **155**:549–556.

Chesney, A. M. 1916. The latent period in the growth of bacteria. *J. Exp. Med.* **24**:387–418.

Cheung, H.-Y., Vitkovic, L, and Freese, E. 1983. Rates of peptidoglycan turnover and cell growth of *Bacillus subtilis* are correlated. *J. Bacteriol.* **156**:1099–1106.

Chung, K. L., Hawirko, R. Z., and Isaac, P. K. 1964. Cell wall replication. II. Cell wall growth and cross wall formation of *Escherichia coli* and *Streptococcus faecalis*. *Can. J. Microbiol.* **10**:473–482.

Churchward, G., and Bremer, H. 1977. Determination of deoxyribonucleic acid replication time in exponentially growing *Escherichia coli* B/r. *J. Bacteriol.* **130**:1206–1213.

Churchward, G., and Holland, I. B. 1976. Envelope synthesis during the cell cycle in *Escherichia coli* B/r. *J. Mol. Biol.* **105**:245–261.

Churchward, G., Bremer, H., and Young, R. 1982. Macromolecular composition of bacteria. *J. Theor. Biol.* **94**:651–670.

Churchward, G., Estiva, E., and Bremer, H. 1981. Growth rate-dependent control of chromosome replication initiation in *Escherichia coli*. *J. Bacteriol.* **145**:1232–1238.

Clark, D. J. 1968. The regulation of DNA replication and cell division in *Escherichia coli* B/r. *Cold Spring Harbor Symp. Quant. Biol.* **33**:823–838.

Clark, D. J., and Maaløe, O. 1967. DNA replication and the division cycle in *Escherichia coli*. *J. Mol. Biol.* **23**:99–112.

Clark, P. F., and Ruehl, W. H. 1919. Morphological changes during the growth of bacteria. *J. Bacteriol.* **4**:615–629.

Cochran, B. H., Reffel, A. C., and Stiles, C. D. 1983. Molecular cloning of gene sequences regulated by platelet-derived growth factor. *Cell* **33**:939–947.

Cole, R. M. 1964. Cell wall replication in *Salmonella typhosa*. *Science* **143**:820–822.

Cole, R. M. 1965. Symposium on the fine structure and replication of bacteria and their parts. III. Bacterial cell-wall replication followed by immunofluorescence. *Bact. Rev.* **29**:326–344.

Cole, R. M., and Hahn, J. J. 1962. Cell wall replication in *Streptococcus pyogenes*: Immunofluorescent methods applied during growth show that new wall is formed equatorially. *Science* **135**:722–724.

Collins, J., and Pritchard, R. H. 1973. Relationship between chromosome replication and F*lac* episome replication in *Escherichia coli*. *J. Mol. Biol.* **78**:143–155.

Collins, J., and Richmond, M. 1962. Rate of growth of *Bacillus cereus* between divisions. *J. Gen. Microbiol.* **28**:15–33.

Condemine, G, and Smith, C. L. 1990. Genetic mapping using large-DNA technology: Alignment of *Sfi*I and *Avr*II sites with the *Not*I genomic restriction map of *Escherichia coli* K12. *In* The Bacterial Chromosome. K. Drlica and M. Riley (eds.), pp. 53–60. American Society for Microbiology, Washington D.C.

Cook, W. R., Kalb, V. F., Jr., Peace, A. A., and Bernlohr, R. W. 1980. Is cyclic guanosine 3',5'-monophosphate a cell-cycle regulator? *J. Bacteriol.* **141**:1450–1453.

Cook, W. R., Kepes, F., Joseleau-Petit, D., MacAlister, T. J., and Rothfield, L. I. 1987. A proposed mechanism for the generation and localization of new division sites during the division cycle of *Escherichia coli*. *Proc. Natl. Acad. Sci. USA* **84**:7144–7148.

Cooper, S. 1969. Cell division and DNA replication following a shift to a richer medium. *J. Mol. Biol.* **43**:1–11.

Cooper, S. 1970. A model for the determination of growth rate. *J. Theor. Biol.* **28**:151–154.

Cooper, S. 1972. Relationship of F*lac* replication and chromosome replication. *Proc. Natl. Acad. Sci. USA* **69**:2706–2710.

Cooper, S. 1974. On a criterion for using chloramphenicol to define different processes in the initiation of DNA synthesis in bacteria. *J. Theor. Biol.* **46**:117–127.

Cooper, S. 1979. A unifying model for the G1 period of prokaryotes and eukaryotes. *Nature (London)* **280**:17–19.

Cooper, S. 1981. The central dogma of cell biology. *Cell Biol. Int. Rep.* **5**:539–551.

Cooper, S. 1982a. The continuum model: Implications for G1-arrest and G(0). *In* Cell Growth. C. Nicolini (ed.), pp. 315–336. Plenum, New York.

Cooper, S. 1982b. The continuum model: Statistical implications. *J. Theor. Biol.* **94**:783–800.

Cooper, S. 1984a. Application of the continuum model to the clock model of the cell cycle. *In* Cell Cycle Clocks. L. Edmunds (ed.), pp. 209–218. Marcel Dekker, New York.

Cooper, S. 1984b. The continuum model as a unified description of the division cycle of eukaryotes and prokaryotes. *In* The Microbial Cell Cycle. P. Nurse and E. Streiblova (eds.), pp. 7–18. CRC Press, Boca Raton, FL.

Cooper, S. 1987. On G(0) and cell cycle controls. *Bioessays* **7**:220–223.

Cooper, S. 1988a. Leucine uptake and protein synthesis are exponential during the division cycle of *Escherichia coli* B/r. *J. Bacteriol.* **170**:436–438.

Cooper, S. 1988b. Rate and topography of cell wall synthesis during the division cycle of *Salmonella typhimurium*. *J. Bacteriol.* **170**:422–430.

Cooper, S. 1988c. The continuum model and c-*myc* regulation. *J. Theor. Biol.* **135**:393–400.

Cooper, S. 1988d. What is the bacterial growth law during the division cycle? *J. Bacteriol.* **170**:5001–5005.

Cooper, S. 1989. The constrained hoop: An explanation of the overshoot in cell length during a shift-up of *Escherichia coli*. *J. Bacteriol.* **171**:5239–5243.

Cooper, S. 1990a. Comparison of backwards and forward methods methods of cell cycle analysis. *FEMS Microbiol. Lett.* **66**:1–4.

Cooper, S. 1990b. The *Escherichia coli* cell cycle. *Res. Microbiol.* **141**:17–29.

Cooper, S. 1991. Conjectures on the mathematics of the cell cycle. *In* Mathematical Population Dynamics. O. Arino, D. E. Axelrod, and M. Kimmel (eds.), pp. 539–546. Marcel Dekker, New York.

Cooper, S., and Helmstetter, C. E. 1968. Chromosome replication and the division cycle of *Escherichia coli* B/r. *J. Mol. Biol.* **31**:519–540.

Cooper, S., and Hsieh, M.-L. 1988. The rate and topography of cell wall synthesis during the division cycle of *Escherichia coli* using N-acetylglucosamine as a peptidoglycan label. *J. Gen. Microbiol.* **134**:1717–1721.

Cooper, S., and Metzger, N. 1986. Efficient and quantitative incorporation of diaminopimelic acid into the peptidoglycan of *Salmonella typhimurium*. *FEMS Microbiol. Lett.* **36**:191–194.

Cooper, S., and Ruettinger, T. 1973. Replication of deoxyribonucleic acid during the division cycle of *Salmonella typhimurium*. *J. Bacteriol.* **114**:966–973.

Cooper, S., and Ruettinger, T. 1975. Temperature-dependent alteration in bacterial protein composition. *Biochem. Biophys. Res. Commun.* **62**:584–586.

Cooper, S., and Weinberger, M. 1977. Medium-dependent variation of deoxyribonucleic acid segregation in *Escherichia coli*. *J. Bacteriol.* **130**:118–127.

Cooper, S., and Wuesthoff, G. 1971. Comment on the use of chloramphenicol to study the initiation of deoxyribonucleic acid synthesis. *J. Bacteriol.* **106**:709–711.

Cooper, S., Hsieh, M.-L., and Guenther, B. 1988. Mode of peptidoglycan synthesis in *Salmonella typhimurium*. *J. Bacteriol.* **170**:3509–3512.

Cooper, S., Schwimmer, M., and Scanlon, S. 1978. Probabilistic behavior of DNA segregation in *Escherichia coli*. *J. Bacteriol.* **134**:60–65.

Coplans, M. 1909–1910. Influences affecting the growth of micro-organisms—Latency: Inhibition: Mass action. *J. Path. Bact.* **14**:1–27.

Cullum, J., and Vicente, M. 1978. Cell growth and length distribution in *Escherichia coli*. *J. Bacteriol.* **134**:330–337.

Cummings, D. J. 1965. Macromolecular synthesis during synchronous growth of *Escherichia coli* B/r. *Biochim. Biophys. Acta* **95**:341–350.

Cummings, D. J. 1970. Synchronization of *E. coli* K12 by membrane (elution) selection. *Biochem. Biophys. Res. Commun.* **41**:471–476.

Cutler, R. G., and Evans, J. E. 1966. Synchronization of bacteria by a stationary-phase method. *J. Bacteriol.* **91**:469–476.

Cutler, R. G., and Evans, J. E. 1967. Relative transcription activity of different segments of the genome through the cell division cycle of *Escherichia coli*. The mapping of ribosomal and transfer RNA and the determination of the direction of replication. *J. Mol. Biol.* **26**:91–105.

Dadd, A. H., and Paulton, R. J. L. 1968. The cell wall of *Bacillus subtilis* during synchronous growth. *J. Gen. Microbiol.* **54**:iii.

Daniels, M. J. 1969a. Aspects of membrane synthesis during the cell cycle of *Bacillus megaterium* and *Escherichia coli*. *J. Gen. Microbiol.* **58**:iv.

Daniels, M. J. 1969b. Lipid synthesis in relation to the cell cycle of *Bacillus megatherium* KM and *Escherichia coli*. *Biochem. J.* **115**:697–701.

Davis, D. B., and Helmstetter, C. E. 1973. Control of F*lac* replication in *Escherichia coli*. *J. Bacteriol.* **114**:294–299.

Degnen, S. T., and Newton, A. 1972. Dependence of cell division on the completion of chromosome replication in *Caulobacter crescentus*. *J. Bacteriol.* **110**:852–856.

Dennis, P. P. 1971. Regulation of stable RNA synthesis in *Escherichia coli*. *Nature (London) New Biol.* **232**:43–48.

Dennis, P. P. 1972. Stable ribonucleic acid synthesis during the cell division cycle in slowly growing *Escherichia coli* B/r. *J. Biol. Chem.* **247**:204–208.

De Boer, W. R., Kruyssen, F. J., and Wouters, J. T. M. 1981. Cell wall turnover in batch and chemostat cultures of *Bacillus subtilis*. *J. Bacteriol.* **145**:50–60.

De Chastellier, C., Frehel, C., and Ryter, A. 1975. Cell wall growth of *Bacillus megaterium*: Cytoplasmic radioactivity after pulse-labeling with tritiated diaminopimelic acid. *J. Bacteriol.* **123**:1197–1207.

De Chastellier, C., Hellio, R., and Ryter, A. 1975. Study of cell wall growth in *Bacillus megaterium* by high-resolution autoradiography. *J. Bacteriol.* **123**:1184–1196.

De Jonge, B. L. M. 1989. Peptidoglycan synthesis during the cell cycle of *Escherichia coli*: Structure and mode of insertion. Thesis, Universiteit van Amsterdam, Netherlands.

De Jonge, B. L. M., Wientjes, F. B., Jurida, I., Driehuis, F., Wouters, J. T. M., and Nanninga, N. 1989. Peptidoglycan synthesis during the cell cycle of *Escherichia coli*: Composition and mode of insertion. *J. Bacteriol.* **156**:136–140.

De Massey, B., Bejar, S., Louarn, J., Louarn, J.-M., and Bouche, J.-P. 1987. Inhibition of replication forks exiting the terminus region of the *Escherichia coli* chromosome occurs at two loci separated by 5 min. *Proc. Natl. Acad. Sci. USA* **84**:1759–1763.

De Pedro, M. A., and Schwarz, U. 1981. Heterogeneity of newly inserted and preexisting murein in the sacculus of *Escherichia coli*. *Proc. Natl. Acad. Sci. USA* **78**:5856–5860.

Dicker, D. T., and Higgins, M. L. 1987. Cell cycle changes in the buoyant density of exponential-phase cells of *Streptococcus faecium*. *J. Bacteriol.* **169**:1200–1204.

Dietzel, I., Kolb, V., and Boos, W. 1978. Pole cap formation in *Escherichia coli* following induction of the maltose binding protein. *Arch. Microbiol.* **118**:207–218.

Dingwall, A., and Shapiro, L. 1989. Rate, origin, and bidirectionality of *Caulobacter* chromosome replication as determined by pulsed-field gel electrophoresis. *Proc. Natl. Acad. Sci. USA* **86**:119–123.

Donachie, W. D. 1965. Control of enzyme steps during the bacterial cell cycle. *Nature (London)* **205**:1084–1086.

Donachie, W. D. 1968. Relationship between cell size and time of initiation of DNA replication. *Nature (London)* **219**:1077–1079.

Donachie, W. D. 1981. The cell cycle of *Escherichia coli*. *In* The cell cycle, P. C. L. John (ed.), pp. 63–83. Cambridge University Press, Cambridge.

Donachie, W. D., and Begg, K. 1990. Genes and the replication cycle of *Escherichia coli*. *Res. Microbiol.* **141**:64–75.

Donachie, W. D., and Begg, K. J. 1970. Growth of the bacterial cell. *Nature (London)* **227**:1220–1224.

Donachie, W. D., and Masters, M. 1966. Evidence for polarity of chromosome replication in F⁻strains of *Escherichia coli*. *Genet. Res.* **8**:119–124.

Donachie, W. D., and Masters, M. 1969. Temporal control of gene expression in bacteria. *In* The Cell Cycle. Gene-Enzyme Interactions. G. M. Padilla, G. L. Whitson, and I. L. Cameron (eds.), pp. 37–76. Academic Press, New York.

Donachie, W. D., and Robinson, A. C. 1987. Cell division: Parameter values and the process, *In* Escherichia coli and Salmonella typhimurium: Cellular and Molecular Biology. F. C. Neidhardt, J. L. Ingraham, K. B. Low, B. Magasanik, M. Schaechter, and H. E. Umbarger (eds.), pp. 1578–1593. American Society for Microbiology, Washington, D.C.

Donachie, W. D., Begg, K. J., and Sullivan, N. F. 1984. Morphogenes of *Escherichia coli*. *In* Microbial Development, R. Losick, and L. Shapiro (eds.), pp. 27–62. Cold Spring Harbor Monograph Series, Cold Spring Harbor, N. Y.

Donachie, W. D., Begg, K. J., and Vicente, M. 1976. Cell length, cell growth, and cell division. *Nature (London)* **264**:328–333.

Doudney, C. 1960. Inhibition of nucleic acid synthesis by chloramphenicol in synchronized cultures of *Escherichia coli*. *J. Bacteriol.* **79**:122–124.

Doudney, C. O. 1965. Ultraviolet light effects on deoxyribonucleic acid replication. *In*

Cellular Radiation Biology, M. D. Anderson Hospital Symposium, pp. 120–138. Williams and Wilkins, Baltimore.

Doyle, R. J., Chaloupka, J., and Vinter, V. 1988. Turnover of cell walls in microorganisms. *Microbiol. Rev.* **52**:554–597.

Driehuis, F., and J. T. M. Wouters. 1985. Effect of diaminopimelic acid-limited growth on the peptidoglycan composition of *Escherichia coli* W7. *Antonie Van Leeuwenhoek* **51**:556.

Driehuis, F., and J. T. M. Wouters. 1987. Effect of growth rate and cell shape on the peptidoglycan composition of *Escherichia coli*. *J. Bacteriol.* **169**:97–101.

Dwek, R. D., Kobrin, L. H., Grossman, N., and Ron, E. Z. 1980. Synchronization of cell division in microorganisms by percoll gradients. *J. Bacteriol.* **144**:17–21.

D'Ari, R., and Bouloc, P. 1990. Logic of the *Escherichia coli* cell cycle. *Trends Biochem. Sci.* **15**:191–194.

D'Ari, R., Maguin, E., Bouloc, P., Jaffe, A., Robin, A., Liebart, J.-C., and Joseleau-Petit, D. 1990. Aspects of cell cycle regulation. *Res. Microbiol.* **141**:9–16.

Eberle, H., and Lark, K. G. 1966. Chromosome segregation in *Bacillus subtilis*. *J. Mol. Biol.* **22**:183–186.

Ecker, R. E., and Kokaisl, G. 1969. Synthesis of protein, ribonucleic acid, and ribosomes by individual bacterial cells in balanced growth. *J. Bacteriol.* **98**:1219–1226.

Edelstein, E. M., Rosenzweig, M. S., Daneo-Moore, L., and Higgins, M. L. 1980. Unit cell hypothesis for *Streptococcus faecalis*. *J. Bacteriol.* **143**:499–505.

Edwards, C. 1981. The microbial cell cycle. American Society for Microbiology, Washington, D.C.

Elliott, S. G., and McLaughlin, C. S. 1978. Rate of macromolecular synthesis through the cell cycle of the yeast *Saccharomyces cerevisiae*. *Proc. Natl. Acad. Sci. USA* **75**:4384–4388.

Elliott, S. G., Warner, J. R., and McLaughlin, C. S. 1979. Synthesis of ribosomal proteins during the cell cycle of the yeast *Saccharomyces cerevisiae*. *J. Bacteriol.* **137**:1048–1050.

Ely, B., and Shapiro, L. 1984. Regulation of cell differentiation in *Caulobacter crescentus*. *In* Microbial Development. R. Losick and L. Shapiro (eds.), pp. 1–26. Cold Spring Harbor Laboratory, New York.

Ely, B., and Shapiro, L. 1989. The molecular genetics of differentiation. *Genetics* **123**:427–429.

Ephrati-Elizur, E., and Borenstein, S. 1971. Velocity of chromosome replication in thymine-requiring and -independent strains of *Bacillus subtilis*. *J. Bacteriol.* **106**:58–64.

Evans, J. E. 1974. Techniques and their applications for synchronization of populations of microorganisms. *In* Handbook of Microbiology, vol. 4. A. I. Laskin and H. A. Lechevalier (eds.), pp. 813–853. CRC Press, Boca Raton, Florida.

Falcone, G., and Szybalski, W. 1956. Biochemical studies on the induction of synchronized cell division. *Exp. Cell Res.* **11**:486–489.

Fantes, P. 1986. Growth factors, G0 and cell cycle controls. *Bioessays* **4**:32–33.

Fantes, P. 1987. (untitled) *Bioessays* **7**:222–223.

Fantes, P. A. 1984a. Cell cycle control in *Schizosaccharomyces pombe*. *In* The Microbial Cell Cycle. P. Nurse and E. Streiblova, E. (eds.), pp. 109–125. CRC Press, Boca Raton, Florida.

Fantes, P. A. 1984b. Temporal control of the *Schizosaccharomyces pombe* cell cycle. *In* Cell Cycle Clocks. L. N. Edmunds (ed.), pp. 233–252, Marcel Dekker, New York and Basel.

Fantes, P. A., and Nurse, P. 1981. Division timing: controls, models and mechanisms. *In* The Cell Cycle. P. C. L. John (ed.), pp. 11–33. Cambridge University Press, Cambridge.

Fan, D. P., Beckman, B. E., and Gardner-Eckstrom, H. L. 1975. Mode of cell wall synthesis in gram-positive bacilli. *J. Bacteriol.* **123**:1157–1162.

Fielding, P., and Fox, C. F. 1970. Evidence for stable attachment of DNA to membrane at the replication origin of *Escherichia coli. Biochem. Biophys. Res. Commun.* **415**:157–162.

Figdor, C. G., Olijhoek, A. J. M., Klencke, S., Nanninga, N., and Bont, W. S. 1981. Isolation of small cells from an exponential growing culture of *Escherichia coli* by centrifugal elutriation. *FEMS Microbiol. Lett.* **10**:349–352.

Finkelstein, M., and Helmstetter, C. E. 1977. Cell cycle analysis of F*lac* replication in *Escherichia coli* B/r. *J. Bacteriol.* **132**:884–895.

Forro, F., Jr. 1965. Autoradiographic studies of bacterial chromosome replication in amino-acid deficient *Escherichia coli* 15T⁻. *Biophys. J.* **5**:629–649.

Forro, F., Jr., and Wertheimer, S. A. 1960. The organization and replication of deoxyribonucleic acid in thymine-deficient strains of *Escherichia coli. Biochim. Biophys. Acta* **40**:9–21.

Frehel, C., and Ryter, A. 1979. Peptidoglycan turnover during growth of a *Bacillus megaterium* Dap⁻ Lys⁻ mutant. *J. Bacteriol.* **137**:947–955.

Frey, J., Chandler, M., and Caro, L. 1981. The initiation of chromosome replication in a *dna*Ats46 and a *dna*A+ strain at various temperatures. *Mol. Gen. Genet.* **182**:364–366.

Frey, J., Chandler, M., and Caro, L. 1984. Overinitiation of chromosome and plasmid replication in a *dna*Acis mutant of *Escherichia coli* K12: Evidence for *dna*A-*dna*B interactions. *J. Mol. Biol.* **179**:171–183.

Fujisawa, T., and Eisenstark, A. 1973. Bi-directional chromosomal replication in *Salmonella typhimurium. J. Bacteriol.* **115**:168–176.

Fuller, R. S., and Kornberg, A. 1983. Purified *dna*A protein in initiation of replication at the *Escherichia coli* chromosomal origin of replication. *Proc. Natl. Acad. Sci. USA* **80**:5817–5821.

Fuller, R. S., Funnell, B. E., and Kornberg, A. 1984. The *dna*A complex with the *E. coli* chromosomal replication origin (*ori*C) and other DNA sites. *Cell* **38**:889–900.

Ganesan, A. T., and Lederberg, J. 1965. A cell-membrane bound fraction of bacterial DNA. *Biochem. Biophys. Res. Commun.* **18**:824–834.

Gausing, K. 1972. Efficiency of protein and messenger RNA synthesis in bacteriophage T4 infected cells of *Escherichia coli. J. Mol. Biol.* **71**:529–545.

Gibson, C. W., Daneo-Moore, L., and Higgins, M. L. 1983b. Cell wall assembly during inhibition of DNA synthesis in *Streptococcus faecium. J. Bacteriol.* **155**:351–356.

Gibson, C. W., Daneo-Moore, L., and Higgins, M. L. 1983. Initiation of wall assembly sites in *Streptococcus faecium. J. Bacteriol.* **154**:573–579.

Gibson, C. W., Daneo-Moore, L., and Higgins, M. L. 1984. Analysis of initiation of sites of cell wall growth in *Streptococcus faecium* during a nutritional shift. *J. Bacteriol.* **160**:935–942.

Glaser, D., and Higgins, M. 1989. Buoyant density, growth rate, and the cell cycle in *Streptococcus faecium. J. Bacteriol.* **171**:669–673.

Glaser, L., and Lindsay, B. 1977. Relation between cell wall turnover and cell growth in *Bacillus subtilis. J. Bacteriol.* **130**:610–619.

Glauner, B. 1986. Das murein von *E. coli.* Thesis, Eberhard-Karls-Universitat Tubingen, F.R.D.

Glauner, B. 1988. Separation and quantification of muropeptides with high-performance liquid chromatography. *Anal. Biochem.* **172**:451–464.

Glauner, B., and Schwarz, U. 1988. Peptidoglycan. Investigation of murein structure and metabolism by high–pressure liquid chromatography. *In* Bacterial Cell Surface Techniques. I. C. Hancock and I. R. Poxton (eds.), pp. 158–174. John Wiley and Sons, New York.

Glauner, B., Höltje, J.-V., and Schwarz, U. 1988. The composition of the murein of *Escherichia coli. J. Biol. Chem.* **263**:10088–10095.

Gmeiner, J., Sarnow, E., and Milde, K. 1985. Cell cycle parameters of *Proteus mirabilis*: Interdependence of the biosynthetic cell cycle and the interdivision cycle. *J. Bacteriol.* **164**:741–748.

Gomes, L., and Shapiro, L. 1984. Differential expression and positioning of chemotaxis proteins in *Caulobacter*. *J. Mol. Biol.* **178**:551–568.

Goodell, E. W. 1983. Peptide crossbridges in the murein of *Escherichia coli* are broken and reformed as the bacterium grows. *In* The Target of Penicillin. R. Hakenbeck, J.-V. Höltje, and H. Labischinski (eds.), pp. 129–134. Walter de Gruyter, Berlin-New York.

Goodell, E. W. 1985. Recycling of murein by *Escherichia coli*. *J. Bacteriol.* **163**:305–310.

Goodell, E. W., and Asmus, A. 1988. Recovery of gram-negative rods from mecillinam: Role of murein hydrolases in morphogenesis. *In* Antibiotic Inhibition of Bacterial Cell Surface Assembly and Function. P. Actor, L. Daneo-Moore, M. L. Higgins, M. R. J. Salton, and G. D. Shockman (eds.), pp. 224–230. American Society for Microbiology, Washington, D. C.

Goodell, E. W., and Schwarz, U. 1983. Cleavage and resynthesis of peptide crossbridges in *Escherichia coli* murein. *J. Bacteriol.* **156**:136–140.

Goodell, E. W., and Schwarz, U. 1985. Release of cell wall peptides into culture medium by exponentially growing *Escherichia coli*. *J. Bacteriol.* **162**:391–397.

Goodell, E. W., Schwarz, U., and Teather, R. M. 1974. Cell envelope composition of *Escherichia coli* K-12: A comparison of cell poles and the lateral cell wall. *Eur. J. Biochem.* **47**:567–572.

Goodwin, B. C. 1966. An entrainment model for timed enzyme synthesis in bacteria. *Nature (London)* **209**:479–481.

Goodwin, B. C. 1969. Synchronization of *Escherichia coli* in a chemostat by periodic phosphate feeding. *Eur. J. Biochem.* **10**:511–514.

Greco, A., Ittmann, M., and Basilico, C. 1987. Molecular cloning of a gene that is necessary for G1 progression in mammalian cells. *Proc. Natl. Acad. Sci. USA* **84**:1565–1569.

Green, E. W., and Schaechter, M. 1972. The mode of segregation of the bacterial cell membrane. *Proc. Natl. Acad. Sci. USA* **69**:2312–2316.

Grossman, N., Ron, E. Z., and Woldringh, C. L. 1982. Changes in cell dimensions during amino acid starvation of *Escherichia coli*. *J. Bacteriol.* **152**:35–41.

Grover, N. B. 1988. Surface-limited growth: A model for the synchronization of a growing bacterial culture through periodic starvation. *J. Theor. Biol.* **134**:77–87.

Grover, N. B., Naaman, J., Ben-Sasson, S., and Doljanski, F. 1969. Electrical sizing of particles in suspension. I. Theory. *Biophys. J.* **9**:1398–1414.

Grover, N. B., Woldringh, C. L., and Koppes, L. J. 1987. Elongation and surface extension of individual cells of *Escherichia coli* B/r: Comparison of theoretical and experimental size distributions. *J. Theor. Biol.* **129**:337–348.

Grover, N. B., Woldringh, C. L., Zaritsky, A., and Rosenberger, R. F. 1977. Elongation of rod-shaped bacteria. *J. Theor. Biol.* **67**:181–193.

Grover, N. B., Zaritsky, A., Woldringh, C. L., and Rosenberger, R. F. 1980. Dimensional rearrangement of rod-shaped bacteria following nutritional shift-up. I. Theory. *J. Theor. Biol.* **86**:421–439.

Grummt, F. 1983. Diadenosine tetraphosphate (Ap$_4$A): A putative chemical messenger of cell proliferation control and inducer of DNA replication. *Plant Mol. Biol.* **2**:41–44.

Gudas, L. J., and Pardee, A. B. 1974. Deoxyribonucleic acid synthesis during the division cycle of *Escherichia coli*: A comparison of strains B/r, K-12, 15, and 15T⁻ under conditions of slow growth. *J. Bacteriol.* **117**:1216–1223.

Guiguet, M., and Cooper, S. 1981. A critique of the use of DNA synthesis as a measure of the effect of mitogens on lymphocytes. *Biosci. Rep.* **2**:91–98.

Gustafsson, P., and Nordstrom, K. 1975. Random replication of the stringent plasmid R1 in *Escherichia coli* K-12. *J. Bacteriol.* **123**:443–448.

Gustafsson, P., Nordstrom, K., and Perram, J. W. 1978. Selection and timing of replication of plasmids R1*drd-19* and F*lac* in *Escherichia coli*. *Plasmid* **1**:187–203.

Hakenbeck, R., and Messer, W. 1977a. Activity of murein hydrolases in synchronized cultures of *Escherichia coli*. *J. Bacteriol.* **129**:1239–1244.

Hakenbeck, R., and Messer, W. 1977b. Oscillations in the synthesis of cell wall components in synchronized cultures of *Escherichia coli*. *J. Bacteriol.* **129**:1234–1238.

Halvorson, H. O., Carter, B. L. A., and Tauro, P. 1971. Synthesis of enzymes during the cell cycle. *Adv. Microb. Physiol.* **6**:47–106.

Hanawalt, P. C., Maaløe, O., Cummings, D. J., and Schaechter, M. 1961. The normal DNA replication cycle. II. *J. Mol. Biol.* **3**:156–165.

Hanks, M. C., and Masters, M. 1987. Transductional analysis of chromosome replication time. *Mol. Gen. Genet.* **210**:288–293.

Hann, S. R., Thompson, C. B., and Eisenman, R. N. 1985. C-*myc* oncogene protein synthesis is independent of the cell cycle in human and avian cells. *Nature* (London) **314**:366–369

Hansen, F. G., and Rasmussen, K. V. 1977. Regulation of the *dna*A product in *Escherichia coli*. *Mol. Gen. Genet.* **155**:219–225.

Hartwell, L. H., and Unger, M. 1977. Unequal division in *Saccharomyces cerevisiae* and its implications for the control of cell division. *J. Cell Biol.* **75**:422–435.

Harvey, J. D. 1972a. Parameters of the generation time distribution of *Escherichia coli* B/r. *J. Gen. Microbiol.* **70**:109–114.

Harvey, J. D. 1972b. Synchronous growth of cells and the generation time distribution. *J. Gen. Microbiol.* **70**:99–107.

Harvey, J. D. 1983. Mathematics of microbial age and size distributions. *In* Mathematics in Microbiology. M. Bazin (ed.), pp. 1–35. Academic Press, New York.

Harvey, R. J., and Marr, A. G. 1966. Measurement of size distribution of bacterial cells. *J. Bacteriol.* **92**:805–811.

Harvey, R. J., Marr, A. G., and Painter, P. R. 1967. Kinetics of growth of individual cells of *Escherichia coli* B/r and *Azotobacter agilis*. *J. Bacteriol.* **93**:605–617.

Hayes, W. 1957. The kinetics of the mating process in *E. coli*. *J. Gen. Microbiol.* **16**:97–119.

Hegarty, C. P. 1939. Physiological youth as an important factor in adaptive enzyme formation. *J. Bacteriol.* **37**:145–152.

Hegarty, C. P., and Weeks, O. B. 1940. Sensitivity of *Escherichia coli* to cold shock during logarithmic growth phase. *J. Bacteriol.* **39**:475–484.

Helmstetter, C. E. 1967. Rate of DNA synthesis during the division cycle of *E. coli* B/r. *J. Mol. Biol.* **24**:417–427.

Helmstetter, C. E. 1968. Origin and sequence of chromosome replication in *Escherichia coli* B/r. *J. Bacteriol.* **95**:1634–1641.

Helmstetter, C. E. 1969. Methods for studying the microbial division cycle. *In* Methods in Microbiology, vol. I, J. R. Norris and D. W. Ribbons (eds.), pp. 327–363. Academic Press, New York and London.

Helmstetter, C. E., and Cooper, S. 1968. DNA synthesis during the division cycle of rapidly growing *E. coli* B/r. *J. Mol. Biol.* **31**:507–518.

Helmstetter, C. E., and Cummings, D. J. 1963. Bacterial synchronization by selection of cells at division. *Proc. Natl. Acad. Sci. USA* **50**:767–774.

Helmstetter, C. E., and Cummings, D. J. 1964. An improved method for the selection of bacterial cells at division. *Biochim. Biophys. Acta* **82**:608–610.

Helmstetter, C. E., and Leonard, A. C. 1987a. Coordinate initiation of chromosome and minichromosome replication in *Escherichia coli*. *J. Bacteriol.* **169**:3489–3494.

Helmstetter, C. E., and Leonard, A. C. 1987b. Mechanism for chromosome and minichromosome segregation in *Escherichia coli*. *J. Mol. Biol.* **197**:295–204.

Helmstetter, C. E., and Leonard, A. C. 1990. Involvement of cell shape in the replication and segregation of chromosomes in *Escherichia coli*. *Res. Microbiol.* **141**:30–39.

Helmstetter, C. E., and Pierucci, O. 1968. Cell division during inhibition of deoxyribonucleic acid synthesis in *Escherichia coli*. *J. Bacteriol.* **95**:1627–1633.

Helmstetter, C. E., and Pierucci, O. 1976. DNA synthesis during the division cycle of three substrains of *Escherichia coli* B/r. *J. Mol. Biol.* **102**:477–486.

Helmstetter, C. E., and Uretz, R. B. 1963. X-ray and ultraviolet sensitivity of synchronously dividing *Escherichia coli*. *Biophys. J.* **3**:35–47.

Helmstetter, C. E., Cooper, S., Pierucci, O., and Revelas, L. 1968. The bacterial life sequence. *Cold Spring Harbor Symp. Quant. Biol.* **33**:809–822.

Helmstetter, C. E., Pierucci, O., Weinberger, M., Holmes, M., and Tang, M. S. 1979. Control of cell division in *Escherichia coli*, In The Bacteria, vol. VII. J. R. Sokatch, and L. N. Ornston (eds.), pp. 517–579. Academic Press, New York.

Hendrickson, W. G., Kusano, T., Yamaki, H., Balakrishnan, R., King, M., Murchie, J., and Schaechter, M. 1982. Binding of the origin of replication of *Escherichia coli* to the outer membrane. *Cell* **30**:915–923.

Henning, U. 1975. Determination of cell shape in bacteria. *Annu. Rev. Microbiol.* **29**:45–60.

Henning, U., and Schwarz, U. 1973. Determinants of cell shape. *In* Bacterial Membranes and Walls. L. Leive, (ed.), pp. 413–438. Marcel Dekker, New York.

Henrici, A. T. 1923-24a. Influence of age of parent culture on size of cells of *Bacillus megatherium*. *Proc. Soc. Exp. Biol. Med.* **21**:343–345.

Henrici, A. T. 1923-24b. Influence of concentration of nutrients on size of cells of *Bacillus megatherium*. *Proc. Soc. Exp. Biol. Med.* **21**:345–346.

Henrici, A. T. 1925. On cytomorphosis in bacteria. *Science* **61**:644–647.

Henrici, A. T. 1928. Morphologic Variation and Rate of Growth of Bacteria. Microbiology Monographs. Bailliere, Tindall, and Cox, London.

Hershey, A. D. 1938. Factors limiting bacterial growth. II. Growth without lag in *Bacterium coli* cultures. *Proc. Soc. Exp. Biol. Med.* **38**:127–128.

Hershey, A. D. 1939. Factors limiting bacterial growth. IV. The age of the parent culture and the rate of growth of transplants of *Escherichia coli*. *J. Bacteriol.* **37**:285–299.

Hershey, A. D. 1940. Factors limiting bacterial growth. VI. Equations describing the early periods of increase. *J. Gen. Physiol.* **23**:11–19.

Hershey, A. D., and Bronfenbrenner, J. 1937. On factors limiting bacterial growth. I. *Proc. Soc. Exp. Biol. Med.* **36**:556–561.

Hershey, A. D., and Bronfenbrenner, J. 1938. Factors limiting bacterial growth. III. Cell size and "physiologic youth" in *Bacterium coli* cultures. *J. Gen. Physiol.* **21**:721–728.

Hewitt, R., and Billen, D. 1965. Reorientation of chromosome replication after exposure to ultraviolet light of *Escherichia coli*. *J. Mol. Biol.* **13**:40–53.

Higgins, M. L. 1976. Three-dimensional reconstruction of whole cells of *Streptococcus faecalis* from thin sections of cells. *J. Bacteriol.* **127**:1337–1345.

Higgins, M. L., and Shockman, G. D. 1970. Model for cell wall growth of *Streptococcus faecalis*. *J. Bacteriol.* **101**:643–648.

Higgins, M. L., and Shockman, G. D. 1976. Study of a cycle of cell wall assembly in *Streptococcus faecalis* by three-dimensional reconstructions of thin sections of cells. *J. Bacteriol.* **127**:1346–1358.

Higgins, M. L., Daneo-Moore, L., Boothby, D., and Shockman, G. D. 1974. Effect of inhibition of deoxyribonucleic acid and protein synthesis on the direction of cell wall growth in *Streptococcus faecalis*. *J. Bacteriol.* **118**:681–692.

Higgins, M. L., Glaser, D., Dicker, D., and Zito, E. 1989. Chromosome and cell wall segregation in *Streptococcus faecium* ATCC 9790. *J. Bacteriol.* **171**:349–352.

Higgins, M. L., Koch, A. L., Dicker, D. T., and Daneo-Moore, L. 1986. Autoradiographic studies of chromosome replication during the cell cycle of *Streptococcus faecium*. *J. Bacteriol.* **168**:541–547.

Hill, T. M., Henson, J. M., and Kuempel, P. L. 1987. The terminus region of the *Escherichia coli* chromosome contains two separate loci that exhibit polar inhibition of replication. *Proc. Natl. Acad. Sci. USA* **84**:1754–1758.

Hinks, R. P., Daneo-Moore, L., and Shockman, G. D. 1978a. Relationship between cellular autolytic activity, peptidoglycan synthesis, separation, and the cell cycle in synchronized populations of *Streptococcus faecium*. *J. Bacteriol.* **134**:1074–1080.

Hinks, R. P., Daneo-Moore, L., and Shockman, G. D. 1978b. Approximation of the cell cycle in synchronized populations of *Streptococcus faecium*. *J. Bacteriol.* **134**:1188–1191.

Hirota, Y., Mordoh, J., and Jacob, F. 1970. On the process of cellular division in *Escherichia coli*. III. Thermosensitive mutants of *Escherichia coli* altered in the process of DNA initiation. *J. Mol. Biol.* **53**:369–387.

Hirota, Y., Ryter, A., and Jacob F. 1968. Thermosensitive mutants of *E. coli* affected in the process of DNA synthesis and cellular division. *Cold Spring Harbor Symp. Quant. Biol.* **33**:677–693.

Hirota, Y., Yasuda, S., Yamada, M., Nishimura, A., Sugimoto, K., Sugisaki, H., Oka, A., and Takanami, M. 1979. Structural and functional properties of the *Escherichia coli* origin of DNA replication. *Cold Spring Harbor Symp. Quant. Biol.* **43**:129–138.

Hirschhorn, R. R., Aller, P., Yvan, Z., Gibson, C. W., and Baserga, R. 1984. Cell-cycle–specific cDNAs from mammalian cells temperature sensitive for growth. *Proc. Natl. Acad. Sci. USA* **81**:6004–6008.

Hitchins, A. D. 1975. Polarized relationship of bacteria spore loci to the "old" and "new" ends of sporangia. *J. Bacteriol.* **121**:518–523.

Hitchins, A. D. 1976. Patterns of spore locations in pairs of *Bacillus cereus* sporangia. *J. Bacteriol.* **125**:366–368.

Hitchins, A. D. 1978. Polarity and topology of DNA segregation and septation in cells and sporangia of the bacilli. *Adv. Microb. Physiol.* **18**:105–176.

Hitchins, A. D. 1980. Polarity and topology of DNA segregation and septation in cells and sporangia of the bacilli. *Can. J. Microbiol.* **24**:1103–1134.

Hochhauser, S. W., Stein, J. L., and Stein, G. S. 1981. Gene expression and cell cycle regulation. *Int. Rev. Cytol.* **71**:95–243.

Hoffman, B., Messer, W., and Schwarz, U. 1972. Regulation of polar cap formation in the life cycle of *Escherichia coli*. *J. Supramol. Struct.* **1**:29–37.

Hoffman, H., and Frank, M. E. 1965. Time-lapse photomicrography of cell growth and division in *Escherichia coli*. *J. Bacteriol.* **89**:212–216.

Holland, I. B. 1987. Genetic analysis of the *E. coli* division clock. *Cell* **48**:361–362.

Holland, I. B., Casaregola, S., and Norris, V. 1990. Cytoskeletal elements and calcium: Do they play a role in the *Escherichia coli* cell cycle? *Res. Microbiol.* **141**:131–136.

Holley, R. W., Baldwin, J. H., and Kiernan, J. A. 1974. Control of growth of a tumor cell by linoleic acid. *Proc. Natl. Acad. Sci. USA* **71**:3976–3978.

Holley, R. W., and Kiernan, J. A. 1974a. Control of initiation of DNA synthesis in 3T3 cells: Low molecular weight nutrients. *Proc. Natl. Acad. Sci. USA* **71**:2942–2945.

Holley, R. W., and Kiernan, J. A. 1974b. Control of initiation of DNA-synthesis in 3T3 cells: Serum factors. *Proc. Natl. Acad. Sci. USA* **71**:2908–2911.

Holmes, M., Rickert, M., and Pierucci, O. 1980. Cell division cycle of *Bacillus subtilis*: Evidence of variability in period D. *J. Bacteriol.* **142**:254–261.

Hotchkiss, R. D. 1954. Cyclical behavior in pneumococcal growth and transformability occasioned by environmental changes. *Proc. Natl. Acad. Sci. USA* **40**:49–55.

Howard, A., and Pelc, S. R. 1951a. Nuclear incorporation of P^{32} as demonstrated by autoradiographs. *Exp. Cell Res.* **2**:178–187.

Howard, A., and Pelc, S. R. 1951b. Synthesis of nucleoprotein in bean root cells. *Nature* (*London*) **167**:599–600.

Höltje, J.-V., and Glauner, B. 1990. Structure and metabolism of the murein sacculus. *Res. Microbiol.* **141**:75–89.

Höltje, J.-V., and Nanninga, N. 1984. The intracellular concentration of cyclic adenosine 3',5'-monophosphate is constant throughout the cell cycle of *Escherichia coli*. *FEMS Microbiol. Lett.* **22**:189–192.

Hughes, R. C., and E. Stokes. 1971. Cell wall growth in *Bacillus licheniformis* followed by immunofluoresence with mucopeptide-specific antiserum. *J. Bacteriol.* **106**:694–696.

Huguenel, E., and Newton, A. 1982. Localization of surface structures during prokaryotic differentiation: Role of cell division in *Caulobacter crescentus*. *Differentiation* **21**:71–78.

Hunter-Szybalska, M., Szybalski, W., and DeLamater, E. 1956. Temperature synchronization of nuclear and cellular division of *Bacillus megatherium*. *J. Bacteriol.* **71**:17–24.

Huzyk, L., and Clark, D. J. 1971. Nucleoside triphosphate pools in synchronous cultures of *Escherichia coli*. *J. Bacteriol.* **108**:74–81.

Iba, H., Fukuda, A., and Okada, Y. 1977. Chromosome replication in *Caulobacter crescentus* growing in a nutrient broth. *J. Bacteriol.* **129**:1192–1197.

Ingraham, J. L., Maaløe, O., and Neidhardt, F. C. 1983. Growth of the Bacterial Cell. Sinauer Associates, Sunderland, Mass.

Ishiguro, E. E., and Ramey, W. D. 1976. Stringent control of peptidoglycan biosynthesis in *Escherichia coli* K-12. *J. Bacteriol.* **127**:1119–1126.

Ishiguro, E. E., and Ramey, W. D. 1978. Involvement of the *rel*A gene product and feedback inhibition in the regulation of UDP-N-acetyl-muramyl-peptide synthesis in *Escherichia coli*. *J. Bacteriol.* **135**:766–774.

Ittman, M., Greco, A., and Basilico, C. 1987. Isolation of the human gene that complements a temperature sensitive cell cycle mutation in BHK cells. *Mol. Cell. Biol.* **7**:3386–3393.

Ivarie, R. D., and Pene, J. J. 1973. Association of many regions of the *Bacillus subtilis* chromosome with the cell membrane. *J. Bacteriol.* **114**:571–576.

Jacob, F., and Wollman, E. L. 1961. Sexuality and the Genetics of Bacteria. Academic Press, New York.

Jacob, F., Brenner, S., and Cuzin, F. 1963. On the regulation of DNA replication in bacteria. *Cold Spring Harbor Symp. Quant. Biol.* **28**:329–347.

Jacq, A., and Kohiyama, M. 1980. A DNA-binding protein specific for the early replicated region of the chromosome obtained from the *Escherichia coli* membrane fractions. *Eur. J. Biochem.* **105**:25–31.

Jacq, A., Kohiyama, M., Lother, H., and Messer, W. 1983. Recognition sites for a membrane-derived DNA binding protein preparation in the *E. coli* replication origin. *Mol. Gen. Genet.* **191**:460–465.

Jaffe, A., and D'Ari, R. 1985. Growth of the *Escherichia coli* cell envelope. *Biochemie* **67**:141–144.

James, R., and Gudas, L. J. 1976. Cell-cycle–specific incorporation of lipoprotein into the outer membrane of *Escherichia coli*. *J. Bacteriol.* **125**:374–375.

Johnston, L. H., Eberly, S. L., Chapman, J. W., Araki, H., and Sugino, A. 1990. The product of the *Saccharomyces cerevisiae* cell cycle gene DBF2 has homology with protein kinases and is periodically expressed in the cell cycle. *Mol. Cell. Biol.* **10**:1358–1366.

Jolliffe, L. K., Doyle, R. J., and Streips, U. N. 1980. Extracellular proteases modify cell wall turnover in *Bacillus subtilis*. *J. Bacteriol.* **141**:1199–1208.

Jones, C., and Holland, I. B. 1985. Role of the SulB (FtsZ) protein in division inhibition during the SOS response in *E. coli*: FtsZ stabilizes the inhibitor SulA in maxicells. *Proc. Natl. Acad. Sci. USA* **81**:4490–4494.

Joseleau-Petit, D., Kepes, F., and Kepes, A. 1984. Cyclic changes of the rate of phospholipid synthesis during synchronous growth of *Escherichia coli*. *Eur. J. Biochem.* **139**:605–611.

Joseleau-Petit, D., Kepes, F., Peutat, L., D'Ari, R., and Kepes, A. 1987. DNA replication initiation, doubling of rate of phospholipid synthesis and cell division in *Escherichia coli*. *J. Bacteriol.* **169**:3701–3706.

Kaguni, M. M., Fuller, R. S., and Kornberg, A. 1982. Enzymatic replication of *E. coli* chromosomal origin is bidirectional. *Nature (London)* **296**:623–627.

Kellenberger-Gujer, G., Podhajska, A. J., and Caro, L. 1978. A cold sensitive *dna*A mutant of *E. coli* which overinitiates chromosome replication at low temperature. *Mol. Gen. Genet.* **162**:9–16.

Kelly, K., Cochran, B., Stiles, C., and Leder, P. 1984. The regulation of *c-myc* by growth signals. M. Potter, F. Melchers, and M. Weigert, (eds.), *Curr. Top. Microbiol. Immunol.* **113**:117–126.

Kendall, D. G. 1948. On the role of variable generation time in the development of a stochastic birth process. *Biometrika* **35**:316–330.

Kendall, D. G. 1952. On the choice of mathematical models to represent normal bacterial growth. *J. R. Stat. Soc. B* **14**:41–44.

Kennett, R. H., and Sueoka, N. 1971. Gene expression during outgrowth of *Bacillus subtilis* spores: Relationship between gene order on chromosome and temporal sequence of enzyme synthesis. *J. Mol. Biol.* **60**:31–44.

Kepes, A., and Autissier, F. 1972. Topology of membrane growth in bacteria. *Biochim. Biophys. Acta* **265**:443–469.

Kepes, F., and D'Ari, R. 1987. Involvement of FtsZ protein in shift-up–uced division delay in *Escherichia coli*. *J. Bacteriol.* **169**:4036–4040.

Kepes, F., and Kepes, A. 1980. Synchronisation automatique de la croissance de *Escherichia coli*. *Ann. Microbiol. (Paris)* **131A**:3–16.

Kepes, F., and Kepes, A. 1981. Long-lasting synchrony of the division of enteric bacteria. *Biophys. Res. Commun.* **99**:761–767.

Kepes, F., and Kepes, F. 1985. Postponement of cell division by nutritional shift-up in *Escherichia coli*. *J. Gen. Microbiol.* **131**:677–685.

Keynan, A., Berns, A. A., Dunn, G., Young, M., and Mandelstam, J. 1976. Resporulation of outgrowing *Bacillus subtilis* spores. *J. Bacteriol.* **128**:8–14.

Kirkwood, T. B. L., and Burdett, I. D. J. 1988. Estimating the growth pattern of microorganisms in distinct stages of the cell cycle. *J. Theor. Biol.* **130**:255–273.

Kjeldgaard, N. O., Maaløe, O., and Schaechter, M. 1958. The transition between different physiological states during balanced growth of *Salmonella typhimurium*. *J. Gen. Microbiol.* **19**:607–616.

Knaysi, G. 1941. A morphological study of *Streptococcus faecalis*. *J. Bacteriol.* **42**:575–586.

Kobayashi, T., Hidaka, M., and Horiuchi, T. 1989. Evidence of a *ter* specific binding protein essential for the termination reaction of DNA replication in *Escherichia coli*. *EMBO J.* **8**:2435–2441.

Koch, A. L. 1966a. Distribution of cell size in growing cultures of bacteria and the applicability of the Collins–Richmond principle. *J. Gen. Microbiol.* **45**:409–417.

Koch, A. L. 1966b. On evidence supporting a deterministic process of bacterial growth. *J. Gen. Microbiol.* **43**:1–5.

Koch, A. L. 1977. Does initiation of chromosome replication regulation cell division? *Adv. Microb. Physiol.* **16**:49–98.

Koch, A. L. 1980. The inefficiency of ribosomes functioning in *Escherichia coli* growing at moderate rates. *J. Gen. Microbiol.* **116**:165–171.

Koch, A. L. 1983a. The surface stress theory of microbial morphogenesis. *Adv. Microb. Physiol.* **24**:301–366.

Koch, A. L. 1983b. The shapes of gram-negative organisms: Variable-T models. *In* The Target of Penicillin. R. Hakenbeck, J.-V. Höltje, and H. Labischinski, (eds.), pp. 99–104. Walter de Gruyter, Berlin-New York.

Koch, A. L. 1985. Bacterial wall growth and division or life without actin. *Trends Biochem. Sci.* **10**:11–14.

Koch, A. L. 1986. The basis of synchronization by repetitive dilution of a growing culture. *J. Theor. Biol.* **123**:333–346.

Koch, A. L. 1987. The variability and individuality of the bacterium. *In Escherichia coli* and *Salmonella typhimurium:* Cellular and Molecular Biology, F. C. Neidhardt, J. L. Ingraham, K. B. Low, B. Magasanik, M. Schaechter, and H. E. Umbarger (eds.), pp. 1606–1614. American Society for Microbiology, Washington, D. C.

Koch, A. L. 1988a. Biophysics of bacterial walls viewed as stress bearing fabric. *Microbiol. Rev.* **52**:337–353.

Koch, A. L. 1988b. Parition of autolysins between the medium, the internal part of the wall, and the surface of gram-positive rods. *J. Theor. Biol.* **134**:463–472.

Koch, A. L. 1989. The origin of the rotation of one end of a cell relative to the other end during growth of gram-positive rods. *J. Theor. Biol.* **141**:931–402.

Koch, A. L. 1990. The surface-stress theory for the case of *Escherichia coli*: The paradoxes of gram-negative growth. *Res. Microbiol.* **141**:119–130.

Koch, A. L., and Blumberg, G. 1976. Distribution of bacteria in the velocity gradient centrifuge. *Biophys. J.* **16**:389–405.

Koch, A. L., and Burdett, I. D. J. 1984. The variable-T model for gram-negative morphology. *J. Gen. Microbiol.* **130**:2325–2338.

Koch, A. L., and Doyle, R. J. 1986. Inside-to-outside growth and the turnover of the gram-positive rod. *J. Theor. Biol.* **117**:137–157.

Koch, A. L., and Pinette, M. F. S. 1987. Nephelometric determination of turgor pressure in growing gram-negative bacteria. *J. Bacteriol.* **169**:3654–3663.

Koch, A. L., and Schaechter, M. 1962. A model for the statistics of the cell division process. *J. Gen. Microbiol.* **29**:435–454.

Koch, A. L., Higgins, M. L., and Doyle, R. J. 1981. Surface tension–like forces determine bacterial shapes: *Streptococcus faecium. J. Gen. Microbiol.* **123**:151–161.

Koch, A. L., Higgins, M. L., and Doyle, R. J. 1982. The role of surface stress in the morphology of microbes. *J. Gen. Microbiol.* **128**:927–945.

Koch, A. L., Mobley, H. L. T., Doyle, R. J., and Streips, U. N. 1981. The coupling of wall growth and chromosome replication in gram-positive rods. *FEMS Microbiol. Lett.* **12**:201–208.

Koch, A. L., Verwer, R. W. H., and Nanninga, N. 1982. Incorporation of diaminopimelic acid into the old poles of *Escherichia coli. J. Gen. Microbiol.* **128**:2893–2898.

Kolter, R., and Helinski, D. R. 1979. Regulation of initiation of DNA replication. *Annu. Rev. Genet.* **13**:335–391.

Koppes, L. J. H., and Nanninga, N. 1980. Positive correlation between size at initiation of chromosome replication in *Escherichia coli* and size at initiation of cell constriction. *J. Bacteriol.* **143**:89–99.

Koppes, L. J. H., and Nordstrom, K. 1985. Insertion of an R1 into the origin of replication of the *Escherichia coli* chromosome: Random timing of replication of the hybrid chromosome. *Cell* **44**:117–124.

Koppes, L. J. H., and von Meyenburg, K. 1987. Nonrandom minichromosome replication in *Escherichia coli* K-12. *J. Bacteriol.* **169**:430–433.

Koppes, L. J. H., Meyer, M., Oonk, H. B., De Jong, M. A., and Nanninga, N. 1980. Correlation between size and age at different events in the cell division cycle of *Escherichia coli*. *J. Bacteriol.* **143**:1241–1252.

Koppes, L. J. H., Overbeeke, N., and Nanninga, N. 1978. DNA replication pattern and cell wall growth in *Escherichia coli* PAT84. *J. Bacteriol.* **133**:1053–1061.

Koppes, L. J. H., Woldringh, C. L., and Nanninga, N. 1978. Size variations and correlation of different cell-cycle events in slow-growing *Escherichia coli*. *J. Bacteriol.* **134**:423–433.

Koppes, L. J., Woldringh, C. L., and Grover, N. B. 1987. Predicted steady-state cell size distributions for various growth models. *J. Theor. Biol.* **129**:325–335.

Kornberg, A. 1974. DNA Synthesis. W. H. Freeman, San Francisco.

Kornberg, A. 1980. DNA Replication. W. H. Freeman, San Francisco.

Kornberg, A. 1982. Supplement to DNA Replication. W. H. Freeman, San Francisco.

Kornberg, A. 1983. Mechanisms of replication of the *Escherichia coli* chromosome. *Eur. J. Biochem.* **137**:377–382.

Kornberg, A. 1984. Enzyme studies of replication of the *Escherichia coli* chromosome. *Adv. Exp. Med. Biol.* **179**:3–16.

Kornberg, A. 1988. DNA replication. *J. Biol. Chem.* **263**:1–4.

Kubitschek, H. E. 1958. Electronic counting and sizing of bacteria. *Nature (London)* **182**:234–235.

Kubitschek, H. E. 1960. Electronic measurement of particle size. *Research* **13**:234–235.

Kubitschek, H. E. 1962a. Loss of resolution in Coulter Counters. *Rev. Sci. Instrum.* **33**:576.

Kubitschek, H. E. 1962b. Normal distribution of cell generation rate. *Exp. Cell Res.* **26**:439–450.

Kubitschek, H. E. 1966. Normal distribution of cell generation rates. *Nature (London)* **209**:1039–1040.

Kubitschek, H. E. 1968a. Linear cell growth in *Escherichia coli*. *Biophys. J.* **8**:792–804.

Kubitschek, H. E. 1968b. Constancy of uptake during the cell cycle of *Escherichia coli*. *Biophys. J.* **8**:1401–1412.

Kubitschek, H. E. 1970a. Evidence for the generality of linear cell growth. *J. Theor. Biol.* **28**:15–29.

Kubitschek, H. E. 1970b. Introduction to Research with Continuous Cultures. Prentice-Hall, Englewood Cliffs, New Jersey.

Kubitschek, H. E. 1971. Distortion in the Coulter Counter and applicability of the Collins–Richmond principle. *Biophys. J.* **11**:124–126.

Kubitschek, H. E. 1974. Estimation of the D period from residual division after exposure of exponential phase bacteria to chloramphenicol. *Mol. Gen. Genet.* **135**:123–130.

Kubitschek, H. E. 1981. Bilinear cell growth of *Escherichia coli*. *J. Bacteriol.* **148**:730–733.

Kubitschek, H. E. 1986. Increase in cell mass during the division cycle of *Escherichia coli* B/rA. *J. Bacteriol.* **168**:613–618.

Kubitschek, H. E. 1987. Buoyant density variation during the cell cycle in microorganisms. *CRC Crit. Rev. Microbiol.* **14**:73–97.

Kubitschek, H. E. 1990a. Cell volume increase in *Escherichia coli* after shifts to richer media. *J. Bacteriol.* **172**:94–101.

Kubitschek, H. E. 1990b. Cell growth and abrupt doubling of membrane proteins in *Escherichia coli* during the division cycle. *J. Gen. Microbiol.* **136**:599–606.

Kubitschek, H. E., and Freedman, M. L. 1971. Chromosome replication and the division cycle of *Escherichia coli* B/r. *J. Bacteriol.* **107**:95–100.

Kubitschek, H. E., and Friske, J. A. 1986. Determination of bacterial cell volume with the Coulter Counter. *J. Bacteriol.* **168**:1466–1467.

Kubitschek, H. E., and Newman, C. N. 1978. Chromosome replication during the division cycle in slowly growing, steady-state cultures of three *Escherichia coli* B/r strains. *J. Bacteriol.* **136**:179–190.

Kubitschek, H. E., and Pai, S. R. 1988. Variation in precursor pool size during the division cycle of *Escherichia coli*: Further evidence for linear cell growth. *J. Bacteriol.* **170**:431–435.

Kubitschek, H. E., and Woldringh, C. L. 1983. Cell elongation and division probability during the *Escherichia coli* growth cycle. *J. Bacteriol.* **153**:1379–1387.

Kubitschek, H. E., Baldwin, W. W., and Graetzer, R. 1983. Buoyant density constancy during the cell cycle of *Escherichia coli*. *J. Bacteriol.* **155**:1027–1032.

Kubitschek, H. E., Baldwin, W. W., Schroeter, S. J., and Graetzer, R. 1984. Independence of buoyant cell density and growth rate in *Escherichia coli*. *J. Bacteriol.* **158**:296–299.

Kubitschek, H. E., Bendigkeit, H. E., and Loken, M. R. 1967. Onset of DNA synthesis during the cell cycle in chemostat cultures. *Proc. Natl. Acad. Sci. USA* **57**:1611–1617.

Kubitschek, H. E., Freedman, M. L., and Silver, S. 1971. Potassium uptake in synchronous and synchronized cultures of *Escherichia coli*. *Biophys. J.* **11**:787–797.

Kuempel, P. L., Masters, M., and Pardee, A. B. 1965. Bursts of enzyme synthesis in the bacterial duplication cycle. Biochem. *Biophys. Res. Comm.* **18**:858–867.

Kuempel, P. L., Pelletier, A. J., and Hill, T. M. 1989. Tus and the terminators: The arrest of replication in prokaryotes. *Cell* **59**:581–583.

Kung, F. C., Raymond, J., and Glaser, D. A. 1976. Metal ion content of *Escherichia coli* versus cell age. *J. Bacteriol.* **126**:1089–1095.

Kunicki-Goldfinger, W., and Mycielski, R. 1966. Sequential UV-induced mutations in synchronized cultures of *Escherichia coli*. *Acta Microbiol. Pol.* **15**:133.

Kurn, N., and Shapiro, L. 1975. Regulation of the *Caulobacter* cell cycle. *In* Current Topics in Cellular Regulation. B. L. Horecker and E. R. Stadtman (eds.), vol. 9, pp. 41–62. Academic Press, New York

Kusano, T., Steinmetz, D., Hendrickson, W. G., Murchie, J., King, M., Benson, A., and Schaechter, M. 1984. Direct evidence for specific binding of the replicative origin of the *Escherichia coli* chromosome to the membrane. *J. Bacteriol.* **158**:313–316.

Lane-Claypon, J. E. 1909. Multiplication of bacteria and the influence of temperature and some other conditions thereon. *J. Hyg.* **9**:239–248.

Lane, D., and Denhardt, D. T. 1974. The *rep* mutation. III. Altered structure of the replicating *Escherichia coli* chromosome. *J. Bacteriol.* **120**:805–814.

Lane, H. E. D., and Denhardt, D. T. 1975. The *rep* mutation. IV. Slower movement of replication forks in *Escherichia coli rep* strains. *J. Mol. Biol.* **97**:99–112.

Lark, K. G. 1958. Variation during the cell-division cycle in the penicillin induction of protoplast-like forms of *Alcaligenes fecalis*. *Can. J. Microbiol.* **4**:179–189.

Lark, K. G. 1966. Regulation of chromosome replication and segregation in bacteria. *Bact. Rev.* **30**:3–32.

Lark, K. G. 1973. Initiation and termination of bacterial deoxyribonucleic acid replication in low concentrations of chloramphenicol. *J. Bacteriol.* **113**:1066–1069.

Lark, K. G., and Lark, C. A. 1960. Changes during the division cycle in bacterial cell wall synthesis, volume and ability to concentrate free amino acids. *Biochim. Biophys. Acta* **43**:520–530.

Lark, K. G., and Lark, C. A. 1965. Regulation of chromosome replication in *Escherichia coli*: Alternate replication of two chromosomes at slow growth rates. *J. Mol. Biol.* **13**:105–126.

Lark, K. G., and Maaløe, O. 1954. The induction of cellular and nuclear division in *Salmonella typhimurium* by means of temperature shifts. *Biochim. Biophys. Acta* **15**:345–356.

Lark, K. G., and Maaløe, O. 1956. Nucleic acid synthesis and the division cycle of *Salmonella typhimurium*. *Biochim. Biophys. Acta* **21**:448–457.

Lark, K. G., and Renger, H. 1969. Initiation of DNA replication in *Escherichia coli* 15T⁻: Chronological dissection of three physiological processes required for initiation. *J. Mol. Biol.* **42**:221–235.

Lark, K. G., Repko, T., and Hoffman, E. J. 1963. The effect of amino acid deprivation on subsequent deoxyribonucleic acid replication. *Biochim. Biophys. Acta* **76**:9–24.

Larsson, O., Zetterberg, A., and Engstrom, W. 1985. Consequences of parental exposure to serum-free medium for progeny cell division. *J. Cell. Sci.* **75**:259–268.

Lau, L. F., and Nathans, D. 1985. Identification of a set of genes expressed during the G0/G1 transition of cultured mouse cells. *EMBO J.* **4**:3145–3151.

Lederberg, J., Cavalli, L. L., and Lederberg, E. M. 1952. Sex compatibility in *E. coli*. *Genetics* **37**:720–730.

Ledingham, J. C. G., and Penfold, W. J. 1914. Mathematical analysis of the lag-phase in bacterial growth. *J. Bacteriol.* **14**:242–260.

Leof, E. B., Wharton, W., Van Wyk, J. J., and Pledger, W. J. 1982. Epidermal growth factor (EGF) and somatomedin-C regulate G1 progression in competent BALB/C-3T3 cells. *Exp. Cell Res.* **141**:107–115.

Leonard, A. C., and Helmstetter, C. E. 1986. Cell cycle–specific replication of *Escherichia coli* minichromosomes. *Proc. Natl. Acad. Sci. USA* **83**:5101–5105.

Leonard, A. C., and Helmstetter, C. E. 1988. Replication patterns of multiple plasmids coexisting in *Escherichia coli*. *J. Bacteriol.* **170**:1380–1383.

Leonard, A. C., Theisen, P. W., and Helmstetter, C. E. 1990. Replication timing and copy number control of *ori*C plasmids. *In* The Bacterial Chromosome. K. Drlica and M. Riley (eds.), pp. 279–286. American Society for Microbiology, Washington D.C.

Leonard, A. C., Weinberger, M., Munson, B. R., and Helmstetter, C. E. 1980. The effects of *ori*C-containing plasmids on host cell growth. *ICN-UCLA Symp. Mol. Cell. Biol.* **19**:171–180.

Lin, E. C. C., Hirota, Y., and Jacob, F. 1971. On the process of cellular division in *Escherichia coli*. VI. Use of a methocel autoradiographic method for the study of cellular division in *Escherichia coli*. *J. Bacteriol.* **108**:375–385.

Lindmo, T., and Steen, H. B. 1979. Characterization of a simple, high-resolution flow cytometer based on a new flow configuration. *Biophys. J.* **28**:33–34.

Liskay, R. M. 1974. A mammalian somatic "cell cycle" mutant defective in G1. *J. Cell Physiol.* **84**:49–56.

Liskay, R. M., and Prescott, D. M. 1978. Genetic analysis of the G1 period: Isolation of mutants (or variants) with a G1 period from a Chinese hamster cell line lacking G1. *Proc. Natl. Acad. Sci. USA* **75**:2873–2877.

Lloyd, D., John, L., Edwards, C., and Chagla, A. H. 1975. Synchronous cultures of microorganisms: Large-scale preparation by continuous-flow size selection. *J. Gen. Microbiol.* **88**:153–158.

Lloyd, D., John, L., Hammill, M., Phillips, C. A., Kader, J., and Edwards, S. W. 1975. Continuous-flow cell cycle fractionation of eukaryotic microorganisms. *J. Gen. Microbiol.* **99**:223–227.

Lloyd, D., Poole, R. K., and Edwards, S. W. 1982. The Cell Division Cycle: Temporal Organization and Control of Cellular Growth and Reproduction. Academic Press, New York.

Longsworth, L. G. 1936. The estimation of bacterial populations with the aid of a photoelectric densitometer. *J. Bacteriol.* **32**:307–328.

Lott, T., Ohta, N., and Newton, A. 1987. Order of gene replication in *Caulobacter crescentus*; use of *in vivo* labeled genomic DNA as a probe. *Mol. Gen. Genet.* **210**:543–550.

Louarn, J., Funderburgh, M., and Bird, R. E. 1974. More precise mapping of the replication origin in *Escherichia coli* K-12. *J. Bacteriol.* **120**:1–5.

Løbner-Olesen, A., Atlung, T., and Rasmussen, K. V. 1987 Stability and replication control of *Escherichia coli* minichromosomes. *J. Bacteriol.* **169**:2835–2842.

Løbner-Olesen, A., Skarstad, K., Hansen, F. G., von Meyenburg, K., and Boye, E. 1989. The *dna*A protein determines the initiation mass of *Escherichia coli* K12. *Cell* **57**:881–889.

Lutkenhaus, J. 1983. Coupling of DNA replication and cell division: *sul*B is an allele of *fts*Z. *J. Bacteriol.* **154**:1339–1346.

Lutkenhaus, J. F., Moore, B. A., Masters, M., and Donachie, W. D. 1979. Individual proteins are synthesized continuously throughout the *Escherichia coli* cell cycle. *J. Bacteriol.* **138**:352–360.

Maaløe, O. 1962. Synchronous growth. *In* The Bacteria. I. C. Gunsalus and R. Y. Stanier (ed.), vol. 4, pp. 1–32. Academic Press, New York.

Maaløe, O., and Bentzon, M. W. 1985. Promoter collectives in bacteria. *In* The Molecular Biology of Bacterial Growth. M. Schaechter, F. C. Neidhardt, J. L. Ingraham, and N. O. Kjeldgaard (eds.), pp. 48–61. Jones and Bartlett, Boston.

Maaløe, O., and Hanawalt, P. C. 1961. Thymine deficiency and the normal DNA replication cycle. *J. Mol. Biol.* **3**:144–155.

Maaløe, O., and Kjeldgaard, N. O. 1966. Control of Macromolecular Synthesis. W. A. Benjamin, New York.

Maaløe, O., and Rasmussen, K. V. 1963. On the *in vivo* replication of bacterial DNA. *Colloq. Intern. Centre. Natl. Rech. Sci. (Paris)* **124**:165–168.

MacAlister, T. J., Cook, W. R., Weigand, R., and Rothfield, L. I. 1987. Membrane-murein attachment at the leading edge of the division septum: A second membrane-murein structure associated with morphogenesis of the gram-negative bacterial division septum. *J. Bacteriol.* **169**:3945–3951.

MacAlister, T. J., MacDonald, B., and Rothfield, L. I. 1983. The periseptal annulus: An organelle associated with cell division. *Proc. Natl. Acad. Sci. USA* **80**:1372–1376.

Mandelstam, J., and Higgs, J. A. 1974. Induction of sporulation during synchronized chromosome replication in *Bacillus subtilis*. *J. Bacteriol.* **120**:38–42.

Manor, H., and Haselkorn, R. 1967. Size fractionation of exponentially growing *Escherichia coli*. *Nature (London)* **214**:983–986.

Manor, H., Deutscher, M. P., and Littauer, U. Z. 1971. Rates of DNA chain growth in *E. coli*. *J. Mol. Biol.* **61**:503–524.

Manor, H., Goodman, H., and Stent, G. S. 1969. RNA and chain growth rates in *Escherichia coli*. *J. Mol. Biol.* **39**:1–29.

Mansour, J. D., Henry, S., and Shapiro, L. 1980. Differential membrane phospholipid synthesis during the cell cycle of *Caulobacter crescentus*. *J. Bacteriol.* **141**:262–269.

Margalit, H., and Grover, N. B. 1987. Initiation of chromosome replication in bacteria: Analysis of an inhibitor control model. *J. Bacteriol.* **169**:5231–5240.

Margolis, S., and Cooper, S.. 1971. Simulation of bacterial growth, cell division, and DNA synthesis. *Comp. Biomed. Res.* **4**:427–443.

Marians, K. 1984. Enzymology of DNA replication in prokaryotes. *Crit. Rev. Biochem.* **17**:153–215.

Marr, A. G., Harvey, R. J., and Trentini, W. C. 1966. Growth and division of *Escherichia coli. J. Bacteriol.* **91**:2388–2389.

Marr, A. G., Painter, P. R., and Nilson, E. H. 1969. Growth and division of individual bacteria. *Symp. Soc. Gen. Microbiol.* **19**:237–261.

Marsh, R. C., and Worcel, A. 1977. A DNA fragment containing the origin of replication of the *Escherichia coli* chromosome. *Proc. Natl. Acad. Sci. USA* **74**:2720–2724.

Martinez-Salas, E., Martin, J. A., and Vicente, M. 1981. Relationship of *Escherichia coli* density to growth rate and cell age. *J. Bacteriol.* **147**:97–100.

Marunouchi, T., and Messer, W. 1973. Replication of a specific terminal chromosome segment in *Escherichia coli* which is required for cell division. *J. Mol. Biol.* **78**:211–228.

Maruyama, Y. 1956. Biochemical aspects of cell growth of *Escherichia coli* as studied by the method of synchronous culture. *J. Bacteriol.* **72**:821–826.

Maruyama, Y., and Lark, K. G. 1959. Periodic synthesis of RNA in synchronized cultures of *Alcaligenes faecalis. Exp. Cell Res.* **18**:389–391.

Maruyama, Y., and Lark, K. G. 1962. Periodic nucleotide synthesis in synchronous cultures of bacteria. *Exp. Cell Res.* **26**:382–394.

Maruyama, Y., and Yanagita, T. 1956. Physical methods for obtaining synchronous culture of *Escherichia coli. J. Bacteriol.* **71**:542–546.

Masters, M. 1970. Origin and direction of replication of the chromosome of *Escherichia coli* B/r. *Proc. Natl. Acad. Sci. USA* **65**:601–608.

Masters, M. 1975. Strains of *Escherichia coli* diploid for the chromosomal origin of DNA replication. *Mol. Gen. Genet.* **143**:105–111.

Masters, M., and Broda, P. 1971. Evidence for the bidirectional replication of the *Escherichia coli* chromosome. *Nature (London) New Biol.* **232**:137–140.

Masters, M., and Donachie, W. D. 1966. Repression and the control of cyclic enzyme synthesis in *Bacillus subtilis. Nature (London)* **209**:476–479.

Masters, M., and Pardee, A. B. 1965. Sequence of enzyme synthesis and gene replication during the cell cycle of *Bacillus subtilis. Proc. Natl. Acad. Sci. USA* **54**:64–70.

Masters, M., Kuempel, P. L., and Pardee, A. B. 1964. Enzyme synthesis in synchronous cultures of bacteria. *Biochem. Biophys. Res. Commun.* **15**:38–42.

Matney, T. S., and Suit, J. C. 1966. Synchronously dividing bacterial cultures. I. Synchrony following depletion and resupplementation of a required amino acid in *Escherichia coli. J. Bacteriol.* **92**:960–966.

Matsuhashi, M., Wachi, M., and Ishino, F. 1990. Machinery for cell growth and division: Penicillin-binding proteins and other proteins. *Res. Microbiol.* **141**:89–103.

Matsushita, R., O'Sullivan, A., White, K., and Sueoka, N. 1974. Chromosome replication in *Bacillus subtilis. In* Mechanism and Regulation of DNA Replication. A. A. Kolber and M. Kohiyama (eds.), pp. 241–251. Plenum Press, New York.

Mauck, J., and Glaser, L. 1970. Turnover of the cell wall of *Bacillus subtilis* W-23 during logarithmic growth. *Biochem. Biophys. Res. Commun.* **39**:699–706.

Mauck, J., and Glaser, L. 1972. On the mode of *in vivo* assembly of the cell wall of *Bacillus subtilis. J. Biol. Chem.* **247**:1180–1187.

Mauck, J., Chan, L., and Glaser, L. 1971. Turnover of the wall of gram-positive bacteria. *J. Biol. Chem.* **246**:1820–1827.

Mauck, J., Chan, L., and Glaser, L., and Williamson, J. 1972. Mode of cell-wall growth of *Bacillus megaterium. J. Bacteriol.* **109**:373–378.

May, J. W. 1963. The distribution of cell-wall label during growth and division of *Salmonella typhimurium. Exp. Cell Res.* **31**:217–220.

McFall, E., and Stent, G. S. 1959. Continuous synthesis of deoxyribonucleic acid in *Escherichia coli. Biochim. Biophys. Acta* **34**:580–582.

McKenna, W. G., and Masters, M. 1972. Biochemical evidence for the bidirectional replication of DNA in *Escherichia coli. Nature (London)* **240**:536–539.

McMacken, R., Silver, L., and Georgopoulos, C. 1987. DNA replication. *In Escherichia coli and Salmonella typhimurium:* Cellular and Molecular Biology. F. C. Neidhardt, J. L. Ingraham, K. B. Low, B. Magasanik, M. Schaechter, and H. E. Umbarger (eds.), pp. 564–612. American Society for Microbiology, Washington, D.C.

Meacock, P. A., Pritchard, R. H., and Roberts, E. M. 1978. Effect of thymine concentration on cell shape in thy *Escherichia coli* B/r. *J. Bacteriol.* **133**:320–328.

Meijer, M., and Messer, W. 1980. Functional analysis of minichromosome replication: Bidirectional and unidirectional replication from the *Escherichia coli* replication origin, oriC. *J. Bacteriol.* **143**:1049–1053.

Meijer, M., Beck, E., Hansen, F. G., Bergmans, H. E. N., Messer, W., von Meyenberg, K., and Schaller, H. 1979. Nucleotide sequence of the origin of replication of the *Escherichia coli* K-12 chromosome. *Proc. Natl. Acad. Sci. USA* **76**:580–584.

Melchers, F., and Lernhardt, W. 1985. Three restriction points in the cell cycle of activated murine B lymphocytes. *Proc. Natl. Acad. Sci. USA* **82**:7681–7685.

Mendelson, N. H. 1976. Helical growth of *Bacillus subtilis:* A new model for cell growth. *Proc. Natl. Acad. Sci. USA* **73**:1740–1744

Mendelson, N. H. 1978. Helical *Bacillus subtilis* macrofibers: Morphogenesis of a bacterial multicellular macroorganism. *Proc. Natl. Acad. Sci. USA* **75**:2478–2482.

Mendelson, N. H. 1982. Bacterial growth and division: Genes, structures, forces, and clocks. *Microbiol. Rev.* **46**:341–375.

Mendelson, N. H., Favre, D., and Thwaits, J. J. 1984. Twisted states of *Bacillus subtilis* macrofiber reflect structural states of the cell wall. *Proc. Natl. Acad. Sci. USA* **81**:3562–3566.

Meselson, M., and Stahl, F. W. 1958. The replication of DNA in *Escherichia coli. Proc. Natl. Acad. Sci. USA* **44**:671–682.

Messer, W. 1972. Initiation of deoxyribonucleic acid replication in *Escherichia coli* B/r: Chronology of events and transcriptional control of initiation. *J. Bacteriol.* **112**:7–12.

Messer, W. 1987. Initiation of DNA replication in *Escherichia coli. J. Bacteriol.* **169**:3395–3399.

Messer, W., Bergmans, H. E. N., Meijer, M., Womack, J. E., Hansen, F. G., and von Meyenburg, K. 1978. Minichromosomes: Plasmids which carry the *E. coli* replication origin. *Mol. Gen. Genet.* **162**:269–275.

Messer, W., Meijer, M., Bergmans, H. E. N., Hansen, F. G., von Meyenburg, K., Beck, E., and Schaller, H. 1979. Origin of replication, oriC, of the *Escherichia coli* K12 chromosome: Nucleotide sequence. *Cold Spring Harbor Symp. Quant. Biol.* **43**:139–145.

Meyer, M., De Jong, M. A., Demets, R., and Nanninga, N. 1979. Length growth of two *Escherichia coli* B/r substrains. *J. Bacteriol.* **138**:17–23.

Minnich, S. A., and Newton, A. 1987. Promoter mapping and cell-cycle regulation of flagellin gene transcription in *Caulobacter crescentus. Proc. Natl. Acad. Sci. USA* **84**:1142–1146.

Mitchison, J. M. 1964. Markers in the cell cycle. *In* The Cell Cycle, Gene-Enzyme Interactions. G. M. Padilla, G. L. Whitson, and I. L. Cameron, (eds.), pp. 361–372. Academic Press, New York.

Mitchison, J. M. 1969. Enzyme synthesis in synchronous cultures. *Science* **165**:657–663.

Mitchison, J. M. 1971. The Biology of the Cell Cycle. Cambridge University Press, Cambridge.

Mitchison, J. M. 1989. Cell cycle growth and periodicities. *In* Molecular Biology of the Fission Yeast. A. Nasim, P. Young, and B. F. Johnson (eds.), pp. 205–242. Academic Press, New York and London.

Mitchison, J. M., and Vincent, W. S. 1965. Preparation of synchronous cell cultures by sedimentation. *Nature (London).* **205**:987–989.

Mobley, H. L., Koch, A. L., Doyle, R. J., and Streips, U. N. 1984. Insertion and fate of the cell wall in *Bacillus subtilis. J. Bacteriol.* **158**:169–179.

Mulder, E., and Woldringh, C. L. 1989. Actively replicating nucleoids influence positioning of division sites in *Escherichia coli* filaments forming cells lacking DNA. *J. Bacteriol.* **171**:4303–4314.

Muller, M. 1895. Ueber den Einflus von Fiebertemperaturen auf die Wachsthumsgeschwindigkeit und die Virulenz des Typhus-Bacillus. *Z. Hyg. Infectionskrakh.* **20**:245–280.

Murray, A. W., and Kirschner, M. W. 1989. Dominoes and clocks: The union of two views of the cell cycle. *Science* **246**:614–621.

Nagata, T. 1962. Polarity and synchrony in the replication of DNA molecules of bacteria. *Biochem. Biophys. Res. Commun.* **8**:348–351.

Nagata, T. 1963. The molecular synchrony and sequential replication of DNA in *Escherichia coli. Proc. Natl. Acad. Sci. USA* **49**:551–559.

Nanninga, N., and Woldringh, C. L. 1985. Cell growth, genome duplication, and cell division, *In* Molecular Cytology of *Escherichia coli*. N. Nanninga (ed.), pp. 259–318. Academic Press, London.

Nanninga, N., den Blaauwen, T., Nederlof, P. M., and de Boer, P. 1983. Premature division in *E. coli* in the presence of tris-EDTA. *In* The Target of Penicillin. R. Hakenbeck, J.-V. Höltje, and H. Labischinski (eds.), pp. 147–152. Walter de Gruyter, Berlin.

Nanninga, N., den Blaauwen, T., Voskuil, J., and Wientjes, F. 1985. Stimulation and inhibition of cell division in synchronized *Escherichia coli. Ann. Inst. Pasteur Microbiol.* **136A**:139–145.

Nanninga, N., Koppes, L. J. H., and de Vries-Thijssen, F. C. 1979. The cell cycle of *Bacillus subtilis* as studied by electron microscopy. *Arch. Microbiol.* **123**:173–181.

Nanninga, N., Woldringh, C. L., and Koppes, L. J. H. 1982. Growth and division of *Escherichia coli. In* Cell Growth. C. Nicolini (ed.), pp. 225–270. Plenum Publishing, New York.

Nathan, P., Osley, M. A., and Newton, A. 1982. Circular organization of the DNA synthetic pathway in *Caulobacter crescentus. J. Bacteriol.* **151**:503–506.

Neidhardt, F. C. 1987. Chemical composition of *Escherichia coli. In Escherichia coli* and *Salmonella typhimurium:* Cellular and Molecular Biology. F. C. Neidhardt, J. L. Ingraham, K. B. Low, B. Magasanik, M. Schaechter, and H. E. Umbarger (eds.), pp. 3–6. American Society for Microbiology, Washington, D. C.

Neidhardt, F. C., Ingraham, J. L., Low, K. B., Magasanik, B., Schaechter, M., and Umbarger, H. E. 1987. *Escherichia coli* and *Salmonella typhimurium*: Cellular and Molecular Biology. American Society for Microbiology, Washington, D.C.

Neidhardt, F. C., Ingraham, J., and Schaechter, M. 1990. The Physiology of the Bacterial Cell: A Molecular Approach. Sinauer Associates, Sunderland, Mass.

Newton, A. 1984. Temporal and spatial control of the *Caulobacter* cell cycle. *In* The Microbial Cell Cycle. P. Nurse and E. Streiblova (eds.), pp. 51–75. CRC Press, Boca Raton, Florida.

Newton, A. 1989. Differentiation in *Caulobacter*: Flagellum development, motility and chemotaxis. *In* Genetics of Bacterial Diversity. D. A. Hopwood and K. F. Chater (eds.), pp. 199–220. Academic Press, London.

Nishioka, Y., and Eisenstark, A. 1970. Sequence of genes replicated in *Salmonella typhimurium* as examined by transduction techniques. *J. Bacteriol.* **102**:320–333.

Nishi, A., and Hirose, S. 1966. Further observations on the rhythmic variation in peptidase activity during the cell cycle of various strains of *Escherichia coli*. *J. Gen. Appl. Microbiol.* (*Tokyo*) **12**:293–297.

Nishi, A., and Horiuchi, T. 1966. *β*-galactosidase formation controlled by episomal gene during the cell cycle of *Escherichia coli*. *J. Biochem* (*Tokyo*) **60**:338–340.

Nishi, A., and Kogoma, T. 1965. Protein turnover in the cell cycle of *Escherichia coli*. *J. Bacteriol.* **90**:884–890.

Norris, V. 1989. Phospholipid flip-out control the cell cycle of *Escherichia coli*. *J. Theor. Biol.* **139**:117–128.

Norris, V., Seror, S. J., Casaregola, S., and Holland, I. B. 1988. A single calcium flux triggers chromosome replication, segregation, and septation in bacteria: A model. *J. Theor. Biol.* **134**:341–350.

Ogden, G. B., and Schaechter, M. 1985. Chromosomes, plasmids, and the bacterial cell envelope. *In* Microbiology-1985. L. Leive (ed.), pp. 282–286. American Society for Microbiology, Washington, D. C.

Ohki, M. 1972. Correlation between metabolism of phosphatidylglycerol and membrane synthesis in *Escherichia coli*. *J. Mol. Biol.* **68**:249–264.

Ohta, N., Chen, L.-S., Swanson, E., and Newton, A. 1985. Transcriptional regulation of a periodically controlled flagellar gene operon in *Caulobacter crescentus*. *J. Mol. Biol.* **186**:107–115.

Oka, A., Sugimoto, K., Takanami, M., and Hirota, Y. 1980. Replication origin of the *Escherichia coli* K-12 chromosome: The size and structure of the minimum DNA segment carrying the information for autonomous replication. *Mol. Gen. Genet.* **178**: 9–20.

Okuda, A., and Cooper, S. 1989. The continuum model: An experimental and theoretical challenge to the G1-model of cell cycle regulation. *Exp. Cell Res.* **185**:1–7.

Olijhoek, A. J. M., Klencke, S., Pas, E., Nanninga, N., and Schwarz, U. 1982. Volume growth, murein synthesis, and murein cross-linkage during the division cycle of *Escherichia coli* PA3092. *J. Bacteriol.* **152**:1248–1254.

Osley, M. A., and Newton, A. 1974. Chromosome segregation and development in *Caulobacter crescentus*. *J. Mol. Biol.* **90**:359–370.

Osley, M. A., Sheffery, M., and Newton, A. 1977. Regulation of flagellin synthesis in the cell cycle of *Caulobacter*: Dependence on DNA replication. *Cell* **12**:393–400.

O'Sullivan, A., and Sueoka, N. 1972. Membrane attachment of the replication origins of a multifork (dichotomous) chromosome in *Bacillus subtilis*. *J. Mol. Biol.* **69**:237–248.

O'Sullivan, M. A., Howard, K., and Sueoka, N. 1975. Location of a unique replication terminus and genetic evidence for partial bidirectional replication of the *Bacillus subtilis* chromosome. *J. Mol. Biol.* **91**:15–38.

Pachler, P. F., Koch, A. L., and Schaechter, M. 1965. Continuity of DNA synthesis in *Escherichia coli*. *J. Mol. Biol.* **11**:650–653.

Padilla, G. M., and Cameron, I. L. 1964. Synchronization of cell division in *Tetrahymena pyriformis* by a repetitive temperature cycle. *J. Cell. Comp. Physiol.* **64**:303–308.

Painter, P. R. 1974. The relative numbers of different genes in exponential microbial cultures. *Genetics* **76**:401–410.

Painter, P. R., and Marr, A. G. 1968. Mathematics of microbial populations. *Annu. Rev. Microbiol.* **22**:519–548.

Paolozzi, L., Nicosia, A., Liebart, J. C., and Ghelardini, P. 1989. Synchronous division induced in *Escherichia coli* K12 by gemts mutants of phage Mu. *Mol. Gen. Genet.* **218**:13–17.

Pardee, A.B. 1968. Control of cell division: Models from microorganisms. *Cancer Res.* **28**:1802–1809.

Pardee, A. B. 1974. A restriction point for control of normal animal cell proliferation. *Proc. Natl. Acad. Sci. USA* **71**:1286–1290.

Pardee, A. B. 1989. G1 events and regulation of cell proliferation. *Science* **246**:603–608.

Pardee, A. B., Dubrow, R., Hamlin, J. L. and Kletzien, R. F. 1978. Animal cell cycle. *Annu. Rev. Biochem.* **47**:725–750.

Park, J.T. 1987a. Murein biosynthesis. *In Escherichia coli* and *Salmonella typhimurium:* Cellular and Molecular Biology. F. C. Neidhardt, J. L. Ingraham, K. B. Low, B. Magasanik, M. Schaechter, and H. E. Umbarger (eds.), pp. 663–671. American Society for Microbiology, Washington, D.C.

Park, J. T. 1987b. The murein sacculus. *In Escherichia coli* and *Salmonella typhimurium:* Cellular and Molecular Biology. F. C. Neidhardt, J. L. Ingraham, K. B. Low, B. Magasanik, M. Schaechter, and H. E. Umbarger (eds.), pp. 23–30. American Society for Microbiology, Washington, D.C.

Park, J. T., and Burman, L. G. 1985. Elongation of the murein sacculus of *Escherichia coli. Ann. Inst. Pasteur. Microbiol.* **136A**:51–58.

Pato, M. L. 1975. Alterations of the rate of movement of deoxyribonucleic acid replication forks. *J. Bacteriol.* **123**:272–277.

Pato, M. L., and Glaser, D. A. 1968. The origin and direction of replication of the chromosome of *Escherichia coli* B/r. *Proc. Natl. Acad. Sci. USA* **60**:1268–1274.

Paulton, R. J. L. 1970a. Analysis of the multiseptate potential of *Bacillus subtilis. J. Bacteriol.* **104**:762–767.

Paulton, R. J. L. 1970b. Cell septation during synchronous growth of *Bacillus subtilis. Nature (London)* **227**:517–518.

Paulton, R. J. L. 1971. The synchronization of *Bacillus subtilis* by the relaxation of a division control system. *Can. J. Microbiol.* **17**:119–122.

Penfold, W. J. 1914. On the nature of bacterial lag. *J. Hyg.* **14**:215–241.

Perry, R. P. 1958. On some properties of growth in a synchronized population of *Escherichia coli. Exp. Cell Res.* **17**:414–419.

Pierucci, O. 1969. Regulation of cell division. *Biophys. J.* **9**:90–112.

Pierucci, O. 1972. Chromosome replication and cell division in *Escherichia coli* at various temperatures of growth. *J. Bacteriol.* **109**:848–854.

Pierucci, O. 1978. Dimensions of *Escherichia coli* at various growth rates: Model for envelope growth. *J. Bacteriol.* **135**:559–574.

Pierucci, O. 1979. Phospholipid synthesis during the cell division cycle of *Escherichia coli. J. Bacteriol.* **138**:453–460.

Pierucci, O., and Helmstetter, C. E. 1976. Chromosome segregation in *Escherichia coli* B/r at various growth rates. *J. Bacteriol.* **128**:708–716.

Pierucci, O., and Zuchowski, C. 1973. Nonrandom segregation of DNA Strands in *Escherichia coli* B/r. *J. Mol. Biol.* **80**:477–503.

Pierucci, O., Melzer, M., Querini, C., Rickert, M., and Krajewski, C. 1981. Comparison among patterns of macromolecular synthesis in *Escherichia coli* B/r at growth rates of less and more than one doubling per hour at 37°C. *J. Bacteriol.* **148**:684–696.

Pinette, M. F. S., and Koch, A. L. 1987. Variability of the turgor pressure of individual cells of the gram-negative heterotroph *Ancylobacter aquaticus. J. Bacteriol.* **169**:4737–4742.

Plank, L. D., and Harvey, J. D. 1979. Generation time statistics of *Escherichia coli* B measured by synchronous culture techniques. *J. Gen. Microbiol.* **115**:69–77.

Plateau, P., Fromant, M., Kepes, F., and Blanquet, S. 1987. Intracellular 5′,5‴-dinucleoside polyphosphate levels remain constant during the *Escherichia coli* cell cycle. *J. Bacteriol.* 169:419–422.

Pledger, W. J., Stiles, C. D., Antoniades, H. N., and Scher, C. D. 1977. Induction of DNA synthesis in BALB/c 3T3 cells by serum components: Reevaluation of the commitment process. *Proc. Natl. Acad. Sci. USA* 74:4481–4485.

Poindexter, J. 1964. Biological properties and classification of the *Caulobacter* group. *Bact. Rev.* 28:231–295.

Poole, R. K. 1977a. Fluctuations in buoyant density during the cell cycle of *Escherichia coli* K12: Significance for the preparation of synchronous cultures by age selection. *J. Gen. Microbiol.* 98:177–186.

Poole, R. K. 1977b. Preparation of synchronous cultures of microorganisms by continuous flow selection: Which cells are selected. *FEMS Microbiol. Lett.* 1:305–307.

Pooley, H. M. 1976a. Layered distribution, according to age, within the cell wall of *Bacillus subtilis. J. Bacteriol.* 125:1139–1147

Pooley, H. M. 1976b. Turnover and spreading of old wall during surface growth of *Bacillus subtilis. J. Bacteriol.* 125:1127–1138.

Powell, E. O. 1955. Some features of the generation times of individual bacteria. *Biometrika* 42:16–44.

Powell, E. O. 1956. Growth rate and generation time of bacteria with special reference to continuous culture. *J. Gen. Microbiol.* 15:492–511.

Powell, E. O. 1958. An outline of the pattern of bacterial generation times. *J. Gen. Microbiol.* 18:382–417.

Powell, E. O., and Errington, F. P. 1963. Generation times of individual bacteria: Some corroborative measurements. *J. Gen. Microbiol.* 31:315–327.

Prentki, P., Chandler, M., and Caro, L. 1977. Replication of the prophage P1 during the cell cycle of *Escherichia coli. Mol. Gen. Genet.* 152:71–76.

Prescott, D. M. 1976. Reproduction of Eukaryotic Cells. Academic Press, New York.

Prescott, D. M., and Kuempel, P. L. 1972. Bidirectional replication of the chromosome in *Escherichia coli. Proc. Natl. Acad. Sci. USA* 69:2842–2845.

Pritchard, R. H. 1968. Control of DNA synthesis in bacteria. *Heredity* 23:472–473.

Pritchard, R. H. 1974. On the growth and form of a bacterial cell. *Philos. Trans. R. Soc. Lond.* B267:303–336.

Pritchard, R. H., and Lark, K. G. 1964. Induction of replication by thymine starvation at the chromosome origin in *Escherichia coli. J. Mol. Biol.* 9:288–307.

Pritchard, R. H., and Zaritsky, A. 1970. Effect of thymine concentration on the replication velocity of DNA in a thymineless mutant of *Escherichia coli. Nature (London)* 226: 126–131.

Pritchard, R. H., Barth, P. T., and Collins, J. 1969. Control of DNA synthesis in bacteria. *Symp. Soc. Gen. Microbiol.* 19:263–297.

Pritchard, R. H., Chandler, M., and Collins, J. 1975. Independence of F replication and chromosome replication in *Escherichia coli. Mol. Gen. Genet.* 138:143–155.

Pritchard, R. H., Meacock, P. A., and Orr, E. 1978. Diameter of cells of a thermosensitive *dnaA* mutant of *Escherichia coli* cultivated at intermediate temperatures. *J. Bacteriol.* 135:575–580.

Pucci, M. J., Hinks, E. T., Dicker, D. T., Higgins, M. L., and Daneo-Moore, L. 1986. Inhibition by β-lactam antibiotics at two different times in the cell cycle of *Streptococcus faecium* ATCC 9790. *J. Bacteriol.* 165:682–688.

Rahn, O. 1906. Ueber den einfluss der stoffwechselproducte auf das Wachstum der Bacterien. *Zentr. Bact. Parasitenk.* **16**:417–429, 609–617.

Rahn, O. 1931. A chemical explanation of the variability of the growth rate. *J. Gen. Physiol.* **15**:257–277.

Ramey, W. D., and Ishiguro, E. E. 1978. Site of inhibition of peptidoglycan biosynthesis during the stringent response in *Escherichia coli*. *J. Bacteriol.* **135**:71–77.

Rapaport, E., Zamecnik, P. C., and Baril, E. F. 1983. Association of diadenosine 5′,5′″-P^1,P^4-tetraphosphate binding protein with DNA polymerase α. *J. Biol. Chem.* **256**:12148–12151.

Reuter, S. H., and Shapiro, L. 1987. Asymmetric segregation of heat-shock proteins upon cell division of *Caulobacter crescentus*. *J. Mol. Biol.* **194**:653–662.

Ricciuti, C. P. 1972. Synchronized division in *Escherichia coli*: An integral portion of culture growth. *J. Bacteriol.* **112**:643–645.

Robin, A., Joseleau-Petit, D., and D'Ari, R. 1990. Transcription of the *ftsZ* gene and cell division in *Escherichia coli*. *J. Bacteriol.* **172**:1392–1399.

Rodriguez, R. L., and Davern, C. I. 1976. Direction of deoxyribonucleic acid replication in *Escherichia coli* under various conditions of cell growth. *J. Bacteriol.* **125**:346–352

Rodriguez, R. L., Dalbey, M. S., and Davern. C. I. 1973. Autoradiographic evidence for bidirectional DNA replication in *Escherichia coli*. *J. Mol. Biol.* **74**:599–604.

Ron, E. Z., Grossman, N., and Helmstetter, C. E. 1977. Control of cell division in *Escherichia coli*: Effect of amino acid starvation. *J. Bacteriol.* **129**:569–573.

Ron, E. Z., Rozenhak, S., and Grossman, N. 1975. Synchronization of cell division in *Escherichia coli* by amino acid starvation: Strain specificity. *J. Bacteriol.* **123**:374–376.

Rosenberg, B. H., Cavalieri, L. F., and Ungers, G. 1969. The negative control mechanism for *E. coli* DNA replication. *Proc. Natl. Acad. Sci. USA* **63**:1410–1417.

Rosenberger, R. F., Grover, N. B., Zaritsky, A., and Woldringh, C. L. 1978a. Control of microbial surface-growth by density. *Nature (London)* **271**:244–245.

Rosenberger, R. F., Grover, N. B., Zaritsky, A., and Woldringh, C. L. 1978b. Surface growth in rod-shaped bacteria. *J. Theor. Biol.* **73**:711–721.

Rothfield, L. I., DeBoer, P., and Cook, W. R. 1990. Localization of septation sites. *Res. Microbiol.* **141**:57–63.

Rothfield, L., and Cook, W. R. 1988. Periseptal annuli: Organelles involved in the bacterial cell division process. *Microbiol. Sci.* **5**:182–185.

Rubin, H., and Steiner, R. 1975. Reversible alterations in chick embryo mitotic cycle in various states of growth regulation. *J. Cell Physiol.* **85**:261–270.

Rudner, D., Rejman, E., and Chargaff, E. 1965. Genetic implications of periodic pulsations of the rate of synthesis and the composition of rapidly labelled bacterial RNA. *Proc. Natl. Acad. Sci. USA* **54**:904–911.

Ryan, F. J., and Cetrulo, S. D. 1963. Directed mutation in a synchronized bacterial population. *Biochem. Biophys. Res. Commun.* **12**:445–447.

Ryter, A. 1968. Association of the nucleus and the membrane of bacteria: A morphological study. *Bact. Rev.* **32**:39–54.

Ryter, A., Hirota, Y., and Jacob, F. 1968. DNA-membrane complex and nuclear segregation in bacteria. *Cold Spring Harbor Symp. Quant. Biol.* **33**:669–676.

Ryter, A., Hirota, Y., and Schwarz, U. 1973. Process of cellular division in *Escherichia coli*: Growth pattern of *E. coli* murein. *J. Mol. Biol.* **78**:185–195.

Ryter, A., Schuman, H., and Schwarz, U. 1975. Integration of the receptor for bacteriophage lambda in the outer membrane of *Escherichia coli*: Coupling with cell division. *J. Bacteriol.* **122**:295–301.

Sargent, M. 1973. Synchronous cultures of *Bacillus subtilis* obtained by filtration with glass fiber filters. *J. Bacteriol.* **116**:736–740.

Sargent, M. G. 1974. Nuclear segregation in *Bacillus subtilis*. *Nature (London)* **250**:252–254.

Sargent, M. G. 1975a. Control of cell length in *Bacillus subtilis*. *J. Bacteriol.* **123**:7–19.

Sargent, M. G. 1975b. Anucleate cell production and surface extension in a temperature-sensitive chromosome initiation mutant of *Bacillus subtilis*. *J. Bacteriol.* **123**:1218–1234.

Sargent, M. G. 1978. Surface extension and the cell cycle in prokaryotes. *Adv. Microb. Physiol.* **18**:105–176.

Savageau, M. 1989. Are there rules governing patterns of gene regulation? *In* Theoretical Biology: Epigenetic and Evolutionary Order from Complex Systems. B. Goodwin and P. Saunders (eds.), pp. 42–66. Edinburgh University Press, Edinburgh.

Schaechter, M., Bentzon, M. W., and Maaløe, O. 1959. Synthesis of deoxyribonucleic acid during the division cycle of bacteria. *Nature (London)* **183**:1207–1208.

Schaechter, M., Maaløe, O., and Kjeldgaard, N. O. 1958. Dependency on medium and temperature of cell size and chemical composition during balanced growth of *Salmonella typhimurium*. *J. Gen. Microbiol.* **19**:592–606.

Schaechter, M., Williamson, J. P., Hood, J. R., Jr., and Koch, A. L. 1962. Growth, cell, and nuclear divisions in some bacteria. *J. Gen. Microbiol.* **29**:421–434.

Schaus, N., O'Day, K., Peters, W., and Wright, A. 1981. Isolation and characterization of amber mutations in gene *dna*A of *Escherichia coli* K-12. *J. Bacteriol.* **145**:904–913.

Schlaeppi, J.-M., and Karamata, D. 1982. Cosegregation of the cell wall and DNA in *Bacillus subtilis*. *J. Bacteriol.* **152**:1231–1240.

Schlaeppi, J.-M., Schaefer, O., and Karamata, D. 1985. Cell wall and DNA cosegregation in *Bacillus subtilis* studied by electron microscope autoradiography. *J. Bacteriol.* **164**: 130–135.

Schwarz, U., and Glauner, B. 1988. Murein structure data and their relevance for understanding of murein metabolism in *Escherichia coli*. *In* Antibiotic Inhibition of Bacterial Cell-Surface Assembly and Function. P. Actor, L. Daneo-Moore, M. L. Higgins, M. R. J. Salton, and G. D. Shockman (eds.), pp. 33–40. American Society for Microbiology, Washington, D.C.

Schwarz, U., Ryter, A., Rambach, A., Hellio, R., and Hirota, Y. 1975. Process of cellular division in *Escherichia coli*: Differentiation of growth zones in the sacculus. *J. Mol. Biol.* **98**:749–759.

Scott, D. B. M., and Chu, E. 1958. Synchronized division of growing cultures of *Escherichia coli*. *Exp. Cell Res.* **14**:166–174.

Scott, R. I., Gibson, J. F., and Poole, R. K. 1980. Adenosine triphosphatase activity and its sensitivity to ruthenium red oscillates during the cell cycle of *Escherichia coli* K12. *J. Gen. Microbiol.* **120**:183–198.

Sedgwick, E. G., and Paulton, R. J. L. 1974. Dimension control in bacteria. *Can. J. Microbiol.* **20**:231–236.

Shapiro, L. 1976. Differentiation in the *Caulobacter* cell cycle. *Annu. Rev. Microbiol.* **30**: 377–407.

Shapiro, L. 1985. Generation of polarity during *Caulobacter* cell differentiation. *Annu. Rev. Cell Biol.* **1**:173–207.

Shapiro, L., Agabian-Keshishian, N., and Bendis, I. 1971. Bacterial differentiation. *Science* **173**:884–892.

Shapiro, L., and Agabian-Keshishian, N. 1970. Specific assay for differentiation in the stalked bacterium *Caulobacter crescentus*. *Proc. Natl. Acad. Sci. USA* **67**:200–203.

Shapiro, L., Mansour, J., Shaw, P., and Henry, S. 1982. Synthesis of specific membrane proteins is a function of DNA replication and phospholipid synthesis in *Caulobacter crescentus*. *J. Mol. Biol.* **159**:303–322.

Sheffery, M., and Newton, A. 1981. Regulation of periodic protein synthesis in the cell cycle: Control of initiation and termination of flagellar gene expression. *Cell* **24**:49–57.

Shehata, T. A., and Marr, A. G. 1975. Effect of temperature on the size of *Escherichia coli* cells. *J. Bacteriol.* **124**:857–862.

Shen, B. H. P., and Boos, W. 1973. Regulation of the β-methylgalactoside transport system and the galactose-binding protein by the cell cycle of *Escherichia coli. Proc. Natl. Acad. Sci. USA* **70**:1481–1485.

Sherman, J. M., and Albus, W. R. 1923. Physiological youth in bacteria. *J. Bacteriol.* **8**:127–139.

Shields, R. 1977. Transition-probability and the origin of variation in the cell cycle. *Nature (London)* **267**:704–707.

Shields, R. 1978. Further evidence for a random transition in cell cycle. *Nature (London)* **273**:755–758.

Siccardi, A. G., Galizzi, A., Mazza, G., Clivio, A., and Albertini, A. A. 1975. Synchronous germination and outgrowth of fractionated *Bacillus subtilis* spores: Tool for the analysis of differentiation and division of bacterial cells. *J. Bacteriol.* **121**:13–19.

Skarstad, K., Løbner-Olesen, A., Atlung, T., von Meyenburg, K., and Boye, E. 1989. Initiation of DNA replication in *Escherichia coli* after overproduction of the *dna*A protein. *Mol. Gen. Genet.* **218**:50–56.

Skarstad, K., Steen, H. B., and Boye, E. 1983. Cell-cycle parameters of slowly growing *Escherichia coli* B/r studied by flow cytometry. *J. Bacteriol.* **154**:656–662.

Skarstad, K., Steen, H. B., and Boye, E. 1985. *Escherichia coli* DNA distributions measured by flow cytometry and compared with theoretical computer simulations. *J. Bacteriol.* **163**:661–668.

Sloan, J. B., and Urban, J. E. 1976. Growth response of *Escherichia coli* to nutritional shift-up: Immediate division stimulation in slow-growing cells. *J. Bacteriol.* **128**:302–308.

Smith, C. L., and Condemine, G. 1990. New approaches for physical mapping of small genomes. *J. Bacteriol.* **172**:1167–1172.

Smith, D.W., and Hanawalt, P. C. 1967. Properties of the growing point region in the bacterial chromosome. *Biochim. Biophys. Acta* **149**:519–531.

Smith, J. A., and Martin, L. 1973. Do cells cycle? *Proc. Natl. Acad. Sci. USA* **70**:1263–1267.

Snellings, K., and Vermeulen, C. W. 1982. Nonrandom layout of the amino acid loci on the genome of *Escherichia coli. J. Mol. Biol.* **157**:687–688.

Sommer, J., and Newton, A. 1988. Sequential regulation of developmental events during polar morphogenesis in *Caulobacter crescentus*: Assembly of pili on swarmer cells requires cell separation. *J. Bacteriol.* **170**:409–415.

Spratt, B. G. 1975. Distinct penicillin-binding proteins involved in the division, elongation, and shape of *Escherichia coli* K12. *Proc. Natl. Acad. Sci. USA* **72**:2999–3003.

Spratt, B. G., and Rowbury, R. J. 1971. Physiological and genetical studies on a mutant of *Salmonella typhimurium* which is temperature-sensitive for DNA synthesis. *Mol. Gen. Genet.* **114**:35–49.

Staley, J. T., and Jordan, T. L. 1973. Crossbands of *Caulobacter crescentus* stalk serve as indicators of cell age. *Nature (London)* **246**:155–156.

Starka, J., and Koza, J. 1959. Nephelometric determination of cell count in synchronously dividing cultures of bacteria. *Biochim. Biophys. Acta* **32**:261–262.

Staugaard, P., van den Berg, F. M., Woldringh, C. L., and Nanninga, N. 1976. Localization of ampicillin-sensitive sites in *Escherichia coli* by electron microscopy. *J. Bacteriol.* **127**:1376–1381.

Steen, H. B. 1980. Further development of a microscope-based flow cytometer: Light scatter detection and excitation intensity compensation. *Cytometry* **1**:26–31.

Steen, H. B. 1983. A microscope-based flow cytometer. *Histochem. J.* **15**:147–160.

Steen, H. B., and Boye, E. 1980. Bacterial growth studied by flow cytometry. *Cytometry* **1**:32–36.

Steen, H. B., and Lindmo, T. 1979. Flow cytometry: a high-resolution instrument for everyone. *Science* **204**:403–404.

Steinberg, D., and Helmstetter, C. E. 1981. F plasmid replication and the division cycle of *Escherichia coli* B/r. *Plasmid* **6**:342–353.

Steinberg, W., and Halvorson, H. O. 1968. Timing of enzyme synthesis during outgrowth of spores of *Bacillus cereus*. II. Relationship between ordered enzyme synthesis and deoxyribonucleic acid replication. *J. Bacteriol.* **95**:479–489.

Stiles, C. D., Capone, G. T., Scher, C. D., Antoniades, H. N., Van Wyk, J. J., and Pledger, W. J. 1979. Dual control of cell growth by somatomedins and platelet-derived growth factor. *Proc. Natl. Acad. Sci. USA* **76**:1279–1283.

Streiblova, E., and Wolf, A. 1972. Cell-wall growth during the cell cycle of *Schizosaccharomyces pombe*. *Z. Allg. Microbiol.* **12**:673–684.

Sud, I. J., and Schaechter, M. 1964. Dependence of the content of cell envelope on the growth rate of *Bacillus megaterium*. *J. Bacteriol.* **88**:1612–1617.

Sueoka, N., and Quinn, W. C. 1968. Membrane attachment of the chromosome replication origin in *Bacillus subtilis*. *Cold Spring Harbor. Symp. Quant. Biol.* **33**:695–705.

Sugimoto, K., Oka, A., Sugisaki, H., Takanami, M., Nishimura, A., Yasuda, S., and Hirota, Y. 1979. Nucleotide sequence of *Escherichia coli* K12 replication origin. *Proc. Natl. Acad. Sci. USA* **76**:575–579.

Sussman, A. S., and Halvorson, H. O. 1966. Spores, Their Dormancy and Germination. Harper and Row, New York.

Swoboda, U., and Dow, C. S. 1979. The study of homogeneous populations of *Caulobacter* stalked (mother) cells. *J. Gen. Microbiol.* **112**:235–239.

Tax, R. 1978. Age distribution of *Caulobacter* cells in an exponential population. *J. Bacteriol.* **135**:16–17.

Temin, H. M. 1971. Stimulation by serum of multiplication of stationary chicken cells. *J. Cell. Physiol.* **78**:161–170.

Terrana, B., and Newton, A. 1975. Pattern of unequal cell division and development in *Caulobacter crescentus*. *Dev. Biol.* **44**:380–385.

Thompson, C. B., Challoner, P. B., Neiman, P. E. and Groudine, M. 1985. Levels of c-*myc* oncogene mRNA are invariant throughout the cell cycle. *Nature (London)* **314**:363–366.

Toennies, G., Iszard, L., Rogers, N. B., and Shockman, G. D. 1961. Cell multiplication studied with an electronic particle counter. *J. Bacteriol.* **82**:857–866.

Trueba, F. J. 1982. On the precision and accuracy achieved by *Escherichia coli* cells at fission about their middle. *Arch. Microbiol.* **131**:55–59.

Trueba, F. J., and Woldringh, C. L. 1980. Changes in cell diameter during the division cycle of *Escherichia coli*. *J. Bacteriol.* **142**:869–878.

Trueba, F. J., Neijssel, O. M., and Woldringh, C. L. 1982. Generality of the growth kinetics of the average individual cell in different bacterial populations. *J. Bacteriol.* **150**:1048–1055.

Trueba, F. J., Van Spronsen, E. A., Traas, J., and Woldringh, C. L. 1982. Effects of temperature on the size and shape of *Escherichia coli* cells. *Arch. Microbiol.* **131**:235–240.

Tsuchido, T., Van Bogelen, R. A., and Neidhardt, F. C. 1986. Heat-shock response in *Escherichia coli* influences cell division. *Proc. Natl. Acad. Sci. USA* **83**:6959–6963.

Tuomanen, E., and Cozens, R. 1987. Changes in peptidoglycan composition and penicillin-binding proteins in slowly growing *Escherichia coli*. *J. Bacteriol.* **169**:5308–5310.

Tyson, C. B., Lord, P. G., and Wheals, A. E. 1979. Dependency of size of *Saccharomyces cerevisiae* cells on growth rate. *J. Bacteriol.* **138**:92–98.

Tyson, J. J. 1979. Periodic enzyme synthesis: Reconsideration of the theory of oscillatory repression. *J. Theor. Biol.* **80**:27–38.

Tyson, J. J. 1985. The coordination of cell growth and division–intentional or incidental? *Bioessays* **2**:72–77.

Upcroft, P., Dyson, H. J., and Wake, R. G. 1975. Characteristics of a *Bacillus subtilis* W23 mutant temperature sensitive for initiation of chromosome replication. *J. Bacteriol.* **121**:121–127.

Vance, D., Goldberg, I., Mitsuhashi, O., Bloch, K., Omura, S., and Nomura, S. 1972. Inhibition of fatty acid synthesis by the antibiotic cerulenin. *Biochem. Biophys. Res. Commun.* **48**:649–656.

Van Alstyne, D., Grant, G. F., and Simon, M. 1969. Synthesis of bacterial flagella: Chromosomal synchrony and flagella synthesis. *J. Bacteriol.* **100**:283–287.

Van de Putte, P., van Dillewijn, J. E., and Rorsch, A. 1964. The selection of mutants of *E. coli* with impaired cell division at elevated temperature. *Mut. Res.* **1**:121–128.

Van Tubergen, R. P., and Setlow, R. B. 1961. Quantitative radioautographic studies on exponentially growing cultures of *Escherichia coli*. The distribution of parental DNA, RNA, protein, and cell wall among progeny cells. *Biophys. J.* **1**:589–625.

Verwer, R. W. H., and Nanninga, N. 1980. Pattern of *meso*-DL-2,6,-diaminopimelic acid incorporation during the division cycle of *Escherichia coli*. *J. Bacteriol.* **144**:327–336.

Verwer, R. W. H., Beachey, E. H., Keck, W., Stoub, A. M., and Poldermans, J. E. 1980. Oriented fragmentation of *Escherichia coli* sacculi by sonication. *J. Bacteriol.* **141**:327–332.

Verwer, R. W. H., Nanninga, N., Keck, W., and Schwarz, U. 1978. Arrangement of glycan chains in the sacculus of *Escherichia coli* *J. Bacteriol.* **136**:723–729.

von Meyenburg, K., Hansen, F. G., Atlung, T., Boe, L., Clausen, I. G., van Deurs, B., Hansen, E. B., Jorgensen, B. B., Jorgensen, F., Koppes, L., Michelsen, O., Nielsen, J., Pedersen, P. E., Rasmussen, K. V., Riise, E., and Skovgaard, O. 1985. Facets of the chromosomal origin of replication, *oriC*, of *Escherichia coli*. *In* Molecular Biology of Bacterial Growth. M. Schaechter, F. C. Neidhardt, J. L. Ingraham, and N. O. Kjeldgaard (eds.), pp. 260–281. Jones and Bartlett, Boston, Mass.

von Meyenburg, K., Hansen, F. G., Riise, E., Bergmans, H. E. N., Meijer, M., and Messer, W. 1979. Origin of replication, *oriC*, of the *Escherichia coli* K12 chromosome: Genetic mapping and minichromosome replication. *Cold Spring Harbor Symp. Quant. Biol.* **43**:121–128.

Vos-Scheperkeuter, G. H., Hofnung, M., and Witholt, B. 1984. High-sensitivity detection of newly induced LamB protein on the *Escherichia coli* cell surface. *J. Bacteriol.* **159**:440–447.

Wain, W. H. 1971. Synthesis of soluble protein during the cell cycle of the fission yeast *Schizosaccharomyces pombe*. *Exp. Cell Res.* **69**:49–56.

Wain, W. H., and Staatz, W. D. 1973. Rates of synthesis of ribosomal protein and total ribonucleic acid through the cell cycle of the fission yeast *Schizosaccharomyces pombe*. *Exp. Cell Res.* **81**:269–278.

Wake, R. G. 1972. Visualization of reinitiated chromosomes in *Bacillus subtilis*. *J. Mol. Biol.* **68**:501–509.

Wake, R. G. 1973. Circularity of the *Bacillus subtilis* chromosome and further study on its bidirectional replication. *J. Mol. Biol.* **77**:569–575.

Wake, R. G. 1975. Bidirectional replication in *Bacillus subtilis*. *In* DNA Synthesis and Its Regulation. M. Goulian, P. Hanawalt, and C. F. Fox (eds.), pp. 650–676. W. A. Benjamin, Menlo Park, Calif.

Ward, C. B., and Glaser, D. A. 1969a. Analysis of the chloramphenicol-sensitive and chloramphenicol-resistant steps in the initiation of DNA synthesis in *E. coli* B/r. *Proc. Natl. Acad. Sci. USA* **64**:905–912.

Ward, C. B., and Glaser, D. A. 1969b. Origin and direction of DNA synthesis in *Escherichia coli*. *Proc. Natl. Acad. Sci. USA* **62**:881–886.

Ward, C. B., and Glaser, D. A. 1969c. Evidence for multiple growing points on the genome of rapidly growing *Escherichia coli* B/r. *Proc. Natl. Acad. Sci. USA* **63**:800–804.

Ward, C. B., and Glaser, D. A. 1970. Control of initiation of DNA synthesis in *Escherichia coli* B/r. *Proc. Natl. Acad. Sci. USA* **67**:255–262.

Ward, C. B., and Glaser, D. A. 1971. Correlation between rate of cell growth and rate of DNA synthesis in *Escherichia coli* B/r. *Proc. Natl. Acad. Sci. USA* **68**:1061–1064.

Ward, H. M. 1895. On the biology of *Bacillus ramosus* (Fraenkel), a schizomycete of the river Thames. *Proc. R. Soc. Lond.* **58**:265–468.

Weidel, W., and Pelzer, A. 1964. Bagshaped molecules, a new outlook on bacterial cell walls. *Adv. Enzymol.* **26**:196–232.

Weinmann-Dorsch, C., Hedl, A., Grummt, I., Albert, W., Ferdinand, F. J., Friis, R. R., Pierron, G., Moll, W., and Grummt, F. 1984. Drastic rise of intracellular adenosine (5')tetraphospho(5')adenosine correlates with onset of DNA synthesis in eukaryotic cells. *Eur. J. Biochem.* **138**:179–185.

Wharton, W. 1983. Hormonal regulation of discrete portions of the cell cycle: Commitment to DNA synthesis is commitment to cellular division. *J. Cell. Physiol.* **117**:423–429.

White, J. H. M., Green, S. R., Barker, D. G., Dumas, L. B., and Johnston, L. H. 1987. The CDC8 transcript is cell-cycle regulated in yeast and is expressed coordinately with CDC9 and CDC12 at a point preceding histone transcription. *Exp. Cell Res.* **171**:223–231.

Wickner, S. H. 1978. DNA replication proteins of *Escherichia coli*. *Annu. Rev. Biochem.* **47**:1163–1191.

Wientjes, F. B., and Nanninga, N. 1989. Rate and topography of peptidoglycan synthesis during cell division in *Escherichia coli*: Concept of a leading edge. *J. Bacteriol.* **171**:3412–3419.

Wientjes, F. B., Olijhoek, A. J. M., Schwarz, U., and Nanninga, N. 1983. Labeling pattern of major penicillin-binding proteins of *Escherichia coli* during the division cycle. *J. Bacteriol.* **153**:1287–1293.

Wientjes, F. B., Schwarz, U., Olijhoek, A. J. M., and Nanninga, N. 1981. Pattern of penicillin-binding proteins during the life cycle of *Escherichia coli*. *In* The Target of Penicillin. R. Hakenbeck, J.-V. Höltje, and H. Labischinski (ed.), pp. 459–464. Walter de Gruyter, Berlin-New York.

Winslow, C.-E. A., and Walker, H. H. 1939. The earlier phases of the bacterial culture cycle. *Bacteriol. Rev.* **3**:147–186.

Woldringh, C. L. 1976. Morphological analysis of nuclear separation and cell division during the life cycle of *Escherichia coli*. *J. Bacteriol.* **125**:248–257.

Woldringh, C. L., and Nanninga, N. 1985. Structure of the nucleoid and cytoplasm in the intact cell. *In* Molecular Cytology of *Escherichia coli*. N. Nanninga (ed.), pp. 161–197. Academic Press, London.

Woldringh, C. L., Binnerts, J. S., and Mans, A. 1981. Variation in *Escherichia coli* buoyant density measured in percoll gradients. *J. Bacteriol.* **148**:58–63.

Woldringh, C. L., De Jong, M. A., van den Berg, W., and Koppes, L. 1977. Morphological analysis of the division cycle of two *Escherichia coli* substrains during slow growth. *J. Bacteriol.* **131**:270–279.

Woldringh, C. L., Grover, N. B., Rosenberger, R. F., and Zaritsky, A. 1980. Dimensional rearrangement of rod-shaped bacteria following nutritional shift-up. II. Experiments with *Escherichia coli* B/r. *J. Theor. Biol.* **86**:441–454.

Woldringh, C. L., Huls., P., Nanninga, N., Pas, E., Taschner, P. E. M., and Wientjes, F. B. 1988. Autoradiographic analysis of peptidoglycan synthesis in shape and division mutants of *Escherichia coli* MC1400 *In* Antibiotic Inhibition of Bacterial Cell-Surface Assembly and Function. P. Actor, L. Daneo-Moore, M. L. Higgins, M. R. J. Salton, and G. D. Shockman (eds.), pp. 66–78. American Society for Microbiology, Washington, D.C.

Woldringh, C. L., Huls, P., Pas, E., Brakenhoff, G. J., and Nanninga, N. 1987. Topography of peptidoglycan synthesis during elongation and polar cap formation in a cell division mutant of *Escherichia coli* MC4100. *J. Gen. Microbiol.* **133**:575–586.

Woldringh, C. L., Mulder, E., Valkenburg, J. A. C., Wientjes, F. B., Zaritsky, A., and Nanninga, N. 1990. Role of the nucleoid in the toporegulation of division. *Res. Microbiol.* **141**:39–49.

Woldringh, C. L., Valkenburg, J. A. C., Pas, E., Taschner, P. E. M., Huls, P., and Wientjes, F. B. 1985. Physiological and geometrical conditions for cell division in *Escherichia coli*. *Ann. Inst. Pasteur Microbiol.* **136A**:131–138.

Wolf, B., Newman, A., and Glaser, D. A. 1968. On the origin and direction of replication of the *Escherichia coli* K12 chromosome. *J. Mol. Biol.* **32**:611–629.

Womble, D. D., and Rownd, R. H. 1986a. Regulation of incFII plasmid DNA replication. A quantitative model for control of plasmid NR1 replication in the bacterial cell-division cycle. *J. Mol. Biol.* **192**:529–548.

Womble, D. D., and Rownd, R. H. 1986b. Regulation of λdv plasmid DNA replication. A quantitative model for control of plasmid λdv replication in the bacterial cell-division cycle. *J. Mol. Biol.* **191**:367–382.

Womble, D. D., and Rownd, R. H. 1987. Regulation of mini-F plasmid DNA replication. A quantitative model for control of plasmid min-F replication in the bacterial cell-division cycle. *J. Mol. Biol.* **195**:99–113.

Wood, N. B., and Shapiro, L. 1975. Morphogenesis during the cell cycle of the prokaryote, *Caulobacter crescentus*. *In* Cell Cycle and Cell Differentiation. J. Reinert and H. Holtzer (eds.), pp. 133–149. Springer-Verlag, New York.

Wraight, C. A., Lueking, D. R., and Kaplan, S. 1978. Synthesis of photopigments and electron transport components in synchronous phototrophic cultures of *Rhodopseudomonas sphaeroides*. *J. Biol. Chem.* **253**:465–471.

Wraight, C. A., Lueking, D. R., Fraley, R. T., and Kaplan, S. 1978. Synthesis of photopigment and electron transport components in synchronous phototrophic cultures of *Rhodopseudomonas sphaeroides*. *J. Biol. Chem.* **253**:465–471.

Wynford-Thomas, D., LaMontagne, A., Marin, G., and Prescott, D. 1985. Location of the isoleucine arrest point in CHO and 3T3 cells. *Exp. Cell Res.* **158**:525–532.

Yanagita, T., Maruyama, Y., and Takebe, I. 1958. Cellular response to deleterious agents during the course of synchronous growth of *Escherichia coli*. *J. Bacteriol.* **75**:523–529.

Yasuda, S., and Hirota, Y. 1977. Cloning and mapping of the replication origin of *Escherichia coli*. *Proc. Natl. Acad. Sci. USA* **74**:5458–5462.

Yoshikawa, H., and Haas, M. 1968. On the regulation of the initiation of DNA replication in bacteria. *Cold Spring Harbor Symp. Quant. Biol.* **33**:843–855.

Yoshikawa, H., and Sueoka, N. 1963a. Sequential replication of the *Bacillus subtilis* chromosome. I. Comparison of marker frequencies in exponential and stationary growth phases. *Proc. Natl. Acad. Sci. USA* **49**:559–566.

Yoshikawa, H., and Sueoka, N. 1963b. Sequential replication of the *Bacillus subtilis* chromosome. II. Isotopic transfer experiments. *Proc. Natl. Acad. Sci. USA* **49**:806–813.

Yoshikawa, H., O'Sullivan, A., and Sueoka, N. 1964. Sequential replication of the *Bacillus subtilis* chromosome. III. Regulation of initiation. *Proc. Natl. Acad. Sci. USA* **52**:973–980.

Young, I. E., and Fitz-James, P. C. 1959. Pattern of synthesis of deoxyribonucleic acid in *Bacillus cereus* growing synchronously out of spores. *Nature (London)* **183**:372–373.

Young, R., and Bremer, H. 1976. Polypeptide chain elongation rate in *Escherichia coli* B/r as a function of growth rate. *Biochem. J.* **160**:195–194.

Zaitseva, G. N. 1963. Some data on protein biosynthesis in synchronous culture of *Azotobacter vinelandii*. *Biochemistry* **28**:653–662.

Zamecnik, P. 1983. Diadenosine $5',5'''$-P^1,P^4-tetraphosphate (Ap_4A): Its role in cellular metabolism. *Anal. Biochem.* **134**:1–10.

Zamecnik, P. C., Rapaport, E., and Baril, E. F. 1982. Priming of DNA synthesis by diadenosine $5',5'''$-P^1,P^4-tetraphosphate with a double-stranded octadecamer as a template and a DNA polymerase α. *Proc. Natl. Acad. Sci. USA* **79**:1791–1794.

Zaritsky, A. 1975a. On dimensional determination of rod-shaped bacteria. *J. Theor. Biol.* **54**:243–248.

Zaritsky, A. 1975b. Rate stimulation of deoxyribonucleic acid synthesis after inhibition. *J. Bacteriol.* **122**:841–846.

Zaritsky, A., and Pritchard, R. H. 1971. Replication time of the chromosome in thymineless mutants of *E. coli*. *J. Mol. Biol.* **60**:65–74.

Zaritsky, A., and Pritchard, R. H. 1973. Changes in cell size and shape associated with changes in the replication time of the chromosome of *Escherichia coli*. *J. Bacteriol.* **114**:824–837.

Zaritsky, A., and Woldringh, C. L. 1978. Chromosome replication rate and cell shape in *Escherichia coli*: Lack of coupling. *J. Bacteriol.* **135**:581–587.

Zetterberg, A. and Larsson, O. 1985. Kinetic analysis of regulatory events in G_1 leading to proliferation or quiescence of Swiss 3T3 cells. *Proc. Natl. Acad. Sci. USA* **82**:5365–5369.

Zeuthen, E. 1958. Artificial and induced periodicity in living cells. *Adv. Biol. Med. Phys.* **6**:37–73.

Zeuthen, E. 1964. The temperature-induced synchrony in *Tetrahymena*. *In* Synchrony and Cell Division and Growth. E. Zeuthen (ed.), pp. 99–158. Wiley/Interscience, New York.

Zeuthen, E., and Scherbaum, O. H. 1954. Synchronous division in mass cultures of the ciliate protozoon *Tetrahymena pyriformis*, as induced by temperature changes. *Colston Pap.* **7**:141–155.

Zeuthen, J., and Pato, M. L. 1971. Replication of the F*lac* sex factor in the cell cycle of *Escherichia coli*. *Mol. Gen. Genet.* **111**:242–255.

Zeuthen, J., Morozow, E., and Pato, M. L. 1972. Pattern of replication of a colicin factor during the cell cycle of *Escherichia coli*. *J. Bacteriol.* **112**:1425–1427.

Zyskind, J. W., Cleary, J. M., Brusilow, W. S. A., Harding, N. E., and Smith, D. W. 1983. Chromosomal replication origin from the marine bacterium *Vibrio harveyi* functions in *Escherichia coli*: *ori*C consensus sequence. *Proc. Natl. Acad. Sci. USA* **80**:1164–1168.

Author Index

Abbo, F. E., 40, 60, 61, 91, 170, *437*
Adler, H. I., 61, *437*
Adolf, E. F., 61, 375, 387, *437, 438*
Agabian, N., 338, *437*
Agabian-Keshishian, N., 60, 337, *465*
Albert, W., 93, *469*
Albertini, A. A., 373, *466*
Albus, W. R., 317, *465*
Aller, P., 427, *450*
Altenbern, R. A., 60, *437*
Anagnostopoulos, G. D., 60, *437*
Anderson, A. J., 372, *437*
Anderson, P. A., 60, *437*
Andresdottir, V., 175, *437*
Antoniades, H. N., 427, 428, *462, 467*
Araki, H., 428, *451*
Archibald, A. R., 372, 373, *437*
Asmus, A., 242, *447*
Atlung, T., 176, 312, *437, 457, 466, 468*
Augenlicht, L., 427, *438*
Autissier, F., 312, *438, 452*

Balakrishnan, R., 311, *449*
Baldwin, J. H., 427, *450*
Baldwin, W. W., 60, 242, 252, *438, 455*
Barber, M. A., 317, *438*
Baril, E. F., 93, *464, 470*
Barker, D. G., 428, *469*
Barner, H. D., 60, *438*
Barth, P. T., 159, 171, 175, 245, *463*
Baserga, R., 426, 427, *438, 450*
Basilico, C., 410, 427, *447, 451*
Bayne-Jones, S., 61, 375, 387, *437, 438*
Beacham, I. R., 172, *438*
Beacham, K., 172, *438*
Beachey, E. H., 241, 242, *438, 468*
Beck, B. D., 244, *438*
Beck, E., 176, 311, *459*
Becker, W., 426, *438*
Beckman, B. E., 373, *445*

Beckwith, J., 174, *438*
Begg, K. J., 189, 218, 228, 236, 242, 245, 246, 312, *438, 444*
Bejaar, S., 176, *444*
Bellino, F. L., 75, 92, *438*
Ben-Sasson, S., 61, *448*
Bendigkeit, H. E., 60, 173, *455*
Bendis, I., 337, *465*
Bennett, D. C., 428, *440*
Benson, A., 311, *455*
Bentzon, M. W., 33, 59, 93, 170, *457, 465*
Bergmans, H. E. N., 176, 311, 312, *459, 468*
Bernlohr, R. W., 93, 244, *442*
Berns, A. A., 373, *452*
Bezanson, G., 167, 175, *438*
Billen, D., 175, *449*
Binkley, S. B., 60, *440*
Binnerts, J. S., 61, 252, *469*
Bird, R. E., 61, 113, 171, 173, 174, *439, 441, 457*
Blanquet, S., 93, *462*
Bleecken, S., 172, *439*
Bloch, K., 357, *468*
Blumberg, G., 243, 252, *453*
Blumenthal, L. K., 59, *439*
Boe, L., 176, *468*
Bolen, J. B., 427, *439*
Bonhoeffer, F., 105, 171, *439*
Boos, W., 92, *444, 465*
Boothby, D., 355, 357, *449*
Borenstein, S., 373, *445*
Botsford, J. L., 93, *439*
Bouche, J.-P., 176, *444*
Bouloc, P., 90, *445*
Bourdeau, P., 356, 357, *439*
Boyd, A., 62, 245, *439*
Boye, E., 61, 91, 171, 176, 388, *439, 466*
Boyle, J. V., 175, *439*
Brakenhoff, G. J., 61, 196, 200, 242, 246, *469*

Bramhill, D., 176, *439*
Braun, R. E., 176, *439*
Braun, V., 241, *439*
Bremer, H., 61, 93, 134, 172, 173, 277, *439,*
　440, 441, 470
Brenner, S., 241, 280, 311, *451*
Brock, T. D., 142, 173, *440*
Broda, P., 174, *458*
Bronfenbrenner, J., 17, 317, *449*
Brooks, R. F., 428, *440*
Brostrom, M. A., 60, *440*
Brusilow, W. S. A., 311, *471*
Buchanan, R. E., 17, 317, *440*
Buchner, H., 313, 317, *440*
Burdett, I. D. J., 244, 371, 373, 376, 388,
　440, 452, 453
Burman, L. G., 242, 244, 246, *440*
Burns, V. W., 60, *440*

Cairns, J., 137, 154, 171, 172, *440*
Cameron, I. L., 59, *461*
Campbell, A., 17, 59, *440*
Canepari, P., 61, 357, *440*
Canovas, J. L., 301, 311, *440*
Capone, G. T., 427, 428, *467*
Carl, P. L., 176, *440*
Caro, L., 61, 173, 175, 176, *441, 446, 452,*
　463
Carter, B. L. A., 90, 159, 175, *448*
Carty, C. E., 212, 244, *440*
Casaregola, S., 93, 243, *450, 461*
Cavalier-Smith, T., 312, *441*
Cavalieri, L. F., 175, *464*
Cavalli, L. L., 174, *456*
Cerda-Olmeda, E., 173, *441*
Cetrulo, S. D., 61, *464*
Chai, N.-C., 277, 311, *441*
Chakraborty, T., 176, *441*
Challoner, P. B., 407, 408, 427, *467*
Chaloupka, J., 243, *441, 445*
Chan, L., 372, 373, *458*
Chandler, M., 61, 173, 174, 175, 176, *439,*
　441, 446, 463
Chang, C.-F., 93, *441*
Chapman, J. W., 428, *441, 451*
Chargaff, E., 60, 90, 92, *464*
Chen, D. J.-C., 428, *441*
Chen, L.-S., 338, *461*
Chesney, A. M., 317, *441*

Cheung, H.-Y., 372, *441*
Chu, E., 59, 60, *465*
Chuang, L., 277, *439*
Chung, K. L., 241, 372, *441*
Churchward, G., 61, 134, 172, 173, 244,
　440, 441
Clark, D. J., 61, 93, 171, 173, *441, 451*
Clark, P. F., 314, 317, *441*
Clausen, E. S., 176, *437*
Clausen, I. G., 176, *468*
Cleary, J. M., 311, *471*
Clivio, A., 373, *466*
Coapes, H. E., 373, *437*
Cochran, B. H., 407, 427, *441, 452*
Cohen, S. S., 60, *438*
Cole, R. M., 242, 355, 372, *438, 441, 442*
Collins, J., 45, 61, 74, 91, 159, 165, 171,
　175, 245, 373, 375, 388, *442, 463*
Condemine, G., 171, *442*
Cook, T. M., 175, *439*
Cook, W. R., 93, 244, *442, 457, 464*
Cooper, S., 24, 62, 72, 91, 92, 93, 170, 171,
　172, 173, 175, 242, 243, 244, 245, 246,
　277, 278, 311, 338, 339, 388, 426, 427,
　428, 431, *442, 443, 447, 448, 458, 461*
Coplans, S., 317, *443*
Cozens, R., 244, *467*
Cullum, J., 277, 377, 388, *443*
Cummings, D. J., 40, 41, 61, 125, 172, *443,*
　448
Cutler, R. G., 35, 59, 92, *443*
Cuzin, F., 241, 280, 311, *451*

Dadd, A. H., 372, *443*
Dalbey, M. S., 154, 174, *464*
Daneo-Moore, L., 61, 355, 356, 357, *439,*
　440, 446, 449, 450, 463
Daniels, M. J., 212, 244, *443*
D'Ari, R., 60, 90, 172, 244, 312, *445, 451,*
　452, 464
Davern, C. I., 154, 174, *464*
Davis, D. B., 175, *443*
de Boer, P., 246, *460*
De Boer, W. R., 372, *444*
De Chastellier, C., 361, 373, *444*
De Jong, M. A., 61, 62, 172, 217, 242, 244,
　245, *454, 459, 469*
De Jonge, B. L. M., 60, 244, *444*
De Massey, B., 176, *444*

De Pedro, M. A., 241, *444*
de Vries-Thijssen, F. C., 277, *460*
DeBoer, P., 244, *464*
Degnen, S. T., 337, 338, 339, *443*
Del Mar Lléo, M., 61, 357, *440*
DeLamater, E., 59, *451*
DeLoach, J. R., 244, *438*
Demets, R., 242, *459*
den Blaauwen, T., 246, *460*
Denhardt, D. T., 61, 173, *455*
Dennis, P. P., 35, 60, 91, 92, *443*
Deutscher, M. P., 173, *457*
Dicker, D. T., 356, 357, *439, 444, 450, 463*
Dietzel, I., 92, *444*
Dingwall, A., 337, *444*
Doljanski, F., 61, *448*
Donachie, W. D., 75, 91, 92, 171, 175, 189,
    218, 228, 236, 242, 245, 246, 312, 366,
    373, *438, 444, 457*
Doudney, C. O., 59, 175, *444*
Dow, C. S., 338, *467*
Doyle, R. J., 243, 312, 357, 360, 370, 372,
    373, *445, 452, 453, 460*
Driehuis, F., 203, 243, 244, *444, 445*
Dubrow, R., 426, *462*
Dumas, L. B., 428, *469*
Dunn, G., 373, *452*
Dwek, R. D., 246, *445*
Dyson, H. J., 373, *467*

Eberle, H., 311, *445*
Eberly, S. L., 428, *451*
Ecker, R. E., 91, 356, *445*
Edelstein, E. M., 356, *445*
Edwards, S. W., 5, 60, *457*
Eisenman, R. N., 407, 408, 427, *448*
Eisenstark, S., 174, *461*
Elliot, S. G., 428, *445*
Ely, B., 337, *445*
Engstrom, W., 427, *456*
Ephrati-Elizur, E., 373, *445*
Errington, F. P., 255, 277, *463*
Estiva, E., 172, *441*
Evans, J. E., 35, 59, 92, *443, 445*
Evinger, M., 338, *437*

Falcone, G., 59, *445*
Fan, D. P., 373, *445*

Fantes, P., 427, 428, *440, 445*
Favre, D., 372, *459*
Ferdinand, F. J., 93, *469*
Fielding, P., 311, *445*
Figdor, C. G., 59, 60, *446*
Finkelstein, M., 175, *446*
Fitz-James, P. C., 60, *470*
Fontana, R., 61, 357, *440*
Forro, F., Jr., 105, 171, 265, 277, 280, 311,
    446
Fox, C. F., 311, *445*
Fraley, R. T., 92, *470*
Frank, M. E., 61, *450*
Freedman, M. L., 93, 146,
    173, *455*
Freese, E., 372, *441*
Frehel, C., 372, 373, *444, 445*
Frey, J., 176, *446*
Friis, R. R., 93, *469*
Friske, J. A., 61, *455*
Fromant, M., 93, *462*
Fukuda, A., 337, 339, *451*
Fuller, R. S., 174, 176, 311, *446, 452*
Funderburgh, M., 173, 174, *441, 457*
Funnell, B. E., 176, *446*

Galizzi, A., 373, *466*
Ganesan, A. T., 311, *446*
Gardner-Eckstrom, H. L., 373, *445*
Gausing, K., 93, *446*
Georgopoulos, C., 176, *459*
Ghelardini, P., 60, *461*
Gibson, C. W., 356, 357, 427, *446, 450*
Gibson, J. F., 92, *465*
Gierer, A., 105, 171, *439*
Glaser, D. A., 92, 152, 153, 173, 175, 357,
    446, 455, 462, 468, 469, 470
Glaser, L., 372, 373, *446, 458*
Glauner, B., 241, 243, 244, 246, *446, 451,*
    465
Gmeiner, J., 61, 374, *447*
Gnirke, H., 241, *439*
Goldberg, I., 357, *468*
Gomes, L., 338, *447*
Goodell, E. W., 242, 243, 244, *447*
Goodman, H., 93, *457*
Goodwin, B. C., 36, 60, 92, 373, *447*
Goss, W. A., 175, *439*
Graetzer, R., 242, 252, *455*

Grant, G. F., 92, *468*
Greco, A., 410, 427, 447, *451*
Green, E. W., 312, *447*
Green, R. S., 372, *437*
Green, S. R., 428, *469*
Grossman, N., 60, 225, 246, *445, 447, 464*
Groudine, M., 407, 408, 427, *467*
Grover, N. B., 36, 60, 61, 91, 175, 230, 242, 245, 246, 252, 376, 377, 388, *447, 457, 464, 469*
Grummt, F., 93, *447, 469*
Grummt, I., 93, *469*
Gudas, L. J., 60, 173, 212, 244, *447, 451*
Guenther, B., 244, *443*
Guiguet, M., 428, *447*
Gustafsson, P., 175, *448*

Haas, M., 175, 373, *470*
Hahn, J. J., 355, *442*
Hakenbeck, R., 244, *448*
Halvorson, H. O., 60, 90, 159, 175, 373, *448, 466, 467*
Hamlin, J. L., 426, *462*
Hammill, M., 60, *457*
Hanawalt, P. C., 6, 125, 154, 172, 173, 174, 311, *441, 448, 457, 466*
Hanks, M. C., 173, *448*
Hann, S. R., 407, 408, 427, *448*
Hansen, E. B., 176, *468*
Hansen, F. G., 176, 311, 312, *437, 448, 457, 459, 468*
Hardigree, A. A., 61, *437*
Harding, N. E., 311, *471*
Hartwell, L. H., 339, *448*
Harvey, J. D., 277, *448, 462*
Harvey, R. J., 61, 277, 377, 388, *448, 458*
Haselkorn, R., 60, *457*
Hayes, W., 174, *448*
Hedl, A., 93, *469*
Hegarty, C. P., 59, 317, *448*
Helinski, D. R., 176, *453*
Hellio, R., 241, 242, 361, 373, *444, 465*
Helmstetter, C. E., 41, 44, 60, 61, 62, 91, 142-143, 170, 171, 173, 175, 209, 218, 244, 245, 273, 278, 289, 299, 311, 312, 339, 388, 426, 427, *443, 446, 448, 449, 456, 462, 464, 466*
Hendrickson, W. G., 311, *449, 455*
Henning, U., 241, 246, *439, 449*
Henrici, A. T., 12, 13, 17, 314, 317, *449*

Henry, S., 338, 339, *457, 465*
Henson, J. M., 176, *450*
Hershey, A. D., 17, 314, 317, *449*
Hewitt, A. D., 175, *449*
Hidaka, M., 176, *452*
Higgins, M. L., 355, 356, 357, 370, 373, *439, 444, 446, 449, 450, 453, 463*
Higgs, J. A., 373, *457*
Hill, T. M., 176, *450, 455*
Hinks, E. T., 357, *463*
Hinks, R. P., 355, 356, 357, *450*
Hirose, S., 92, *461*
Hirota, Y., 176, 225, 241, 242, 243, 246, 286, 288, 311, 312, *450, 456, 461, 464, 465, 467, 470*
Hirschhorn, R. R., 427, *450*
Hitchins, A. D., 374, *450*
Hochhauser, S. W., 426, 427, *450*
Hoffman, B., 241, 242, *450*
Hoffman, E. J., 61, 154, 157, 174, *456*
Hoffman, H., 61, *450*
Hofnung, M., 312, *468*
Holland, I. B., 60, 62, 93, 243, 244, 245, *439, 441, 450, 452, 461*
Holley, R. W., 427, *450*
Holmes, M., 209, 244, 367, 373, *449, 450*
Höltje, J.-V., 93, 241, 243, 246, *446, 451*
Hood, J. R., Jr., 241, 242, 277, *465*
Horiuchi, T., 175, 176, *452, 461*
Hotchkiss, R. D., 4, 6, 33, 59, *451*
Howard, A., 33, 59, 170, 428, *451*
Howard, K., 373, *461*
Hsieh, M.-L., 242, 244, 246, *443*
Hughes, R. C., 372, *451*
Huguenel, E., 338, *451*
Huls, P., 61, 196, 200, 242, 246, *469, 470*
Hunt, T., 428, *440*
Hunter-Szybalska, M., 59, *451*
Huzyk, L., 93, *451*

Iba, H., 337, 339, *451*
Ingraham, J. L., 1, 5, 93, *451, 460*
Ingram, L. O., 212, 244, *440*
Isaac, P. K., 241, 372, *441*
Ishiguro, E. E., 243, *451, 463*
Ittmann, M., 410, 427, 447, *451*

Jacob, F., 174, 176, 241, 243, 280, 286, 288, 311, *450, 451, 456, 464*

Jacq, A., 311, *451*
Jaffe, A., 312, *451*
James, R., 212, 244, *451*
John, L., 60, *457*
Johnston, L. H., *428*, *441*, *451*, *469*
Jolliffe, L. K., 372, *452*
Jones, C., 244, *452*
Jordan, T. L., 338, *466*
Jorgensen, B. B., 176, *468*
Jorgensen, F., 176, *468*
Joseleau-Petit, D., 60, 172, 212, 244, 245, *442*, *452*, *464*
Jurida, I., 244, *444*

Kader, J., 60, *457*
Kaguni, M. M., 174, *452*
Kalb, V. F., Jr., 93, 244, *442*
Kaplan, S., 92, *470*
Karamata, D., 374, *465*
Kawirko, R. Z., 241, 372, *441*
Keck, W., 241, *468*
Kellenberger-Gujer, G., 176, *452*
Kelly, K., 407, 427, *452*
Kendall, D. G., 277, *452*
Kennett, R. H., 93, 373, *452*
Kepes, A., 36, 37, 60, 212, 244, 245, 312, *438*, *452*
Kepes, F., 36, 37, 60, 93, 172, 212, 244, 245, *442*, *452*, *462*
Keynan, A., 373, *452*
Kiernan, J. A., 427, *450*
King, M., 311, *449*, *455*
Kirkwood, T. B. L., 373, 376, 388, *440*, *452*
Kirschner, M. W., *428*, *460*
Kjeldgaard, N. O., 4, 6, 13, 14, 17, 127, 130, 141, 171, 172, 217, 245, 246, 278, *452*, *457* *465*
Klencke, S., 61, 241, 242, *461*
Kletzien, R. F., 426, *462*
Knaysi, G., 375, 387, *452*
Kobayashi, T., 176, *452*
Kobrin, L. H., 246, *445*
Koch, A. L., 59, 60, 61, 93, 170, 183, 185, 186, 200, 203, 231, 241, 242, 243, 244, 246, 252, 262, 264, 265, 276, 277, 278, 312, 338, 356, 357, 360, 370, 371, 372, 373, 388, *440*, *450*, *452*, *453*, *460*, *461*, *462*, *465*
Kogoma, T., 60, 61, *461*
Kohiyama, M., 311, *451*

Kokaisl, G., 91, 356, *445*
Kolb, V., 92, *444*
Kolter, R., 176, *453*
Koppes, L. J. H., 61, 62, 91, 172, 173, 175, 176, 217, 242, 243, 244, 245, 276, 277, 312, 376, 377, 388, 447, *454*, *460*, *468*, *469*
Kornberg, A., 169, 174, 176, 311, 436, *439*, *446*, *452*, *454*
Koza, J., 59, *466*
Krajewski, C., 244, *462*
Kruyssen, F. J., 372, *443*
Kubitschek, H. E., 60, 61, 92, 93, 146, 173, 242, 243, 245, 246, 252, 277, 388, *438*, *454*, *455*
Kuempel, P. L., 60, 92, 154, 174, 176, 361, 373, *450*, *455*, *458*, *463*
Kung, F. C., 92, *455*
Kunicki-Goldfinger, W., 59, *455*
Kurn, N., 337, *455*
Kusano, T., 311, *449*, *455*

LaMontagne, A., *428*, *470*
Lane, D., 61, 173, *455*
Lane, H. E. D., 173, *455*
Lane-Claypon, J. E., 313, 317, *455*
Lark, C. A., 60, 106, 171, *456*
Lark, K. G., 33, 59, 60, 61, 93, 106, 154, 157, 160, 170, 171, 172, 173, 174, 175, 277, 311, 324, 338, *439*, *441*, *445*, *455*, *456*, *458*, *463*
Larsson, O., 403, 427, *456*, *471*
Lau, L. F., 427, *456*
Leder, P., 407, 427, *452*
Lederberg, E. M., 174, *456*
Lederberg, J., 174, 311, *446*, *456*
Ledingham, J. C. G., 313, 317, *456*
Leof, E. B., 427, *456*
Leonard, A. C., 175, 218, 245, 311, 312, *448*, *449*, *456*
Lernhardt, W., 427, *459*
Leuking, D. R., 92, *470*
Liebart, J. C., 60, *461*
Lin, E. C. C., 243, 286, 288, 311, *456*
Lindmo, T., 61, *466*
Lindsay, B., 372, *446*
Liskay, R. M., 397, 401, 426, 427, *456*
Littauer, U. Z., 173, *457*

Lloyd, D., 5, 60, 457
Løbner-Olesen, A., 176, 312, 457, 466
Loken, M. R., 60, 173, 455
Longard, K., 313, 317, 440
Longsworth, L. G., 17, 317, 457
Lopez-Saez, J. F., 301, 311, 440
Lord, P. G., 339, 428, 467
Lother, H., 176, 311, 441, 451
Lott, T., 337, 457
Louarn, J., 113, 171, 173, 174, 176, 439, 444, 457
Louarn, J.-M., 176, 444
Low, K. B., 5, 460
Lutkenhaus, J., 75, 91, 244, 457

Maaløe, O., 1, 4, 5, 6, 13, 14, 17, 33, 59, 61, 93, 125, 127, 130, 141, 154, 170, 171, 172, 174, 175, 246, 271, 278, 441, 448, 451, 452, 456, 457, 465
MacAlister, T. J., 244, 442, 457
MacDonald, B., 244, 457
Magasanik, B., 5, 460
Mandelstam, J., 373, 452, 457
Manor, H., 60, 93, 173, 457
Mans, A., 61, 252, 469
Mansour, J. D., 338, 339, 457, 465
Margalit, H., 175, 457
Margolis, S., 172, 458
Marians, K., 176, 458
Marin, G., 428, 470
Marr, A. G., 61, 109, 171, 259, 270, 277, 377, 388, 448, 458, 461, 465
Marsh, R. C., 155, 174, 458
Martin, J. A., 242, 252, 458
Martin, L., 428, 466
Martinez-Salas, E., 242, 252, 458
Martuscelli, J., 113, 171, 174, 439
Marunouchi, T., 173, 458
Maruyama, Y., 40, 60, 61, 93, 458, 470
Masters, M., 60, 75, 91, 92, 155, 173, 174, 175, 361, 373, 437, 444, 448, 455, 457, 458, 459
Matney, T. S., 60, 458
Matsuhashi, M., 241, 458
Mauck, J., 372, 373, 458
May, J. W., 242, 459
Mazza, G., 373, 466
McFall, E., 170, 459
McKenna, W. G., 155, 174, 459

McLaughlin, C. S., 428, 445
McMacken, R., 176, 459
Meacock, P. A., 245, 246, 459, 463
Meijer, M., 174, 176, 311, 312, 459, 468
Melchers, F., 427, 459
Melzer, M., 244, 462
Mendelson, N. H., 372, 459
Meselson, M., 106, 156, 171, 174, 459
Messer, W., 173, 174, 175, 176, 241, 242, 244, 311, 312, 441, 448, 450, 451, 458, 459, 468
Metzger, N., 243, 443
Meyer, M., 62, 242, 244, 454, 459
Michelsen, O., 176, 468
Milde, K., 61, 374, 447
Minnich, S. A., 338, 459
Mitchison, J. M., 2, 5, 38, 60, 420, 428, 459, 460
Mitsuhashi, O., 357, 468
Mobley, H. L. T., 312, 372, 453, 460
Moll, W., 93, 469
Moore, B. A., 75, 91, 457
Mordoh, J., 176, 450
Mulder, E., 217, 228, 244, 245, 312, 460, 470
Muller, M., 313, 317, 460
Munson, B. R., 312, 456
Murchie, J., 311, 449, 455
Murray, A. W., 428, 460
Mycielski, R., 59, 455

Naaman, J., 61, 447
Nagata, T., 40, 60, 61, 157, 174, 460
Nanninga, N., 61, 62, 75, 91, 93, 173, 196, 200, 224, 241, 242, 243, 244, 246, 263, 276, 277, 338, 451, 453, 454, 459, 460, 461, 466, 468, 469
Nathan, P., 338, 460
Nathans, D., 427, 456
Navarette, M. H., 301, 311, 440
Nederlof, P. M., 246, 460
Neidhardt, F. C., 1, 5, 244, 388, 460, 467
Neijssel, O. M., 61, 377, 388, 467
Neiman, P. E., 407, 408, 427, 467
Newman, A., 153, 173, 470
Newman, C. N., 61, 173, 455
Newton, A., 337, 338, 339, 443, 457, 459, 460, 461, 465, 466
Nicosia, A., 60, 461

Nielsen, J., 176, *468*
Nilson, E. H., 259, 277, *458*
Nishi, A., 60, 61, 92, 175, *461*
Nishimura, A., 176, *450, 467*
Nishioka, Y., 174, *461*
Nomura, S., 357, *468*
Nordstrom, K., 175, *448, 454*
Norris, V., 93, 243, 244, *450, 461*

O'Day, K., 176, *439, 465*
Ogden, G. B., 311, *461*
Ohki, M., 212, 244, *461*
Ohta, N., 337, 338, *457, 461*
Oka, A., 176, *450, 461, 467*
Okuda, A., 426, *461*
Olijhoek, A. J. M., 59, 60, 61, 75, 91, 241, 242, *446, 461, 469*
Omura, S., 357, *468*
Oonk, H. B., 62, 242, 244, *454*
Orr, E., 246, *463*
Osley, M. A., 338, 339, *460, 461*
O'Sullivan, M. A., 311, 373, *458, 461, 470*
Overbeeke, N., 242, *454*

Pachler, P. F., 59, 61, 170, *461*
Padilla, G. M., 59, *461*
Pai, S. R., 92, *455*
Painter, P. R., 173, 259, 270, 277, *458, 461*
Paolozzi, L., 60, *461*
Pardee, A. B., 40, 60, 61, 91, 92, 170, 173, 361, 373, 411, 412, 426, 427, 428, *437, 447, 455, 458, 462*
Park, J. T., 241, 242, 244, 246, *438, 440, 462*
Parker, G., 338, *437*
Pas, E., 61, 196, 200, 241, 242, 246, *461, 469, 470*
Pato, M. L., 152, 173, 175, *462, 471*
Paulton, R. J. L., 372, 373, *443, 462, 465*
Peace, A. A., 93, 244, *442*
Pedersen, P. E., 176, *468*
Pelc, S. R., 33, 59, 170, 428, *451*
Pelletier, A. J., 176, *455*
Pelzer, A., 241, *469*
Pene, J. J., 311, *451*
Penfold, W. J., 313, 317, *456, 462*
Perram, J. W., 175, *448*
Perry, R. P., 91, *462*

Peters, W., 176, *465*
Pettijohn, D. E., 60, *437*
Phillips, C. A., 60, *457*
Pierron, G., 93, *469*
Pierucci, O., 62, 142–143, 171, 172, 173, 175, 189, 209, 242, 244, 245, 273, 278, 289, 299, 311, 367, 373, 426, 427, *449, 450, 462*
Pinette, M. F. S., 246, 252, *453, 462*
Plank, L. D., 277, *462*
Plateau, P., 93, *462*
Pledger, W. J., 427, 428, *456, 462, 467*
Podhajska, A. J., 176, *452*
Poindexter, J., 337, *463*
Poldermans, J. E., 241, *468*
Poole, R. K., 5, 60, 92, 252, *457, 463, 465*
Pooley, H. M., 372, *463*
Powell, E. O., 46, 61, 255, 277, *463*
Prentki, P., 175, *463*
Prescott, D., 428, *470*
Pritchard, R. H., 135, 154, 159, 165, 171, 172, 173, 174, 175, 242, 243, 245, 246, 252, 397, 426, 427, *438, 442, 456, 459, 463, 471*
Pucci, M. J., 357, *463*

Querini, C., 244, *462*
Quinn, W. C., 311, *467*

Rahn, O., 277, 317, *463*
Raichler, J., 242, *440*
Rambach, A., 241, 242, *465*
Ramey, W. D., 243, *451, 463*
Rapaport, E., 93, *464, 470*
Rasmussen, K. V., 175, 176, 312, *437, 448, 457, 468*
Raymond, J., 92, *455*
Reffel, A. C., 427, *441*
Rehn, K., 241, *439*
Rejman, E., 60, 90, 92, *464*
Renger, H., 160, 175, 324, 338, *456*
Repko, T., 61, 154, 157, 174, *456*
Reuter, S. H., 312, 339, *464*
Revelas, L., 426, *448*
Ricciuti, C. P., 60, *464*
Richmond, M., 45, 61, 74, 91, 373, 375, 388, *442*
Rickert, M., 244, *462*

Riedlin, G., 313, 317, *440*
Riise, E., 176, *468*
Roberts, E. M., 245, *459*
Robin, A., 60, 172, 244, *464*
Robinson, A. C., 218, 234, 245, *444*
Rodriguez, R. L., 154, 174, *464*
Rogers, N. B., 356, *467*
Ron, E. Z., 60, 225, 246, *445, 447, 464*
Rorsch, A., 246, *468*
Rosenberg, B. H., 175, *464*
Rosenberger, R. F., 230, 242, 245, 246, 252, *447, 464, 469*
Rosenzweig, M. S., 356, *445*
Rothfield, L. I., 244, *442, 457, 464*
Rowbury, R. J., 172, *466*
Rownd, R. H., 176, *470*
Rozenhak, S., 60, *464*
Rubin, H., 427, *464*
Rudner, D., 60, 90, 92, *464*
Ruehl, W. H., 314, 317, *441*
Ruettinger, T., 171, 244, *443*
Ryan, F. J., 61, *464*
Ryter, A., 225, 241, 242, 246, 311, 312, 361, 372, 373, *444, 446, 464, 465*

Sargent, M., 60, 242, 368, 372, 373, *464*
Sarnow, E., 61, 374, *447*
Savageau, M., 174, *465*
Scanlon, S., 245, 311, *443*
Schaechter, M., 1, 4, 5, 6, 13, 14, 17, 33, 59, 61, 125, 127, 130, 141, 170, 171, 172, 217, 241, 242, 245, 246, 262, 277, 278, 311, 312, *447, 448, 449, 452, 453, 455, 460, 461, 465, 467*
Schaller, H., 176, 311, *459*
Schaus, N., 176, *465*
Scher, C. D., 427, 428, *462, 467*
Scherbaum, O. H., 59, *471*
Schlaeppi, J.-M., 374, *465*
Schroeter, S., 242, 252, *455*
Schuman, H., 312, *464*
Schwarz, U., 61, 75, 91, 225, 241, 242, 243, 244, 246, 312, *444, 446, 447, 449, 450, 461, 464, 465, 468, 469*
Schwimmer, M., 245, 311, *443*
Scott, D. B. N., 59, 60, *465*
Scott, R. I., 92, *465*
Sedgwick, E. G., 373, *465*
Seror, S. J., 93, 243, *461*

Setlow, R. B., 104, 170, 200, 242, 279, 308, 310, *468*
Shapiro, L., 60, 337, 338, 339, *444, 445, 447, 455, 457, 465, 470*
Shaw, P., 338, *465*
Sheffery, M., 338, 339, *461, 465*
Shehata, T. A., 109, 171, *465*
Shen, B. H. P., 92, *465*
Sherman, J. M., 317, *465*
Shields, R., 428, *466*
Shockman, G. D., 61, 355, 356, 357, *440, 450, 467*
Shuman, H., 93, *441*
Siccardi, A. G., 373, *466*
Silver, L., 176, *459*
Silver, S., 93, *455*
Simon, M., 92, *468*
Skarstad, K., 61, 91, 176, 312, 388, *439, 457, 466*
Skovgaard, O., 176, *468*
Sloan, J. B., 172, *466*
Smith, C. L., 171, *442*
Smith, D. W., 311, *466, 471*
Smith, G. L., 427, *439*
Smith, J. A., 428, *440, 466*
Snellings, K., 156, 174, *466*
Somlyo, A., 93, *441*
Sommer, J., 338, *466*
Spratt, B. G., 92, 172, *466*
Staatz, W. D., 428, *468*
Stahl, F. W., 106, 156, 171, 174, *459*
Staley, J. T., 338, *466*
Starka, J., 59, *466*
Staugaard, P., 277, *466*
Steen, H. B., 47, 61, 91, 171, 388, *438, 466*
Stein, G. S., 426, 427, *450*
Stein, J. L., 426, 427, *450*
Steinberg, D., 175, *466*
Steinberg, W., 373, *466*
Steiner, R., 427, *464*
Steinmetz, D., 311, *455*
Stent, G. S., 93, 170, *457, 459*
Stiles, C. D., 407, 427, 428, *441, 452, 462, 467*
Stockman, G. D., 355, 357, *449*
Stokes, E., 372, *451*
Stoub, A. M., 241, *468*
Streiblova, E., 312, *467*
Streips, U. N., 312, 372, *452, 453, 460*
Strnadova, M., 243, *441*

Author Index

Sturman, A. J., 372, 437
Sud, I. J., 246, 467
Sueoka, N., 93, 311, 373, 452, 458, 461, 467, 470
Sugimoto, K., 176, 450, 461, 467
Sugino, A., 428, 451
Sugisaki, H., 176, 450, 467
Suit, J. C., 60, 458
Sullivan, N. F., 236, 246, 444
Sussman, A. S., 60, 467
Swanson, E., 338, 461
Swoboda, U., 338, 467
Szybalski, W., 59, 445, 451

Takanami, M., 176, 450, 461, 467
Takebe, I., 61, 470
Tang, M. S., 209, 244, 449
Taschner, P. E. M., 242, 469, 470
Tauro, P., 90, 159, 175, 448
Tax, R., 334, 339, 467
Teather, R. M., 244, 447
Temin, H. M., 427, 467
Terrana, B., 337, 467
Thompson, C. B., 407, 408, 427, 448, 467
Thwaits, J. J., 372, 459
Toennies, G., 356, 467
Traas, J., 245, 277, 311, 467
Trentini, W. C., 277, 458
Tresguerres, E. F., 301, 311, 440
Trueba, F. J., 61, 222, 224, 245, 277, 311, 377, 388, 467
Tsuchido, T., 244, 467
Tuomanen, E., 244, 467
Tyson, C. B., 339, 428, 467
Tyson, J. J., 277, 373, 467

Umbarger, H. E., 5, 460
Unger, M., 339, 448
Ungers, G., 175, 464
Upcroft, P., 373, 467
Urban, J. E., 172, 466
Uretz, R. B., 61, 449

Valkenburg, J. A. C., 217, 228, 242, 244, 245, 312, 470
Van Alstyne, D., 92, 468
Van Bogelen, R. A., 244, 467

Van de Putte, P., 246, 468
van den Berg, F. M., 277, 466
van den Berg, W., 61, 172, 217, 245, 469
van Deurs, B., 176, 468
van Dillewijn, J. E., 246, 468
Van Spronsen, E. A., 245, 277, 311, 467
Van Tubergen, R. P., 104, 170, 200, 242, 279, 308, 310, 468
Van Wyk, J. J., 427, 428, 456, 467
Vance, D., 357, 468
Vermeulen, C. W., 156, 174, 466
Verwer, R. W. H., 241, 242, 263, 277, 312, 338, 453, 468
Vicente, M., 189, 218, 228, 242, 245, 246, 252, 277, 377, 388, 443, 444, 458
Vincent, W. S., 38, 60, 460
Vinter, V., 243, 445
Vitkovic, L., 372, 441
von Meyenburg, K., 176, 311, 312, 454, 457, 459, 466, 468
Vos-Scheperkeuter, G. H., 312, 468
Voskuil, J., 246, 460

Wain, W. H., 428, 468
Wake, R. G., 373, 467, 468
Wang, R. J., 428, 441
Ward, C. B., 92, 152, 173, 175, 468, 469
Ward, H. M., 6, 375, 387, 469
Weeks, O. B., 59, 448
Wegener, W. S., 60, 438
Weidel, W., 241, 469
Weigand, R., 244, 457
Weinberger, M., 209, 244, 245, 311, 312, 443, 449, 456
Weinmann-Dorsch, C., 93, 469
Wertheimer, S. A., 105, 171, 265, 277, 280, 311, 446
Whalley, J. B., 373, 440
Wharton, W., 427, 456
Wheals, A. E., 339, 428, 467
Wheatley, D., 428, 440
White, J. H. M., 428, 469
White, K., 373, 458
Wicker, S. H., 176, 241, 469
Wientjes, F. B., 75, 91, 217, 224, 228, 242, 243, 244, 245, 246, 312, 444, 460, 469, 470
Williamson, J., 373, 458
Williamson, J. P., 241, 242, 277, 465

Winslow, C.-E. A., 314, 317, *469*
Witholt, B., 312, *468*
Woldringh, C. L., 61, 62, 91, 142, 172,
  173, 196, 200, 211, 217, 222, 224, 225,
  228, 230, 242, 243, 244, 245, 246, 252,
  276, 277, 311, 312, 376, 377, 388, *447,*
  454, 460, 464, 466, 467, 469, 470, 471
Wolf, A., 312, *467*
Wolf, B., 153, 173, *470*
Wollman, E. L., 174, *451*
Womack, J. E., 176, 312, *459*
Womble, D. D., 176, *470*
Wood, N. B., 337, *470*
Worcel, A., 155, 174, *458*
Wouters, J. T. M., 203, 243, 244, 372, *443,*
  444, 445
Wraight, C. A., 92, *470*
Wright, A., 176, *439, 465*
Wuesthoff, G., 175, 338, 427, *443*
Wynford-Thomas, D., 428, *470*

Yamada, M., 176, *450*
Yamaki, H., 311, *449*

Yanagita, T., 60, 61, *458, 470*
Yasuda, S., 176, 311, 312, *450, 467, 470*
Yoshikawa, H., 175, 373, *470*
Yoshinaga, K., 176, *441*
Young, I. E., 60, *470*
Young, M., 373, *452*
Young, R., 93, 172, *441, 470*
Yousif, A. E., 301, 311, *440*
Yvan, Z., 427, *450*

Zahler, S. A., 59, *439*
Zaitseva, G. N., 59, *470*
Zamecnik, P. C., 93, *464, 470*
Zaritsky, A., 134, 135, 172, 173, 174, 217,
  228, 230, 242, 244, 245, 246, 252, *438,*
  447, 463, 464, 469, 470, 471
Zetterberg, A., 427, *456*
Zeuthen, E., 59, *471*
Zeuthen, J., 175, *471*
Zuchowski, C., 311, *462*
Zyskind, J. W., 311, *471*

Subject Index

Abbo, F. E., filtration and
synchronization, 40–41, 91n.7
Acceptors
defined, 241n.16
maturing of peptidoglycan, 206
Adolph, E. F., early studies of bacterial
growth, 375
Age
distribution and study of bacterial
growth, 9–12
inverse distribution, 265–266, 269–270
law of conservation of cell order, 429–430
and size distributions of *Caulobacter*,
334–335
and size structure of bacterial cultures,
273, 275–276
Aggregation problem, bacterial division
cycle, 19–22
Alcaligenes fecalis, nucleotide synthesis, 82
Aldea, M., variation of cell diameter, 224
Alteration of generations forbidden, 302
Amino acids
cell width changes after starvation, 225
peptidoglycan structure, 178–179
thymine starvation and
synchronization, 35
Ancylobacter aquaticus, measurement of
turgor pressure, 251
Antibiotics
minichromosomes and resistance,
305–306
nonsynchrony methods, 44
Streptococcus and chromosome
replication, 351
Aspartate transcarbamylase, cycle-specific
synthesis, 76
Autolysins, variability of growth and
filamentation in *Bacillus subtilis*, 362
Autoradiography
analysis of cell wall synthesis, 188–190,
200–201, 224–225, 227

analysis of number of growing points in
cell, 113
analysis of protein synthesis during
division cycle, 71–72
differential methods for total population
analysis, 46
DNA chain extension and methods for
measuring C period, 136–137
early studies on patterns of DNA
replication, 104–105
slow-growing cells, 149

Baby Machine, 435–436
Bacillus, strains and membrane-elution
method, 57
Bacillus cereus
enzyme synthesis, 364
general pattern of growth, 365
size distribution and Collins–Richmond
method, 375–376
size distribution of exponential
populations, 74
Bacillus subtilis
cytoplasm synthesis during division
cycle, 363–365
DNA segregation, 289
DNA synthesis during division cycle,
365–370
flagella synthesis, 80
nonrandom segregation of DNA, 280
order of genome replication, 157
protein synthesis during division cycle,
84–85
stability and turnover of peptidoglycan,
203
surface growth during division cycle of,
358–363
Backwards analysis
eukaryotic division cycle, 424–425

methodological implications of
  Caulobacter growth, 334
nonsynchronization methods, 49–58
variability and theory of, 270–271, 273
Baconian tradition, 18
Balanced growth
  DNA synthesis, 24n.2
  life cycle of bacteria, 13
  rate of DNA synthesis during division
    cycle of *Bacilli*, 366–370
  study of bacterial growth, 8–9
Basilico, C., cloning and identification of
  G1 genes, 410–411
Batch treatment
  law of conservation of age order, 430
  law of conservation of size distribution,
    431
Bacteriophage λ
  models for sequential pathways during
    division cycle, 67
Bayne–Jones, S., early studies of bacterial
  growth, 375
Becker, W., regulation of eukaryotic
  division cycle, 390
Begg, K. J.
  cell-wall synthesis and zonal growth,
    189
  classification of mutants affecting cell
    surface and division, 236
Bellino, F. L., enzyme synthesis, 76
Bentzon, M. W., perturbation of cell
  growth and division, 34
Bezanson, G., resistance plasmids, 167
β-galactosidase
  doubling of enzyme production, 81
  negative regulation of DNA synthesis,
    158
  phenomenon of induction, 77
Biology
  Baconian tradition, 18
  Cartesian approach, 18
  conservation laws, 429
  general argument on direction of
    research, 336–337
  theory and experiment in, 4–5
  unity of as principle, 3
  unity with regard to division cycle,
    432–433
Biosynthesis, during division cycle,
  433–434
Bird, R. E., bidirectional genome
  replication, 153–154

Bleecken, S., early studies of DNA
  synthesis, 172n.44
Bonhoeffer, F., number of growing points
  in chromosome, 105
B period, bacterial division cycle, 151
Brakenhoff, G. J., location of newly
  synthesized peptidoglycan, 200–201
Bremer, H.
  gene-frequency analysis, 138
  increment-of-DNA synthesis, 134
  replicon model, 280
Brock, T. D., history of bacterial
  nucleoids, 142
Bromouracil, labeling of DNA, 105
Buchanan, R. E., early studies of bacterial
  life cycle, 313–314
Buchner, H., early studies of bacterial life
  cycle, 313
Burdett, I. D. J.
  Collins–Richmond method and growth
    rate, 376
  surface-stress model and *Bacilli*, 371
  variable-T model for pole synthesis, 211

Cairns, J., autoradiographic analysis of
  DNA chain extension, 137
Calcium
  regulation of bacterial-division cycle,
    423
  soluble components during division
    cycle, 83
Canovas, J. L., nonrandom segregation of
  DNA, 301
Caro, L., bidirectional genome replication,
  153–154
Carter, B. L. A., positive or negative
  regulation of synthesis, 159
Cartesian approach, 18
Carty, C. E., lipid or membrane synthesis,
  212
Cations, soluble components during
  division cycle, 82
Caulobacter crescentus
  applications of division cycle, 334–336
  D period compared to *Escherichia coli*,
    337n.4
  flagella synthesis, 80
  as gram-negative rod, 336–337
  growth and division pattern of,
    318–320, 322–327, 329–333

heat-shock proteins and nonrandom segregation, 308
Cefoxitin, *Streptococcus* and chromosome replication, 351
Centrifugation
cell density studies, 249
selective methods of synchronization, 38–39
Cephalothin, *Streptococcus* and chromosome replication, 351
Cerulenin, *Streptococcus* and chromosome replication, 351
Chai, N.-C., co-segregation of DNA and cell envelope, 280
Challoner, P. B., c-*myc* synthesis, 407
Chemostat, studies of bacterial division cycle, 150–151
Chloramphenicol
Caulobacter and DNA regulation, 325
DNA fork movement, 140
DNA synthesis in *Streptococcus faecium*, 343–344
initiation and restriction points, 161
measurement of D period, 143
Chromosomes
a priori considerations on regulation of replication, 114–115, 117–119
a priori considerations on segregation, 281, 283–286
number of growing points and study of DNA replication, 105
Churchward, G.
gene-frequency analysis, 138
increment-of-DNA-synthesis method, 134
Clark, D. J., early studies of bacterial life cycle, 314
Cloning, identification of G1 genes, 410–411
C-*myc* synthesis
analysis of during division cycle, 406–408
criteria to demonstrate cell-cycle regulation, 408–410
Cochran, B. H., c-*myc* synthesis, 407
Colicinogenic plasmids, replication during division cycle, 167
Collins, J.
growth rate of cells and cell size, 45, 74
plasmid replication, 165
static analysis of cell growth, 375

Collins–Richmond method, 72–74, 375–377
Competence, mitogenesis processes, 417, 419
Conservation of age order, 430
Conservation of size distribution, 431
Constrained-hoop model
constant cell width during division cycle, 220
synthesis of cell surface, 230–232
Constriction, regulation of in rod-shaped cells, 209–212
Continuum model
eukaryotic cell cycle, 392, 394–395
restriction point experiments, 412–414
variability of interdivision times, 416
Cooper, S., cytoplasm and vision of cell cycle, 90n.4
Copenhagen school, study of bacterial growth, 16
Coplans, M., early studies of bacterial life cycle, 313
Coulter Counter, investigator bias and synchronization, 28
C period
constant at moderate growth rates, 109
constant at slow growth rates, 146–147, 148
defined, 98
determination of in *Bacillus subtilis*, 366–368
DNA replication and slow-growing cells, 144
DNA synthesis in *Streptococcus faecium*, 343, 345
marker ratios during slow growth, 147–148
methods for measuring, 133–140
time variation, 273
variable at slow growth rates, 109
variation in replication-segregation sequence, 257–258
Criteria for synchronization methods, 26–29
Cross-linking
maturing of peptidoglycan, 206–207, 209
peptidoglycan chains, 184
Cullum, J., exponential growth law, 377
Cummings, D. J.
filtration and conflicting results, 40
membrane-elution technique, 41, 44

Cutler, R. G., starvation and RNA
   synthesis, 35
Cuzin, F., replicon model, 280
Cytogenesis, alternate analysis of
   eukaryotic division cycle, 417
Cytoplasm
   biosynthetic pattern, 433–434
   conservation of size distribution, 431
   definition, 63
   events during division cycle, 89–90
   exponential growth law, 380
   periodicities in *Schizosaccharomyces
      pombe*, 421–424
   regulation of DNA synthesis at
      initiation, 381, 383
   segregation, 308
   synthesis
      alternate proposals for during
         division cycle, 76–84
      a priori considerations of, 64–68
      Bacillus subtilis division cycle, 363–365
      categories of molecules during
         bacterial division cycle, 19, 20–21
      experimental analysis of during
         division cycle, 69–76
      history of analysis and psychological
         aspects of science, 84
      interrelationships between DNA
         synthesis, surface synthesis, and,
         22–23

Daniels, M. J., lipid or membrane
   synthesis, 212
Dennis, P. P., starvation and RNA
   synthesis, 35
Density, cell
   during division cycle, 247–250
   experimental support for pressure
      model of cell-surface synthesis, 198,
      200
   filamentation, 362
   regulatory mechanisms, 252
   Streptococcus, 353
Density-shift analysis, DNA replication
   during division cycle, 48
Deoxythymidine triphosphate (TTP),
   cycle-specific variations in enzyme
   synthesis, 68

Determinism, cell regulation, 276
Diameter, cell
   apparent variation during division
      cycle, 222, 224
   variation or constancy of during
      division cycle, 202
Diaminopimelic acid, peptidoglycan
   structure, 179–180
Differential methods, cell-cycle analysis,
   30–32
Differentiation
   Caulobacter and developmental
      pathways, 330
   Caulobacter as model for division cycle,
      319
Division
   observed variation during, 253–255
   relationship to DNA synthesis, 264–265
   relationship to size at initiation, 264
   remaining problems of bacterial division
      cycle, 435
Division cycle
   aggregation problem, 19–22
   Caulobacter as model for differentiation,
      319
   conservation laws of, 429–431
   experimental analysis of, 25–26
   field of studies, 1–3
   history of studies, 4
   logic of, 24, 434
   passive and independent regulation, 23
   unity throughout biology, 432–433
DNA
   biosynthetic patterns, 434
   cell component density, 247
   concentration as function of growth
      rate, 152
   distinction between replication and
      synthesis, 95
   formula for cell content, 111
   replication
      bidirectional, 171n.13
      concept of the fundamental cell, 348,
         350
      continuum of division-cycle patterns,
         395
      dichotomous in *Bacillus subtilis*,
         365–366
      during division cycle of *Caulobacter*,
         319–320, 322–325

during division cycle of slow-growing
cells, 144, 146–150
early studies on pattern of, 103–106
further analysis of during division
cycle, 133–143
icon for, 114
icon for regulation of, 133
membrane-elution technique and
analysis, 107, 109–113
segregation
Caulobacter crescentus, 335–336
and cell wall in *Bacillus*, 371
methods of analyzing, 286, 288
nonrandom, 288–289, 291–292
random, 288
Streptococcus, 353
step times as measure of C period,
139–140
synthesis
age-size structure of bacterial
cultures, 276
balanced and exponential growth,
24n.2
categories of molecules during
bacterial division cycle, 19, 21–22
cell growth and continuum model, 417
confined to central portion of division
cycle in animal cells, 59n.9
continuum model, 395
during division cycle of *Bacillus
subtilis*, 365–370
during division cycle of *Streptococcus
faecium*, 343–345
early studies on pattern, 187
enzyme synthesis in *Bacilli*, 364
events leading to initiation, 160–162
experimental analysis of regulation,
119–127, 129–133
fast and moderate growth rates,
94–95, 97–103
increase in oldest cells, 173n.83
increment and methods of measuring
C period, 134
initiation and termination, 169–170
interrelationships between cytoplasm
synthesis, surface synthesis, and,
22–23
newborn cells, 39
periodicities in growth of
Schizosaccharomyces pombe, 420

rate of during division cycle, 107, 109
regulation of initiation, 157–160, 381,
383
relationship to cell division, 264–265
restriction point experiments, 412–414
Schaechter, Maaloe, Kjeldgaard
experiment, 385
synchronized cells, 104
time variation and choice of methods,
271
unity of cell biology, 432–433
Donachie, W. D.
analysis of specific protein synthesis
during division cycle, 75
cell length at different growth rates, 218
cell-wall synthesis and zonal growth,
189
classification of mutants affecting cell
surface and division, 236
C period in *Bacillus subtilis*, 366
critical-length hypothesis, 234
Donors
defined, 241n.16
maturing of peptidoglycan, 206
D period
in *Bacillus*, 368
concept of the fundamental cell, 348,
350
constant at slow growth rates, 146–147,
148
DNA replication and determination of,
142–143
DNA replication and slow-growing
cells, 144
DNA synthesis in *Streptococcus faecium*,
343, 345
Escherichia coli and *Caulobacter*
compared, 337n.4
time variation, 273
variation in *Escherichia coli*, 261–262
variation in replication–segregation
sequence, 257–258
Driehuis, F., stability and turnover of
peptidoglycan, 203

Eberle, H., nonrandom segregation of
DNA, 280
Economics, aggregation problem, 19

Einstein, A., theory and experiment in
    physics, 4
Eisenman, R. N., c-*myc* synthesis, 407
Electron microscopy, nucleoids, 142
Elutriation, selective methods of
    synchronization, 40
Enterobacteriaceae
    size and initiation of DNA, 355
    study of division cycles, 3
Enterococcus hirae. See Streptococcus faecium
Entrainment, nonselective methods for
    synchronization, 36–37
Enzymes
    cell-cycle–specific synthesis, 66–67,
        68, 76
    cytoplasm synthesis and exponential
        growth, 20
    positive and negative regulation,
        158–159, 174–175n.122
    Schizosaccharomyces pombe and
        periodicities, 421
    synthesis during division cycle of *Bacilli*,
        363–365
    synthesis under steady and nonsteady
        growth conditions, 90–91n.6
Equipartition mechanism, segregation of
    plasmids, 303, 305–306
Escherichia coli
    bacterial division cycle and *Caulobacter
        crescentus*, 318
    bidirectional genome replication,
        153–154
    calculation of cell mass at time of
        division, 118
    cell-cycle–specific protein synthesis, 325
    cell density, 198
    cell growth pattern of *Bacilli*, 365
    chromosome replication in *Caulobacter*,
        323
    Collins–Richmond method and growth
        rate, 376
    C values in strain B/r, 134
    division cycles of *Schizosaccharomyces
        pombe*, 420
    D period compared to *Caulobacter*,
        337n.4
    elutriation and synchronization, 40
    fundamental cell, 348
    general model of bacterial growth, 372
    icon for *Caulobacter* division cycle, 333
    nucleotide synthesis, 82

older and newer poles, 338n.30
pattern of DNA synthesis during
    division cycle, 94–95
pole development compared to
    Caulobacter, 329–330
poles and asymmetric segregation of
    DNA, 293
ratio-of-rates method and cell-surface
    synthesis, 194
Schaechter, Maaløe, Kjeldgaard
    experiment, 387
short-term labeling experiments,
    242n.56
stability and turnover of peptidoglycan,
    202–203
strains and membrane-elution
    method, 57
stress model of surface growth, 362
study of division cycles, 3
synchronization by phage infection, 37
theoretic maximal growth rate, 88
unit-cell models of cell growth and
    division, 309
variation in D period, 261–262
Ethylenediaminetetraacetate (EDTA), cell
    width changes after treatment with,
    225
Eukaryotic cells
    alternate analysis of division cycle,
        391–392, 394–417, 419
    backwards analysis of division cycle,
        424–425
    division cycle compared to bacterial,
        389–391
    icon of division cycle, 426
    Schizosaccharomyces pombe as model
        system, 420
    terminology of division cycle, 425–426
Evans, J. E., starvation and RNA
    synthesis, 35
Evolution, periodicities, 422
Execution points, nonsynchrony methods
    and total population analysis, 48–49
Exponential growth
    cellular growth law, 379–380
    criteria for synchronization, 27, 28
    cytoplasm synthesis, 20–21
    DNA synthesis, 24n.2
    inverse correlation between length and
        width of cells, 91n.11
    mass synthesis, 171n.31

periodicities in *Schizosaccharomyces pombe*, 421–422
size distribution of population, 72–74
study of bacterial growth, 7–8, 11–12
synthesis of mass in *Streptococcus faecium*, 346
Exponential synthesis
  definition, 196
  modes of cytoplasm increase, 65

Filamentation, variability of and growth in *Bacillus subtilis*, 362
Filtration, selective methods of synchronization, 40–41, 91n.7
F*lac*, plasmid replication during division cycle, 163–167
Flagella
  alternate proposals for cytoplasm synthesis during division cycle, 80–81
  Caulobacter crescentus, 319, 325, 332, 338n.19
Flow cytometry
  differential methods for total population analysis, 46–47
  DNA content of individual cells, 111–113
  slow-growing cells, 149
Fluorescence, flow cytometry method, 112
Forro, F., Jr.
  autoradiography and DNA replication, 105
  segregation in microcolonies, 280
Freedman, M. E., constant C and D at slow growth rates, 146
Frehel, C., zonal surface growth, 361
Fundamental cell, proposal of the, 347–348, 350
Fundamental experiment of bacterial physiology described, 13–15
  reanalyzed, 383–387

Genes
  dosage as source of variation in enzyme synthesis, 68
  order of in genome, 155–156
Gene-frequency analysis
  DNA replication analysis during division cycle, 48

measurement of C period, 137–138
Generations, alternation forbidden principle, 302
Genetics, analysis of division cycle, 234–236, 239
Genome
  bidirectional replication, 153–155
  early work on order of replication, 152–153
  order of genes in, 155–156
  unique order of replication, 156–157
Gierer, A., number of growing points in chromosome, 105
Glaser, D., order of genome replication, 152, 153
G(0) cells
  analysis of, 402–403
  model of Zetterberg and Larsson, 403–406
G1 cell, definition, 391–392
G1 genes, cloning and identification, 410–411
G1 period
  analysis of arrest, 399–400
  analysis of variation, 397–399
  eukaryotic division cycle, 390
  release from arrest, 400–401
Goodwin, B. C., starvation and synchronization, 36
Gram-negative bacteria
  cell-cycle–specific protein synthesis, 326
  early studies on macromolecule segregation in, 279–280
  icon for *Caulobacter* division cycle, 333
  shape of rod-shaped cells, 213–215, 217–222, 224–225, 227–228, 230
  structure of cell surface, 177–183
Gram-negative rods, *Caulobacter* as, 336–337
Greco, A., cloning and identification of G1 genes, 410–411
Grossman, N., cell width and amino acid starvation, 225
Groudine, M., c-*myc* synthesis, 407
Grover, N. B.
  cell length overshoot during shift-up, 230–231
  Collins–Richmond method and growth rate, 376
  entrainment and synchronization, 36

Growth
  comments on pattern of *Streptococcus*, 353–355
  fundamental experiment of bacterial physiology, 13–17
  law of cellular, 375–380
  law of length, 380–381
  macromolecular synthesis and *Bacillus subtilis*, 365
  multiseptate in *Bacillus subtilis*, 362–363
  observed variation during, 253–255
  population and pattern, 1
  rates
    cell mass at different, 119–121, 369–370
    constant shape of cells at different, 215, 217–218
    determination by medium, 85–89
    DNA concentration as function of, 152
    DNA content at different, 110–111
    DNA synthesis at moderate and fast, 94–95, 97–103
    segregation of DNA, 297
    theoretic maximal rate, 88–89
  and regulation during division cycle of *Bacillus*, 371–372
  Schizosaccharomyces pombe and periodicities, 420–424
  study of bacterial, 7–13
  unified view of bacterial during division cycle, 372
  variability of and filamentation in *Bacillus subtilis*, 362
Gudas, L. J., lipid or membrane synthesis, 212

Haemocytometer, investigator bias and synchronization, 28
Halvorson, H. O., positive or negative regulation of synthesis, 159, 160
Hanawalt, P. C., mechanism of replication of chromosome replication, 125–126
Hann, S. R., c-*myc* synthesis, 407, 408
Harvey, J. D., exponential growth law, 377
Heat shocks
  Caulobacter and protein synthesis, 332
  DNA synthesis in synchronized cells, 104

  nonselective methods for synchronization, 33–34
Heat treatments, nonselective methods for synchronization, 33–34
Helmstetter, C. E.
  cell shape at different growth rates, 218
  F*lac* and plasmid replication, 166
  membrane-elution technique, 41, 44
  nucleoid-occlusion model, 209
  order of genome replication, 152–153
  residual cell division after inhibition of DNA synthesis, 142–143
  segregation model, 292–294, 297, 299–301
Henrici, A. T.
  classic life cycle of bacteria, 12–13
  early studies of bacterial life cycle, 314
Hershey, A. D., early studies of bacterial life cycle, 314–315
Hirota, Y.
  central zone surface synthesis, 225
  methocel method, 286, 288
History of cell-cycle studies, 4
Hoffman, B., unique order of genome replication, 157
Holmes, M., *Bacillus subtilis* replication pattern, 367–368
Hook protein gene, synthesis of flagella-related protein during *Caulobacter* division cycle, 325
Hotchkiss, R. D.
  history of cell-cycle studies, 4
  temperature and synchronization, 33
Howard, A., midcycle DNA synthesis in eukaryotic cells, 33
Huls, P., location of newly synthesized peptidoglycan, 200–201
Hydroxyurea, *Caulobacter* and DNA regulation, 323–324

Indexing synchrony experiments, 29–30
Inertia, nonrandom DNA segregation, 289
Ingram, L. O., lipid and membrane synthesis, 212
Inhibitors
  effect on rate of DNA replication, 140
  nonselective methods of synchronization, 37
Initiation
  regulation of DNA synthesis, 381, 383

relationship to size at division, 264
remaining problems of bacterial division
    cycle, 434
Schaechter, Maaløe, Kjeldgaard
    experiment, 385
variation in mass, 258–259
Integral methods, compared to
    differential, 31–32
Invagination
    DNA replication and start of, 211–212
    remaining problems of bacterial division
        cycle, 435
Investigator, bias and synchronization, 28
I period, defined, 171n.34
Isodensity
    meaning of, 250
    regulatory mechanisms and cell density,
        248
Ittmann, M., cloning and identification of
    G1 genes, 410–411

Jacob, F.
    methocel method, 286, 288
    replicon model, 280
    theory and experiment in biology, 5, 18
James, R., lipid and membrane synthesis,
    212
Joseleau-Petit, D., membrane synthesis,
    212

Kaufmann, W., theory and experiment in
    physics, 4
Kelly, K., c-myc synthesis, 407, 409
Kepes, A., automatic phosphate
    starvation method, 37
Kepes, F.
    automatic phosphate starvation
        method, 37
    starvation and synchronization, 36
Kirkwood, T. B. L., Collins–Richmond
    method and growth rate, 376
Kjeldgaard, N. O.
    chromosome replication and
        implications of experiment, 117–118
    fundamental experiment reanalyzed,
        383–387
    influence on cell-cycle studies, 4
    nucleoid content during steady-state
        growth and transitions, 141–142

problem of DNA replication during
    division cycle, 94
steady-state growth, 13–14
Knaysi, G., early studies of bacterial
    growth, 375
Koch, A. L.
    age distribution, 265
    arrangement of peptidoglycan strands
        on cell surface, 183
    Bacillus subtilis and peptidoglycan
        release, 370
    cell diameter and larger cell size,
        231–232
    inside-to-outside growth of
        peptidoglycan, 360
    Lark and Lark model of DNA
        replication, 106
    mechanical segregation models, 308
    peptidoglycan and mass synthesis, 185,
        186
    relationship between size at initiation
        and size at division, 264
    relationship of DNA synthesis and cell
        division, 264–265
    stability and turnover of peptidoglycan,
        203
    starvation and synchronization, 36
    surface-stress model and Bacilli, 371
    variable-T model for pole synthesis, 211
Koppes, L. J. H., Collins–Richmond
    method and growth rate, 376
Kornberg, A., unity of biochemistry, 433
Kubitschek, H. E.
    constant C and D at slow growth rates,
        146
    linear synthesis during division cycle,
        77, 78, 93n.38
Kuempel, P. L., bidirectional genome
    replication, 154–155

Labeling, radioactive, membrane-elution
    method, 56
Lactobacillus acidophilus, co-segregation of
    DNA and cell envelope, 280
Lag phase
    early studies of bacterial life cycle,
        313–315
    as shift-up, 315
Lark, C. A., model of DNA replication,
    106

Lark, K. G.
  Caulobacter and DNA regulation,
    324–325
  events leading to initiation of DNA
    synthesis, 160, 161
  heat treatment and synchronization, 33
  model of DNA replication, 106
  patterns of DNA segregation, 280
  unique order of genome replication, 157
Larsson, O., G(0) model, 403–406
Law of Conservation of age order, 430
Law of Conservation of Energy, theory
    and experiment in physics, 4
Law of Conservation of size distribution,
    431
Leading-edge model, pole growth,
    201–202
Leder, P., c-*myc* synthesis, 407
Ledingham, J. C. G., early studies of
    bacterial life cycle, 313
Length, cell
  constant mass within culture, 220–222
  distribution of newborn cells, 227–228
  growth law, 380–381
  minimal or critical, 232–234
  rate of growth, 200
Leonard, A. C.
  cell shape at different growth rates, 218
  F*lac* and plasmid replication, 166
  segregation model, 292–294, 297,
    299–301
Life cycle
  bacterial as shift-ups and shift-downs,
    315–317
  classic of bacteria, 12–13
  short history of the bacterial, 313–315
Lin, E. C. C., methocel method, 286, 288
Linear synthesis
  alternate proposals for cytoplasm
    synthesis, 77–80
  cytoplasm, 21
Lipids
  cell component density, 247
  synthesis during division cycle, 212–213
Liskay, R. M.
  arrest of G1 phase, 401
  variation in G1 phase, 397
Load-bearing layer, cell surface structure,
    180
Logarithmic phase, study of bacterial cell
    cycle, 317

Logic, of division cycle, 24, 434
Longard, K., early studies of bacterial life
    cycle, 313
Lopez-Saez, J. F., nonrandom segregation
    of DNA, 301
Louarn, J., bidirectional genome
    replication, 153–154
Lutkenhaus, J. F., analysis of specific
    protein synthesis during division
    cycle, 75

Maaløe, O.
  chromosome replication and
    implications of experiment, 117–118
  heat treatment and synchronization, 33
  fundamental experiment reanalyzed,
    383–387
  influence on cell-cycle studies, 4
  mechanism of regulation of
    chromosome replication, 125–126
  nucleoid content during steady-state
    growth and transitions, 141–142
  perturbation of cell growth and
    division, 34
  problem of DNA replication during
    division cycle, 94
  steady-state growth, 13–14
Maaløe–Hanawalt experiment, 4, 125–126
McKenna, W. G., bidirectional genome
    replication, 155
Macromolecular synthesis, cell growth
    during division cycle of *Bacillus
    subtilis*, 365
Make-before-break
  internal pressure of cell, 186
  load-bearing structure of cell, 180
Marker ratios, C value during slow
    growth, 147–148
Marr, A. G.
  cumulative and noncumulative
    variation, 259–260
  exponential growth law, 377
  inverse age distribution, 270
Marsh, R. C., bidirectional genome
    replication, 155
Martuscelli, J., bidirectional genome
    replication, 153–154
Maruyama, Y., filtration and conflicting
    results, 40

Mass, cell
  determination of at different growth
    rates, 369–370
  initiation of replication, 114–115, 117,
    121
Mass synthesis
  cell growth and continuum model, 417
  exponential growth, 171n.31
  exponentiality in *Streptococcus*, 356n.23
  exponential rate of, 64
  icon for during division cycle, 89
  inhibition and regulation of DNA
    synthesis, 121–123
  linear rate of, 65
  progression and competence factors,
    419
  relationship to peptidoglycan synthesis,
    184–187
  Streptococcus faecium, 346–347
  variation in rate of, 258
Mass-to-DNA ratios, methods for
    measuring C period, 135–136
Masters, M.
  analysis of specific protein synthesis
    during division cycle, 75
  bidirectional genome replication, 155
Membrane, cell
  segregation of, 309–310
  synthesis during division cycle, 212–213
Membrane-elution technique
  analysis of DNA replication, 107,
    109–113
  analysis of specific protein synthesis
    during division cycle, 74–76
  backwards methods of cell-cycle
    analysis, 50–58
  criteria for method, 55
  DNA synthesis during division cycle of
    slow-growing cells, 144, 146
  Flac and plasmid replication, 165–166
  importance to understanding of
    bacterial division cycle, 435–436
  individual growth laws, 380
  measurement of DNA segregation, 292
  methodological implications of
    Caulobacter growth, 334
  methods of analyzing segregation of
    DNA, 288
  ratio of peptidoglycan synthesis to
    protein synthesis during division
    cycle, 197

selective methods of synchronization,
    41, 44
time variation and choice of methods,
    271, 273
total protein synthesis during division
    cycle, 69–71
variable D period within culture, 368
Meselson –Stahl experiment
  DNA replication, 106
  unique order of genome replication,
    156–157
Metabolism, cell, *Schizosaccharomyces pombe*
    and periodicities, 421
Methicillin, *Streptococcus* and chromosome
    replication, 351
Methocel method, analysis of DNA
    segregation, 286, 288
Methods
  experiments and discussions of, 25
  implications of *Caulobacter* growth,
    334
Microbiology, understanding of bacterial
    cell division, 2
Microcolonies, segregation in, 280
Minichromosomes
  copy number and stability, 306–307
  directed nonrandom segregation, 307
  origin, replication, and significance,
    302–303
  paradox of high copy number, 303
  replication during division cycle, 166,
    167
  resistance to antibiotics, 305–306
Mitchison, J. M.
  periodicities in growth of
    Schizosaccharomyces pombe, 420
  proposal of quasi-linear synthesis, 421
Mitogenesis
  alternate analysis of eukaryotic division
    cycle, 417
  competence and progression, 417, 419
Mitomycin, DNA synthesis in *Streptococcus
    faecium*, 344
Mitosis
  definition of G1 cell, 391
  method of labeled and backward
    analysis, 424
Molecular mechanisms, shape
    determination, 228, 230
Moore, B. A., analysis of specific protein
    synthesis during division cycle, 75

Morphogenes, classification of mutants
    affecting cell surface and division, 238
Muller, M., early studies of bacterial life
    cycle, 313
Murein. *See* Peptidoglycan
Mutations
    classification of mutants affecting cell
        surface and division, 238, 239
    simple and complex processes, 235–236

Nagata, T.
    filtration and conflicting results, 40
    unique order of genome replication, 157
Nanninga, N.
    cell width variability and peptidoglycan
        synthesis, 224
    leading-edge model of pole growth, 201
    location of newly synthesized
        peptidoglycan, 200–201
    synthesis of penicillin-binding
        proteins, 75
    unit-cell model, 189, 263
Navarette, M. H., nonrandom segregation
    of DNA, 301
Neijssel, O. M., exponential growth law,
    377–378
Newborn cells
    cell length distribution, 227–228
    centrifugation and synchronization,
        38–39
    criteria for synchronization methods, 26
    size variation, 254
Newman, A., order of genome
    replication, 153
N-formimidoyl thienamycin, *Streptococcus*
    and chromosome replication, 351
Nieman, P. E., c-*myc* synthesis, 407
Nilson, E. H., cumulative and
    noncumulative variation, 259–260
Nonselective methods for
    synchronization, 33–38
Nonsynchronization methods
    backwards methods, 49–58
    compared to synchronization, 25–26
    total population analysis, 44–49
Normality, perturbation of cell growth
    and division, 34
Nucleoid-occlusion model
    DNA replication and start of
        invagination, 211

    regulation of constriction in rod-shaped
        cells, 209
Nucleoids
    concentrations of DNA in bacterial cell,
        17n.7
    division in relation to DNA synthesis,
        140–142
    electron microscopy, 142
    segregation of DNA, 307–308
Nucleotides, soluble components during
    division cycle, 82
Nutrients, *Caulobacter* and DNA synthesis,
    323

Ohki, M., lipid and membrane synthesis,
    212
Olijhoek, A. J. M., synthesis of
    penicillin-binding protein, 75
Oncogenes, analysis of c-*myc* synthesis
    during division cycle, 406–408
*Ori*C, initiation of DNA replication, 169
Osmotic pressure, sources of internal
    pressure in cell, 251

Pachler, P. F., Lark and Lark model of
    DNA replication, 106
Painter, P. R.
    cumulative and noncumulative
        variation, 259–260
    exponential growth law, 377
    inverse age distribution, 270
Pardee, A. B.
    filtration and synchronization, 40–41,
        91n.7
    restriction points and eukaryotic cells,
        411–412
Pas, E., location of newly synthesized
    peptidoglycan, 200–201
Passive regulation, 23
Pato, M. L.
    Flac and chromosome replication, 164
    order of genome replication, 152
Pelc, S. R., midcycle DNA treatments in
    eukaryotic cells, 33
Penfold, W. J., early studies of bacterial
    life cycle, 313
Penicillin
    Caulobacter and periodic synthesis of
        protein, 332

rate of synthesis of binding proteins, 75–76

Pentapeptide
synthesis of precursor, 183–184
transfer of to growing chain, 184

Peptidoglycan
amount of on cell, 179–181
arrangement of strands on cell surface, 183
composition and cross-linking of subunits, 178–179, 184
heterogeneity of structure, 181–182
inside-to-outside growth of in *Bacillus subtilis*, 359–360
maturing of, 206–207, 209
segregation of bacterial surface, 308–310
structure of *Bacilli* cell wall, 358
synthesis
biochemistry of, 183–184
central zone of increased, 225
early studies on rate of during division cycle, 190–191
rate and topography of during division cycle, 191–192, 194–198, 200–203
relationship of mass synthesis to, 184–187
stringent regulation, 203–204
turnover of in *Bacillus subtilis*, 358–359

Periodicities, growth of *Schizosaccharomyces pombe*, 420–424

Periseptal annulus, regulation and constriction in rod-shaped cells, 209–210

Phage infection, nonselective methods of synchronization, 37–38

Phosphate, starvation and synchronization, 36–37

Physics
conservation laws, 429
theory and experiment in, 4–5

Physiology, experiments and balanced growth, 9

Pierucci, O.
Bacillus subtilis replication pattern, 367–368
DNA segregation, 299
cell-wall synthesis and zonal growth, 189
pattern of DNA segregation, 289

phospholipid synthesis, 213
residual cell division after inhibition of DNA synthesis, 142–143

Plasmids
replication during division cycle, 162–169
segregation of, 302–303, 305–307

Platelet extract, competence factors, 417, 419

Platelet-poor plasma, progression factors, 417, 419

Pneumococcus, temperature and synchronization, 33

Poisson distribution, explained, 310n.2

Poles
leading-edge model, 201–202
location of newly synthesized peptidoglycan, 201
pattern of development in *Caulobacter*, 326–327, 329–333
pattern of synthesis in *Streptococcus*, 340–341
Schaechter, Maaløe, Kjeldgaard experiment, 385
synthesis and regulation of constriction in rod-shaped cells, 211

Polyvinylpyrrolidone (PVP), cell density studies, 249

P1 prophage, replication during division cycle, 167

Pools, membrane-elution method, 56–57

Population
growth and pattern, 1
total analysis and nonsynchrony methods, 44–49

Powell, E. O.
temporal variation, 255
time-lapse analysis, 46

Prescott, D. M.
bidirectional genome replication, 154–155
variation in G1 phase, 397

Pritchard, R. H.
cell-wall synthesis and zonal growth, 189
entrainment and synchronization, 36
plasmid replication, 165
positive or negative regulation of initiation, 159
rate stimulation method, 135

Probability, cell regulation, 276
Probability and determinism, 276
Progression, mitogenesis processes, 417, 419
Prokaryotes, terminology of division cycle, 425–426
Protein synthesis
  autoradiographic analysis during division cycle, 71–72
  Caulobacter and linear model of regulation of periodic, 331–333
  division cycle of *Schizosaccharomyces pombe*, 422
  D period and residual cell division after inhibition of, 143
  functional model for cell-cycle regulation of specific, 84–85
  membrane during division cycle, 213
  membrane-elution method and analysis of specific, 74–76
  pattern of pole development in *Caulobacter*, 326–327, 329–333
  ratio of peptidoglycan synthesis during division cycle using membrane-elution, 197
  Schizosaccharomyces pombe and periodicities, 420
  specific during division cycle of *Caulobacter*, 325–326
  total during division cycle analyzed by membrane-elution method, 69–71
  variations in capacity for specific, 81–82
  variations in specific during division cycle, 80
Protozoans, temperature changes in synchronizing, 59n.8
Pseudosynchronous culture, synchronization methods, 33

Quasi-linear synthesis, periodicities and proposal of, 421–422
Quiescence, model of and restriction point, 412

Radioactive decay, DNA replication during division cycle, 48

Rate maintenance
  defined, 15
  deviations from and regulation of DNA synthesis, 130–132
Rate stimulation method, measurement of C period, 134–135
Ratio-of-rates method
  cell-surface synthesis, 194
  ratio of peptidoglycan synthesis to protein synthesis, 197
Regulation, cell, probability and determinism, 276
Renger, H.
  Caulobacter and DNA regulation, 324–325
  events leading to initiation of DNA synthesis, 160, 161
Repko, T., unique order of genome replication, 157
Replication-segregation sequence, variation in C and D periods, 257–258
Replicon model, segregation in microcolonies, 280
Rep mutation, modifications of rate of DNA replication, 140
Resistance plasmids, replication during division cycle, 167
Restriction points
  eurkaryotic division cycle, 411–414
  proposal of bacterial, 160–162
Ribosomes, cytoplasm synthesis and exponential growth, 20
Richmond, M.
  growth rate of cells and cell size, 45, 74
  static analysis of cell growth, 375
Rickert, M., *Bacillus subtilis* replication pattern, 367–368
Riedlin, G., early studies of bacterial life cycle, 313
Rifampin, DNA fork movement, 140
RNA
  cell component density, 247
  level of c-*myc* and ribosomal, 409
  Schaechter, Maaløe, Kjeldgaard experiment, 386
  Schizosaccharomyces pombe and periodicities, 420
  specific synthesis during division cycle, 82
  synthesis and membrane-elution method, 72

Robinson, A. C.
cell length at different growth rates, 218
critical-length hypothesis, 234
Rod-shaped cells
cell-cycle–specific protein synthesis, 326
icon for *Caulobacter* division cycle, 333
regulation of constriction in, 209–212
shape of, 213–215, 217–222, 224–225,
227–228, 230
unit-cell model of growth, 263
Ron, E. Z., cell width after amino acid
starvation, 225
Rosenberger, R. F., cell length overshoot
during shift-up, 230–231
Ruehl, W. H., early studies of bacterial
life cycle, 314
Ryter, A.
diffuse surface growth, 361
central zone surface synthesis, 225

Saccharomyces cerevisiae, division cycle and
periodic synthesis, 423–424
Salmonella
bidirectional genome replication, 154
short-term labeling experiments,
242n.56
strains and membrane-elution
method, 57
Salmonella typhimurium
bacterial division cycle and *Caulobacter
crescentus*, 318
calculation of cell mass at time of
division, 118
constant cell width during division
cycle, 224
constant shape of cells, 217, 219
DNA content at different rates of
growth, 111
heat treatment and synchronization, 33
measurement of D period, 141
ratio-of-rates method and cell-surface
synthesis, 194
Schaechter, Maaløe, Kjeldgaard
experiment, 387
stability and turnover of peptidoglycan,
203, 243n.69
Sargent, M. G.
D period in *Bacillus*, 368
linear synthesis, 365

zonal surface growth, 360
Savageau, M., positive or negative
regulation of initiation, 158–159,
174n.122
Schaechter, M.
chromosome replication and
implications of experiment,
117–118
fundamental experiment reanalyzed,
383–387
influence on cell-cycle studies, 4
nucleoid content during steady-state
growth and transitions, 141–142
perturbation of cell growth and
division, 34
problem of DNA replication during
division cycle, 94
steady-state growth, 13–14
Schizosaccharolyces pombe
division cycle of, 420–424
unit-cell hypothesis, 309
Schwarz, U.
central zone surface synthesis, 225
synthesis of penicillin-binding
proteins, 75
Science
Cartesian and deductive, 435
history of analysis of cytoplasm
synthesis and psychological
aspects, 84
study of division cycle, 2
theories of and biology, 18
Segregation
alternate model, 301–302
a priori considerations on chromosome,
281, 283–286
bacterial surface, 308–310
cell wall and DNA in *Bacillus*, 371
cytoplasm, 308
early studies on macromolecule in
gram-negative bacteria, 279–280
Helmstetter–Leonard model, 292–294,
297, 299–301
mechanical models, 307–308
methods of analyzing DNA, 286, 288
observation and explanation of
nonrandom patterns, 288–289,
291–292
plasmids, 302–303, 305–307
see also DNA, segregation

Selective methods for synchronization,
38–44
Setlow, R. B.
autoradiography and DNA replication,
104–105
macromolecule segregation in
gram-negative bacteria, 279–280
segregation of cytoplasm, 308
segregation of peptidoglycan, 200
Shape
constant and DNA segregation data,
300
rod-shaped gram-negative bacteria,
213–215, 217–222, 224–225,
227–228, 230
Shift-down, experimental analysis of
regulation of DNA synthesis, 132–133
Shift-up
experimental analysis of in *Streptococcus
faecium*, 347
experimental analysis of regulation of
DNA synthesis, 127, 129–130, 132
lag phase as, 315
Schaechter, Maaløe, Kjeldgaard
experiment, 387
sharp break in rate of cell increase, 273
studies of bacterial growth, 15–16
Size, cell
and age distributions of *Caulobacter*, 334
and age structure of bacterial cultures,
273, 275–276
conservation of distribution, 431
direct measurements and nonsynchrony
methods, 44–46
distribution and study of bacterial
growth, 375–376
distribution of exponential populations
and membrane-elution method,
72–74
homeostasis during balanced growth,
259
relationship between at initiation and
division, 264
relative of swarmer and stalked cells at
initiation of DNA replication in
Caulobacter, 322
variations, 254
Slow-growing cells
DNA replication during division cycle
of, 144, 146–150

DNA synthesis in *Streptococcus faecium*,
344–345
physiological state, 13
Snellings, K., order of genes in genome,
156
Spore germination, nonselective methods
of synchronization, 37
Stalked cells
DNA synthesis and swarmer cells,
319–320, 322–323
pattern of pole development in
Caulobacter, 326–327, 329–333
Starvation
analysis of G(0) cells, 402
cell width changes after starvation, 225
nonselective methods of
synchronization, 34–35
thymine and premature initiation of
DNA synthesis, 126
Stationary phase, as historical artifact,
315–316
Stationary-phase cells, physiological
state, 13
Steady-state growth, nucleoid content
during, 141–142
Steen, H. B., flow cytometry methods, 47
Stiles, G. D., c-*myc* synthesis, 407
Strand-inertia model
alternate-segregation model, 302
nonrandom DNA segregation, 289,
291–292
Streptococcus
comments on growth pattern, 353–355
density during division cycle, 353
division pattern, 340–347
events in growth, 350–352
mass synthesis and exponentiality in,
356n.23
proposal of the fundamental cell,
347–348, 350
segregation of DNA in, 353
surface-stress model and growth, 352
zonal growth pattern, 187
Streptolydigin, DNA fork movement, 140
Stringent regulation of wall synthesis,
203–204
Sucrose gradient synchronization,
measurement of C period, 138–139
Sullivan, N. F., classification of mutants
affecting cell surface and division, 236

Surface, cell
  biosynthetic patterns, 434
  diffuse growth in *Bacilli*, 360–361
  growth during division cycle of *Bacillus subtilis*, 358–363
  growth in *Streptococcus*, 340–342, 350–352
  regulation of synthesis at initiation, 383
  segregation of bacterial, 308–310
  structure of gram-negative bacteria, 177–183
  synthesis
    categories of molecules during bacterial division cycle, 22
    central zone during division cycle, 225, 227
    early studies on pattern of, 187–191
    experimental support for pressure model, 197–198, 200–203
    interrelationships between cytoplasm synthesis, DNA synthesis, and, 22–23
    laws, critical tests, predictions, and exceptions, 239–240
    quantitative analysis, 194–196
    topography of growth in *Bacillus subtilis*, 362
Surface-stress model
  Bacillus growth, 371
  gram-positive rods, 362
  gram-negative rods, 184–187
  Streptococcal growth, 352
Surface tension (T), formal concept, 211
Swarmer cells
  DNA synthesis and stalked cells, 319–320, 322–323
  pattern of pole development in *Caulobacter*, 327
Symmetry, variation in at division, 259
Synchronization methods
  compared to nonsynchronization, 25, 26
  comparison of selective and nonselective methods, 32–33
  criteria for, 26–29
  experiments demonstrating periodicities, 421
  exponential increase in pattern and faulty experimental situations, 104
  indexing of experiments, 29–30
  measurement of C period, 138–139

  membrane-elution apparatus, 57
  nonselective methods, 33–38
  selective methods, 38–41, 44
  time variation and choice of methods, 271, 273
Synchronous division, inhibition of DNA synthesis, 124

Tauro, P., positive or negative regulation of synthesis, 159
Tax, R., asymmetrically dividing yeast and *Caulobacter* age distributions, 334–335
Temperature
  degree of segregation at different, 300
  DNA replication at different, 109–110
  release of sensitive cells from G1 arrest, 401
  sensitive mutants of *Streptococcus*, 351–352
  shifts and synchronization, 34
Termination, remaining problems of bacterial division cycle, 434–435
Terminology, division cycle in prokaryotes and eukaryotes, 425–426
Tetrahymena, temperature changes in synchronizing, 59n.8
Thompson, C. B., c-*myc* synthesis, 407, 408
Thymine
  modifications of rate of DNA replication, 140
  starvation and premature initiation of DNA synthesis, 126
  starvation and synchronization, 35
Time
  criteria for synchronization, 27
  interdivision
    animal cells, 277n.10
    Caulobacter and cell-division patterns, 320, 322
    Caulobacter and E period, 330
    correlations between different, 263–264
    transition-probability model, 414–416
    and variability, 254–255
    variability and age distribution, 266, 269–270
    variation in total, 259

T period, bacterial division cycle, 151–152
Transductional analysis, measurement of
    C period, 139
Transition-probability model, eukaryotic
    division cycle, 414–416
Trueba, F. J.
    exponential growth law, 377–378
    variation of cell diameter, 222, 224
Turgor
    during division cycle, 250–251
    regulatory mechanisms, 252
Turnover of peptidoglycan, 202–203

Unit cell model
    cell growth and division, 309
    evidence against, 309
    zonal growth pattern, 189–190
Unity of Biology, 3
Unity of Cell Biology, 432–433, 436n.2
U period, bacterial division cycle, 152

Van Tubergen, R. P.
    autoradiography and DNA replication,
        104–105
    macromolecule segregation in
        gram-negative bacteria, 279–280
    segregation of cytoplasm, 308
    segregation of peptidoglycan, 200
Variability
    analysis of G1 phase, 397–399
    correlations among different variables,
        263–265
    elements of during division cycle, 255,
        257–262
    equality of division, 262–263
    growth and filamentation in *Bacillus*
        subtilis, 362
    inverse age distribution, 265–266,
        269–270
    observed during cell growth and
        division, 253–255
    predictions of continuum model, 416
    theory of backwards analysis, 270–271,
        273
Variable-T model, regulation of
    constriction in rod-shaped cells, 211
Vermeulen, C. W., order of genes in
    genome, 156

Verwer, R. W. H., unit cell model of
    growth, 189, 263
Vibrio cholerae asiaticae, early studies of
    bacterial life cycle, 313
Vicente, M.
    cell-wall synthesis and zonal growth,
        189
    exponential growth law, 377

Walker, H. H., early studies of bacterial
    life cycle, 314
Wall, cell
    control mechanisms for synthesis,
        203–206
    experimental analysis of *Streptococcus*
        growth, 342
    experimental support for zonal growth,
        188–189
    icon for growth of *Streptococcus faecium*,
        343
    icon for synthesis during division, 240
    quantitative analysis of growth during
        division cycle, 194–196
    segregation in *Bacillus*, 371
    structure of the *Bacillus subtilis*, 358
Ward, C. B.
    early studies of bacterial growth, 375
    order of genome replication, 153
Wertheimer, S. A., segregation in
    microcolonies, 280
Width, cell
    changes after EDTA treatment or amino
        acid starvation, 225
    constant mass within culture, 220–222
    variability and peptidoglycan synthesis,
        224–225
    variability within culture, 219–220
Wientjes, E. B.
    cell width variability and peptidoglycan
        synthesis, 224
    leading-edge model of pole growth, 201
    synthesis of penicillin-binding
        proteins, 75
Winslow, C. -E. A., early studies of
    bacterial life cycle, 314
Woldringh, C. L.
    cell length overshoot during shift-up,
        230–231
    cellular growth law, 377–378

cell width after amino acid starvation, 225
Collins–Richmond method and growth rate, 376
constant shape of cells, 217–218
electron microscopy of nucleoids, 142
invagination and DNA replication, 211
location of newly synthesized peptidoglycan, 200–201
pattern of cell-surface growth during division cycle, 196
peptidoglycan synthesis at start of invagination, 224
variation of cell diameter, 222, 224
Wolf, B., order of genome replication, 153
Worcel, A., bidirectional genome replication, 155
Wouters, J. T. M., stability and turnover of peptidoglycan, 203

Yanagita, T., filtration synchronization method, 41
Yeast
as ymmetrically dividing and age distribution of Caulobacter, 334–335
see also Schizosaccharomyces pombe

Yousif, A. E., nonrandom segregation of DNA, 301

Zaritsky, A.
cell length overshoot during shift-up, 230–231
constant shape of cells, 217
rate stimulation method, 134, 135, 324
Zetterburg, A., G(0) model, 403–406
Zeuthen, E., Flac and chromosome replication, 164
Zinc, soluble components during division cycle, 82
Zonal growth
diffuse surface growth in Bacilli, 360–361
early studies on location of cell surface synthesis, 187–188
evidence against, 190
experimental support for in wall synthesis, 188–189
regulatory mechanisms of gram-negative rods and Streptococci, 342
Zuchowski, C., pattern of DNA segregation, 289